The Urban Book Series

The Urban Book Series is a resource for urban studies and geography research worldwide. It provides a unique and innovative resource for the latest developments in the field, nurturing a comprehensive and encompassing publication venue for urban studies, urban geography, planning and regional development.

The series publishes peer-reviewed volumes related to urbanization, sustainability, urban environments, sustainable urbanism, governance, globalization, urban and sustainable development, spatial and area studies, urban management, transport systems, urban infrastructure, urban dynamics, green cities and urban landscapes. It also invites research which documents urbanization processes and urban dynamics on a national, regional and local level, welcoming case studies, as well as comparative and applied research.

The series will appeal to urbanists, geographers, planners, engineers, architects, policy makers, and to all of those interested in a wide-ranging overview of contemporary urban studies and innovations in the field. It accepts monographs, edited volumes and textbooks.

Indexed by Scopus.

Ayyoob Sharifi · Pourya Salehi
Editors

Resilient Smart Cities

Theoretical and Empirical Insights

Springer

Editors
Ayyoob Sharifi
Graduate School of Humanities and Social Sciences
Hiroshima University
Higashihiroshima, Japan

Pourya Salehi
ICLEI—Local Governments for Sustainability (World Secretariat)
Bonn, Nordrhein-Westfalen, Germany

ISSN 2365-757X ISSN 2365-7588 (electronic)
The Urban Book Series
ISBN 978-3-030-95036-1 ISBN 978-3-030-95037-8 (eBook)
https://doi.org/10.1007/978-3-030-95037-8

This Springer imprint is published by the registered company Springer Nature Switzerland AG
The registered company address is: Gewerbestrasse 11, 6330 Cham, Switzerland

Foreword by Gino Van Begin

Today, we are experiencing the impacts of climate change in both the Global South and Global North, which reinforces the need to respond to the global climate emergency. Local and regional governments are at the forefront of the climate crisis; they need to ensure that basic services are provided to citizens and businesses, and at the same time manage the local impacts and aftermath of the increasingly frequent and extreme weather events, while dealing with other complexities such as population growth and urbanization. The COVID-19 pandemic has brought into focus that the range of challenges faced today by local and regional governments is expanding as they are being forced to respond to a wide range of uncertainties and increasingly complex issues such as social inequity and local economic growth while identifying and responding to local priorities to guide a just and inclusive transition. Planning and action today have the potential to set a course to determine the reality of everyday life for generations to come. This not only presents local leaders and decision makers with a powerful opportunity, but also a great responsibility to make the right choices and make more informed decisions.

ICLEI—Local Governments for Sustainability (ICLEI), as the leading global network of cities, towns, and regions, and active in more than 125 countries—is committed to sustainable development. Our technical and policy guidance is supported by robust knowledge generation and brokering, addressing a diverse range of themes relevant to sustainable development. This includes the two key interconnected concepts of resilient development and smart cities. As part of our concerted efforts, we work hand in hand with our research and innovation partners to generate, broker, and use actionable knowledge. This means not only exploring new groundbreaking theoretical foundations and principles, but also making sure practical knowledge is accessible by showcasing case studies that illustrate application and facilitate improved guidance over time. This publication embraces that approach, helping unpack and distill challenging concepts, and provide actionable knowledge for decision and policy-makers.

The notable outcomes derived from thought-provoking discussions at ICLEI's Daring Cities Forum 2021, and the Innovate4Cities 2021 Conference where ICLEI was a core partner, and also having research and innovation as one of the main foci at the Glasgow Climate Change Conference (also known as COP 26), reiterated the importance of addressing knowledge and innovation gaps to accelerate sustainable development. In line with this perspective, equipping our leaders and policy-makers at national, regional, and local level with cutting-edge knowledge and guidance that is built on robust research, to engage in multi-level action, and enhance collaboration across all tiers of government is needed, now more than ever.

We remain optimistic that smart solutions, digital innovation, and technologies are advancing remarkably and will help address societal challenges and accelerate resilient development. This book pioneers the approach of capturing synergies generated by combining resilient and smart solutions. It also explores how resilient development and smart cities concepts intertwine, and investigates how subnational governments can embed resilience thinking in the design of their climate action plans and benefit from smart solutions to enhance resilience development. With this in mind, I sincerely hope this illuminative publication and the featured contributions can inform cities' work on the transition toward smart resilient systems globally.

Gino Van Begin
Secretary General
ICLEI—Local Governments for Sustainability
Bonn, Germany

Foreword by Frank Cownie

Cities across the globe are dealing with a wide range of stressors and adverse impacts caused by climate change, which are projected to get worse if we don't respond. Many cities are already feeling the burden of climate change at the ground level, so we must act with urgency and purpose in order to re-route the current trajectory of our planet.

Bolstering urban resilience and thinking is essential to prepare for climate shocks and minimize potential human and economic losses. The conclusions drawn from this book form a thorough assessment of the literature that points toward resilience being a process rather than a goal. As a process, resilience is dynamic and constantly evolving, a characteristic shared by smart initiatives. The development of new technologies shapes the advancement of smart cities.

Resilience and smart cities are terms that are increasingly being referred to in the sustainability arena; however, sometimes they are used as buzzwords and lose their meaning. This book attempts to remedy this by tethering resilience to practical examples and case studies, and explores the academic context behind the terms.

By linking resilience to smart cities, this literature artfully guides the reader through the intricacies of both concepts, as well as highlighting the synergies created when they are blended. The recommendations presented in this volume illuminate a clear pathway for local and subnational governments wanting to leverage the promise of smart solutions to increase resilience in their communities.

As the first of its kind to deep dive into the theoretical underpinnings of resilient smart cities, I am delighted that this book offers robust actionable knowledge for city leaders. The featured case studies from around the world present ways in which the concept of resilient smart cities can be achieved. I am convinced that this volume will draw more attention to smart city resilience, thereby contributing to the achievement of SDG 11 and the goals of the Sendai Framework.

Honorable Mayor Frank Cownie
President of ICLEI
Des Moines, USA

Preface

The convergence of climate change and urbanization has caused significant unprecedented challenges for cities globally. The emergence of other stressors such as the COVID-19 pandemic has further complicated the issues. As the frequency and intensity of stressors and disasters induced by climate change and other factors are expected to significantly increase in the future, a top priority for cities and communities is to build up on their resilience and they need scientific support toward this goal. Therefore, it is essential to develop evidence-based scientific solutions to improve the capacities to prepare for, absorb, recover from, and adapt to disastrous events. This requires not only a better understanding of urban complexities, but also enhanced predictive abilities to reduce uncertainties and to avoid being overwhelmed by extreme events. To deal with these challenges, scientists cannot only rely on conventional methods and need to develop disruptive and transformative approaches. Accordingly, the rapid advances in ICT-enabled smart cities that rely on big data analytics provide manifold new possibilities for scientists to better understand the complexities of urban systems and subsystems, to provide decision makers with better and more regularly updated information on human activities that may relate to climate change adaptation and mitigation, to facilitate online monitoring of risks, to inform different stakeholders on how to enhance their preparation and predicative abilities, and to develop methods that enable real-time response to disasters.

Overall, there is now reasonable consensus on the utility of smart city solutions for resilience and climate change adaptation/mitigation. However, the underlying principles of the concept of smart city resilience are not well explored. Also, limited knowledge exists on the actual and/or potential contributions of smart solutions. By filling these gaps, this edited book will support urban researchers, planners, and decision makers in their efforts toward developing smart and sustainable cities.

The main question that this book will address is 'do smart city projects contribute to urban climate resilience?'. Other noteworthy questions are 'what are the indicators of smart city resilience?', 'what procedures should be taken to improve efficacy of smart city solutions?', and 'what are the opportunities and challenges for promoting smart city resilience and for integrating resilience thinking into smart city planning?'.

The book has two parts, one focused on the theoretical aspects of the concept of smart city resilience and the other on the empirical aspects. The first part discusses theoretical insights by synthesizing the state of the art of research on smart city and resilience. To make these theoretical aspects more tangible, through detailed analyses of selected smart city initiatives around the world, the second part provides insights on how to harness smart technologies for climate resilience and sustainability planning. In addition, performance of selected smart cities around the world will be evaluated via several case-study analyses.

We would like to thank all colleagues who gave up their time to contribute to this project. We would also like to appreciate the partial financial support from the Asia-Pacific Network for Global Change Research. We hope you will enjoy reading this book and hope that the strategies proposed in this book will be used to bolster resilience and sustainability of cities in the face of climate change and other threats.

Higashihiroshima, Japan — Ayyoob Sharifi
Bonn, Germany — Pourya Salehi

Contents

Contributors

Teslim Abdul-Kareem Georgia Department of Natural Resources, Environmental Protection Division, Atlanta, GA, USA

Ismaila Rimi Abubakar College of Architecture and Planning, Imam Abdulrahman Bin Faisal University, Dammam, Saudi Arabia

Yusuf A. Aina Department of Geomatics Engineering Technology, Yanbu Industrial College, Yanbu, Saudi Arabia

Asad Asadzadeh University of Bonn, Bonn, Germany

Abdul-Lateef Balogun Geospatial Analysis and Modelling (GAM) Research Laboratory, Department of Civil & Environmental Engineering, Universiti Teknologi PETRONAS (UTP), Seri Iskandar, Perak, Malaysia

Paulina Budryte Faculty of Social Sciences, Arts and Humanities, Kaunas University of Technology, Kaunas, Lithuania

Armelle Cibaka Knowledge Management and Youth Regional Coordinator, ICLEI South America, Sao Paulo, Brazil

Alison Grovert Department of Landscape Architecture, School of Architecture and Environment, University of Oregon, Eugene, OR, USA

Anina Hartung ICLEI—Local Governments for Sustainability (World Secretariat), Bonn, Germany

Bao-Jie He School of Architecture and Urban Planning, Chongqing University, Chongqing, China;
Key Laboratory of New Technology for Construction of Cities in Mountain Area, Ministry of Education, Chongqing University, Chongqing, China

Leila Irajifar School of Architecture and Urban Design, RMIT University, Melbourne, Australia

Ewa Jackson ICLEI – Local Governments for Sustainability (Canada Office), Toronto, Canada

Pedro Jacobi President of the Regional Directive Council, ICLEI South America, Sao Paulo, Brazil

Amy Jones ICLEI – Local Governments for Sustainability (World Secretariat), Bonn, Nordrhein-Westfalen, Germany

Prakash Kamtam University of Hertfordshire, Hatfield, UK

Yesuel Kim KAIST Urban Design Lab, KAIST Smart City Research Center, Department of Civil and Environmental Engineering, KAIST, Daejeon, South Korea

Youngchul Kim KAIST Urban Design Lab, KAIST Smart City Research Center, Department of Civil and Environmental Engineering, KAIST, Daejeon, South Korea

Yekang Ko Department of Landscape Architecture, School of Architecture and Environment, University of Oregon, Eugene, OR, USA

Nae-Wen Kuo Department of Geography, National Taiwan Normal University, Taipei, Taiwan

Sunghee Lee KAIST Urban Design Lab, KAIST Smart City Research Center, Department of Civil and Environmental Engineering, KAIST, Daejeon, South Korea

Chong-En Li Department of Geography, National Taiwan Normal University, Taipei, Taiwan

Hasan Masrur Faculty of Engineering, University of the Ryukyus, Nishihara, Okinawa, Japan

Briana Meier History, Anthropology, and Science and Technology Studies, Massachusetts Institute of Technology, Cambridge, MA, USA

Roman Serdar Mendle ICLEI—Local Governments for Sustainability (European Secretariat), Freiburg, Germany

Melasutra Md Dali Centre for Sustainable Urban Planning and Real Estate (SUPRE), University of Malaya, Malaya, Malaysia;
Department of Urban and Regional Planning, Faculty of Built Environment, University of Malaya, Kuala Lumpur, Malaysia

Yasmin Mohd Adnan Department of Real Estate, Faculty of Built Environment, University of Malaya, Kuala Lumpur, Malaysia

Mariana Nicolletti Low Carbon and Resilience Regional Manager, ICLEI South America, Sao Paulo, Brazil

Olasunkanmi Habeeb Okunola Federal University of Technology, Institute for Land and Community Resilience, Minna, Nigeria;
Global Change Institute, University of Witwatersrand, Johannesburg, South Africa

Shrutika Parihar Global Centre for Environment and Energy, Ahmedabad University, Gujarat, India

Rodrigo Perpétuo Regional Director, ICLEI South America, Sao Paulo, Brazil

Maria Rebecca Quintero School of Architecture and Built Environment, University of Newcastle, Newcastle, NSW, Australia

Mustapha A. Raji Department of Urban Planning and Administration, University of Seoul, Seoul, South Korea

Pourya Salehi ICLEI—Local Governments for Sustainability (World Secretariat), Bonn, Nordrhein-Westfalen, Germany

Carmela Sambo Department of Landscape Architecture, School of Architecture and Environment, University of Oregon, Eugene, OR, USA

Ayyoob Sharifi Graduate School of Humanities and Social Sciences, Hiroshima University, Higashihiroshimashi, Japan;
Graduate School of Advanced Science and Engineering, Hiroshima University, Hiroshima, Japan

Himanshu Shekhar Institute for Environment and Human Security (UNU-EHS), United Nations University, UN Campus, Bonn, Germany

Nehmat Singh IHS, Erasmus University Rotterdam, Rotterdam, The Netherlands

Rhea Srivastava The Energy and Resources Institute (TERI), New Delhi, India

Abdulwaheed Tella Geospatial Analysis and Modelling (GAM) Research Laboratory, Department of Civil & Environmental Engineering, Universiti Teknologi PETRONAS (UTP), Seri Iskandar, Perak, Malaysia;
Earth, Environment and Space Division, Foresight Institute of Research and Translation, Ibadan, Nigeria

Ruchi Tomar School of Habitat Studies, Tata Institute of Social Sciences, Mumbai, India

Khanh N. Vu Centre for Urban Transitions, Swinburne University of Technology, Melbourne, VIC, Australia

Ke Xiong School of Architecture and Urban Planning, Chongqing University, Chongqing, China;
Key Laboratory of New Technology for Construction of Cities in Mountain Area, Ministry of Education, Chongqing University, Chongqing, China

Shamsudeen T. Yekeen Geospatial Analysis and Modelling (GAM) Research Laboratory, Department of Civil & Environmental Engineering, Universiti Teknologi PETRONAS (UTP), Seri Iskandar, Perak, Malaysia;
Department of Geography, Environment, and Geomatics, University of Guelph, Guelph, ON, Canada

Part I
Theoretical Insights

Chapter 1
Cities in the Context of Global Change: Challenges and the Need for Smart and Resilient Cities

Ayyoob Sharifi

Abstract Cities are now home to over 4.3 billion people, more than 56% of world population, and further growth in urbanization trends is projected for the coming decades. According to United Nations, 68% of world population is projected to live in urban areas by 2050. Traditionally, people have migrated to cities in search for better livelihood opportunities, better access to services and amenities, and enhanced quality of life. These aspirations may not always be fully realized due to various factors such as ineffective urban development and management policies and practices. As a result, externalities and problems such as social inequality, crime, environmental pollution, and traffic jam are common in many urban areas, especially those in developing countries. Climate change and the recent COVID-19 pandemic have reignited the debates over cities and their future. On the one hand, there are many concerns over the vulnerability of cities to the impacts of climate change and other stressors and extreme events such as pandemics. On the other hand, it is argued that effective urban management policies and practices can provide solutions for addressing the increasing challenges that cities are facing and contribute to mitigating global climate change. Reliance on conventional approaches and strategies may, however, not be sufficient if cities want to be part of the solution to climate change and other challenges. Therefore, there has been increasing emphasis on adopting innovative and disruptive solutions that are transformative and can accelerate transition toward creating cities that are more resilient and sustainable. This has led to growing interest and investment in smart solutions and technologies enabled by advances in information and communication technologies. Based on an overview of the existing literature, in this chapter, I first discuss some of the major challenges that cities are now facing. Results show that major challenges are related to ecological degradation, unregulated urban expansion, climate change adaptation and mitigation, resource management, fragmented urban management, air pollution, housing, and transportation. Next, I briefly discuss potential contributions of smart city solutions and technologies to overcoming these challenges. Finally, I provide a summary

A. Sharifi (✉)
Graduate School of Humanities and Social Sciences, Hiroshima University, Higashihiroshima, Japan
e-mail: sharifi@hiroshima-u.ac.jp

A. Sharifi and P. Salehi (eds.), *Resilient Smart Cities*, The Urban Book Series,
https://doi.org/10.1007/978-3-030-95037-8_1

of this edited volume and its contributions to advancing knowledge on smart and resilient cities.

Keywords Smart cities · Global change · Climate change · Urbanization · Sustainability · Resilience

1.1 Introduction

Cities are now home to over 4.3 billion people,more than 56% of world population, and further growth in urbanization trends is projected for the coming decades. According to United Nations, 68% of world population is projected to live in urban areas by 2050 (UN 2018). This will be an almost twofold increase compared to the percentage reported for 1960 and shows the growing interest in cities as centers of economic growth, innovation, and cultural activities. In fact, cities already account for more than 80% of global Gross Domestic Product (GDP), and this share is likely to further increase as more people will move into cities in the coming decades (Dobbs et al. 2011). It is arguably due to this concentration of economic activities that, traditionally, people have migrated to cities in search for better livelihood opportunities, better access to services and amenities, and enhanced quality of life. These aspirations may, however, not always be fully realized due to various factors such as ineffective urban development and management policies and practices. As a result, externalities and problems such as social inequality, crime, environmental pollution, and traffic jam are common in many urban areas, especially those in developing countries. For instance, currently about 30% of world urban population is living in slums (WB 2021). In developing countries timeline of urban development plans and programs is often not compatible with the rapid rates of urbanization (Sharifi et al. 2014). In other words, urban planners and policy makers fail to keep up with the rapid urbanization rates, and this results in unbalanced urban development that is not conducive to achieving sustainable development goals. Urbanization challenges, however, are not only unique to developing countries and socio-economic and environmental issues are also common in the cities of developed countries. For instance, currently, more than 91% of global population is exposed to PM 2.5 air pollution (WB 2021), and this issue is more severe in cities. According to World Bank, data mortality rate (per 100,000 population) attributed to household and ambient air pollution is more than 114 (WB 2021).

Climate change and the recent COVID-19 pandemic have reignited the debates over cities and their future. On the one hand, there are many concerns over the vulnerability of cities to the impacts of climate change and other stressors and extreme events such as pandemics. According to some estimates, the annual direct loss attributable to natural disasters in cities is more than USD 314 billion (WBG 2016). As climate change is expected to further increase the frequency and intensity of extreme events, in the absence of efforts to enhance urban resilience, the amount of annual loss to disasters may increase significantly in the coming decades. As will be further discussed in Chapters 2 and 4 of this volume, resilience is a multi-dimensional property and enhancing it requires concerted efforts across different sectors, actors, and stakeholders. While there is no universal definition for the term resilience, it can be described as the ability to plan and prepare for, absorb, recover from, and adapt to adverse events (Cutter et al. 2013; Sharifi 2020b). Enhancing resilience is also needed for better capacity to deal with other stressors such as the recent COVID-19 pandemic. The pandemic has hit cities around the world hard and has exposed many vulnerabilities that need to be addressed in order to ensure better resilience to similar future adverse events (Sharifi 2021b; Sharifi and Khavarian-Garmsir 2020).

In addition to being exposed to the impacts of climate change, cities are also major drivers of climate change given the large share of energy consumed in cities. It is estimated that cities account for over 70 of global energy consumption and associated CO_2 emissions (WB 2010). This is a clear indication of the significance of urban climate action plans for addressing climate change issues. Investment in clean and renewable-based energy technologies not only allows cities to reduce their emissions, but also provides them with opportunities to achieve multiple co-benefits related to health, equity, air quality, and energy resilience (Sharifi 2021a; Sharifi and Yamagata 2016).

Recognizing the significance of cities for addressing climate change and other societal challenges, many international policy frameworks have highlighted the need for taking urgent actions in cities. For instance, the significance of cities and urban climate action plans has been underscored in the Paris Agreement (UNFCCC 2015) and the New Urban Agenda (Habitat 2017). Also, the United Nations Sustainable Development Goals (different goals, particularly Goal 11) emphasize the need for taking urgent actions in cities to address multiple social, economic, and environmental concerns (UNSDG 2015). As many cities are also exposed to various types of hazards, the Sendai Framework for Disaster Risk Reduction has also underscored the need for taking actions at the city level (UNISDR 2015). In response, many cities around the world have taken lead in efforts aimed at tackling climate change and other societal challenges. For instance, there are now alliances and networks of cities such as the C40 cities[1] and the Global Covenant of Mayors[2] that are focused on such activities. Indeed, the fact that policy making at the city scale is less complicated compared to the level of nation-states allows cities to take more nimble actions toward addressing climate change and other contemporary challenges. Based on this, it is argued that effective urban management policies and practices can provide solutions for addressing the increasing challenges that cities are facing and contribute to mitigating global climate change. Reliance on conventional approaches and strategies may, however, not be sufficient if cities want to be part of the solution to climate change and other societal challenges. Therefore, there has been increasing emphasis on adopting innovative and disruptive solutions that are transformative and can accelerate transition toward creating cities that are more resilient and sustainable. Such innovative solutions and technologies are also expected to contribute to achievement

[1] https://www.c40.org/.

[2] https://www.globalcovenantofmayors.org/.

of the Sustainable Development Goals (SDGs). Cities have traditional been centers of innovation and the demand for developing innovative solutions has led to growing interest and investment in smart solutions and technologies enabled by advances in information and communication technologies. There is also a vast body of literature on smart city solutions and technologies and their contribution to sustainability and, to some extent, to climate change adaptation and mitigation. However, existing literature is fragmented and there is, relatively, limited knowledge of the contributions of smart city solutions and technologies to resilience. This is a major gap that this volume aims to fill. To set the stage for the next chapters, here, based on an overview of the existing literature, I first discuss some of the major challenges that cities are now facing (Sect. 1.3). In Sect. 1.4, I briefly discuss potential contributions of smart city solutions and technologies to overcoming these challenges. Finally, in Sect. 1.5, I provide a summary of this edited volume and its contributions to advancing knowledge on smart and resilient cities.

1.2 Methods

There has been substantial increase in the number of academic publications in the past decade or so. Researchers now find it challenging to keep pace with the rapid rate of academic publishing. The large number of annually published articles makes conducting systematic reviews difficult, if not impossible. Further, given time limitations, researchers will not be able to follow all newly published studies. This challenge has been, to some extent, addressed by using scientometrics and bibliometric analysis techniques and tools that allow obtaining macroscopic overviews of the state of knowledge and structure of academic fields (Sharifi 2021c; Van Eck and Waltman 2013). Various tools such as VOSviewer, CiteSpace, and SciMAT have been developed for this purpose. These tools are capable of highlighting thematic transitions as well as major authors, documents, journals, countries, and institutions that have contributed to the development and evolution of academic fields. VOSviewer is particularly suitable for mapping research keywords (terms) and their patterns of co-occurrence. It can, therefore, be used to identify major thematic areas related to a field and determine how different terms/clusters are linked to one another. Such results can be used to find answers to broad questions that require a macroscopic overview. For instance, in the case of this work, it allows identifying major challenges that cities are facing and potential contributions of smart city solutions and technologies to overcoming these challenges.

Input data for bibliometric analysis in VOSviewer can be obtained from databases such as the Web of Science (WoS) or Scopus that archive bibliographic data of academic publications. Given its reputation for archiving quality peer-reviewed research, WoS was selected for the purpose of this chapter. To retrieve relevant articles for term co-occurrence analysis, two different search strings related to the two major objectives of the chapter were developed. First, a string was developed to identify key challenges facing cities (search string #1 in Table 1.1). This string

Table 1.1 Search strings used for retrieving relevant articles

Purpose	Search string
Identifying key challenges that cities are facing	Search string #1: TS=((“urbanization” OR “urban areas” OR “cities”) AND (“challenges” OR “problems” OR “issues”))
Identifying potential contributions of smart solutions and technologies discussed in the literature	Search string #2: TS=((“sustainability” OR “sustainable development” OR “climate change” OR “resilience” OR “air pollution” OR “air quality” OR “food security” OR “mobility” OR “accessibility” OR “water resources” OR “waste management” OR “ecosystem services” OR “housing” OR “health” OR “COVID-19” OR “energy” OR “vulnerability” OR “vulnerabilities” OR “sdgs” OR “urban regeneration” OR “biodiversity” OR “urban heat”) AND (“digitali*ation” OR “digital technolog*” OR “Information and communication technolog*” OR “ict” OR “information technolog*” OR “internet of things” OR “iot” OR “artificial intelligence” or “AI” or “machine learning” OR “blockchain” OR “virtual reality” OR “VR” OR “augmented reality” OR “AR” OR “3-d print*” OR “three-dimensional printing” OR “cloud computing” OR “big data” OR “5G” OR “6G” OR “Smart technolog*” OR “smart home*” OR “smart house*” OR “smart cit*” OR “home energy management system*” OR “industry 4*” OR “society 5*” OR “robotic*” OR “automation” OR “unmanned aerial vehicle*” OR “UAV*” OR “smart meter*” OR “smart grid” OR “vehicle-to-vehicle communication” OR “machine-to-machine communication”) AND (“urbanization” OR “urban areas” OR “cities”))

includes terms related to urbanization and its challenges. Searching for the string in the “titles, abstracts, and keywords” field of WoS, for an indefinite period, returned 34,765 articles on September 12, 2021. An initial check showed that some of these articles are not directly related to the subject matter of this study. Therefore, articles belonging to irrelevant fields (e.g., surgery, linguistics, etc.) were excluded from the database. At the end, 21,529 articles remained in the database and were used for identifying major challenges that cities are facing using the term co-occurrence analysis function of VOSviewer. To ensure synonyms and different variants of a term are not treated differently in the analysis, a thesaurus file was developed and used in the analysis (e.g., the term “urbanisation” was replaced with “urbanization” in the analysis).

After the key challenges were identified, the second search string (Table 1.1) was developed to explore potential contributions of smart city solutions and technologies to overcoming them. As can be seen from the table, search string #2 in a combination of different terms related to smart solutions and technologies, urbanization/cities, and urban issues and challenges. Again, the string was used to search for relevant articles in the "title, abstracts, and keywords" field of the WoS, for an indefinite period, and returned 5,881 articles on September 13, 2021. The retrieved articles were used as the input data for term co-occurrence analysis in VOSviewer. As can be seen in the next sections, the outputs are presented as a network (graph) of nodes and links. Each node refers to a highly occurred term and the node size is proportional to the frequency of the term's co-occurrence with other terms. The links between terms indicate the state of connection between them and link width is proportional to the strength of connection. Terms that have co-occurred more frequently and are strongly linked to each other form clusters that are presented in different colors. These clusters often represent major thematic focus areas related to academic fields and can be used to better understand the research structure and dominant topics. Results of the first analysis on the key challenges facing cities are presented in the following section.

1.3 Major Challenges Facing Cities

The output of the term co-occurrence analysis (Fig. 1.1) provides interesting insights into major challenges that have been discussed in existing research. It is clear that issues related to sustainability and climate change have received more attention. As mentioned in the Introduction section, existing climate change is widely recognized as the most important challenge that cities need to address in the future decades. Also, the compounded effects of climate change and urbanization may render transition toward sustainability challenging.

Four major clusters can be identified from Fig. 1.1. The cluster in green includes more frequently occurred terms that are strongly linked to terms within the cluster and terms from other clusters. This cluster is mainly focused on the impacts of unregulated and uncontrolled urban expansion (i.e., dispersed urban development/urban sprawl) on biodiversity and ecosystems. There is a lot of evidence showing how unregulated urban growth results in substantial land cover and land use changes, transforming ecological assets (e.g., forests and agricultural lands) into built up areas (Kafy et al. 2021; Sharifi et al. 2014). This often leads to biodiversity loss and ecosystem degradation with major implications for food security, urban microclimate regulation, and stormwater management. As the term map indicates, strategies and solutions such as urban green infrastructure and urban agriculture have been proposed to compensate for some of the ecological losses. In terms of linkages to smart solutions and technologies, the term map implies that smart city solutions have been rarely discussed in connection with biodiversity and ecological losses in cities and this could be highlighted as a gap to be addressed in the future. It is worth noting, however, that some of the techniques mentioned at the bottom of the figure (i.e., GIS and remote sensing) have benefited from smart solutions and big data analytics, thereby facilitating a

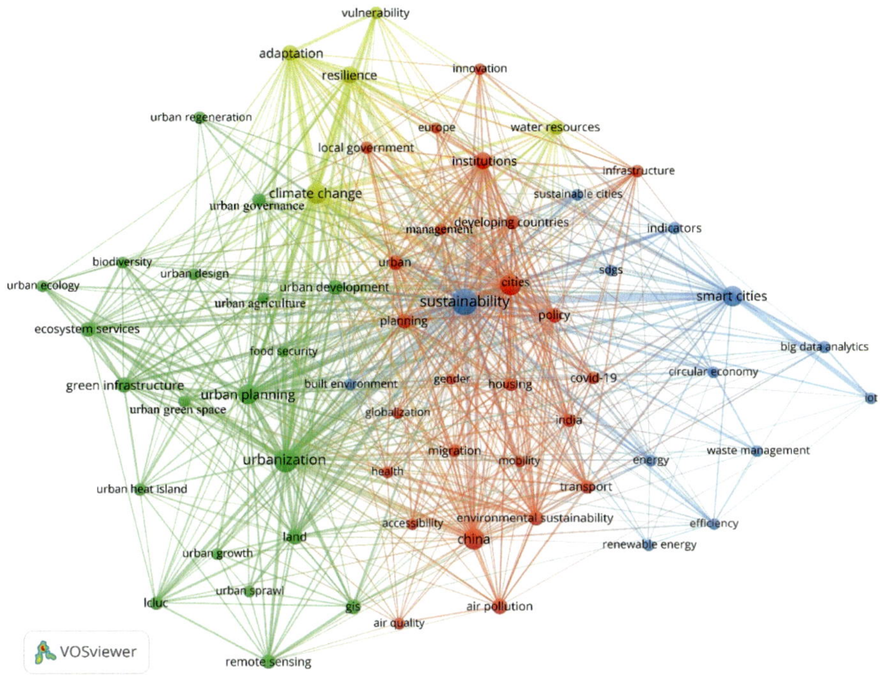

Fig. 1.1 Results of the term co-occurrence analysis to identify major challenges facing cities

better understanding of the past, current, and future patterns of land use and land cover change and ecological loss in urban areas (Xie et al. 2017).

The second major cluster (presented in red in Fig. 1.1) is focused on various issues related to urban planning and management, housing, and mobility. The most noteworthy point about this cluster is that it is located in between the smart cities cluster (in blue) and the clusters on ecological impacts (in green) and climate change adaptation (in yellow). Accordingly, it can, for instance, be argued that smart city solutions and technologies applied to the management, housing, and mobility sectors can indirectly impact ecological systems of cities and their hinterland areas. For example, using smart technologies to enhance energy efficiency of housing and mobility sectors could contribute to mitigating environmental impacts. Indeed, smart solutions and technologies can be utilized to develop integrated planning and assessment systems that could contribute to better energy performance across different sectors (Wang et al. 2021).

In terms of major challenges that can be highlighted, at the lower part of the cluster, terms such as air quality and air pollution are noteworthy. These are closely linked with other terms such as transport and health, indicating that addressing issues related to public health is a major challenge that could be addressing by focusing on the transportation sector.

The terms "gender," "housing," and "migration" are also highlighted and closely linked to each other. These could be linked to the challenges that cities are facing regarding provision of suitable housing to different societal groups. Access to affordable housing could particularly be challenging for females and those migrating from rural areas. Such challenges may also have ramifications for public health as was demonstrated during the recent pandemic. In many places such as the slums of Indian cities, the limited per-capital housing space and lack of access to services (e.g., clean water) made it difficult to comply with social distancing and hygiene measures recommended for controlling the spread of COVID-19 (Sharifi and Khavarian-Garmsir 2020). While smart solutions alone would not be sufficient for dealing with such complicated issues, they may provide opportunities for better access to services, or in the case of health crises such as the recent pandemic smart solutions can provide innovative and effective solutions for controlling the spread of the virus (Sharifi et al. 2021).

At the upper part of the red cluster terms such as "institutions," "local governments," "management," and "innovation" are dominant. Indeed, given the fact that cities are complex systems, and their management requires engagement with multiple stakeholders, urban governance has always been challenging. Traditionally, different urban sectors have operated in silos, resulting in inefficient urban management. The use of smart solutions and technologies will provide added value and could offer opportunities for integrated urban management that can, among other things, enhance efficiency, maximize co-benefits and synergies, and minimize conflicts and trade-offs (Jiang et al. 2021; Sharifi 2020a, 2021a).

The third major cluster (in blue) is focused on smart cities. Terms related to the energy sector are dominant in this cluster. Ensuring availability, accessibility, affordability, and acceptability (i.e., environmental sustainability) of energy resources in cities has always been a major challenge (Sharifi and Yamagata 2016). To address these challenges, smart solutions and technologies such as home energy management systems, smart metering systems, and smart grids have been widely used (Masrur et al. 2021). Waste management is another challenge that has been highlighted in this section. Smart solutions are also increasingly used to overcome this challenge in cities (Ali et al. 2020). Collectively, efforts aimed at enhancing efficiency of resource management can contribute to promoting circular economy in cities.

Finally, the cluster in yellow is focused on resilience to climate change impacts. As discussed earlier, climate change is expected to increase the frequency and intensity of adverse events in cities, and this is likely to lead to major challenges for urban planning and management. For instance, a large share of world urban population is living in coastal areas that are exposed to hazards such as sea level rise and flooding. Other common hazards that may affect quality of life in cities are extreme heat, drought, and storms. Therefore, urgent actions are needed to minimize urban vulnerabilities through developing and implementing resilience plans (Sharifi and Yamagata 2018). In this regard, smart solutions and technologies can contribute by, for instance, facilitating development of early warning systems or real-time monitoring systems that could lead to behavioral changes for better climate change adaptation.

1.4 Potential Contributions of Smart Solutions and Technologies

Some of the major challenges that cities are facing were highlighted in the previous section. Also, as "smart cities" was one of the major thematic clusters that emerged from the analysis reported in the previous section, potential contributions of smart solutions to addressing the challenges were briefly discussed. In this section, potential avenues for contribution will be further discussed based on the results of the term co-occurrence analysis that was specifically conducted for this purpose.

In the previous section, it was discussed that environmental degradation is one of the major challenges that cities need to deal with. Results of the term co-occurrent analysis conducted to highlight potential avenues for the contribution of smart city solutions to addressing urban challenges (Fig. 1.2) show that interlinkages between smart cities and sustainability is one of the most dominant clusters (in red). The terms "sustainability" and "smart cities" are strongly linked to each other, indicating that exploring implications of smart city solutions and technologies for sustainability is of high interest to researchers. It is widely acknowledged that smart cities should also be able to contribute to sustainability and this has led to the emergence and spread

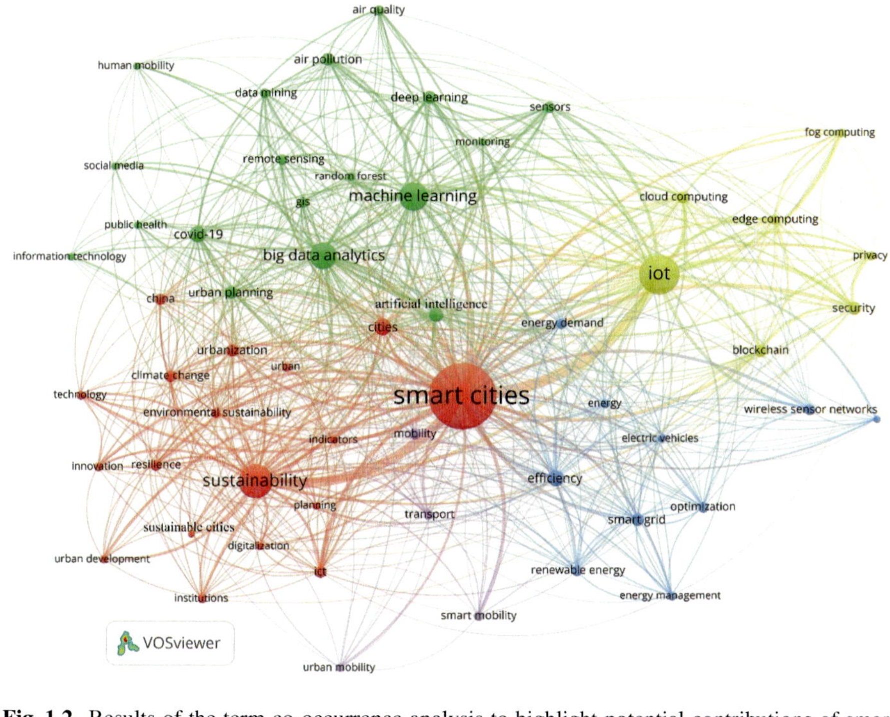

Fig. 1.2 Results of the term co-occurrence analysis to highlight potential contributions of smart city solutions and technologies to addressing urban challenges

of the concept of "smart sustainable cities" (Bibri and Krogstie 2017). According to the literature, smart city initiatives have been successful in enhancing sustainability awareness and have led to the development of more ambitious sustainability targets (Haarstad and Wathne 2019). However, there are also concerns that inappropriate implementation of energy-intensive smart solutions could have negative impacts on environmental sustainability. The red cluster also includes terms such as climate change and resilience, indicating that smart solutions and technologies can contribute to increasing resilience against climate change adaptation and mitigation (Allam 2020). More specific examples of such contributions will be provided when describing the other clusters below.

The second major cluster that is closely linked to the red cluster is the green cluster. As shown in Fig. 1.2, this cluster is mainly focused on smart city solutions and technologies such as machine learning, deep learning, data mining, sensors, and big data analytics. These terms are closely linked to the terms "air pollution" and "air quality," indicating that a lot of research exists on using such technologies for dealing with the issue of air pollution in cities. For instance, smart solutions and technologies can be used for enhanced monitoring, modeling, and prediction of air pollution (Ma et al. 2019; Zhu et al. 2018). This will allow planners and policy makers take more effective actions toward improving urban air quality. The term "COVID-19" is also a frequently used term in this cluster, indicating that a large body of research has been conducted on it over the past two years. On the one hand, a lot of pandemic-related publications are focused on the environmental implications of COVID-19 and methods such as big data analytics and monitoring have been used to examine how urban environmental quality has changed following measured developed to control the pandemic (Sharifi and Khavarian-Garmsir 2020). On the other hand, there is extensive evidence showing how mart solutions have contributed to better resilience against the pandemic by strengthening planning, absorption, recovery, and adaptation capacities in cities (Hassankhani et al. 2021; Sharifi et al. 2021). Such contributions will be further discussed in Chapter 7 of this volume.

The clusters in yellow and blue are closely linked to each other and are mainly focused on the energy sector and energy resource management. Various technologies such as IoT and blockchain have been used for enhancing efficiency of energy resource management (Tomazzoli et al. 2020). ICT and smart solutions have enabled technologies and systems such as home energy management systems, smart meters, vehicle to grid systems, and decentralized energy systems. There is a lot of evidence showing that such smart technologies provide multiple benefits for improving efficiency of energy supply and demand through, among other things, scheduling optimization, monitoring performance, real-time communication with users, and minimizing the mismatch between supply and demand (Ding and Wu 2019; Yusri and Nashiruddin 2020).

The yellow cluster also includes the terms "privacy" and "security." As smart solutions involve collection of large volumes of, sometimes, sensitive information, there are concerns over the privacy implications of smart technologies (Al-Turjman et al. 2021). However, efforts have been made to design secure architecture for ensuring security of smart technologies such as IoT, cloud computing, and fog computing (Al-Turjman et al. 2021). Appropriate design and implementation of blockchain-based systems can, for instance, contribute to overcoming security and privacy concerns (Shi et al. 2020).

The last cluster (in purple) is focused on mobility and transport. Among other things, IoT and big data analytics have contributed to better operation and management of transportation systems (Muthuramalingam et al. 2019), enhancing time management, scheduling, and energy efficiency of transportation systems (Kyriazis et al. 2013), more effective and efficient parking management (Saarika et al. 2017), and improved transportation safety through safety sensors or alcohol detection systems (Uzairue et al. 2018).

Overall, this macroscopic overview showed that smart solutions and technologies have a lot of potential to contribute to addressing urban challenges. While addressing these challenges can also directly or indirectly contribute to enhancing resilience, the term maps show that "resilience" has not been the main focus of smart cities literature. Therefore, further research is needed to better understand contributions of smart city solutions and technologies to resilience. As will be discussed in the next section, this volume aims to fill this gap by providing more theoretical and empirical insights on the resilience of smart cities.

1.5 Final Remarks

Based on macroscopic overview of the existing literature, in this chapter, I first discussed some of the major challenges that cities are now facing. Results showed that major challenges are related to ecological degradation, unregulated urban expansion, climate change adaptation and mitigation, resource management, fragmented urban management, air pollution, housing, and transportation. Next, I briefly discussed potential contributions of smart city solutions and technologies to overcoming these challenges. It was discussed that smart solutions and technologies can contribute to solving issues related to different domains such as sustainability, public health, air quality, energy management, transportation, and security. It was, however, found that limited attention has been paid to the contributions of smart cities to resilience. To address this issue, in this volume various issues related to the theoretical and empirical aspects of smart city planning, development, and implementation are discussed.

The first part is focused on theoretical underpinnings of the concept of resilient smart cities. In the next chapter, recent advances in smart cities and urban resilience will be presented and the need for resilient smart cities will be discussed. Chapters 3 and 4 will focus on the multiple concepts and underlying principles and characteristics of smart cities and resilient cities, respectively. Following this, the concept of resilient smart cities and its underlying principles will be discussed in Chapter 5. To complement this, results of a survey on the principles and characteristics that should be considered for achieving smart city resilience will be presented in Chapter 6 of the volume.

In the second part of the volume, different empirical results related to resilient smart cities will be presented. The recent pandemic provided an unprecedented opportunity to examine actual and/or potential contributions of smart solutions and technologies to resilience. Some such contributions are discussed in Chapter 7. In Chapter 8, results of a case-based literature survey to understand the general typology and focus of smart city projects around the world are presented. Following this, there will be more specific case studies from Taiwan (Chapter 9), Malaysia (Chapter 10), Australia (Chapter 11), India (Chapter 12), the United States (Chapter 13), South Korea (Chapter 14), Europe (Chapter 15), general cases from around the world (Chapter 16), and South America (Chapter 17). These cases are focused on multiple issues such as urban management, to climate change, energy systems, and transportation systems and will contribute to gaining better insights into actual and/or potential contributions of smart city solutions and technologies to urban resilience. They show that adopting smart solutions will put cities in a stronger position to deal with adverse events. Hopefully, planners and policy makers will find these insights useful for planning and developing cities that are more sustainable and resilient.

References

Al-Turjman F, Zahmatkesh H, Shahroze R (2021) An overview of security and privacy in smart cities' IoT communications. Trans Emerg Telecommun Technol n/a(n/a):e3677. https://doi.org/10.1002/ett.3677

Ali T, Irfan M, Alwadie AS, Glowacz A (2020) IoT-based smart waste bin monitoring and municipal solid waste management system for smart cities. Arab J Sci Eng 45(12):10185–10198. https://doi.org/10.1007/s13369-020-04637-w

Allam Z (2020) The rise of autonomous smart cities: Technology, economic performance and climate resilience. Springer Nature

Bibri SE, Krogstie J (2017) Smart sustainable cities of the future: An extensive interdisciplinary literature review. Sustain Cities Soc 31:183–212. https://doi.org/10.1016/j.scs.2017.02.016

Cutter SL, Ahearn JA, Amadei B, Crawford P, Eide EA, Galloway GE, Goodchild MF, Kunreuther HC, Li-Vollmer M, Schoch-Spana M, Scrimshaw SC, Stanley E.M, Whitney G, Zoback ML (2013). Disaster resilience: A national imperative. Environ: Sci Policy Sustain Dev 55(2):25–29. https://doi.org/10.1080/00139157.2013.768076

Ding X, Wu J (2019) Study on energy consumption optimization scheduling for Internet of Things. IEEE Access 7:70574–70583. https://doi.org/10.1109/ACCESS.2019.2919769

Dobbs R, Smit S, Remes J, Manyika J, Roxburgh C, Restrepo, A (2011) Urban world: Mapping the economic power of cities, 62. McKinsey Global Institute

Haarstad H, Wathne MW (2019). Are smart city projects catalyzing urban energy sustainability? Energy Policy 129:918–925. https://doi.org/10.1016/j.enpol.2019.03.001

Habitat U (2017) New urban agenda. Accessed on 5 March 2020. Retrieved from http://habitat3.org/wp-content/uploads/NUA-English.pdf

Hassankhani M, Alidadi M, Sharifi A, Azhdari A (2021). Smart city and crisis management: Lessons for the COVID-19 pandemic. Int J Environ Res Public Health 18(15):7736. Retrieved from https://www.mdpi.com/1660-4601/18/15/7736

Jiang H, Geertman S, Witte P (2021) Smartening urban governance: An evidence-based perspective. Reg Sci Policy & Pract 13(3):744–758. https://doi.org/10.1111/rsp3.12304

Kafy AA, Dey NN, Al Rakib A, Rahaman ZA, Nasher NMR, Bhatt A (2021) Modeling the relationship between land use/land cover and land surface temperature in Dhaka, Bangladesh using CA-ANN algorithm. Environ Chall 4:100190. https://doi.org/10.1016/j.envc.2021.100190

Kyriazis D, Varvarigou T, White D, Rossi A, Cooper J (2013, June 4–7). Sustainable smart city IoT applications: Heat and electricity management & Eco-conscious cruise control for public transportation. Paper presented at the 2013 IEEE 14th International Symposium on "A World of Wireless, Mobile and Multimedia Networks" (WoWMoM)

Ma J, Cheng JCP, Lin C, Tan Y, Zhang J (2019) Improving air quality prediction accuracy at larger temporal resolutions using deep learning and transfer learning techniques. Atmos Environ 214:116885. https://doi.org/10.1016/j.atmosenv.2019.116885

Masrur H, Sharifi A, Islam MR, Hossain MA, Senjyu T (2021) Optimal and economic operation of microgrids to leverage resilience benefits during grid outages. Int J Electr Power Energy Syst 132:107137. https://doi.org/10.1016/j.ijepes.2021.107137

Muthuramalingam S, Bharathi A, Rakesh kumar S, Gayathri N, Sathiyaraj R, Balamurugan B (2019) IoT based intelligent transportation system (IoT-ITS) for global perspective: A case study. In: Balas VE, Solanki VK, Kumar R, Khari M (eds) Internet of things and big data analytics for smart generation. Springer International Publishing, Cham, pp 279–300

Saarika PS, Sandhya K, Sudha T (2017) Smart transportation system using IoT. Paper presented at the 2017 International Conference on Smart Technologies for Smart Nation (SmartTechCon), 17–19 August

Sharifi A (2020a) Trade-offs and conflicts between urban climate change mitigation and adaptation measures: A literature review. J Clean Prod 276:122813. https://doi.org/10.1016/j.jclepro.2020.122813

Sharifi A (2020b) Urban resilience assessment: Mapping knowledge structure and trends. Sustainability (Switzerland) 12(15). https://doi.org/10.3390/SU12155918

Sharifi A (2021a) Co-benefits and synergies between urban climate change mitigation and adaptation measures: A literature review. Sci Total Environ 750:141642. https://doi.org/10.1016/j.scitotenv.2020.141642

Sharifi A (2021b) The COVID-19 pandemic: Lessons for urban resilience. In: Linkov I, Keenan JM, Trump BD (eds) COVID-19: Systemic risk and resilience. Springer International Publishing, Cham, pp 285–297

Sharifi A (2021c) Urban sustainability assessment: An overview and bibliometric analysis. Ecol Indic 121:107102. https://doi.org/10.1016/j.ecolind.2020.107102

Sharifi A, Chiba Y, Okamoto K, Yokoyama S, Murayama A (2014) Can master planning control and regulate urban growth in Vientiane, Laos? Landsc Urban Plan 131:1–13. https://doi.org/10.1016/j.landurbplan.2014.07.014

Sharifi A, Khavarian-Garmsir AR (2020) The COVID-19 pandemic: Impacts on cities and major lessons for urban planning, design, and management. Sci Total Environ 749. https://doi.org/10.1016/j.scitotenv.2020.142391

Sharifi A, Khavarian-Garmsir AR, Kummitha RKR (2021) Contributions of smart city solutions and technologies to resilience against the COVID-19 pandemic: A literature review. Sustainability 13(14):8018. Retrieved from https://www.mdpi.com/2071-1050/13/14/8018

Sharifi A, Yamagata Y (2016) Principles and criteria for assessing urban energy resilience: A literature review. Renew Sustain Energy Rev 60:1654–1677. https://doi.org/10.1016/j.rser.2016.03.028

Sharifi A, Yamagata Y (2018) Resilience-oriented urban planning. In: Yamagata Y, Sharifi A (eds) Resilience-oriented urban planning: Theoretical and empirical insights. Springer International Publishing, Cham, pp 3–27

Shi S, He D, Li L, Kumar N, Khan MK, Choo K-KR (2020) Applications of blockchain in ensuring the security and privacy of electronic health record systems: A survey. Comput & Secur 97:101966. https://doi.org/10.1016/j.cose.2020.101966

Tomazzoli C, Scannapieco S, Cristani M (2020) Internet of Things and artificial intelligence enable energy efficiency. J Ambient Intell Humlzed Comput. https://doi.org/10.1007/s12652-020-02151-3
UN (2018) World urbanization prospects: The 2018 revision, online edition. Department of Economic and Social Affairs PD New York, NY
UNFCCC (2015) Paris Agreement, United Nations Framework Convention on Climate Change. Accessed on 24 May 2020. Retrieved from https://unfccc.int/files/essential_background/convention/application/pdf/english_paris_agreement.pdf
UNISDR (2015) Sendai framework for disaster risk reduction 2015–2030. Paper presented at the Proceedings of the 3rd United Nations World Conference on DRR, Sendai, Japan
UNSDG (2015) About the Sustainable Development Goals, United Nations. Accessed on 30 October 2019. Retrieved from http://www.un.org/sustainabledevelopment/sustainable-development-goals/
Uzairue S, Ighalo J, Matthews VO, Nwukor F, Popoola SI (2018) IoT-enabled alcohol detection system for road transportation safety in smart city, Cham
Van Eck NJ, Waltman L (2013) VOSviewer manual. Leiden: Univeristeit Leiden 1(1):1–53
Wang C, Gu J, Sanjuán Martínez O, González Crespo R (2021) Economic and environmental impacts of energy efficiency over smart cities and regulatory measures using a smart technological solution. Sustain Energy Technol Assess 47:101422. https://doi.org/10.1016/j.seta.2021.101422
WB (2010) Cities and Climate Change: An Urgent Agenda (Urban Development Series; Knowledge Papers No. 10). Washington, DC. © World Bank. License: CC BY 3.0 IGO. Retrieved from https://openknowledge.worldbank.org/handle/10986/17381
WB (2021) World Bank Open Data. https://data.worldbank.org/. Accessed on 13 September 2021
WBG (2016) Investing in urban resilience: Protecting and promoting development in a changing world. World Bank Group, Washington, DC. © World Bank. License: CC BY 3.0 IGO. Retrieved from https://openknowledge.worldbank.org/handle/10986/25219
Xie H, He Y, Xie X (2017) Exploring the factors influencing ecological land change for China's Beijing–Tianjin–Hebei Region using big data. J Clean Prod 142:677–687. https://doi.org/10.1016/j.jclepro.2016.03.064
Yusri A, Nashiruddin MI (2020) LORAWAN Internet of Things Network Planning for Smart Metering Services. Paper presented at the 2020 8th International Conference on Information and Communication Technology (ICoICT), 24–26 June
Zhu D, Cai C, Yang T, Zhou X (2018) A machine learning approach for air quality prediction: Model regularization and optimization. Big Data Cogn Comput 2(1):5. Retrieved from https://www.mdpi.com/2504-2289/2/1/5

Chapter 2
Recent Advances in Smart Cities and Urban Resilience and the Need for Resilient Smart Cities

Ayyoob Sharifi, Rhea Srivastava, Nehmat Singh, Ruchi Tomar, and Mustapha A. Raji

Abstract Cities around the world have traditionally dealt with a wide array of natural and human-made risks and hazards. Annually, this results in significant human and economic losses in urban areas. As climate change is expected to further increase the frequency and intensity of adverse events, and other adverse events such as the recent COVID-19 pandemic may also hit cities again in the future, cities around the world increasingly recognize the importance of building on urban resilience to minimize vulnerabilities and enhance resistance, absorption, recovery and adaptation capacities. The rapid advances in smart city solutions enabled by information and communication technologies have also provided cities with more tools and opportunities to deal with adverse events. There is a vast body of literature on both smart city and urban resilience. However, the concept of smart city resilience has received limited attention in the literature. To fill this gap, in this chapter, we first provide overviews of the underlying principles of the smart city and urban resilience concepts. Next, we explain how adopting integrated approaches that simultaneously consider both smartness and resilience can help cities take more effective and efficient efforts toward dealing with adverse events, enhancing quality of life and ensuring transition toward sustainable development.

Keywords Smart cities · Urban resilience · Smart and resilient cities · Urban planning

A. Sharifi (✉)
Graduate School of Humanities and Social Sciences, Hiroshima University, Higashihiroshima, Japan
e-mail: sharifi@hiroshima-u.ac.jp

R. Srivastava
The Energy and Resources Institute (TERI), New Delhi, India

N. Singh
IHS, Erasmus University Rotterdam, Rotterdam, The Netherlands

R. Tomar
School of Habitat Studies, Tata Institute of Social Sciences, Mumbai, India

M. A. Raji
Department of Urban Planning and Administration, University of Seoul, Seoul, South Korea

A. Sharifi and P. Salehi (eds.), *Resilient Smart Cities*, The Urban Book Series,
https://doi.org/10.1007/978-3-030-95037-8_2

2.1 Introduction

In 2017, the United Nations Conference on Housing and Sustainable Urban Development (Habitat III) developed a strategic action plan to address the urban challenges, which was adopted by all of its member states (Kalapurakal 2018). This action plan and many other international policy frameworks have increased the recognition of the significance of cities for addressing global challenges such as climate change, increased inequality and public health threats. Cities can be described as a complex form of human system and urban ecosystem with powerful pulling effects on the people, organizations and creative innovations (Singh 2015). Modern cities are expected to be the catalyst to urban challenges. Presently, 55% of the world's population lives in the urban areas and this share is projected to increase to 68% by 2050. One of the most significant change is from Latin America and the Caribbean with 81.2% living in the cities, increasing from 41.3% in the 50s (Buchholz 2020). The role of cities in the economic growth and development, and social and environmental issues has been in global focus in recent years (McCormick et al. 2013). Thus, these projections are to enable experts and policy makers in planning and implementing actions to mitigate urban challenges.

Cities continue to play a prevailing role in global production and consumption of energy and food (Negev et al. 2009). They act in an interconnected network of functions and do not operate in isolation (McCormick et al. 2013). Thus, they are faced with all sorts of challenges that will impact their growth and development. The continuous growth of population and change in geographical patterns in cities often lead to a number of urban problems that have serious impacts on the built and natural environment. Among other things, noteworthy challenges that cities are facing include environmental degradation, inequitable access to resources and services, ineffective mobility infrastructure, increasing vulnerability to disasters and increasing air pollution (Chan et al. 2019). In terms of disasters, as the impacts of climate change continue to emerge, cities are increasingly dealing with disasters such as floods and extreme weather events. As a case in point, in 2021, many cities around the world faced challenges caused by torrential rains, storms and record floods. The compounding effects of other events such as the COVID-19 pandemic have also made it more challenging for cities to overcome risks. Annually, in the absence of appropriate urban resilience programs, disasters result in major human and economic losses in cities and the impacts are expected to further increase in the future.

While most experts and organizations focus on environmental issues, socio-economic and institutional dimensions also deserve attention. Inequality is a major issue in our cities and inequality in the provision of basic social infrastructure and resources can greatly cause harm or destabilize the society (Chan et al. 2019). Equal distribution of resources and services is needed to enhance socio-economic and environmental justice. Among other things, this would require actions such as the redistribution of urban infrastructure, controlling the use of green space to increase agricultural production and prevent food security issues consequently, and proper use of modern and smart technologies. These, in turn, could contribute to improving the livelihoods of people in the cities.

It is generally known that the industrial revolution of the nineteenth century was a result of technological innovations in the agricultural sectors and industries. Consequently, it contributed to the increase in population and gave rise to change in urban geographical patterns. Subsequently, it catalyzed transformation in cities. While technology is perceived to have immensely contributed to environmental degradation, researchers have argued that technology can also improve the living standards.

The application of urban transformation approach toward repositioning and building resilience cities requires the adoption of the sustainable development strategies which is the UN Framework that was agreed upon by member nations in tackling the urban issues. Sustainable development strategies are goals designed to uplift the social and economic development of nations and as such attempt to resolve the urban issues that are constraints to the overall growth of our cities. These goals also recognize the role of technology and its importance for facilitating transition toward sustainable development. Sustainable urban transformation necessitates the comprehension that cities are sources of possibilities and potentials, promoting active partnership among diverse stakeholders, integrating different concepts and ideas, and stimulating experimentation with different solutions, and these approaches are necessary in dealing with critical urban issues in our modern times.

While cities are influenced in diverse ways by large-scale transformation processes, the vulnerability and opportunities for cities may also differ due to internal factors, such as the local economic structure as well as external relations and geographic location. These are challenges toward the implementation of transformative approaches to urban issues, and all the stakeholders are to take critical measures while addressing local issues in our cities. It is hoped that smart cities enabled by advances in information and communication technologies could provide opportunities to not only accelerate transition toward sustainable development, but also make cities more resilient against the increasing hazards and disasters.

In the remainder of this chapter, we first provide brief descriptions of urban resilience and smart cities. Next, we discuss the need for developing smart resilient cities.

2.2 Urban Resilience

2.2.1 A Brief History of Resilience

The word resilience is used particularly in relation to adaptation from a period of great stress. Since decades, resilience has been adapted by the researchers from various fields of ecology, public health, psychology, disaster management, urban planning and development, and business. The use of the term in different study domains resulted in different conceptions of the term. Even with emerging differences in defining and measuring the concept, resilience is commonly understood as bouncing and rebounding from difficult situations (Sharifi et al. 2017). A resilient system should also feature other characteristics such as resistance and elasticity (Olsson et al. 2015).

Resilience has been deeply rooted in psychology than in any other field of study (Olsson et al. 2015). According to the American Psychological Association (2014, p. 2), resilience is "the process of adapting well in the face of adversity, trauma, tragedy, threats or even significant sources of stress". In psychology, resilience is often understood either as a personal trait, or the process of adaptation or as an outcome now known as emergent resilience. Norman Garzemy is considered as a pioneer in resilience research focusing on growth and resistance under unfavorable circumstances (Kolar 2011).

In ecology, the term resilience became famous in the 1960s and 1970s. Ecologists interpreted resilience in two different ways: first, to measure the recovery time of a system from the disturbance; second, to know the extent of the disturbance to displace the system from its original state (Perrings 2006). However, Perrings (1998) considers resilience in a much broader sense and defines it as a measure of the ability of a system to withstand stresses and shocks—its ability to persist in an uncertain world.

With the origin of the term in ecology, the concept of resilience has diffused to other related fields, such as disaster management and cities. In disaster management, resilience refers to the ability of an individual or a community to adapt and recover from any hazard or stresses and also be able to restore and transform its structure (Combaz 2014). In regard to socio-ecological systems, interdisciplinary scientists define resilience as the "capacity to adapt or transform in the face of change in social-ecological systems, particularly unexpected change, in ways that continue to support human well-being" (Folke et al. 2016, para. 7). Hence, resilience as a concept has an acknowledged history with the meaning of the term being malleable, allowing stakeholders from various fields to accept a common terminology without having to agree on an exact definition.

2.2.2 Emergence and Characteristics of Urban Resilience

The theory of resilience has recently become a significant concept in the context of cities. As cities continue to urbanize, there are various uncertainties and challenges being faced, such as overexploitation of natural resources, climate change, water security, economic uncertainties and social conflicts. To solve these challenges, cities need to be designed, planned and managed while fostering urban resilience. Although the definition of 'urban resilience' as a term remains unclear, yet it is commonly accepted as the capacity of a city to resist or absorb shocks and stresses.

Urban resilience is traditionally linked with the theory of engineering resilience, which considers city resilience as the ability of physical infrastructure to withstand shocks and restore to their state prior to the shock (Martin-Moreau and Ménascé

2018). This meaning of resilience sees the city as a stable entity, where only certain disturbances can be addressed at a time, since risks are not considered to be related to each other. Further, this explanation directly focuses on the material element of the urban fabric, such as infrastructure, while ignoring non-material elements like social capital (UN-Habitat 2017).

Meerow et al. (2016, p. 39) define urban resilience as "the ability of an urban system and all its constituent socio-ecological and socio-technical networks across temporal and spatial scales-to maintain or rapidly return to desired functions in the face of a disturbance, to adapt to change, and to quickly transform systems that limit current or future adaptive capacity". This definition is dynamic and identifies the importance of temporal and spatial scales with having a general adaptability rather than being specific.

In literature, the definitions presented by various scholars on urban resilience embrace it as a positive attribute (Leichenko 2011; Brown et al. 2012). However, there is an ongoing debate on the equilibrium state described in the definitions by various scholars (Meerow et al. 2016). Urban resilience does not make it necessary for a system to return to the previous equilibrium path after stresses and shocks; rather, the previous equilibrium may have disappeared for any reason and an alternate or a new track may appear, all of which can indeed change the ambit of the system (Tasan-Kok et al. 2012). Some describe urban resilience as a state to go back to the 'normal' or 'same' state, but the question arises whether the normal state with conditions like poverty is desirable.

According to Gleeson (2008), urban resilience doesn't have a definite 'blueprint'; that is, the structure of a resilient city has its own path of evolution and adaptive capacity, and no two cities would have the same resilience. Further, the literature also determines three pathways to urban resilience, which are, *persistence* indicating that a system should be able to resist any disturbances; *transition* or adapt from one state to another; and *transform* radically or extremely.

Different researchers and institutes have defined different characteristics of urban resilience. According to Tasan-Kok et al. (2012, p. 46), scholars like Godschalk and Fleischhauer attempted to combine characteristics that measure the resilience of a city; they are, "*redundancy, diversity, efficiency, autonomy, strength, interdependence, adaptability*, and *collaboration*". However, scholars like Walker and Salt consider these characteristics from a qualitative perspective along with adding social dimensions like diversity, social capital, innovation, tight feedback, ecosystem services and overlap in governance (Tasan-Kok et al. 2012, p. 46).

Specific to addressing disaster management of cities at broader capacity, Jha et al. (2013) have divided urban resilience into four components: economic, institutional, infrastructural and social. Although it is essential to understand that a city might not necessarily undergo a sudden transition, a city can also adapt itself subject to slow burning processes (Meerow et al. 2016).

However, Martin-Moreau and Ménascé (2018) consider one serious limitation of urban resilience as being too multifaceted. In their paper, they mention the two schools of thoughts that have originated surrounding the notion of city resilience:

the notion is more useful when used in an inclusive sense and that it is considered as an ill-defined concept without any specificity toward the term.

2.2.3 Recent Developments

As a theoretical concept, resilience is continuously being conceptualized. There is vast academic literature focusing on the complexity of the notion and recommending the use of policy frameworks that need to be implemented in its practical form to be able to manage better complex issues of urban resilience (UN-Habitat 2017). With the increasing number of disasters and weather-related incidents taking place due to the climate change crisis, the discussion about resilience in cities is growing. Likewise, the spread of COVID-19 in the year 2020 further highlighted the need for and importance of better preparation of urban systems during the times of contagion.

There are several definitions and frameworks available in the publications on urban resilience. However, to enhance the concept even further, implementation of the same is necessary. As international and national organizations move forward with developing policies, frameworks and indicators for measuring resilience, it will impact how the concept is further appraised and advanced in the cities. The implementation of policy frameworks can be viewed from three approaches: top-down, bottom-up and mixed approach (Shamsuddin 2020).

In his paper, Shamsuddin (2020) focuses on the need to evolve the connection of urban resilience between theory and applicability for sealing the implementation gap. The implementation process, though, poses some challenges of its own in terms of coordination, adaptability of the concept and its diverse outcomes. But as a transformative concept, it is imperative to address resilience qualities such as redundancy and flexibility in new planning approaches (Frantzeskaki and Kabisch 2016).

Urban resilience is not only about preparing for natural or economic disasters, but also preventing the disasters from happening in the first place. It should be considered as a shift in mindset and policy priorities for a more integrated, forward-looking approach (Frantzeskaki and Kabisch 2016). In achieving urban resilience, both horizontal and vertical links in regard to the governance structure need to be strengthened for cities to be able to mitigate and adapt to any possible change (Tasan-Kok et al. 2012). Trust and transparency are considered as good principles to be practiced by members of different departments in the public administration when working on the topic of urban resilience projects (Frantzeskaki and Kabisch 2016).

Many international organizations have developed quantitative tools to measure the resilience of an urban system (Sharifi 2016). The International Organization for Standardization developed a set of indicators known as "Sustainable development of communities – Indicators for city services and quality of life" in 2014, which is considered to be applied at international scale in cities (UN-Habitat 2017). The Standard for Sustainable and Resilient Infrastructure (SuRe) has developed criteria for infrastructure at a global scale across social, governance and environmental aspects.

While measurement tools are being produced and developed by international organizations, the private sector is also engaging in the same. For example, the 100 Resilient Cities project by the Rockefeller Foundation is one such index developed for evaluating urban resilience in cities for physical, social and economic shocks and stresses. Ma et al. (2020) state that although every study has developed an urban resilience evaluation framework under their respective focus, yet, the components of economy, environment, social organization and infrastructure have been recognized widely.

In 2010, ICLEI launched the Resilient Cities Congress to raise awareness and provide support to the local and regional governments in challenges related to resilience. ICLEI laid special focus on the multi-level governance structure for adopting a resilient approach in cities. ICLEI encourages the government sector to collaborate with the private sector at local and regional project levels for various services, such as technology, consultancies, think tanks and research institutes, and investors, which would further enhance resilience in the cities (ICLEI 2019).

Another concept of urban resilience is implemented through a place-based experiment known as Urban Living Labs. It is a new type of government intervention which considers urban living labs as spaces for local experimental projects through intense collaboration by bringing together different stakeholders that are, public, private and civil society and among different sectoral divisions within a city. The advantage of urban living labs is that they engage into participatory planning by giving citizens an equal importance with private companies, research institutions and public organizations. Urban living labs give immense importance to dealing with sustainability threats in a city, where involvement of multiple actors facilitates new ways to deal with sustainability challenges (Frantzeskaki and Kabisch 2016). The concept of urban living labs is being preferred since the experimentation is placed and tested in a real-life setting, which enables the needs of citizens to come to equal grounds with the policies of the public actors to achieve desirable outcomes (Frantzeskaki and Kabisch 2016). Thus, the entire process of urban living labs ensures more efficiency in the final version of the project.

The concept of urban resilience can be integrated into all urban sectors, especially water supply and management, waste management, energy, communications and transportation (Jha et al. 2013). It is essential to give particular attention for the quick recovery of these urban sectors as they play an imperative role in the lives of the community members and their well-being. Jha et al. (2013) consider four essential strategies for building resilience in these urban sectors:

1. Locational approach: It is considered to be effective as a long-term approach that addresses risks particularly related to infrastructure development.
2. Structural approach: It is considered to be effective as a medium-term approach, most suitable for existing infrastructure.
3. Operational mitigation: Under this mitigation approach, alternative plans of actions need to be prepared in case of a system failure.
4. Financial approach: This approach undertakes short- and long-term financial needs of certain urban sectors.

Since cities are understood as complex, integrated socio-ecological systems, it is important to understand that having resilience in different domains, such a, water supply, health care and climate change, would not contribute to a city's overall resilience; rather, incorporating resilience in a system as a whole would make a larger impact (UN-Habitat 2017). For supporting urban resilience through the above-mentioned four strategies, institutional arrangements will have to promote interconnectedness, redundancy and flexibility—which can be achieved through collaboration between public, private and civil actors (Frantzeskaki et al. 2014).

To conclude, urban resilience differs from one city to another depending on temporal and spatial scales as well as the priority areas and hazards (UN-Habitat 2017). It is important for urban communities and its environment to be resilient taking into consideration every kind of disaster and disturbances. The major key to making urban resilience more efficient and integrated is by connecting to various other individuals and organizations (Coaffee et al. 2018). For this, it is essential for countries to develop new and effective quantitative and qualitative tools to measure urban resilience at local and regional level. Implementing urban resilience challenges the traditional functioning of public administration and is proving to be difficult to change the institutional set up in accordance with "resilience" (Coaffee et al. 2018). Nevertheless, with limited resources and capacity, the public and the humanitarian sector worldwide have started to recognize the importance of partnerships for additional resources and skills from the private sector that are necessary for making cities resilient.

Along with the advancement of the research and practice of urban resilience, the concept of smart city has gained traction in science and policy circles. However, as will be discussed later, these too concepts have rarely been discussed together. It is expected that integration of smart technologies would provide opportunities for better implementation of urban resilience plans and programs.

2.3 Smart Cities

2.3.1 A Brief History of Smart Cities

The concept of smart cities has been around since a long time and its usage has gained momentum in recent years, often being associated with the future of urban management via technological means. Despite this, there is no universally agreed definition of smart cities yet (Angelidou 2015; Ahvenniemi et al. 2017; Ismagilova et al. 2019). The disparity among smart city definitions can be attributed to the varying and complementary perspectives with which they have been described throughout literature over time. Initial conceptions, such as 'digital cities', 'information cities', 'ubiquitous cities' and 'intelligent cities' were regarded as the vision for what cities would look like in the future (Angelidou 2015). Nonetheless, they were seen from a technological perspective only and had a dominant focus on the use of ICT along with other modern technologies for innovative functioning of physical infrastructure systems (e.g., energy, transport, waste, communications, etc.) (Batty, 2013; Albino et al. 2015; Ahvenniemi et al. 2017). Contrastingly, other paradigms used such as the 'knowledge city' and 'creative city' incorporated social and human capital as well, like education, social innovation, knowledge economy and creativity, while actively demonstrating and elevating the socio-economic aspects of cities (Nam and Pardo 2011; Kitchin 2014; Angelidou 2015). Angelidou (2015, p. 99) particularly asserts that knowledge and innovation economy has played a crucial role in the smart city discourse and the emergence of the idea of smart cities in the current world scenario.

The emergence of urbanization, globalization and neoliberal ideologies along with the fast-paced technological advancements that accompanied them in cities dramatically aided the integration of the above-mentioned discourses in a holistic way (Kitchin 2015; Angelidou 2015). In fact, the inculcation of technologies in the management of city services and the socio-economic progress (in the form of privatization, collaboration, etc.) enabled by them has resulted in more efficient governance, facilitated innovation and a sustainable approach toward urban life (Kitchin 2015; Angelidou 2015; Silva et al. 2018).

2.3.2 Dimensions of Smart Cities

Due to the complexity of smart cities with several aspects involved, many researchers demarcated the concept into different dimensions/principles for simplification. According to Nam and Pardo (2011, p. 286), the core components of smart cities are technology, people and institutions. The authors highlight the interconnection between these components, mentioning that investments in technology (IT infrastructure) and social capital encourage sustainable growth while increasing the quality of life with a citizen-centric participatory governance. Similarly, Silva et al. (2018) also maintained physical, social, institutional and economic infrastructure as the four pillars of a generic smart city, where institutional infrastructure embodied transparent and participatory governance, physical infrastructure comprised of natural resources (green spaces, waterways, etc.) and other manufactured infrastructure such as buildings and energy systems, social infrastructure encompassed community engagement and citizen awareness while economic infrastructure focused on smart economy with the utilization of applications such as e-commerce for better economic management.

Filling the gap, Giffinger and Gudrun (2010), Neirotti et al. (2014) and Albino et al. (2015) in their literature review deemed the natural environment as one of the major components of a smart city too, which was missed out by others. Giffinger and Gudrun (2010, p. 14) identified smart economy, smart mobility, smart people, smart environment, smart governance and smart living as the fundamental components to smart cities. They associate smart economy with flexible labor markets and ICT industries, smart mobility with the use of ICT in transport infrastructure, smart people with learning, creativity and participation, smart environment with sustainable resource management and pollution reduction using renewables, smart governance with optimal, collaborative and transparent administration and finally smart living with access to technical and social infrastructure. Meanwhile, Albino et al. (2015) shortlisted four common underlying characteristics, incorporating social inclusion, natural environment, attention on business-led urban development and networked infrastructure.

On the whole, even with separated dimensions, it is important to note the systemic connection of each component with one another that also entails the essence of smart cities. Far away from a much debated technocentric approach, smart cities encompass a fuller picture, where technology acts more as an enabler, aiming for a lower environmental impact and improved quality of life through information processes and innovation. Caragliu et al. (2011, p. 50) appropriately put together all components mentioned above and described "a city to be smart when investments in human and social capital and traditional (transport) and modern (ICT) communication infrastructure fuel sustainable economic growth and a high quality of life, with a wise management of natural resources, through participatory governance".

2.3.3 Recent Developments in Smart City Research

With the current evolution of digital technologies and increasing affordability, smart cities are getting even smarter and transforming urban planning (Yigitcanlar 2015; Talari et al. 2017). Research on smart cities is also expanding, with contributions from the electronics and machine learning community on the application of these technologies in cities. The concept of Internet of Things (IoT), where information is gathered by interconnected digital devices, such as smartphones, sensors and computers, and subsequently stored on cloud services, has expanded smart city activities (Talari et al. 2017; Costa and Duran-Faundez 2018; Muhammad et al. 2021). Correspondingly, the onset of IoT has resulted in increased reliance on digitization and, in turn, a large volume of data generation by city stakeholders, the formation of which is known as big data (Allam and Dhunny 2019). The correlative working of big data, ICT and cloud services simultaneously along with the help of Artificial Intelligence (AI) for data mining, analysis and interpretation facilitates interoperability among sub-systems and fosters informed decision making and efficient policies in cities in many smart city sectors/components (Allam and Dhunny 2019; Batty 2013). Supporting that, Nuaimi et al. (2015) and Hashem et al. (2016) assert that big data analytics, albeit requiring proper ICT infrastructure and smart networks, can play a role in improving urban sectors ranging from health care to education, transportation, agriculture, waste, water, energy and governance. These technological tools not only create synergy through integration and communication among the smart city sectors and other stakeholders, but also reflect better liveability through enhanced user experiences, civic engagement and job opportunities (Allam and Dhunny 2019).

For instance, one of the big data applications on smart cities, the paradigmatic smart grid has enabled better management of energy systems. With the help of smart meters and sensors, production and consumption energy data are used for predicting disruptions, increasing reliability, quality and profits from energy systems as well as making users more aware, thereby reducing environmental impact (Al Nuaimi et al. 2015; Talari et al. 2017). Similarly, the healthcare sector utilizes effective diagnosis and treatment tools for better services, and transportation systems have become more environmentally friendly and optimized with real-time traffic monitoring and route information (via GPS) that can further play a crucial role in disaster and evacuation circumstances (Batty et al. 2012; Talari et al. 2017; Aqib et al. 2020). Furthermore, Angelidou et al. (2017) in their research also bring up the resource consumption aspect, arguing that the usage of the above-mentioned technologies has the potential to drive a lesser material-intensive economy, thus, encouraging eco-economic decoupling.

A crucial advancement can particularly be seen in the smart urban governance component of smart cities. With respect to the technological advancements, big data and IoT are revolutionizing how governments manage cities. More and more open data portals by city governments are emerging to encourage transparency and accountability (Batty et al. 2012). Moreover, additional innovative approaches such as the use of blockchain technology for transactions have proved to be useful in the prevention of bureaucratic lags, disorganization and uncertainties, along with stimulating collaboration between machines, government and people (Allam and Dhunny 2019). Indeed, the citizen-driven collaborative element of smart cities is enhanced with public–private partnerships, maintaining a balance between top-down policies and bottom-up grassroots initiatives (Baccarne et al. 2014). A relevant concept increasingly being used is that of urban living labs. As explained in the previous section, this concept drives innovation by keeping end-users, private industry, academia and other stakeholders involved in real-life testing of new ideas and technologies, thus fostering experiment-based approaches to urban issues (Cosgrave et al. 2013; Baccarne et al. 2014). As they are mostly supervised by governments, living labs have an imminent focus on social value creation, an inherent fundamental of smart cities.

In general, the number of real-time smart city projects is increasing at a drastic speed, with cities and countries drafting elaborate smart strategies that also enable the achievement of the United Nations Sustainable Development Goals (SDGs). The European Union (EU) has been endorsing the smart city concept to generate social and economic value, competitiveness along with a sustainable living for many years. In the realm of urban development, their main focus lies on environmental sustainability and the use of greener technologies to tackle greenhouse gas (GHG) emissions, for which they have set ambitious targets and policies (Ahvenniemi et al. 2017). Correspondingly, smart city projects that get heavy public-sector funding in the EU are designated with the term 'lighthouse city'. Lange and Knieling (2020) claim that there have been as many as 42 lighthouse cities under 15 EU projects since 2014, tackling energy, mobility and other ICT issues. They also argue that the competition for funding has even institutionalized the smart city vision in the EU. As a result, cities like Amsterdam, Copenhagen, Barcelona, etc., are running at the forefront in smart city initiatives currently. Apart from that, the Government of India also launched the Smart Cities Mission in 2015, aiming to develop 100 world class smart cities in India. This mission largely focused on economic development and quality life in existing cities and highlighted the collaborative approach between the government and private actors (Praharaj et al. 2018). In contrast, The United States underpin smart cities with relatively low funding but looks at the concept from an entrepreneurial perspective with a range of start-ups coming up already (Claudel et al. 2015). As a matter of fact, it can be said that smart cities are looked at differently even spatially and not just temporally.

The growth of smart cities has also prompted an increase in the assessment tools to monitor and evaluate smart cities and address any shortcomings (Sharifi 2020). Ahvenniemi et al. (2017) point out in their research the similarities between smart city and sustainable city assessment frameworks, thus directly highlighting their link. These tools majorly encompass the above-discussed components of a smart city, and numerous types of indicators are used to measure them with their emphasis ranging from impacts to outputs and processes (Sharifi 2019). While these assessment tools facilitate governance and policy processes, many tools are also used for comparison among different cities in terms of their smartness (Albino et al. 2015). CITY keys indicators for smart city projects and smart cities by Bosch et al. (2017) is such an example. This tool is used to evaluate smart city project impacts based on five key themes, "people, planet, prosperity, governance and propagation" in order to tackle societal challenges (Bosch et al. 2017, p. 9). Other assessment methods outlined by Caird (2018) also include the Smart City Maturity Model, European Smart Cities Ranking Model, IBM's Smarter City Assessment Tool, Smart City Index Master Indicators (SCIMI) framework, etc.

2.3.4 Challenges Ahead of Smart Cities

Smart city development and implementation come with radical, disruptive changes in the perception of managing cities. Although mainstreaming of emergent technologies in cities is still at a nascent stage, Batty (2013, p. 277) opine that the rise of big data in smart cities is pushing the world into "short-termism", causing a shift from long-term planning (20 or 50 years) to short-term thinking (5 years) about management of cities. At the same time, their implementation also comes with many technical, socio-economic and environmental challenges (Ahad et al. 2020). First and foremost is the issue of security and privacy. With an enormous reliance on data in smart cities, data confidentiality, privacy of citizen data and resistance against cyber-attacks should be a top priority for cities and their governments in order to maintain trust among citizens for successful implementation of advanced technologies (Talari et al. 2017; Ahad et al. 2020). Access and acceptance of these technologies also pose as a big challenge, often causing a digital divide between tech savvy groups and citizens with no access to them or no knowledge of how to use them (Ahad et al. 2020). Additionally, upgradation of existing infrastructure fit for smart technologies also incurs high implementation costs (Al Nuaimi et al. 2015; Hashem et al. 2016; Ahad et al. 2020). As a result of this, there is also the possibility of just developed countries going ahead with the initiatives while the lower income countries back out. Moreover, since data in smart cities is gathered from several sources, data mining, integration, sharing and visualization become much more complex, thus requiring advanced software systems to manage (Al Nuaimi et al. 2015; Hashem et al. 2016). Under the environment realm, e-waste, that is the disposal of devices in an environment friendly way, is one of the most crucial challenges to be faced with (Ahad et al. 2020). Cities are drafting various kinds of policies on e-waste disposal to ensure no harm is being caused to the planet. All in all, while there are many more challenges to smart city implementation, cities are increasingly willing to invest into the complex technologies and innovation for an inclusive, efficient, safe and liveable environment.

Smart cities have the potential to balance the social, economic, institutional and environmental aspects of urbanity. Hence, the importance of a multidisciplinary as well as an interdisciplinary approach (among and within systems) toward smart cities for urban development is reiterated multiple times, both in literature and in practice. In any case, what was once regarded as a distant utopia is now becoming a reality, radically changing the existing notions of urban development. Data platforms are continuously evolving, uncertainties are being eliminated and new tools are opening up a plethora of opportunities for various stakeholders while providing a promising and sustainable future of smart cities.

2.4 The Need for Developing Smart and Resilient Cities

As mentioned earlier, due to increasing urban development and disaster issues, both smart and resilient cities have been studied and practiced over the years across different geographies of the world. As shown, the studies revolving around the concept of resilience and smart city have given new perspectives and requirements for integration in the framework of the built environments. It is realized and established by many studies that both concepts are originated and planned on the similar basis and have similar goals and pathways toward development and resolve similar predicaments (de Jong et al. 2015). Despite this, they have rarely been studied together in an integrated manner.

Preferably, the concepts of smart city and resilient city should be interweaved and/or converged for enhanced understanding and implementation of the projects, while the existing definitions and literature on both of the concepts have proved inversely (van den Buuse and Kolk 2019). Despite many efforts in the past, this complexity is majorly attributed to the fact that both concepts are not defined in a clear way. The real meaning of the terms resilience and smart city is still unclear and ambiguous as the popular notion and trend in climate change and urban development field convey various different interpretations (Bellini et al. 2017). Consequently, in most of the cities or countries, the projects related to smart and resilient city concepts are being implemented in isolation, even though there are many overlaps in their meaning and relationships.

On the other side, the necessity of integrating the smart city and resilience concept at planning and operational levels has lately started to appear with focus on transition of mismanaged, conventional and vulnerable built environments to the ones with inclusion of resilient, smart and sustainable concepts (Jovanović et al. 2019). Interestingly, some of the research projects are exploring the significance of integrating smart city and resilience frameworks in the existing urban system and planning. Some such examples include Bahnstadt, the city quarter of Heidelberg in Germany where practitioners tried to model the vulnerabilities and resilience of the built environment in a smart city context and Florence city from Italy where smart technology was employed to monitor urban resilience levels. At the country and city levels, 187 Chinese smart cities were evaluated on the resilience concept and only four cities, namely Beijing, Shanghai, Shenzhen and Guangzhou, emerged as most successful on the concept of smart and resilient urban development (Zhu et al. 2019). These demonstrated examples are positive steps toward convergence of both the concepts. However, the status quo demonstrates the existence of an imprecise cognitive context that obstructs the integration of both concepts despite proven potential and functioning need for such development. Hence, this section of the chapter will focus on the possible limitations of isolated implementation of both concepts and benefits of integrated approach while outlining the possible opportunities and challenges.

2.4.1 Limitations of Isolated Approaches

Cities are complex systems with many risks and opportunities entangled within their structures. The scale and complexity of these risks make cities more vulnerable and amplify the challenges to achieve sustainable growth. To completely explore the opportunities of integration between smart and resilience in cities, we need to understand the limitations in isolated implementation of both concepts. To understand the possible limitations of isolated implementation of resilient and smart cities, there is need to consider all four dimensions of development, i.e., social, physical, economic and environmental (Moraci et al. 2018).

In physical context, smart city projects are mostly based on the smart solutions by utilizing the Internet of Things (IoT), with the aim to ease the quality of life for citizens. However, without the existence of resilient infrastructure in IoT, the data stream disruptions would not be possible, thus, resulting in negative effects and failure of smart city project implementation. On the other hand, the technological solutions provide opportunities to develop resilient cities by adopting modern ICTs and IoT, such as optimized resource allocation in the energy sector, transport sector, waste, lifestyle and management, and so on. These solutions suggested under smart cities provide better resilience to the city's infrastructure and improve social, economic and environmental development, which directly or indirectly bring better quality of life for its citizens. The isolated implementation of projects has negative economic impacts, as it involves more costs whereas by integrating both the concepts, an economic co-benefit approach is possible and can result in more profitable and sustainable solutions. Along with this, the operational and maintenance costs also increase without the availability of resilient infrastructure (Shah et al. 2019).

The social and environmental context of urban development also gets impacted by the isolated implementation of smart and resilient projects in a city. It is evident from various natural disasters or any natural hazard that the socially and economically weaker section of the society gets mostly affected by the impacts. While smart city projects can bring ease and equitable quality of life for citizens, the inclusion of resilience can improve its infrastructure to minimize the impact of natural disasters. The smart city and resilient city projects in isolation do not provide inclusive approaches in any of the sectors or areas considered in urban systems. This can bring more losses than the expected outcomes of the implemented projects as both smart city and resilient cities are based on many similar characteristics, including efficiency, flexibility, adaptability, diversity, innovation and so on. Hence, the next section will emphasize on possible integration of both the concepts (Zhu et al. 2020).

2.4.2 Convergence of Smart and Resilient Cities

Based on different research findings, there are various conceptual and operational aspects to support the convergence of urban resilience and smart solutions in an integrated and single development framework. Understanding the complexities in the continuous evolving field of urban development, it is needed to include the proportions of protection and adaptation in the resilience context (de Jong et al. 2015). Urban resilience has become prerequisite in designing and operations of any city model, including smart city, which are further defined with specific capacities, like preparedness, learning, adaptation and response system (Gazzola et al. 2019). The compatibility and convergence of resilience principles and smart concepts need an inclusive and integrated approach with focus on these fundamental capacities at the core of urban development framework.

In a practical approach, systematic thinking acts as a common operational urban development framework for resilience and smart city concepts. Conceptual constructs like complex adaptive systems can be employed to promote urban resilience, as it has been used to define smart city development (Colding and Barthel 2017). Moreover, it is easier to identify the rational connection between the two approaches and concepts from the sustainability perspective, and more precisely, from the perspective of vulnerability reduction and efficiency improvement of the overall quality of life in urban systems. Many researchers in the field of smart city that have acknowledged the concept of resilience claim the enhancement of urban system capacities toward resilient performance by the adoption of technology (ITU-T 2014).

Another convergence point between smart and resilience cities depends on the management and awareness creation. The structures in smart urban systems are built on networks of knowledge (ITU-T 2014) and information but the multilayered innovations have proved to be catalyst in this system (Gazzola et al. 2019). However, the capability of knowledge creation to solve complications in adaptation process redirects the major objectives of resilience approach. Furthermore, in a given socio-technical urban system, innovation is considered corresponding to adaptation in complex challenges. Hence, resilience thinking reverberates with the concept of smartness. In the given context of convergence, the augmentation of the concept of smartness results in the improvement of the whole system's resilience (Marsal-Llacuna and Segal 2016).

The resolutions stated by the smart city transformation include some operational methods and tools to attain the essential urban resilience qualities. Smart city concept focuses on the betterment of operations, planning and services in the urban system which fundamentally targets the conversion of processes rather than the certain infrastructural resources (Estevez et al. 2016). In case of resilience, it encompasses further development of various fundamental capacities which can develop as outcome of multiple complexes and codependent processes of urban development system and framework (Estevez et al. 2016). Considering these two points, the concept of smart city has potential to become an effective resource for building essential capacities in the context of urban resilience. With this framework of understanding, it can be implied that resilience is an inherent concept in the context of smart city and vice versa.

The smart technological solutions offered in the development frameworks and operational processes promote better convergence and urban development. Apart from technology, both smart cities and resilient cities can complement each other in developing sustainable designing and management of built environment and governance systems. The information developed and collected while constructing smart cities can further be used to support and enhance urban resilience systems such as development of disaster warning systems. It is believed that rational strategies for designing, planning and governance of resilient and smart cities can be further supported by integrating multidisciplinary sciences, such as engineering, social sciences and others.

2.5 Conclusion

We are living in an era of increasing risks and uncertainties induced by climate change, increased population growth and precarious geopolitical conditions. Annually, these result in major adverse events and cause significant human and economic losses. Being home to the majority of world population, cities are particularly hit hard by adverse events as was demonstrated during the recent pandemic (Sharifi and Khavarian-Garmsir 2020). Given the projected impacts of climate change, in the absence of appropriate strategies to deal with hazards, risks and uncertainties, the amount of human and property losses in cities could significantly increase in the coming decades. Understanding this, over the past two decades or so, enhancing urban resilience has become a priority in many cities around the world. There is now a wealth of knowledge on urban resilience principles and characteristics and many initiatives have also been developed for implementation of urban resilience principles and concepts.

One potentially effective way of enhancing resilience is through adoption of smart city solutions and technologies enabled by advances in information and communication technologies. Despite this, there is still a lack of integrated approaches that simultaneously consider both smartness and resilience. Some works on the integration of these two have been published following the pandemic (Sharifi et al. 2021). However, more research is still needed. Elaborating on the underlying principles of smartness and resilience, in this chapter we discussed that both concepts emphasize common characteristics such as efficiency, flexibility and adaptability. Integrating these two concepts could provide opportunities for achieving synergistic benefits and helps saving resources. In the absence of resilience, smart city initiatives may not be capable to function properly during extreme events such as major floods or storms that could disrupt operations of vulnerable infrastructure. On the other hand, failure to adopt and integrate smart technologies may result in costly, uncoordinated and inefficient urban resilience building plans and programs that are likely to fail to achieve their full potentials. It is, therefore, essential to take more efforts toward integrating the two concepts for creating smart resilient cities.

References

Ahad MA, Paiva S, Tripathi G, Feroz N (2020) Enabling technologies and sustainable smart cities. Sustain Cities Soc 61:102301. https://doi.org/10.1016/j.scs.2020.102301

Ahvenniemi H, Huovila A, Pinto-Seppä I, Airaksinen M (2017) What are the differences between sustainable and smart cities? Cities 60:234–245. https://doi.org/10.1016/j.cities.2016.09.009

Al Nuaimi E, Al Neyadi H, Mohamed N, Al-Jaroodi J (2015) Applications of big data to smart cities. J Internet Serv Appl 6(1):1–15. https://doi.org/10.1186/s13174-015-0041-5

Albino V, Berardi U, Dangelico RM (2015) Smart cities: Definitions, dimensions, performance, and initiatives. J Urban Technol 22(1):3–21. https://doi.org/10.1080/10630732.2014.942092

Allam Z, Dhunny ZA (2019) On big data, artificial intelligence and smart cities. Cities 89:80–91. https://doi.org/10.1016/j.cities.2019.01.032

American Psychological Association (2014) The road to resilience. American Psychological Association. Retrieved from https://www.uis.edu/counselingcenter/wp-content/uploads/sites/87/2013/04/the_road_to_resilience.pdf

Angelidou M (2015) Smart cities: A conjuncture of four forces. Cities 47:95–106. https://doi.org/10.1016/j.cities.2015.05.004

Aqib M, Mehmood R, Alzahrani A, Katib I (2020) A smart disaster management system for future cities using deep learning, GPUs, and in-memory computing. In: Mehmood R, See S, Katib I, Chlamtac I (eds) Smart infrastructure and applications: Foundations for smarter cities and societies (pp. 159–184). Springer International Publishing, Cham. https://doi.org/10.1007/978-3-030-13705-2_7

Baccarne B, Schuurman D, Mechant P, De Marez L (2014) The role of urban living labs in a smart city. In XXV ISPIM Innovation Conference, Proceedings. Presented at the XXV ISPIM Innovation Conference. Retrieved from http://hdl.handle.net/1854/LU-5646684

Batty M (2013) Big data, smart cities and city planning. Dialogues Hum Geogr 3(3):274–279. https://doi.org/10.1177/2043820613513390

Batty M, Axhausen KW, Giannotti F, Pozdnoukhov A, Bazzani A, Wachowicz M, Ouzounis G, Portugali Y (2012) Smart cities of the future. Eur Phys J Spec Top 214(1):481–518. https://doi.org/10.1140/epjst/e2012-01703-3
Bellini E, Nesi P, Bellini P (2017) Information Model and Interoperability. RESOLUTE Project. Florence, Italy: University of Florence. Retrieved from http://www.resolute-eu.org/files/RESOLUTE-D4-4-Information-model-and-interoperability-v9-0.pdf
Bosch P, Jongeneel S, Rovers V, Neumann H, Airaksinen M, Huovila A (2017) CITYkeys indicators for smart city projects and smart cities. CITYkeys Report. Retrieved from https://nws.eurocities.eu/MediaShell/media/CITYkeystheindicators.pdf
Brown A, Dayal A, Rumbaitis Del Rio C (2012) From practice to theory: Emerging lessons from Asia for building urban climate change resilience. Environ Urban 24(2):531–556. https://doi.org/10.1177/0956247812456490
Buchholz K (2020). How has the world's urban population changed from 1950 to today? World Economic Forum, 4 November. Retrieved from https://www.weforum.org/agenda/2020/11/global-continent-urban-population-urbanisation-percent/#:~:text=In%202020%2C%2056.2%20percent%20of,has%20risen%20in%20every%20content.
Caird S (2018) City approaches to smart city evaluation and reporting: Case studies in the United Kingdom. Urban Res & Pract 11(2):159–179. https://doi.org/10.1080/17535069.2017.1317828
Caragliu A, Del Bo C, Nijkamp P (2011) Smart cities in Europe. J Urban Technol 18(2):65–82. https://doi.org/10.1080/10630732.2011.601117
Chan H, Lee KY, Harvey N (2019) 5 big challenges facing big cities of the future. World Economic Forum, 29 October. Retrieved from: https://www.weforum.org/agenda/2018/10/the-5-biggest-challenges-cities-will-face-in-the-future/
Claudel M, Birolo A, Ratti C (2015) Government's role in growing a smart city. In: Araya D (ed) Smart cities as democratic ecologies. Palgrave Macmillan, London, pp 23–34. https://doi.org/10.1057/9781137377203_3
Coaffee J, Therrein MC, Chelleri L, Henstra D (2018) Urban resilience implementation: A policy challenge and research agenda for the 21st century. J Contingencies Cris Manag 26(13):403–410. https://doi.org/10.1111/1468-5973.12233
Colding J, Barthel S (2017) An urban ecology critique on the "smart city" model. J Clean Prod 164:95–101. https://doi.org/10.1016/j.jclepro.2017.06.191
Combaz E (2014) Disaster resilience: Topic guide. GSDRC, University of Birmingham, Birmingham, UK. Retrieved from https://gsdrc.org/topic-guides/disaster-resilience/
Cosgrave E, Arbuthnot K, Tryfonas T (2013) Living labs, innovation districts and information marketplaces: A systems approach for smart cities. Procedia Comput Sci 16:668–677. https://doi.org/10.1016/j.procs.2013.01.070
Costa DG, Duran-Faundez C (2018) Open-source electronics platforms as enabling technologies for smart cities: Recent developments and perspectives. Electronics 7(12):404. https://doi.org/10.3390/electronics7120404
de Jong M, Joss S, Schraven D, Zhan C, Weijnen M (2015) Sustainable–smart–resilient–low carbon–eco–knowledge cities; making sense of a multitude of concepts promoting sustainable urbanization. J Clean Prod 109:25–38. https://doi.org/10.1016/j.jclepro.2015.02.004
Estevez E, Lopes N, Janowski T (2016) Smart sustainable cities: Reconnaissance study. United Nations University Operating Unit on Policy-Driven Electronic Governance. Retrieved from http://collections.unu.edu/eserv/UNU:5825/Smart_Sustainable_Cities_v2final.pdf
Folke C, Biggs R, Norström A, Reyers B, Rockström J (2016) Social-ecological resilience and biosphere-based sustainability science. Ecol Soc 21(3). https://doi.org/10.5751/ES-08748-210341
Frantzeskaki N, Kabisch N (2016) Designing a knowledge co-production operating space for urban environmental governance—Lessons from Rotterdam, Netherlands and Berlin, Germany. Environ Sci & Policy 62:90–98. https://doi.org/10.1016/j.envsci.2016.01.010

Frantzeskaki N, Wittmayer J, Loorbach D (2014) The role of partnerships in 'realising' urban sustainability in Rotterdam's city ports area, the Netherlands. J Clean Prod 65:406–417. https://doi.org/10.1016/j.jclepro.2013.09.023

Gazzola P, Del Campo AG, Onyango V (2019) Going green vs going smart for sustainable development: Quo vadis? J Clean Prod 214:881–892. https://doi.org/10.1016/j.jclepro.2018.12.234

Giffinger R, Gudrun H (2010) Smart cities ranking: An effective instrument for the positioning of the cities? ACE: Arch City Environ 4(12):7–26. https://doi.org/10.5821/ace.v4i12.2483

Gleeson B (2008) Critical commentary. Waking from the dream: An Australian perspective on urban resilience. Sage Journals 45(13):2653–2668. https://doi.org/10.1177/0042098008098198

Hashem IAT, Chang V, Anuar NB, Adewole K, Yaqoob I, Gani A, Ahmed E, Chiroma H (2016) The role of big data in smart city. Int J Inf Manag 36(5):748–758. https://doi.org/10.1016/j.ijinfomgt.2016.05.002

ICLEI (2019) Resilient cities, thriving cities: The evolution of urban resilience. ICLEI, Bonn, Germany. Retrieved from https://e-lib.iclei.org/publications/Resilient-Cities-Thriving-Cities_The-Evolution-of-Urban-Resilience.pdf

Ismagilova E, Hughes L, Dwivedi YK, Raman KR (2019) Smart cities: Advances in research—An information systems perspective. Int J Inf Manag 47:88–100. https://doi.org/10.1016/j.ijinfomgt.2019.01.004

ITU-T (2014) An overview of smart sustainable cities and the role of information and communication technologies. International Telecommunication Union Telecommunication Standardization

Jha AK, Miner TE, Stanton-Geddes Z (2013) Building urban resilience: Principles, tools and practices. Directions in development: Environment and sustainable development. World Bank, Wasington, DC. Retrieved from http://hdl.handle.net/10986/13109

Jovanović A, Jelić M, Rosen T, Klimek P, Macika S, Øien K (2019) Smart Resilience D3.7: The "ResilienceTool" of the SmartResilience project. EU project SmartResilience, Project No. 700621.

Kalapurakal, R (2018) Cities 2030: Implementing the new urban agenda. United Nations Development Programme, 7 February. Retrieved from https://www.undp.org/blog/cities-2030-implementing-new-urban-agenda

Kitchin R (2014) The real-time city? Big data and smart urbanism. GeoJournal 79(1):1–14. https://doi.org/10.1007/s10708-013-9516-8

Kitchin R (2015) Making sense of smart cities: Addressing present shortcomings. Camb J Reg Econ Soc 8(1):131–136. https://doi.org/10.1093/cjres/rsu027

Kolar K (2011) Resilience: Revisiting the concept and its utility for social research. Int J Ment Health Addict 9(4):421–433. https://doi.org/10.1007/s11469-011-9329-2

Lange K, Knieling J (2020) EU smart city lighthouse projects between top-down strategies and local legitimation: The case of Hamburg. Urban Plan 5(1):107–115. https://doi.org/10.17645/up.v5i1.2531

Leichenko R (2011) Climate change and urban resilience. Curr Opin Environ Sustain 3(3):164–168. https://doi.org/10.1016/j.cosust.2010.12.014

Ma F, Wang Z, Sun Q, Yuen KF, Zhang Y, Xue H, Zhao S (2020) Spatial-Temporal evolution of urban resilience and its influencing factors: Evidence from the Guanzhong plain urban agglomeration. Sustain (Basel, Switzerland) 12(7):2593. https://doi.org/10.3390/su12072593

Martin-Moreau M, Ménascé D (2018) Urban resilience: Introducing this issue and summarizing the discussions. Field Actions Sci Rep: 6–11. Retrieved from http://journals.openedition.org/factsreports/4629

Marsal-Llacuna M, Segal ME (2016) The intelligenter method (I) for making "smarter" city projects and plans. Cities 55:127–138. https://doi.org/10.1016/j.cities.2016.02.006

McCormick K, Anderberg S, Coenen L, Neij L (2013) Advancing sustainable urban transformation. J Clean Prod 50:1–11. https://doi.org/10.1016/j.jclepro.2013.01.003

Meerow S, Newell JP, Stults M (2016) Defining urban resilience: A review. Landsc Urban Plan 147:38–49. https://doi.org/10.1016/j.landurbplan.2015.11.011

Moraci F, Errigo M, Fazia C, Burgio G, Foresta S (2018) Making less vulnerable cities: Resilience as a new paradigm of smart planning. Sustain (Basel, Switzerland) 10(3):755. https://doi.org/10.3390/su10030755
Muhammad AN, Aseere AM, Chiroma H, Shah H, Gital AY, Hashem IAT (2021) Deep learning application in smart cities: Recent development, taxonomy, challenges and research prospects. Neural Comput Appl 33(7):2973–3009. https://doi.org/10.1007/s00521-020-05151-8
Nam T, Pardo T (2011) Conceptualizing smart city with dimensions of technology, people, and institutions. Paper presented at the 282–291, 12 June. https://doi.org/10.1145/2037556.2037602. Retrieved from http://dl.acm.org/citation.cfm?id=2037602
Negev M, Garb Y, Biller R, Sagy G, Tal A (2009) Environmental problems, causes, and solutions: An open question. J Environ Educ 41(2):101–115. https://doi.org/10.1080/00958960903295258
Neirotti P, De Marco A, Cagliano AC, Mangano G, Scorrano F (2014) Current trends in smart city initiatives: Some stylised facts. Cities 38:25–36. https://doi.org/10.1016/j.cities.2013.12.010
Olsson L, Jerneck A, Thoren H, Persson J, O'Byrne D (2015) Why resilience is unappealing to social science: Theoretical and empirical investigations of the scientific use of resilience. Sci Adv 1(4):e1400217. https://doi.org/10.1126/sciadv.1400217
Perrings C (1998) Resilience in the dynamics of economy-environment systems. Environ Resource Econ 11(3):503–520. https://doi.org/10.1023/A:1008255614276
Perrings C (2006) Resilience and sustainable development. Environ Dev Econ 11(4):417–427. https://doi.org/10.1017/S1355770X06003020
Praharaj S, Han JH, Hawken S (2018) Urban innovation through policy integration: Critical perspectives from 100 smart cities mission in India. City Cult Soc 12:35–43. https://doi.org/10.1016/j.ccs.2017.06.004
Shah SA, Seker DZ, Rathore MM, Hameed S, Ben Yahia S, Draheim D (2019) Towards disaster resilient smart cities: Can internet of things and big data analytics be the game changers? IEEE Access 7:91885–91903. https://doi.org/10.1109/ACCESS.2019.2928233
Shamsuddin S (2020) Resilience resistance: The challenges and implications of urban resilience implementation. Cities 103:102763. https://doi.org/10.1016/j.cities.2020.102763
Sharifi A (2016) A critical review of selected tools for assessing community resilience. Ecol Ind 69:629–647. https://doi.org/10.1016/j.ecolind.2016.05.023
Sharifi A, Chelleri L, Fox-Lent C, Grafakos S, Pathak M, Olazabal M, Yamagata Y (2017) Conceptualizing dimensions and characteristics of urban resilience: Insights from a co-design process. Sustain (Basel, Switzerland) 9(6):1032. https://doi.org/10.3390/su9061032
Sharifi A (2019) A critical review of selected smart city assessment tools and indicator sets. J Clean Prod 233:1269–1283. https://doi.org/10.1016/j.jclepro.2019.06.172
Sharifi A (2020) A typology of smart city assessment tools and indicator sets. Sustain Cities Soc 53:101936. https://doi.org/10.1016/j.scs.2019.101936
Sharifi A, Khavarian-Garmsir AR (2020) The COVID-19 pandemic: Impacts on cities and major lessons for urban planning, design, and management. Sci Total Environ 749:142391. https://doi.org/10.1016/j.scitotenv.2020.142391
Sharifi A, Khavarian-Garmsir AR, Kummitha RK (2021) Contributions of smart city solutions and technologies to resilience against the COVID-19 pandemic: A literature review. Sustain (Basel, Switzerland) 13(8018):8018. https://doi.org/10.3390/su13148018
Silva BN, Khan M, Han K (2018) Towards sustainable smart cities: A review of trends, architectures, components, and open challenges in smart cities. Sustain Cities Soc 38:697–713. https://doi.org/10.1016/j.scs.2018.01.053
Singh RB (2015) Urban development challenges, risks and resilience in Asian mega cities. Springer, Tokyo
Talari S, Shafie-khah M, Siano P, Loia V, Tommasetti A, Catalão J (2017) A review of smart cities based on the internet of things concept. Energies (Basel) 10(4):421. https://doi.org/10.3390/en10040421
Taşan-Kok T, Stead D, Lu P (2012) Conceptual overview of resilience: History and context. In: Resilience thinking in urban planning. Springer Netherlands, Dordrecht, pp 39–51. https://doi.org/

10.1007/978-94-007-5476-8_3. Retrieved from http://link.springer.com/10.1007/978-94-007-5476-8_3
UN-Habitat (2017) Trends in urban resilience. United Nations Human Settlements Programme, Kenya. Retrieved from https://unhabitat.org/sites/default/files/download-manager-files/Trends_in_Urban_Resilience_2017_smallest.pdf
van den Buuse D, Kolk A (2019) An exploration of smart city approaches by international ICT firms. Technol Forecast Soc Chang 142:220–234. https://doi.org/10.1016/j.techfore.2018.07.029
Yigitcanlar T (2015) Smart cities: An effective urban development and management model? Aust Plan 52(1):27–34. https://doi.org/10.1080/07293682.2015.101975
Zhu S, Li D, Feng H (2019) Is smart city resilient? evidence from china. Sustain Cities Soc 50:101636. https://doi.org/10.1016/j.scs.2019.101636
Zhu S, Li D, Feng H, Gu T, Hewage K, Sadiq R (2020) Smart city and resilient city: Differences and connections. Wiley Interdiscip Reviews Data Min Knowl Discov 10(6):n/a. https://doi.org/10.1002/widm.1388

Chapter 3
Smart Cities: Concepts and Underlying Principles

Rhea Srivastava and Ayyoob Sharifi

Abstract The concept of smart cities has been gaining importance in the academic and policy fields as a means to provide innovative solutions to tackle the rapid urbanization, globalization and climate change challenges faced by cities. However, the concept is still contested and is continually evolving with numerous debates on what it entails altogether. Therefore, this chapter aims at providing a comprehensive understanding of the smart city concept by elaborating on its roots and discussing the major themes, dichotomies and gaps in its definitions across contemporary literature. It also identifies and studies the main dimensions of smart cities that may be pertinent in the identification and evaluation of smart cities. The in-depth literature review intends on acting as a reference point for scholars to get a clearer picture of the smart city research landscape and enable policy makers to develop, implement and monitor smart city solutions for a sustainable future.

Keywords Smart cities · Sustainable urban development · Smart city dimensions · IoT applications · Urban planning

3.1 Introduction

Cities play a dominant and synergistic role in the social, economic and environmental aspects of development worldwide. However, cities are currently facing unprecedented colossal challenges as more than 50% of the world's population now lives in urban areas, with the United Nations estimating an increase of up to 68% in that by 2050 (United Nations 2018; Albino et al. 2015). This rapid urbanization brings along a complex set of risks, concerns and issues with regard to catering to the growing needs and demands of citizens and climate change adversities, while also

R. Srivastava (✉)
The Energy and Resources Institute (TERI), New Delhi, India

A. Sharifi
Graduate School of Humanities and Social Sciences, Hiroshima University, Higashihiroshima, Japan
e-mail: sharifi@hiroshima-u.ac.jp

A. Sharifi and P. Salehi (eds.), *Resilient Smart Cities*, The Urban Book Series,
https://doi.org/10.1007/978-3-030-95037-8_3

maintaining a sustainable approach towards urban development (Nam and Pardo 2011; Sharifi 2019). In order for cities to thrive on urbanization and its associated challenges, smarter, creative and more innovative ways of urban planning, design and management are pre-requisites. Accordingly, the Information and Communication Technology (ICT) uprising has become the most sought after, 'smart' solution of today. Digitization of urban systems through networked ICTs since the past decade has caused a fundamental shift in cities, with the internet playing a major role in citizens' daily lives as well as in governance. In conformity with this, the concept of smart cities has emerged as a model for efficient and sustainable urban development (Chourabi et al. 2012; Angelidou 2015; Yigitcanlar 2015; Ahvenniemi et al. 2017; Allam and Newman 2018).

Smart cities utilize a number of smart technologies, such as big data, Internet of Things (IoT), cloud computing and Artificial Intelligence (AI), to manage infrastructure and urban services, improve the quality of life and enhance governance mechanisms (Kitchin et al. 2019). Since the conception of the smart city paradigm, smart cities have been evolving in every aspect, from small-scale projects to larger city strategies, incorporating an increasing number of characteristics in order to facilitate innovation, achieve a high quality of life and minimize environmental impacts (Yigitcanlar 2015). Using advanced technological interventions, smart cities also contribute to address climate change by aiming for greenhouse gas emission reduction in urban areas, attaining energy efficiency, optimizing resource use and facilitating better and more effective adaptation strategies (Ahvenniemi et al. 2017). Moreover, while the smart city model provides a competitive edge in the global economy to draw more people and businesses, it is also viewed as a key strategy in eliminating socio-economic issues such as that of poverty, unemployment and inequality (Manville et al. 2014; Sharifi 2019).

Overall, with smart cities representing such a multi-disciplinary and evolutionary nature, the main objective of this research is to dig deeper into the concept to gain a greater understanding of what it entails for the future of cities and urban development. To do so, the specific sub-objectives of the study are: (a) to provide an overview of the history of smart city movements; (b) to provide an in-depth investigation of the concepts and definitions related to smart cities; and (c) to elaborate on the dimensions and principles of smart cities. Being explanatory in nature, the insights derived from the theoretical review of literature will build up on the previous and existing discourse on the smart city notion and will aid scholars studying the smart city field in obtaining a clearer picture of the fragmented research on it. Additionally, by laying out the possible application domains where smart solutions can be used, the study also intends on guiding professionals and policy makers towards developing, implementing, evaluating and monitoring a sustainable, smart city in a holistic way.

In particular, the paper is structured as follows. Section 3.2 gives a brief overview of the emergence of the smart city concept along with exploring the conceptual movements related to it, such as information cities, digital cities and intelligent cities. Section 3.3 outlines the various concepts and definitions of smart cities found in literature and highlights the numerous perspectives through which they were studied. Section 3.4 explores the key dimensions of smart cities while analysing the categories and indicators related to each dimension. Finally, the last section offers a discussion of the main findings and draws concluding remarks.

3.2 Background and Historical Overview

The roots of smart cities can be traced back to the post-industrial, neo-liberal era (mid-twentieth century), when new technologies started influencing urban scholars to study their impact on cities and to visualize what cities of the future would look like (Angelidou 2015). Along with the technological boom that enabled the popularization of ICTs in everyday life, Angelidou (2015) asserts that knowledge and innovation economy around which this technological advancement was centred is another significant driver that shaped the smart city concept. Over two decades, the smart city concept built up on its foundational shaping forces and utopias, as it evolved with not just technological developments, but also social and economic, such as a demand for change due to urbanization, an increasing market for smart products and the need to address sustainability (Angelidou 2017). This evolution of the ICT approach in cities has been documented in the form of city models/movements, many of them overlapping with each other in terms of their origin timeline and descriptions. According to Batty et al. (2012), the origin of this ICT approach can be traced back to James Martin's concept of the wired society in 1977 that gave rise to wired cities, a city in which telecommunication technologies provide information services to businesses and households (Strauss et al. 1996). But Hollands (2008) and Batty et al. (2012) also acknowledge that this model was not necessarily smart and needed an additional IT spiral to make cities truly smart with efficient service delivery and computing. Other related conceptions such as the 'information cities' and 'digital cities' also started to develop during that time, possibly arising from wired cities themselves. As a result of this, the narrow one-dimensional perspective of looking at cities started to incrementally evolve to a multi-dimensional smart approach that is the present-day smart cities. The sequential set of major movements preceding the smart city model is discussed in the remainder of this section.

3.2.1 Information Cities

The rise of the information city with the widespread penetration of information technologies triggered a paradigm shift in urban planning (Knox 2010). The second half of the twentieth century (late 1970s and 1980s) focused a great deal on the emergence of the knowledge society, with cities acting as drivers of change (Webster 2001; Stock 2011). The advent of globalization additionally caused quite the 'information revolution', also resulting in a societal transformation (Castells, 1985 in Hepworth 1987; Gillespie and Richardson 2000). According to Castells (1989), innovation in science and technology, capitalism and informational processes reshaping the production, consumption and state control in US cities. This not only made interactions/ decision-making more globalized, but also changed the focus to profitability and industrial organization (Knox 2010). He also emphasized on the importance of information technologies in minimizing the gap between society and economy (Castells 1989 in Knox 2010). All in all, as explained by Nam and Pardo (2011, p. 285), an information city acted as an "urban centre for commerce, social and civic services, and social interactions among people, businesses and government institutions".

Information cities, according to Hepworth (1987), are a combination of two dimensions: the economic and technological dimensions. Building up on Porat's (1977) concept of the information economy, Hepworth considered an information city as a metropolitan economy, one that is heavily specialized in the production, processing and distribution of information. Consequently, the technological dimension dealt with the technical infrastructure from the introduction of new technologies (computers and telecommunications), thereby covering local as well as sub-networks. These not only acted as transport networks but also facilitated electronic production and management systems for the public and private sector (Hepworth 1987, p. 259). In addition to this point of view, Castells (1989) incorporated two spaces in the construct of information cities: the space of places (geographical spaces) and the space of flows (information, power, capital), where technology acted as an essential instrument in knowledge processing, generation and management. Much emphasis is given to the space of flows as fundamental for the characterization of a city.

While the information city during that time increasingly elevated the economy, several downsides were considered as well. For example, Castells (1989, p. 225) asserted that the emergence of the information age ended up differentiating between information-based formal economy and the labour-based informal economy, which also resulted in displacing workers, thus affecting their economic state and widening inequality. Moreover, the space of flows dominated the space of places due to internationalization enabled by the information economy. This, as stated by Castells, posed as an issue for cities from a socio-political stand, as there was a greater risk of vulnerability in the international arena. However, Castells' perspective did not comprehend the massive development surrounding these themes in the current world scenario. While he may have laid the foundation, ample research studying and solving the complexities of the information age followed and are subsequently covered in the next sections.

3.2.2 *Digital Cities*

Shifting from the one-dimensional technology perspective in cities, the concept of digital cities was conceived in the year 1985 but got popularized in the early 1990s, and was based on the use of localized information and communication networks providing daily (web-based) services to citizens for a better quality of life (Yasuoka et al. 2010; Dameri and Cocchia 2013). Yasuoka et al. (2010, p. 940) establish that this concept was originally initiated by three different actors: the not-for-profit electronic community forums, government (as an initiative for information sharing and service delivery) and private companies (for commercial services). To get an aerial perspective, Yovanof and Hazapis's (2009, p. 446) definition of a digital city—commonly used in literature—explains it as "a connected community that combines broadband communications infrastructure; a flexible, service-oriented computing infrastructure based on open industry standards, and innovative services to meet the needs of governments and their employees, citizens and businesses". Therefore, a digital city worked on facilitating an environment where people could have a seamless experience with interactions, connections and knowledge sharing (Ishida 2002).

Despite a common terminology used to describe digital cities, they have different interpretations, goals and architecture, depending upon their social (and spatial) context. Yasuoka et al. (2010) classify the digital cities according to their "socio-technical" and "virtual-physical dimensions". To simplify, the "socio-technical" dimension consists of one trend that goes in the high-tech direction, where state-of-the-art technologies are applied to regional information spaces, and another one that deals with more social aspects, such as citizen participation and better social life using local information. For the "virtual-physical" dimension, they established that while some digital cities feel like a virtual world where functions are carried out to support community of interests, other digital cities have closer ties to real cities where functions are performed to obtain regional information and activities (Yasuoka et al. 2010, p. 940). A noteworthy observation by them is that these two dimensions are more or less interwoven and inseparable, especially as digital cities expanded with wireless networks. Modern digital cities cannot be categorized just as technology test beds of community spaces, but a more integrated approach is applied when looking at them. This also resonates with Besselar et al.'s (2000, p. 19) findings on the many interpretations of digital cities. According to them, some digital cities can be interpreted as a "local social information infrastructure", where information about the 'real city' is disseminated to locals and visitors, thus combining the dimensions of Yasuoka et al.'s (2010) classification. Other interpretations by them included identifying the digital city as a "communication medium", a tool to enhance "local democracy", a resource for organization of daily life, and a free space to "experience and experiment with cyberspace" as well as coordinate social life.

When compared to information cities, Anthopoulos and Tsoukalas (2006) developed a multi-level common architecture for digital cities, which acts as an extension of the previous information city approach. This multi-level architecture consists of four main layers. The 'infrastructure layer', contains facilities such as information systems, broadband networks, network equipment, software systems and other hard or soft infrastructure needed for the deployment of e-services. The 'service layer', being important for the diffusion of digital activities in the community, includes the e-services that the city offers. The 'information layer' contains all the information required, produced and stored in the infrastructure layer (public/private documents, geospatial data, etc.). Finally, the 'user layer' deals with all end users/ stakeholders using e-services. Since this layer combines the whole architecture, it is present at the top as well as bottom, as it concerns both the stakeholders that provide e-services (e-government, e-commerce, etc.), and the end users (citizens, students, local enterprises) who are served by the services.

The emergence of digital cities marked an elemental shift in the channel of interaction and service delivery between the government and citizens, hence, also causing a major change in the regulatory environment with more efficient government operations (Yovanof and Hazapis 2009). Digital cities research also branched out further into other concepts such as virtual cities (digital representation of the city) and ubiquitous cities (ubiquitous accessibility and infrastructure). However, as Aurigi (2000) concluded, most initiatives in a digital city have been designed from a top-down approach, where the government often perceives citizens as 'clients'. He further argues that as citizens just end up following instructions, they should be involved right from the conception stage of the services that are intended for them. Furthermore, Aurigi also points out a lack of pluralism in the digital city visions, stating that many of the cities are predominantly being looked at from the government and the scientists' perspective while missing out on other stakeholders, such as urban planners and architects. Lastly, he discusses the issue of social polarization in the use of advanced technologies, a challenge also found in the information city literature. Going back to the virtual-physical dimension of digital cities, Aurigi (2000) explains that there is inadequate integration between the virtual and real city. Therefore, he suggests that considerable attention must also be given to capacity building in the form of policies and computer literacy campaigns in order to bridge the gap.

3.2.3 Intelligent Cities

A new generation of cities called intelligent cities started emerging around the same time as digital cities, with its oldest reference going back to Batty's (1990) paper on using intelligent cities for a competitive advantage (Komninos 2009). The late 1990s and early 2000s saw more studies on intelligent cities, simultaneously bringing along with them new ways of dealing with development, innovation and smart infrastructure (Komninos 2015). Intelligent cities emerge at the crossing of digital cities and knowledge society (Nam and Pardo 2011; Albino et al. 2015). According to Komninos (2008), intelligent cities have two core components: the first one deals with the innovation systems driving the development of innovations inside the organizations within the system; and the second component is that of the digital applications and tools that aid in knowledge creation, communication, data storage and transfer, etc. To sum up, Komninos (2006, p. 1) describes intelligent cities as spaces "with a high capacity for learning and innovation, which is built in the creativity of their population, their institutions of knowledge creation, and their digital infrastructure for communication and knowledge management". This also shows where intelligent cities differ from digital cities, even though the terms are used interchangeably throughout literature. While digital cities involve functions of the city (work, housing, environment, etc.), intelligent cities facilitate collaborative innovation, social cooperation and learning through digital spaces and ICT. Therefore, every intelligent city has digital components but every digital city is not necessarily an intelligent city (Komninos 2006 2008; Nam and Pardo 2011).

Komninos (2008) further goes on to analyse the three levels in the configuration of intelligent cities. The physical level consists of people (especially knowledge workers) and knowledge intensive activities of the city, where innovation is centred around human skills, creativity and their cooperation with each other. The second level entails the institutional mechanisms and organizations that facilitate innovation. Intelligence at this level is social and collective, as compared to the physical level where it is human. This level, built on social and intellectual capital and collective intelligence, provides infrastructure for collective action. Finally, the third level encapsulates the digital aspects through artificial intelligence by making use of web-based applications, interactive technologies and other infrastructure for digital communication, collaborative product development, process innovation and knowledge management.

However, while intelligent cities are definitely a few steps ahead of digital cities with their incorporation of multi-level intelligence and innovation, Ojo et al. (2015) claim that their scope is still narrow and standalone (as compared to smart cities), and the governance mechanisms to integrate the components are still missing. Hollands (2008) also supports this finding by discussing the fragmented nature of intelligent cities, in which some aspects are technology driven and others deal with human networks and human capital approaches, neglecting a wider and a more sustainable approach towards city development. Nonetheless, the notion of intelligent cities was quite a breakthrough for the twenty-first century and is most analogous to the concept of smart cities that followed.

3.2.4 Smart Cities

Smart cities are a further proliferation of the concepts of information, digital and intelligent cities, with researchers also calling intelligent cities as the first generation smart cities (Yigitcanlar et al. 2018). The concept started getting traction in mid-2000s, when state-of-the-art developments in IT such as ubiquitous computing, satellite TVs, electronic commerce and cloud-based solutions along with their falling prices gave rise to new possibilities in providing the infrastructure for improving urban services and living (Carvalho 2015; Angelidou 2015). In the beginning, the smart city literature remained largely technocentric and top-down in nature. The elements of citizen engagement and human capital (creativity, learning, knowledge) gained momentum gradually. Admittedly, Hollands (2008) considered the intelligence embedded by infrastructures as the main incentive for the development of smart cities, but also their interaction with the city stakeholders (citizens, businesses, government) to foster innovation for social, economic, environmental and cultural growth.

Smart city literature discusses many principles underlying the concept that mainly revolve around sustainability, quality of life, smartness and urbanization (Silva et al. 2018). Physical infrastructure, social infrastructure, institutional infrastructure and economic infrastructure are the four widely accepted pillars on which a smart city stands (Mohanty, 2016 in Silva et al. 2018, p. 699). While the physical infrastructure theme incorporates the natural resources and service infrastructure, the social infrastructure theme deals with the human and social capital such as citizen awareness, the institutional infrastructure entails the governance aspect of smart cities, and the economic infrastructure ensures a smart and stable economy (Silva et al. 2018). The dimensions of smart cities encapsulating all the application domains associated with the concept, namely smart economy, smart people, smart living, smart governance, smart mobility, smart environment and smart data, which are studied in the later sections of this study, are also built upon these four pillars.

Smart city initiatives are becoming more and more crucial in ensuring sustainable urban development and transforming cities for a sustainable future through various innovative technologies and collaborative business models. Research arcs discussing the same under the term 'sustainable smart cities' have also been formulated in the recent history of smart city literature (Kramers et al. 2014). However, proper management of these cities is also growing in importance simultaneously. Challenges that smart cities are currently dealing with, such as the growing complexity of data collection and integration, privacy and ownership of data, cybersecurity as well as knowledge/skill and infrastructure upgradation (Al Nuaimi et al. 2015; Ahad et al. 2020), require solutions to tackle them at hand while bridging the social and cultural gap in the society at the same time.

To summarize, Fig. 3.1 shows a broad timeline of the emergence of the concepts discussed above, ultimately leading to the smart city concept.

3.3 Smart Cities Concepts and Definitions

The term 'smart city' has been defined countless number of times in literature and policy discussions, not just limited to being a concept for managing cities but a movement for growth and development in cities. Despite the increasing number of publications and heavy popularity of the term, there is no single, universally accepted definition for it, which can be attributed to the varying perceptions and viewpoints with which the smart city concept is looked at. Nonetheless, the broad nature of smart city research is dichotomous, incorporating both a technological and holistic view of looking at city development.

Initial conceptions that were also built upon earlier city models (wired city, information city, etc.) had a one-dimensional technology-driven focus. The authors, mostly in the industrial literature, viewed and associated smart cities with the use of advanced technologies like ICT for more efficient urban systems, ranging from energy systems (e.g., smart grid) to water, waste to transportation systems (Harrison

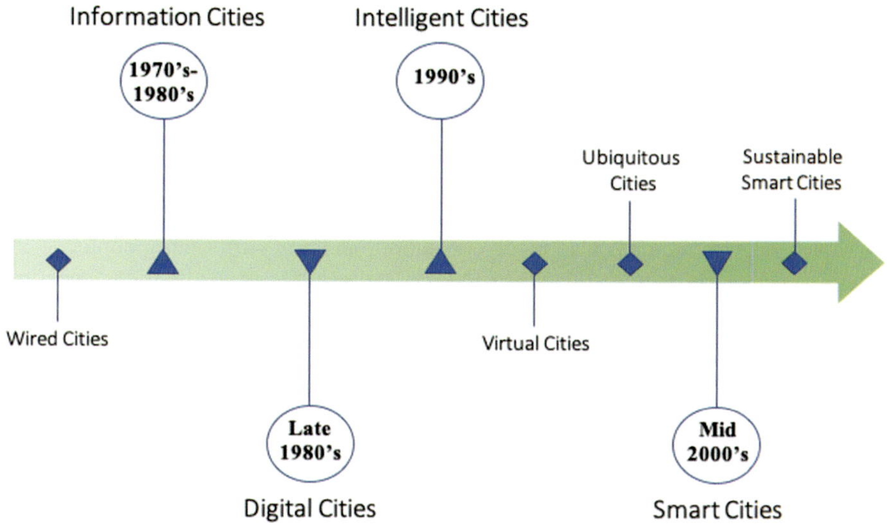

Fig. 3.1 Temporal evolution of smart city movements (*Source* Authors [2021])

et al. 2010; Lombardi et al. 2012). Often seen as a business opportunity, Harrison et al. (2010, p. 1–2) in an IBM document consider cities as systems of systems and argue that the term 'smart city' denotes "an instrumented, interconnected and intelligent city", where "instrumented" referred to digitization for information creation and measurement, "interconnected" implied information communication, and "intelligent" meant incorporation of data analytics and modelling for informed decision-making (Mosannenzadeh and Vettorato 2014). Taking a more utopic stance, Canton (2011) further perceives smart cities to be the hub of advanced technologies, such as artificial intelligence inducing super intelligence in humans and machines, that would ultimately address complex city challenges in the domains of energy, health, safety and commerce. A slightly deeper comprehension of smart cities encompassed the use of advanced technologies but for a better quality of life and economic prosperity. For example, Marsal-Llacuna et al. (2015) in their paper conclude that smart cities not only just improve urban performance using technologies, but also aid in efficient service provision to citizens, increase economic collaborations, and encourage business models among public and private stakeholders (see also Silva et al. 2018).

Correspondingly, this utopian, business-driven outlook towards smart cities was questioned by many, with Giffinger et al. (2007) and Hollands (2008) laying the foundation and opening up the concept to a progressive and wholesome vision, in which technological intervention works in tandem with social, human, cultural and governance considerations (Caragliu et al. 2011; Nam and Pardo 2011; Angelidou 2014; Mora et al. 2019; Praharaj and Han 2019). Denoted as the 'soft domain' by Neirotti et al. (2014), a broad spectrum of research and definitions about smart cities stemmed from this school of thought in varying capacities. Some authors relate smart cities with smart people strongly. Winters (2011) substantiates that an ideal smart city provides several opportunities for higher education that, in turn, increase growth with skilled workforce. Keeping people at the centre of smart city operations, these conceptualizations focus on drivers such as higher education, culture and creativity for people to generate as well as gain from social capital and urban development (Rios 2008; Lombardi et al. 2012; Winters 2011; Yigitcanlar et al. 2018). In fact, this knowledge-based urban development has also given rise to alternative terms such as 'knowledge city' among scholars to explore more about it. In other works, collaborative governance is also given consideration along with smart people. As an example, Nam and Pardo (2011, p. 286) look at smart cities as an organic connection between technology, people and institutions, where investments in the technological and social capital "enhance quality of life through participatory governance", thus also encouraging efficient policies for urban innovation. In addition to the soft domains, significant amount of literature also focuses on the urban environmental sustainability viewpoint towards smart cities, with researchers studying about ways to inculcate sustainability into smart city approaches (Batty et al. 2012; Neirotti et al. 2014; Ahvenniemi et al. 2017). Kramers et al. (2014) use sustainable smart cities as a way to outline initiatives where smart solutions improve sustainability. Further elaborating on the importance of an interdisciplinary sustainable smart city approach in urban development in their paper, Bibri and Krogstie (2017, p. 194) state that "it is high time to link technological progress with the agenda of sustainable development and thus to justify future ICT investments by environmental concerns and socio–economic needs in the context of smart cities". Particularly, the environmental concerns involve elements such as energy efficiency, carbon neutrality, pollution reduction, zero-waste and sustainable transport (p. 193).

These differences in smart city perceptions are also evident spatially. Mora et al. (2018), after performing a bibliometric analysis on the smart city literature, found that Europe is the largest contributor to the research, followed by North America. However, while Europe's authors are predominantly from academic institutions, they support the holistic view of smart city literature discussed above. Contrastingly, North American authors follow the technocentric and innovative interpretation of the concept, with authors coming from both universities and businesses. Israilidis et al. (2021) also found that only 7% of the empirical studies on smart cities have incorporated a multi-country perspective, compared to the initiatives covered from just one country. Rightly so, Hollands (2008), Mosannenzadeh and Vettorato (2014), Angelidou (2017) and Yigitcanlar et al. (2018, p. 3) accurately assert that one of the reasons for inconsistency in the smart city conception is due to the different nature of research by academic, commercial/industrial and governmental organizations that is tailored to correspond to their own perspective, whether it is from a practical, disciplinary, or domain orientation, thus also causing a knowledge gap between theory and practice.

Apart from the technology-led vs. holistic angle in the smart city debate, other metaphors/dichotomies which end up increasing the knowledge gap have been discussed in literature by Mora et al. (2019). They address three more development paths in smart cities: "double or quadruple helix model of collaboration";

"top-down or bottom-up approach"; and "mono-dimensional or integrated intervention logic" (p. 71). In case of the collaborative model to go with, the double helix model of collaboration between technology providers and the local governments works more in favour of corporations, and is therefore supported by them in research. As a counter-approach, many researchers criticize this closed model and call for a broader model of collaboration between industry, institutions, governments and citizens for efficient innovation and development (Mosannenzadeh and Vettorato 2014; Selada 2017; Mora et al. 2019). Meijer and Bolivar (2016, p. 398) in their own definition of smart cities also explained: "the smartness of a city refers to its ability to attract human capital and to mobilize this human capital in collaborations between the various (organized and individual) actors through the use of information and communication technologies". Simultaneously, the debate between top-down and bottom-up approach towards smart cities is also extensive in literature (Mora et al. 2019). While the supporters of a top-down approach justify their stance by suggesting that it provides a long-term vision for proper governance, many scholars point out that this approach is more market oriented and less citizen centric, thereby also ignoring the innovation potential from bottom-up approaches (Ratti and Townsend 2011; Mora et al. 2019). Finally, the last dichotomy deals with the mono vs multi-dimensional outlook towards smart cities. According to Mora et al. (2019), the European Commission has been inclining towards a mono-dimensional vision of smart cities in their research, with emphasis on ICT solutions for smart growth in energy, transport, buildings, etc. However, Mosannenzadeh and Vettorato (2014) and Manville et al. (2014) suggest a multi-dimensional smart city model that integrates domains like environment, mobility, community, governance, etc. Mora et al. (2019) also highlight that this particular integrated approach can now be seen in smart city assessment frameworks, fostering a fuller picture.

Pertaining to all these considerations, many scholars developed comprehensive conceptualizations of smart cities. While the smart city adaptation by Yigitcanlar et al. (2018) focused on the balance between environmental, societal, economic and institutional development, the interpretation of smart cities by Giffinger et al. (2007, p. 14) dealt with economy, mobility, people, environment, governance and living, especially emphasizing on the citizen-centricity of cities with independent and aware citizens. Perhaps a prominent definition of a smart city is the one formulated by Caragliu et al. (2011, p. 70): "We believe a city to be smart when investments in human and social capital and traditional (transport) and modern (ICT) communication infrastructure fuel sustainable economic growth and a high quality of life, with a wise management of natural resources, through participatory governance". This definition of a smart city is considered as the most holistic one among majority of the researchers. All in all, incorporating multiple principles while looking at smart cities enables a strategic transformation while ensuring operational efficiency, better quality of life and sustainability (Sharifi 2019). As a whole, Table 3.1 illustrates the numerous definitions of smart cities covered in literature.

Table 3.1 Smart city definitions throughout literature

Definition	Reference
"A Smart City is a city that gives inspiration, shares culture, knowledge, and life, a city that motivates its inhabitants to create and flourish in their own lives. A smart city is an admired city, a vessel to intelligence, but ultimately an incubator of empowered spaces."	Rios (2008)
"A city well performing in a forward-looking way in economy, people, governance, mobility, environment, and living, built on the smart combination of endowments and activities of self-decisive, independent and aware citizens. Smart city generally refers to the search and identification of intelligent solutions which allow modern cities to enhance the quality of the services provided to citizens."	Giffinger and Gudrun (2010)
"A smart city is a city connecting the physical infrastructure, the IT infrastructure, the social infrastructure, and the business infrastructure to leverage the collective intelligence of the city."	Harrison et al. (2010)
"The Smart City is one that will use advanced technology and sciences – computing, neuroscience, nanoscience, information science – to address the challenges of the future of the city such as energy, health, safety and commerce."	Canton (2011)
"We believe a city to be smart when investments in human and social capital and traditional (transport) and modern (ICT) communication infrastructure fuel sustainable economic growth and a high quality of life, with a wise management of natural resources, through participatory governance."	Caragliu et al. (2011)
"A Smart City has three key components: technology (infrastructures of hardware and software), people (creativity, diversity, and education), and institution (governance and policy). Given the connection between the factors, a city is smart when investments in human/social capital and IT infrastructure fuel sustainable growth and enhance a quality of life, through participatory governance."	Nam and Pardo (2011)
"Smart Cities are metropolitan areas with a large share of the adult population with a college degree."	Winters (2011)
"A Smart City is a synthesis of hard infrastructure with the availability and quality of knowledge communication and social infrastructure. Smart Cities are also instruments for improving competitiveness in such a way that community and quality of life are enhanced."	Batty et al. (2012)

(continued)

(continued)

Definition	Reference
"Smart cities are all urban settlements that make a conscious effort to capitalize on the new Information and Communications Technology (ICT) landscape in a strategic way, seeking to achieve prosperity, effectiveness and competitiveness on multiple socio-economic levels."	Angelidou (2014); Angelidou (2017)
"A Smart City is a city seeking to address public issues via ICT-based solutions on the basis of a multi-stakeholder, municipally based partnership."	Manville et al. (2014)
"Smart Cities initiative tries to improve urban performance by using data, information and information technologies (IT) to provide more efficient services to citizens, to monitor and optimize existing infrastructure, to increase collaboration amongst different economic actors and to encourage innovative business models in both the private and public sectors."	Marsal-Llacuna et al. (2015)
"The smartness of a city refers to its ability to attract human capital and to mobilize this human capital in collaborations between the various (organized and individual) actors through the use of information and communication technologies."	Meijer and Bolivar (2016)
"A smarter city can be described as a city where advanced ICT is combined with physical, infrastructural, architectural, operational, functional, and ecological systems across many spatial scales, as well as with urban planning approaches, with the aim of improving efficiency, sustainability, equity, and livability."	Bibri and Krogstie (2017)
"Smart city is a system that facilitates interoperability among various sub systems to improve the QoL of urban citizens."	Silva et al. (2018)
"Smart cities can be defined through the three 'S' of urban development: smart technology, society and sustainability."	Praharaj and Han (2019)
"The smart city is an urban locality functioning as a healthy system of systems with sustainable and knowledge-based development activities to generate desired outcomes for all humans and non-humans."	Yigitcanlar et al. (2019)

Source Author (2021)

To recapitulate, while these definitions certainly started a new league of smart city discourses, they also further added to the complexity of smart cities leading to the inconsistency in finding a universal smart city definition (Praharaj and Han 2019). The already existing jargons used before smart cities (intelligent cities, digital cities, ubiquitous cities, etc.) add to the confusion due to them getting conflated (Hollands 2008), despite the differences between them that were pointed out in the previous section. Even today, smart city concepts are context/domain dependent with respect to policies, capital, resources, political situations, etc. Mora et al. (2018, p. 20) call attention to the fragmentation and knowledge gap, and argue that in order to address this challenge, greater number of studies as well as exchange among the scholars researching about smart cities is the way forward.

3.4 Dimensions and Principles of Smart Cities

Based on the previous section, it is evident that smart cities aim at improving citizens' quality of life and addressing urban issues by utilizing ICT solutions, that also foster collaboration and partnerships among diverse constituents. The fact that the smart city concept applies to various aspects of urban life and influences different actors and sectors justifies the ambiguity in the smart city dimensions literature. There exists a vast literature traversing the many dimensions of smart cities, either in the form of broad principles or as indicators for smart city assessments. The most commonly adopted framework is developed by that of Giffinger et al. (2007), who propose six dimensions, namely, smart economy, smart governance, smart people, smart mobility, smart environment and smart living. There is ample documentation on this set of dimensions as it has been used as the foundation by many researchers (Lombardi et al. 2012; Manville et al. 2014). Other studies have also proposed smart education, smart waste and water, smart infrastructure, smart health care and smart security systems (Neirotti et al. 2014; Wahab et al. 2020). Moreover, smart technology or smart data are also discussed as an important dimension due to the big role real-time data management plays in smart cities (Yigitcanlar et al. 2018; Huovila et al. 2019; Wahab et al. 2020).

For this study, the smart city framework developed by Sharifi (2019) is used. The framework has been developed based on a synthesis of existing research on smart city dimensions and the seven themes covered in it (Fig. 3.2) are smart economy, smart people, smart governance, smart environment, smart living, smart mobility and smart data. As shown in Fig. 3.2, the indicators under each theme cover a majority of the additional dimensions discussed above, thus forming a multi-disciplinary structure that is integrative of the wider disciplines of urban growth and social, economic and environmental development. Correspondingly, the dimensions identified from this smart city framework are covered below in detail.

3.4.1 Smart Economy

Smart economy is described as one of the main domains of smart cities as it is also one of their principles. Wahab et al. (2020, p. 4) opine from the smart city framework by Giffinger et al. (2007), that the economy dimension entails three major perspectives—one where production and innovation encourage economic growth, followed by the viewpoint of a smart city being an economic driver in itself, and lastly for applying the actual economics behind the smart city. Indeed, Giffinger et al. (2007, p. 12) covered

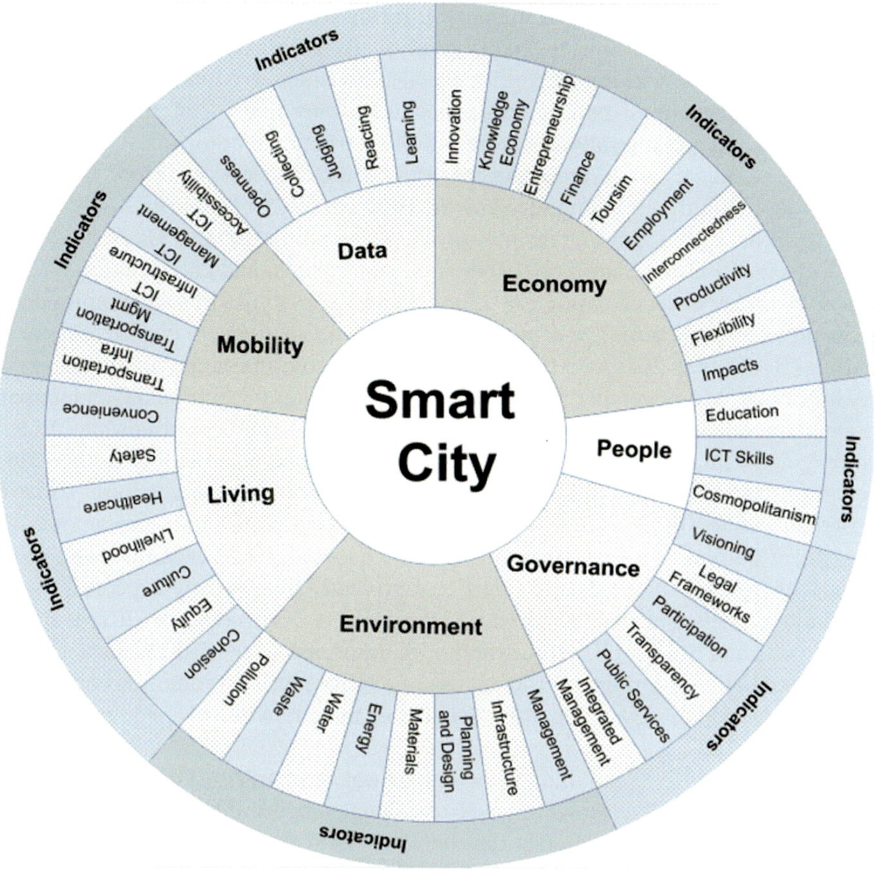

Fig. 3.2 Smart city dimensions and sub-dimensions by Sharifi (2019)

factors such as entrepreneurship, innovative spirit, international embeddedness and ability to transform, economic image and trademarks, productivity and flexibility of labour markets. The innovative spirit dimension, which is also the most recurring factor under economy according to Sharifi's (2019) review of smart city assessment tools, deals with measures such as R&D expenditure, employment rate and other policies to ensure and promote innovation/creativity in cities. Along with innovation, entrepreneurship covered elements such as number of new businesses/start-ups and self-employment rates. A high degree of entrepreneurial activities for developing new business models also aids in increasing productivity as well as the competitiveness with that growing economic image (Bosch et al. 2017; Lu et al. 2019). Additionally, other indicators such as GDP per employed person, Foreign Direct Investments and ICT measures for economic development also come under the dimension of productivity (Giffinger and Gudrun 2010; Sharifi 2019). To explain flexibility in the labour market, Giffinger and Gudrun (2010) refer to unemployment rates, which Bosch et al. (2017) also deem as one of the most informative labour market and economic performance indicators. Sharifi (2019) additionally found that other elements that are concerned with working flexibility are also considered. Moreover, the international interconnectedness between cities and economies in the form of international events, cross city collaborations, migration of international companies and people, etc., clearly indicates cities' attractiveness and competitiveness; therefore, they must be considered highly for economic growth (Bosch et al. 2017; Sharifi 2019).

In addition to the sub-dimensions provided by Giffinger and Gudrun (2010), Bosch et al. (2017) and Sharifi (2019) include employment and knowledge economy indicators under smart economy too. Creation and availability of local employment opportunities act as strong incentives for setting up new smart city/development projects (Bosch et al. 2017), and can significantly improve the employment rate of the city in general. Moreover, in the era of knowledge economy where social and human capitals are valuable assets, provision of green jobs, e-commerce transactions and industry-academia-governance collaboration are tremendously adding to the value creation aspect of smart cities with a simultaneous boost in the economy (Sharifi 2019). Subsequently, Sharifi (2019) also identifies two more indicators among the assessment frameworks, first being finance, which includes market demand and value along with funding methods for smart city projects, followed by aspects such as economic impacts and risks from smart city projects. Neirotti et al. (2014), Bosch et al. (2017) and Sharifi (2019) further highlight access to touristic places, such as cultural and heritage buildings, efficient use of ICT for tourism promotion and impact management, which elevate the attractiveness of the city while also adding value to the local economy.

This comprehensive set of indicators help to look at smart cities from an economic sustainability perspective. However, some challenges such as that of privacy concerns (for transactions or other interactions between industry and citizens) may need further consideration while looking at the economic side of ICT usage in smart cities (Ismagilova et al. 2019).

3.4.2 Smart Governance

An innovative governance approach that facilitates collaboration between the government, citizens and other stakeholders with numerous technological advancements is fundamental for smart cities to become successful. Much emphasis is given to the participatory and collaborative facets in many smart city governance processes (Giffinger et al. 2007; Caragliu et al. 2011; Chourabi et al. 2012; Batty et al. 2012; Lombardi et al. 2012; Meijer and Bolivar 2016; Ruhlandt 2018; Sharifi 2019). Channels of communication between stakeholders (e-governance, e-participation, social media, etc.) in the form of access to documents, transactions, feedbacks, voting for initiatives and other bureaucratic tasks that impart public and social services improve participation and engagement, thereby also increasing the government's transparency, accountability and responsiveness in decision-making/implementation processes (Pereira et al. 2018; Gil et al. 2019; Ismagilova et al. 2019; Lu et al. 2019). Simultaneously, this model of governance enhances the quality of relationship between the government and citizens by building public confidence, commitment and a sense of empowerment in a democracy (Pereira et al. 2018).

Apart from the participation sub-dimension, vision and leadership, both within the government and for the public in order to encourage innovation, are also considered essential for the success of smart cities (Nam and Pardo 2011; Bosch et al. 2017; Sharifi 2019). Smart governance calls for a cross- and multi-disciplinary, coordinated effort that provides a clear strategy as well as the resources (funds, subsidies, etc.) for urban transformation. It ensures interoperability among systems as well as integration of solutions to fulfil the strategic goals mapped out to tackle complexity and uncertainty in urban management much more efficiently (Bolivar 2015; Sharifi 2019). Generally, legal and regulatory frameworks in regard to planning, implementation, user privacy and data ownership are required for the government to create feasible conditions for enabling and supporting their digital urban development vision (Chourabi et al. 2012; Ruhlandt 2018; Sharifi 2019). All factors combined, smart governance represents the interaction between technology, people, policies, resources, public private partnerships, laws and information, that ultimately serves citizens better by improving their quality of life while also making urban processes transparent and efficient.

3.4.3 Smart Environment

The smart environment dimension deals with the use of ICT applications to preserve natural resources, reduce energy and carbon emissions and augment environmental sustainability in general (Lu et al. 2019; Wahab et al. 2020). According to Sharifi (2019), the most recurring sub-dimension in this dimension is that of energy resources, involving renewable energy sources for local energy distribution, smart grids, efficient energy management and GHG emission control. In fact, smart energy technologies such as smart grids, energy monitoring, forecasting and storage are transforming the energy informatics, with Artificial Intelligence (AI) often being referred to as the new electricity for a more sustainable energy outlook (Pieroni et al. 2018; Rolnick et al. 2019). Similarly, the use of ICT measures for the efficient management of water and waste resources is also considered. While advanced technologies such as sensors can predict the level of contamination of water bodies, other monitoring applications like smart meters can effectively help in reducing water consumption (Pradhan et al. 2017). For the waste management challenge, IoT-based applications play a major role in real-time waste collection, disposal, recycling and recovery, thus also improving the environment and quality of life as a whole for citizens (Neirotti et al. 2014; Ismagilova et al. 2019). Additionally, other resources such as food and material consumption and recycling are also given importance in the smart environment literature (Bosch et al. 2017; Sharifi 2019). Furthermore, Lu et al. (2019), Ismagilova et al. (2019) and Sharifi (2019) highlight the reduction of environmental pollution with the help of sensors that inform about air, water, soil and noise pollution levels to safeguard the ecosystem and to ensure sustainability. This information further aids in drafting mitigation strategies for pollution reduction in cities.

In order to promote the use of smart technologies mentioned above, physical and social infrastructure are pre-requisites. For instance, smart energy technologies that reduce energy demand and enable local distribution through renewable sources require decentralized and modular infrastructure in the form of smart grids and green buildings (Sharifi 2019). Chourabi et al. (2012) and Ismagilova et al. (2019) also claim that green infrastructure such as parks, green roofs and other green spaces have a considerate impact on the sustainability as well as the liveability of a city, and therefore must be considered. Lastly, Sharifi (2019) includes constructive strategies, plans and policies to monitor resources as well as the ICT infrastructure being used for sustainable environmental management. Recent advancements have also enabled citizens to be more involved with state-of the-art concepts such as citizen science that allow them to collect environmental data, contribute to research and remain aware about the environmental issues around them (Hunt et al. 2015).

3.4.4 Smart People

As mentioned earlier, an important feature that distinguishes smart city paradigm from its predecessors is the recognition of the central role played by people as the end users of smart city solutions and applications. The dimension of smart people stresses on the role of human capital in urban management and development. According to Nam and Pardo (2011, p. 287), issues arising from urbanization can be solved using "smart solutions", that is the use of "creativity, human capital, cooperation among relevant stakeholders, and their bright scientific ideas". Further supported by Lu et al. (2019, p. 728), Dameri and Rosenthal-Sabroux (2014), and Madakam and Ramaswamy (2015), smart people generate as well as benefit from social capital by creating a 'lifelong learning environment' through ICT innovation. Deemed as a centre for higher education by Winters (2011), smart cities attract creative workers, providing them with multiple opportunities, thus also making the city smarter, economically successful, culturally or ethnically diverse and open minded (Giffinger et al. 2007; Nam and Pardo 2011). Therefore, it is important to have the smart people dimension while considering a smart city as it not only improves the social, economic and human development but also adds flexibility and resilience in the face of adversity (Wahab et al. 2020).

3.4.5 Smart Living

Smart living includes factors that essentially enhance the citizens' quality of life. There are two main viewpoints for smart living in literature. The first one looks at the facilities enabled by ICT that are used to transform office, residential, energy, transportation and other infrastructures into smart environments for a modern living experience (Madakam and Ramaswamy 2015; Tahir and Malek 2016; Lu et al. 2019). Tahir and Malek (2016, p. 7), in relation to this, state that smart living increases citizens' understanding towards "how people and technology interact by combining senses with physical action, social behaviour analysis, data analytics, engineering, technology, and communication". On the other hand, Giffinger et al. (2007), Kumar (2017) and Ismagilova et al. (2019) also involve factors such as health care, livelihood quality, culture preservation, tourism, public safety, equity, convenience and social cohesion, to entail a more integrated outlook towards smart living for a better quality of life. Equity was found to be the most covered factor in smart city assessment frameworks (Sharifi 2019). Han and Kim (2021) in their paper on sustainable smart living opine that the interaction between all citizens and smart environments is what makes a city sustainable and that this acceptance to adopt new technologies by citizens enables smart equity in sustainable smart living. The authors also explain the role of the government in bringing about this equity by taking care of the accessibility, affordability, productivity and efficiency of the services to diverse groups, as they are the main smart service providers of a city. Social cohesion, which embeds solidarity and a sense of belonging in the society, is an inherent characteristic of smart living. ICT measures also aid in improving connectedness and support among communities as well as providing equal opportunities to everyone (disabled, elderly, special needs, etc.) (Giffinger et al. 2007).

Literature further considers the healthcare domain as a crucial measure of quality of life in smart cities (Ismagilova et al. 2019). Researchers like Pramanik et al. (2017) studied the smart health concept in detail and have come up with numerous frameworks that incorporate big data analytics to reconstruct existing healthcare infrastructure for better diagnosis, treatment and monitoring, eventually promoting well-being. As a whole, Pramanik et al. (2017, p. 371) fittingly establish that " smart health defines not only ICT development, but also a state-of-thinking, a way of lifestyle and approach, and a vow for connected entities to improve healthcare facilities in the home, city, country and globe with the aid of a number of intelligent agents". Another significant factor that is discussed in length among researchers is that of safety and security in a smart city. ICT is used in many areas to provide the community with a safe environment. These range from detection of crime hotspots and natural disaster risk areas, vulnerability assessments, shorter response time for ambulance, police and fire services, local citizen involvement and data privacy (Neirotti et al. 2014; Petrova-Antonova and Ilieva 2018; Ismagilova et al. 2019; Sharifi 2019). An additional category of housing/livelihood brought up by Giffinger et al. (2007) discusses the housing quality to ensure affordable and good living conditions. Moreover, according to

Kumar (2017), a smart city with a smart living environment promotes culture, celebrates people, history and art, and motivates people to lead a meaningful and satisfied life by being more involved. Therefore, the presence of good quality cultural infrastructure such as cinema halls, museums, and libraries, and the ICT functions that promote them make up the cultural sub-dimension of smart living.

3.4.6 Smart Mobility

Smart mobility is a multi-faceted prime element for a successful smart city and provides multiple benefits to all its stakeholders. It is important to note that this dimension deals with mobility of people and not just through vehicle transportation. New and innovative technologies satisfy people's needs by providing them with accessibility to real-time information from any location, enhancing performance and attractiveness of the mobility systems, saving time, space and money as well as reducing GHG emissions (Giffinger et al. 2007; Manville et al. 2014; Battarra et al. 2018; Bokolo and Peterson 2019). Sourbati (2020, p. 2) explains smart mobility as a "convergence of movement in the physical space and in data flow" through networked communications. Therefore, there are two major aspects under smart mobility: transport and ICT infrastructure/management/accessibility.

A smart city is said to have an efficient transport infrastructure that promotes multiple modes of transportation. Battarra et al. (2018) argue that smart mobility is essentially sustainable mobility as the transport sector plays a major role in impacting the environmental quality of a city and smart solutions can help them be more compatible with the environment. For instance, using public transport, such as metro, train and bus reduces traffic congestion and air pollution. If the infrastructure for that is accessible, of good quality, diverse and provides connectivity throughout an intended geographical area, more people will shift to using them, thus decreasing the extent of the above-mentioned issues (Petrova-Antonova and Ilieva 2018). Other modes of non-motorized transport (walking, bicycle) are also encouraged, provided that there is availability of infrastructure such as bike lanes and vibrant and safe walkable areas for pedestrians with digital navigation (Petrova-Antonova and Ilieva 2018; Sharifi 2019). Additionally, cities are increasingly adopting innovative green transportation systems with electric vehicles and charging stations that predominantly focus on sustainable, clean energy (Benevolo et al. 2015). Moreover, technologies such as automated vehicles and car sharing services are also covered extensively in the smart mobility literature as concepts which are revolutionizing mobility and improving the service quality, ultimately proving to be economically beneficial for the city as more and more private actors are being involved (Docherty et al. 2018; Lu et al. 2019). With the rise of multi-modal transportation, Benevolo et al. (2015) and Šurdonja et al. (2020) also talk about Intelligent Transportation Systems (ITS) such as traffic management systems, demand control systems and parking guidance systems. Surveillance equipment like sensors and cameras can be very useful to track vehicles, detect deteriorating road conditions and monitor traffic volume for better management of traffic congestion or emergency situations (Petrova-Antonova and Ilieva 2018; Ismagilova et al. 2019; Lu et al. 2019).

Benevolo et al. (2015) also assert that these solutions require the involvement of citizens, their readiness as well as their willingness to use the technologies. Once the people are ready to accept, ICT amplifies proper management of the mobility services, resulting in a modern, usable and sustainable mobility system. Therefore, ICT infrastructure such as broadband, mobile phone network coverage and apps for services (parking, car sharing, etc.) are important factors when looking at smart mobility. In fact, Sourbati (2020) claims that mobile ICTs can be considered as effective means of transport mobility in the form of better connectivity (through navigating apps), awareness (traffic information) and digitalization (transactions for transport fares) (Petrova-Antonova and Ilieva 2018). Proper ICT infrastructure and management can also help in better access to transport systems; however, older groups face some difficulties in accessing/using platforms which can be a cause for increased inequality of digital access (Sourbati 2020). It is evident that the extent of ICT varies in different mobility solutions in a city, but is fundamental for an integrated, sustainable and forward-looking mobility system.

3.4.7 Smart Data

Data are the cornerstone of smart cities. With a wide array of smart city initiatives relying on ICTs, smart cities have become data production and sharing engines (Moustaka et al. 2019). Since such a vast amount of data are constantly produced and collected from technologies—such as mobile devices, seniors, social media platforms and other IoT infrastructure—efficient management, storage and publication of these data play a major role in maintaining consistency in smart cities (Huovila et al. 2017; Moustaka et al. 2019). According to Moustaka et al. (2019, p. 103:6), development and management of open data which are accessible for and available to everyone are crucial for smart cities because of their contribution in improving decision-making, citizen engagement and data economy. Huovila et al. (2017, p. 120) used five characteristics to rate the quality of this open data: data are published in an open, structured format under an open license, and it is in linked data format with URLs and is linked to other data. Furthermore, Sharifi (2019) also states three more factors under the smart data dimension. The judging/ data analytics factors include data harvesting tools/strategies to move the data from production stage to storage, and data mining tools to produce analytics/ publication intelligence from that (Moustaka et al. 2019). This is followed by the decision-making by the government, enterprise and citizens based on the published data, and finally the learning factor that deals with the upgradation impact of the data on the mode of operation, planning processes and interaction (Sharifi 2019).

It is evident that literature provides a multi-disciplinary set of dimensions for a comprehensive understanding of smart cities and their complex nature. However, the current literature also does not fully cover the interconnections between the dimensions themselves that may be necessary to address the complexities arising out of one another. For example, it is the knowledgeable, informed section of society that is willing to use sustainable smart mobility options for a better environment, thus enforcing a connection between smart people and smart mobility. Nonetheless, this smart city framework has the potential to not only contribute in expanding the conceptual understanding of smart cities but also helps in the practical aspect of identifying, evaluating and monitoring them.

3.5 Conclusions

Rapid advances in information and communication technologies and big data analytics, coupled with the need for developing more innovative and efficient solutions for addressing urbanization and global challenges, have led to the increasing development of smart city initiatives around the globe. Interest in smart city initiatives is expected to continue in the coming decades as cities are expected to deal with more severe challenges induced by climate change and continued population growth. In fact, smart city initiatives have already proven effective in helping cities overcome and control major societal challenges such as the recent COVID-19 pandemic, and it is argued that this has further accelerated development and uptake of smart city initiatives and solutions (Hassankhani et al. 2021; Sharifi and Khavarian-Garmsir 2020; Sharifi et al. 2021).

While the smart city concept has been around for several decades and a large body of research has been published on it, there is still a need for a more comprehensive overview of its underlying principles and dimensions. Therefore, this chapter attempted to gain a deeper understanding of the rapidly thriving concept of a smart city through extensive literature review of its history, definitions, dimensions and other related concepts as per the determined sub-objectives of this study. Firstly, the in-depth analysis of smart city roots exhibited a clear ascension of city models with respect to their incorporation of multiple perspectives at every growing stage over the decades, finally giving rise to the present-day holistic version of smart cities. While it did clarify the conflated, interchangeable use of these terms, there is still no conclusion when it comes to defining a smart city. Ranging from an intra-disciplinary, technocentric approach to a cross- and multi-disciplinary approach, smart city definitions vary temporally, conceptually as well as spatially. Current trends and developments in the smart city arena have become highly context dependent, which makes it the work of each city's policy makers to strategize and implement smart solutions/policies according to their local demands rather than following a generic workbook. Moreover, as pointed out in literature, it is also important for diverse stakeholders (researchers, businesses, government) to have a common line of thought regarding the concept in order to reduce inconsistency in its usage. Additionally, the paper covers the seven main dimensions pertinent to the smart cities along with their respective sub-dimensions, that grant a doorway to a stronger understanding of smart cities. This framework also serves as a base for further investigation into smart city frameworks which facilitate development and implementation of evaluation processes, and aid in formulating comprehensive strategies concerning each domain, possibly also achieving the interdisciplinary solutions to address social, economic, environmental and cultural issues on every scale.

References

Ahad MA, Paiva S, Tripathi G, Feroz N (2020) Enabling technologies and sustainable smart cities. Sustain Cities Soc 61:102301. https://doi.org/10.1016/j.scs.2020.102301

Ahvenniemi H, Huovila A, Pinto-Seppä I, Airaksinen M (2017) What are the differences between sustainable and smart cities? Cities 60:234–245. https://doi.org/10.1016/j.cities.2016.09.009

Al Nuaimi E, Al Neyadi H, Mohamed N, Al-Jaroodi J (2015) Applications of big data to smart cities. J Internet Serv Appl 6(1):1–15. https://doi.org/10.1186/s13174-015-0041-5

Albino V, Berardi U, Dangelico RM (2015) Smart cities: Definitions, dimensions, performance, and initiatives. The Journal of Urban Technology 22(1):3–21. https://doi.org/10.1080/10630732.2014.942092

Allam Z, Newman P (2018) Redefining the smart city: Culture, metabolism and governance. Smart Cities 1(1):4–25. https://doi.org/10.3390/smartcities1010002

Angelidou M (2014) Smart city policies: A spatial approach. Cities 41:S3–S11. https://doi.org/10.1016/j.cities.2014.06.007

Angelidou M (2015) Smart cities: A conjuncture of four forces. Cities 47:95–106. https://doi.org/10.1016/j.cities.2015.05.004

Angelidou M (2017) The role of smart city characteristics in the plans of fifteen cities. J Urban Technol 24(4):3–28. https://doi.org/10.1080/10630732.2017.1348880

Anthopoulos LG, Tsoukalas IA (2006) The implementation model of a digital city. the case study of the digital city of trikala, greece. J E-Government 2(2):91–109. https://doi.org/10.1300/J399v02n02_06

Aurigi A (2000). Digital city or urban simulator? Paper presented at the Digital, 33–44

Battarra R, Zucaro F, Tremiterra MR (2018) Smart mobility and elderly people. Can ICT make the city more accessible for everybody? TeMA: J Land Use:23–42. https://doi.org/10.6092/1970-9870/5768

Batty M, Axhausen K, Giannotti F, Pozdnoukhov A, Bazzani A, Wachowicz M, Ouzounis G, Portugali Y (2012) Smart cities of the future. Eur Phys J Spec Top 214(1):481–518. https://doi.org/10.1140/epjst/e2012-01703-3

Batty M (2013) Big data, smart cities and city planning. Dialogues Hum Geogr 3(3):274–279. https://doi.org/10.1177/2043820613513390

Benevolo C, Dameri RP, D'Auria B (2015) Smart mobility in smart city. In: Empowering organizations. Springer International Publishing, Cham, pp 13–28. https://doi.org/10.1007/978-3-319-23784-8_2. Retrieved from http://link.springer.com/10.1007/978-3-319-23784-8_2

Bibri SE, Krogstie J (2017) Smart sustainable cities of the future: An extensive interdisciplinary literature review. Sustain Cities Soc 31:183–212. https://doi.org/10.1016/j.scs.2017.02.016

Bokolo AJ, Petersen SA (2019) A smart city adoption model to improve sustainable living. Retrieved from https://explore.openaire.eu/search/publication?articleId=dedup_wf_001::224eed3db2bc20ce174a9985f0fca465

Bosch P, Jongeneel S, Rovers V, Hans-Martin Neumann, Miimu Airaksinen, Huovila A (2017) CITYkeys indicators for smart city projects and smart cities (Unpublished). https://doi.org/10.13140/rg.2.2.17148.23686. Retrieved from https://search.datacite.org/works/10.13140/rg.2.2.17148.23686

Bolívar, M. P. R. (2015). Smart cities: Big cities, complex governance? In: Transforming city governments for successful smart cities. Springer International Publishing, Cham, pp 1–7. https://doi.org/10.1007/978-3-319-03167-5_1 Retrieved from http://link.springer.com/10.1007/978-3-319-03167-5_1

Canton J (2011) The extreme future of megacities. Significance (Oxford, England) 8(2):53–56. https://doi.org/10.1111/j.1740-9713.2011.00485.x

Caragliu A, Del Bo C, Nijkamp P (2011) Smart cities in europe. J Urban Technol 18(2):65–82. https://doi.org/10.1080/10630732.2011.601117

Carvalho L (2015) Smart cities from scratch? A socio-technical perspective. Camb J Reg Econ Soc 8(1):43–60. https://doi.org/10.1093/cjres/rsu010

Castells M (1989).The informational city: Information technology, economic restructuring, and the urban-regional process. Blackwell, Oxford

Chourabi H, Taewoo Nam, Walker S, Gil-Garcia JR, Mellouli S, Nahon K, Pardo TA, Scholl HJ (2012) Understanding smart cities: An integrative framework. Paper presented at the 2289–2297, January. https://doi.org/10.1109/HICSS.2012.615 Retrieved from https://ieeexplore.ieee.org/document/6149291

Dameri RP, Cocchia A (2013) Smart city and digital city: Twenty years of terminology evolution. Paper presented at the X Conference of the Italian Chapter of AIS, ITAIS, 1–8.

Dameri RP, Rosenthal-Sabroux C (2014) Smart city and value creation. In: Smart city. Springer International Publishing, Cham, pp 1–12. https://doi.org/10.1007/978-3-319-06160-3_1. Retrieved from http://link.springer.com/10.1007/978-3-319-06160-3_1

Docherty I, Marsden G, Anable J (2018) The governance of smart mobility. Transp Research Part A, Policy and Practice 115:114–125. https://doi.org/10.1016/j.tra.2017.09.012

Giffinger R, Pichler-Milanović N (2007) Smart cities: Ranking of European medium-sized cities. Vienna University of Technology, Centre of Regional Science

Giffinger R, Gudrun H (2010) Smart cities ranking: An effective instrument for the positioning of the cities? Centre de Política del Sòl i Valoracions - Universitat Politècnica de Catalunya. https://doi.org/10.5821/ace.v4i12.2483

Gil O, Cortés-Cediel M, Cantador I (2019) Citizen participation and the rise of digital media platforms in smart governance and smart cities. Int J E-Plan Res 8(1):19–34. https://doi.org/10.4018/IJEPR.2019010102

Gillespie A, Richardson R (2000) Teleworking and the city: Myths of workplace transcendence and travel reduction. In: Cities in the telecommunications age: The fracturing of geographies, pp 228–245

Harrison C, Eckman B, Hamilton R, Hartswick P, Kalagnanam J, Paraszczak J, Williams P (2010) Foundations for smarter cities. IBM J Res Dev 54(4):1–16. https://doi.org/10.1147/JRD.2010.2048257

Nikki Han MJ, Kim MJ (2021) A critical review of the smart city in relation to citizen adoption towards sustainable smart living. Habitat Int 108:102312. https://doi.org/10.1016/j.habitatint.2021.102312

Hassankhani M, Alidadi M, Sharifi A, Azhdari A (2021) Smart City and Crisis Management: Lessons for the COVID-19 Pandemic. Int J Environ Res Public Health 18(15):7736. Retrieved from https://www.mdpi.com/1660-4601/18/15/7736

Hepworth ME (1987) The information city. Cities 4(3):253–262

Hollands RG (2008) Will the real smart city please stand up? City (London, England) 12(3):303–320. https://doi.org/10.1080/13604810802479126

Hunt N, O'Grady MJ, Muldoon C, Kroon B, Rowlands T, Wan J, O'Hare GM (2015) Citizen science: A learning paradigm for the smart city? IxD&A 27:44–65

Huovila A, Airaksinen M, Pinto-Seppä I, Piira K, Bosch P, Penttinen T, Neuman HM, Kontinakis N (2017) CITYkeys smart city performance measurement system. Retrieved from https://cris.vtt.fi/en/publications/3b417c09-2a57-4e49-a68c-6cb41a9aacd8

Huovila A, Bosch P, Airaksinen M (2019) Comparative analysis of standardized indicators for smart sustainable cities: What indicators and standards to use and when? Cities 89:141–153. https://doi.org/10.1016/j.cities.2019.01.029

Ishida T (2002) Digital city Kyoto. Commun ACM 45(7):76–81

Ismagilova E, Hughes L, Dwivedi YK, Raman KR (2019) Smart cities: Advances in research—An information systems perspective. Int J Inf Manage 47:88–100. https://doi.org/10.1016/j.ijinfomgt.2019.01.004

Israilidis J, Odusanya K, Mazhar MU (2021) Exploring knowledge management perspectives in smart city research: A review and future research agenda. Int J Inf Manage 56:101989. https://doi.org/10.1016/j.ijinfomgt.2019.07.015

Kitchin R, Coletta C, Evans L, Heaphy L (2019) Creating smart cities. In: Creating smart cities, pp 1–18

Knox H (2010) Cities and organisation: The information city and urban form. Cult Organ 16(3):185–195. https://doi.org/10.1080/14759551.2010.503496

Komninos N (2006) The architecture of intelligent cities: Integrating human, collective and artificial intelligence to enhance knowledge and innovation. Paper presented at the https://doi.org/10.1049/cp:20060620. Retrieved from http://digital-library.theiet.org/content/conferences/10.1049/cp_20060620

Komninos N (2008) Intelligent cities. In: Anttiroiko A (ed) Electronic government: Concepts, methodologies, tools, and applications. IGI Global, pp 4205–4212. https://doi.org/10.4018/978-1-59904-947-2

Komninos N (2009) Intelligent cities: Towards interactive and global innovation environments. Int J Innov Reg Dev 1(4):337–355. https://doi.org/10.1504/IJIRD.2009.022726

Komninos N (2015) The age of intelligent cities: Smart environments and innovation-for-all strategies, 1st edn. Routledge. https://doi.org/10.4324/9781315769349

Kramers A, Höjer M, Lövehagen N, Wangel J (2014) Smart sustainable cities—Exploring ICT solutions for reduced energy use in cities. Environ Model Softw 56:52–62. https://doi.org/10.1016/j.envsoft.2013.12.019

Kumar TMV (2017) Smart economy in smart cities. Springer, Singapore

Lombardi P, Giordano S, Farouh H, Yousef W (2012) Modelling the smart city performance. Innovation (Abingdon, England) 25(2):137–149. https://doi.org/10.1080/13511610.2012.660325

Lu H, Chen C, Yu H (2019) Technology roadmap for building a smart city: An exploring study on methodology. Futur Gener Comput Syst 97:727–742. https://doi.org/10.1016/j.future.2019.03.014

Madakam S, Ramaswamy R (2015) 100 new smart cities (India's smart vision). Paper presented at the 1–6, Febeuary. https://doi.org/10.1109/NSITNSW.2015.7176407. Retrieved from https://ieeexplore.ieee.org/document/7176407

Manville C, Cochrane G, Cave J, Millard J, Pederson JK, Thaarup RK, Kotterink B (2014) Mapping smart cities in the EU. European Parliament; Directorate General for Internal Policies. Policy Department Economic and Scientific policy A

Marsal-Llacuna M, Colomer-Llinàs J, Meléndez-Frigola J (2015) Lessons in urban monitoring taken from sustainable and livable cities to better address the smart cities initiative. Technol Forecast Soc Chang 90:611–622. https://doi.org/10.1016/j.techfore.2014.01.012

Meijer A, Bolívar MPR (2016) Governing the smart city: A review of the literature on smart urban governance. Int Rev Adm Sci 82(2):392–408. https://doi.org/10.1177/0020852314564308

Mora L, Deakin M, Reid A (2018) Smart-city development paths: Insights from the first two decades of research. In: Smart and sustainable planning for cities and regions. Springer International Publishing, Cham, pp 403–427. https://doi.org/10.1007/978-3-319-75774-2_28. Retrieved from http://link.springer.com/10.1007/978-3-319-75774-2_28

Mora L, Deakin M, Reid A (2019) Combining co-citation clustering and text-based analysis to reveal the main development paths of smart cities. Technol Forecast Soc Chang 142:56–69. https://doi.org/10.1016/j.techfore.2018.07.019

Mosannenzadeh F, Vettorato D (2014) Defining smart city. A conceptual framework based on keyword analysis. TeMA: J Land Use. https://doi.org/10.6092/1970-9870/2523

Moustaka V, Vakali A, Anthopoulos L (2019) A systematic review for smart city data analytics. ACM Comput Surv 51(5):1–41. https://doi.org/10.1145/3239566

Nam T, Pardo T (2011) Conceptualizing smart city with dimensions of technology, people, and institutions. Paper presented at the 282–291, June 12. https://doi.org/10.1145/2037556.2037602. Retrieved from http://dl.acm.org/citation.cfm?id=2037602

Neirotti P, De Marco A, Cagliano AC, Mangano G, Scorrano F (2014) Current trends in smart city initiatives: Some stylised facts. Cities 38:25–36. https://doi.org/10.1016/j.cities.2013.12.010

Ojo A, Curry E, Janowski T, Dzhusupova Z (2015) Designing next generation smart city initiatives: The SCID framework. In: Transforming city governments for successful smart cities. Springer International Publishing, Cham, pp 43–67. https://doi.org/10.1007/978-3-319-03167-5_4 Retrieved from http://link.springer.com/10.1007/978-3-319-03167-5_4

Pereira GV, Parycek P, Falco E, Kleinhans RJ (2018) Smart governance in the context of smart cities: A literature review. Information Polity 23(2):143–162. https://doi.org/10.3233/IP-170067

Petrova-Antonova D, Ilieva S (2018) Smart cities evaluation—A survey of performance and sustainability indicators. Paper presented at the 486–493, August. https://doi.org/10.1109/SEAA.2018.00084. Retrieved from https://ieeexplore.ieee.org/document/8498251

Pieroni A, Scarpato N, Di Nunzio L, Fallucchi F, Raso M (2018) Smarter city: Smart energy grid based on blockchain technology. Int J Adv Sci, Eng Inf Technol 8(1):298. https://doi.org/10.18517/ijaseit.8.1.4954

Porat MU (1977) The input-output structure of the information economy/prepared by marc U. Porat and Michael R. Rubin; produced by the U. S. dept of commerce, office of telecommunications. District of Columbia: For sale by the Supt. of Docs., U. S. Govt. Print. Off., 1977

Pradhan MA, Patankar S, Shinde A, Shivarkar V, Phadatare P (2017) IoT for smart city: Improvising smart environment. Paper presented at the 2003–2006, August. https://doi.org/10.1109/ICECDS.2017.8389800. Retrieved from https://ieeexplore.ieee.org/document/8389800

Praharaj S, Han H (2019) Building a typology of the 100 smart cities in india. Smart Sustain Built Environ 8(5):400–414. https://doi.org/10.1108/SASBE-04-2019-0056

Pramanik MI, Lau RYK, Demirkan H, Azad MAK (2017) Smart health: Big data enabled health paradigm within smart cities. Expert Syst Appl 87:370–383. https://doi.org/10.1016/j.eswa.2017.06.027

Ratti C, Townsend A (2011) The social nexus. Sci Am 305(3):42–49. Retrieved from http://www.jstor.org/stable/26002792

Rios P (2008) Creating "The Smart City". Retrieved from http://hdl.handle.net/10429/393

Rolnick D, Donti PL, Kaack LH, Kochanski K, Lacoste A, Sankaran K, Ross AS, Milojevic-Dupont N, Jaques N, Waldman-Brown A, Bengio Y (2019) Tackling climate change with machine learning. Retrieved from https://arxiv.org/abs/1906.05433

Ruhlandt RWS (2018) The governance of smart cities: A systematic literature review. Cities 81:1–23. https://doi.org/10.1016/j.cities.2018.02.014

Selada C (2017) Smart cities and the quadruple helix innovation systems conceptual framework: The case of Portugal. In: The quadruple innovation helix nexus. Palgrave Macmillan, New York, pp 211–244. https://doi.org/10.1057/978-1-137-55577-9_8. Retrieved from http://link.springer.com/10.1057/978-1-137-55577-9_8

Sharifi A (2019) A critical review of selected smart city assessment tools and indicator sets. J Clean Prod 233:1269–1283. https://doi.org/10.1016/j.jclepro.2019.06.172

Sharifi A (2020) A typology of smart city assessment tools and indicator sets. Sustain Cities Soc 53:101936. https://doi.org/10.1016/j.scs.2019.101936

Sharifi A, Khavarian-Garmsir AR (2020) The COVID-19 pandemic: Impacts on cities and major lessons for urban planning, design, and management. Sci Total Environ 749. https://doi.org/10.1016/j.scitotenv.2020.142391

Sharifi A, Khavarian-Garmsir AR, Kummitha RKR (2021) Contributions of smart city solutions and technologies to resilience against the COVID-19 pandemic: A literature review. Sustainability 13(14):8018. Retrieved from https://www.mdpi.com/2071-1050/13/14/8018

Silva BN, Khan M, Han K (2018) Towards sustainable smart cities: A review of trends, architectures, components, and open challenges in smart cities. Sustain Cities Soc 38:697–713. https://doi.org/10.1016/j.scs.2018.01.053

Sourbati M (2020) Age and the city: The case of smart mobility. In: Human aspects of IT for the aged population. Technology and society. Springer International Publishing, Cham, pp 312–326. https://doi.org/10.1007/978-3-030-50232-4_22. Retrieved from http://link.springer.com/10.1007/978-3-030-50232-4_22

Stock WG (2011) Informational cities: Analysis and construction of cities in the knowledge society. J Am Soc Inform Sci Technol 62(5):963–986. https://doi.org/10.1002/asi.21506

Strauss RE, Schoder D, Müller G (1996) Wired cities—Opportunities for small and medium sized cities on the information highway. In:Terashima N, Altman E (eds) Advanced IT tools: IFIP world conference on IT tools 2–6 September 1996, Canberra, Australia. Springer, Boston, MA, pp 3–10. https://doi.org/10.1007/978-0-387-34979-4_1. Retrieved from https://doi.org/10.1007/978-0-387-34979-4_1

Šurdonja S, Giuffrè T, Deluka-Tibljaš A (2020) Smart mobility solutions—Necessary precondition for a well-functioning smart city. Transp Res Procedia 45:604–611. https://doi.org/10.1016/j.trpro.2020.03.051

Tahir Z, Abdul Malek J (2016) Main criteria in the development of smart cities determined using analytical method. Plan Malays J 14(5). https://doi.org/10.21837/pmjournal.v14.i5.179

United Nations (2018) World urbanization prospects; the 2011 revision. Department of Economic and Social Affairs, United Nations, New York. Available at: https://www.un.org/en/events/citiesday/assets/pdf/the_worlds_cities_in_2018_data_booklet.pdf. Accessed 7 May 2021

van den Besselaar P, Melis I, Beckers D (2000) Digital cities: Organization, content, and use. In: Digital cities. Springer, Berlin, Heidelberg, pp 18–32. https://doi.org/10.1007/3-540-46422-0_3. Retrieved from http://link.springer.com/10.1007/3-540-46422-0_3

Wahab NSN, Seow TW, Radzuan ISM, Mohamed S (2020) A systematic literature review on the dimensions of smart cities. IOP Conf Ser: Earth Environ Sci 498:12087. https://doi.org/10.1088/1755-1315/498/1/012087

Webster F (2001) Re-inventing place: Birmingham as an information city? City (London, England) 5(1):27–46. https://doi.org/10.1080/13604810125315

Winters JV (2011) Why are smart cities growing? J Reg Sci 51(2):253–270. https://doi.org/10.1111/j.1467-9787.2010.00693.x

Yasuoka M, Ishida T, Aurigi A (2010) The advancement of world digital cities. In: Handbook of ambient intelligence and smart environments. Springer, Boston, MA, pp 939–958. https://doi.org/10.1007/978-0-387-93808-0_35 Retrieved from http://link.springer.com/10.1007/978-0-387-93808-0_35

Yigitcanlar T (2015) Smart cities: An effective urban development and management model? Aust Plan 52(1):27–34. https://doi.org/10.1080/07293682.2015.1019752

Yigitcanlar T, Kamruzzaman M, Buys L, Ioppolo G, Sabatini-Marques J, da Costa EM, Yun JJ (2018) Understanding 'smart cities': Intertwining development drivers with desired outcomes in a multidimensional framework. Cities 81:145–160. https://doi.org/10.1016/j.cities.2018.04.003

Yovanof G, Hazapis G (2009) An architectural framework and enabling wireless technologies for digital cities & intelligent urban environments. Wirel Pers Commun 49(3):445–463. https://doi.org/10.1007/s11277-009-9693-4

Yigitcanlar T, Kamruzzaman M, Foth M, Sabatini-Marques J, da Costa E, Ioppolo G (2019) Can cities become smart without being sustainable? A systematic review of the literature. Sustain Cities Soc 45:348–365. https://doi.org/10.1016/j.scs.2018.11.033

Chapter 4
Resilient Cities: Concepts and Underlying Principles

Nehmat Singh and Ayyoob Sharifi

Abstract In the era of increasing risks and uncertainties induced by various stressors such as climate change and social and geopolitical conflicts, resilience is high on the agenda of planners, policy makers, and researchers. This is manifested in the increasing number of plans, programs, policies, and frameworks that are developed annually to enhance urban resilience. One potential impediment to the proper design and implementation of resilience plans, programs, policies, and frameworks is the incomplete understanding of the resilience concept itself. This issue becomes even more complicated when considering the fact that resilience is a contested notion and various definitions exist for it depending on the background, field, context, and objectives of the stakeholders. In an effort to better understand different conceptualizations of resilience in the context of urban planning, this chapter elaborates on the genealogy of the resilience concept and its underlying principles and characteristics. It is argued that resilience as a concept has an old history in fields such as physics and psychology but has been introduced to and used in urban studies only since a few decades ago. Urban scholars and practitioners have relied on the vast body of literature from other fields to conceptualize resilience depending on their specific purposes. Three dominant approaches that guide such conceptualizations are, namely, engineering, socio-ecological, and adaptive. The latter one has gained more momentum in the recent years considering the increasing recognition of the concept of living with risk and the need for continuous improvement and evolvement. This chapter concludes by elaborating on various underlying resilience characteristics such as Robustness, redundancy, flexibility, agility, adaptive capacity, modularity, resourcefulness, creativity, equity, foresight capacity, diversity, inclusiveness, connectivity, and efficiency. These characteristics are essential for developing

N. Singh
IHS, Erasmus University Rotterdam, Rotterdam, The Netherlands

A. Sharifi (✉)
Graduate School of Humanities and Social Sciences, Hiroshima University, Higashihiroshima, Japan
e-mail: sharifi@hiroshima-u.ac.jp

Graduate School of Advanced Science and Engineering, Hiroshima University, Hiroshima 739-8511, Japan

A. Sharifi and P. Salehi (eds.), *Resilient Smart Cities*, The Urban Book Series,
https://doi.org/10.1007/978-3-030-95037-8_4

more objective resilience plans, programs, policies, and frameworks. They could also contribute to making the resilience concept more tangible to various stakeholders.

Keywords Resilient cities · Urban resilience · Urban planning · Principles

4.1 Introduction

The world today is experiencing the largest shift from rural to urban areas in the history of settlements. Currently, the world population of 7.8 billion is expected to reach about 9.8 billion in 2050. With this increase in population, the share of humans living in the urban areas is also expected to increase up to 68% by 2050 (United Nations 2018).

While this process of urbanization is creating larger economic opportunities; it is also placing extraordinary stress on natural environment as well as the government institutions. Local authorities and state governments are finding it challenging to meet the demands of the ever-increasing population in cities, especially of the poor inhabitants. The authorities are unable to provide them with basic facilities of housing, sewerage, and water. Therefore, we see urban areas being more prone to social risks as well as natural disasters.

In many cities, the scale of the risks is much higher due to the quality of housing and infrastructure services (Huq et al. 2007). The rural inhabitants who migrate to urban areas seeking for better opportunities end up living in the riskiest zones of the city usually near riverbeds and steep slopes and are trapped into the condition of urban poverty. This increases the vulnerability of the people. The rapid scale of urbanization is leading to urban conflicts and social unrest. Widening inequalities, practice of unfair law and order, and weak urban governance are further escalating the intensity of social risk (Lehmann 2015). In regard to environmental hazards, landslides, catastrophic fires, and massive floods are all increasing at an alarming rate affecting huge parts of the cities. The increasing frequency of these natural disasters is being caused by human-induced activities. For instance, large scale deforestation and mining may cause landslides. Climate change is expected to further increase the frequency and intensity of adverse events in cities. Additionally, the recent pandemic demonstrated that cities could also be highly vulnerable to infectious diseases (Sharifi and Khavarian-Garmsir 2020).

Urban areas have seemed to form a direct relation between climate change and the process of urbanization (Filho et al. 2019). Climate change is leading to depletion in water supplies in cities, and water scarcity issues are further exacerbated due to high rates of evaporation and losses and excessive consumption. Cities are also affected by other climate-induced impacts such as extreme heat. This way, it is impacting the health of the citizens, more among the people residing in the informal settlements in developing and less developed countries.

The two continents of Africa and Asia are identified as regions most vulnerable to the impacts of climate change mostly due to rise in population growth rates and rapid increase in urban slums (Filho et al. 2019). Vulnerability and risks further increase in cities situated in flood plains and coastal areas. For instance, Dhaka, Mumbai, and Shanghai, considered as the largest cities of the world, are becoming more and more prone to floods and storms every year. Many cities from Africa, such as Lagos, Alexandria, and others are prone to higher risks from flooding and storms as well (Huq et al. 2007).

In certain urban regions of the world, there is increase in the frequency and intensity of flooding which might be due to high sea levels, glacial outbursts, and heavier or prolonged rainfall. In other cities, there is a reduction in the average rainfall which has even led to droughts. Hence, all these changing phenomena is likely to be caused due to climate change.

Reckien et al. (2017) state that "climate change is acknowledged as the largest threat to our societies in the coming decades." The water quantity and quality are heavily being impacted and will continue to if significant amount of investment is not done for various interventions. (Huq et al. 2007). Change in urban temperature would continue to rise which is highly influenced by Urban Heat Island (UHI) effect. Furthermore, the climate change will increase the heat stress caused by UHI. This will lead to negative consequences to the ecosystem services as well as on the health of the citizens. (Emilsson and Sang 2017; Sharifi et al. 2021). The shift in temperatures and rainfall patterns, and intensified concentration of CO_2 are projected to have adverse impacts on the ways species interact in an urban space and on the ecosystem services. Warmer urban temperatures will also be a breeding ground for mosquitoes, thus, spreading diseases like malaria and dengue fever (Huq et al. 2007). Climate change has the potential to multiply the effect on these existing stresses, which will expose the cities to more dangerous disasters and risks.

The impact of these risks caused due to the environmental degradation processes and gaps in urban governance structure are felt differently in cities across the world. Future catastrophes would be further exaggerated if right actions are not taken on time. There is a need to develop integrated solutions to address a complex range of vulnerabilities and risks in urban areas by building resilience in cities. Urban resilience doesn't only look into preparing cities for responding to disasters or hazards but rather taking initiatives to prevent them from happening in the first place (Gonçalves 2018). Resilience in cities does not only account for natural hazards but also human conflicts, economic downturns, and governance failures. This makes it even more imperative for cities to be resilient. To achieve resilience an incredible coordination level between different actors in the city, such as government authorities, private institutes, and industries and, the community, is necessary (Gonçalves 2018). Thereafter, resilient cities can handle natural and human-induced disasters, protect the lives of their citizens, and promote well-being and sustainable growth of cities and its people.

This chapter aims to explain the origin of and underlying principles of the concept of resilience as a beneficial strategy for improved urban management. For this, it is essential to first understand the various concepts and characteristics that revolve around urban resilience. Therefore, this chapter is based on literature review about this field.

The remainder of this chapter has the following structure: Sect. 4.2 provides an overview of the historical background of resilience and the building of the concept of urban resilience. In Sect. 4.3, concepts and definitions of urban resilience have been reviewed. Section 4.4 elaborates the main dimensions and characteristics of urban resilience. Lastly, in Sect. 4.5, conclusions on the need for urban resilience are drawn.

4.2 Background and Historical Overview

The word "resilience" is believed to be derived from the Latin verb "resilire" which means to recoil or rebound. "Resilience" or "resiliency" became common terms during the second half of the seventeenth century for explaining any counteractions in physical aspects (Gößling-Reisemann et al. 2018). The concept first originated in the works of Thomas Tredgold from physics of material in the year 1818 (Iliuk and Teperik 2018). He used the words to explain the reason behind the ways certain types of wood could accommodate sudden and strong loads without breaking (Iliuk and Teperik 2018). Through this academic discipline, the concept of resilience was further introduced in English speaking countries (Gößling-Reisemann et al. 2018).

With the outgrowing popularity of the word "resilience," there are different views and applications surrounding the concept. According to a study conducted by Iliuk and Teperik (2018), resilience has a long history across different fields of study. The infographic developed by them illustrates the evolution of the concept in various fields since 1950. The chart[1] explicitly shows that psychology has been in clear lead since the past 40 years, since the term is extensively used in relation to cognitive responses.

According to some scholars in psychology, resilience is considered as a personal trait (Garmezy 1971). According to Olsson et al. (2015), psychologists use the term resilience "to describe an individual's reactions to potentially traumatic events." Norman Garmezy, a clinical psychologist, is considered to be the founder of research in resilience. He defined resilience as being "not necessarily impervious to stress, rather, resilience is designed to reflect the capacity for recovery and maintained adaptive behaviour that may follow initial retreat or incapacity upon initiating a stressful event" (Shean 2015). Other scholars of psychology such as Michael Rutter, Emily Werner, Suniya Luther, and Michael Unger have also defined resilience according to their understanding and field of specialization (Shean 2015).

In ecology, the concept of resilience was introduced by C. S. Holling to express the persistence of an ecosystem going through a change due to natural or anthropogenic causes. Holling argued that the concept is applicable to more fields than just ecosystems but also, socio-ecological and economic systems (Thorén 2014). Ecological resilience will be discussed further in the remainder of this section.

[1] https://public.flourish.studio/visualisation/4795648/.

4.2.1 Major Approaches to Resilience

With regards to an urban system, resilience thinking has three major approaches: (1) Engineering resilience (2) Ecological resilience (3) Socio-ecological resilience. These will be briefly discussed below.

Resilience from an engineering perspective is based on the ability to resist shocks and the speed of recovery of a system after going through a series of disturbances. The sooner the system recovers, the more resilient it is. However, this interpretation considers the system to be in a static state and so, this conceptualization of resilience emphasizes the speed and ability of the system to bounce back to its pre-disturbance stage (UN-Habitat 2017). The definition revolves around the measure of stability. Physics, control system design, and material engineering are some prominent fields which use this definition (Gunderson 2000).

Over time, ecological resilience became prominent in understanding the science and management of ecosystems when pushed beyond their limits for recovering from a situation that involved many stressors on the system (Falk et al. 2019). The term resilience was introduced into ecology by C. S. Holling in 1973. It was to help in understanding the dynamics observed in ecosystems and defined it as "the amount of disturbance that an ecosystem could withstand without changing self-organized processes and structures" (Gunderson 2000). Ecological resilience measures the magnitude of disturbance a system can absorb before "the system redefines its structure by changing the variables and processes that control behaviour" (Gunderson 2000).

The stark difference between engineering and ecological resilience lies in the equilibrium state. Engineering resilience considers only one stable or equilibrium state emphasizes staying within that state or returning to it in a timely manner. Whereas, ecological resilience presumes multiple states for a system and the ability of the system to transition between the states is considered to be important. Gunderson (2000) states some examples of the transition in ecological systems, such as grassland dominated terrain to woodland dominated terrain in Zimbabwe. The multiple states are described in the form of plantation and the stressor or disturbance is grazing.

From 1970s, many social scientists and urbanists started to believe that the social fabric of regions is intrinsically interwoven with the ecosystems they depend on. This conceptual thinking led to the emergence of concepts focused on socio-ecological systems. When resilience was embedded in this concept, it came to be known as socio-ecological resilience, which "incorporates the idea of adaptation, learning and self-organization in addition to the general ability to persist disturbance" (Folke 2006).

Carpenter et al. (2001) interpret socio-ecological resilience as:

1. The amount of disturbance a system can handle and still remain in the same state,
2. The degree of a system to be self-organized,
3. The degree to which the system can build and increase the capacity for learning and adaptation.

This further challenges the equilibrium state of engineering and ecological resilience as this view highlights the ever-evolving state of system without having to go through drastic disturbances. Therefore, Kim and Lim (2016) describe it as "evolutionary" resilience since it examines the evolutionary process of a society. However, the term "evolutionary" mentioned here has a slightly different meaning. It includes flexibility, diversity, and adaptability, thus improving the capacity through a constant process of adaptation (Kim and Lim 2016).

Research on ecosystems has led to a deeper understanding of the important role that biodiversity plays in self-organizing of complex adaptive systems in relation to absorbing a shock and reorganizing of the system after the disturbance or shock (Folke 2006). The adaptive cycle developed by C. S. Holling is applicable in complex adaptive systems in relation to ecosystems and social systems through the process of non-linear changes. It describes the way a complex adaptive system evolves in four stages, namely, rapid growth (r), equilibrium (K), collapse or destruction (Ω), and reorientation (α) (Folke 2006). The adaptive cycle in Fig. 4.1 describes the movement of the system through a three dimensional state: system potential, connectedness, and resilience (Holling, 1986).

The Panarchy (Fig. 4.2) by Gunderson and Holling (2002) shows nested adaptive cycle indicating that the processes that give shape to an ecosystem take place at different spatial and temporal scales. According to Kim and Lim (2016), the adaptive capacity cycle needs to evolve continuously to be able to evolve continuously and experience transition into a state that has greater amount of resilience and allows continuous interactions between various features such as adaptability and transformability (Kim and Lim 2016). Sundstorm and Allen (2019) firmly argue the adaptive cycles to be ubiquitous in complex adaptive systems as they are continuously generating dynamics due to self-organization and evolution processes.

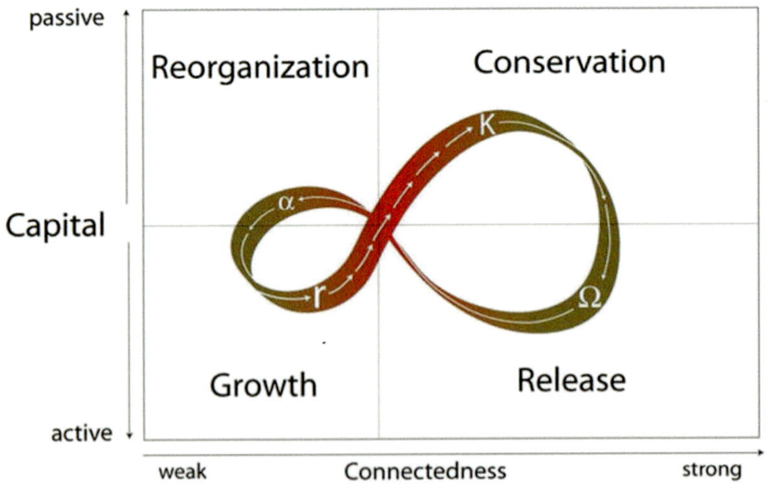

Fig. 4.1 The adaptive cycle (*Source* Holling, 1986 in Garmestani et al., [2009])

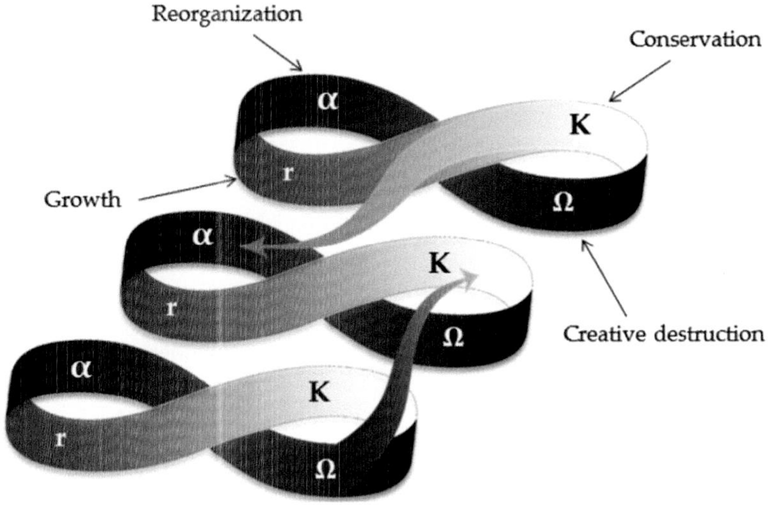

Fig. 4.2 The panarchy adaptive cycle (*Source* Kim and Lim 2016)

Research on socio-ecological systems considers that there is more than just social and ecological systems combined. Most of the literature available on resilience from 2000 onwards adopted the term "socio-ecological system" exclusively considering the social side of it to address global challenges (Berkes and Ross 2013).

4.2.2 Emergence and Spread of Urban Resilience

While the concept of resilience has been addressed and researched upon by different academicians and scientists from different study fields, it was not until the early years of the new millennium that it was widely used in urban studies (Sharifi et al. 2017). Since then, the application of this concept has been constantly changing more along with the changes in the urban environment; hence, it is considered as an "evolving concept" (UN-Habitat 2017). Cities are considered as the part of the problem due to high emissions of greenhouse gases and increase in vulnerability to risks yet; they are also parts of the solutions they can provide opportunities for better and more efficient coping and response. With the rise in catastrophic events, urbanization process, and population density, urban resilience has become a major agenda for urban planners and policy thinkers. Due to the increasing pressure that challenges the well-being of citizens, many cities are adopting resilience frameworks to also accelerate transition toward economic, socio-cultural, and environmental sustainability.

Frantzeskaki (2016) recognizes five benefits of urban resilience in cities at strategic and program level. They are as follows:

Strategic Level

- As an integrated concept, resilience allows connecting goals and actions across various departments for being able to develop a common goal and agenda for achieving it,
- Urban resilience helps in building up solutions that systematically address vulnerabilities and risks,
- As a transformative concept, it requires approaches that address characteristics such as redundancy and flexibility.

Program Level

- It is a multifaceted concept that requires in-depth understanding of socio-cultural, ecological, economic, and institutions factors along with assets and vulnerabilities.
- It is an empowering concept for community engagement as it looks into understanding risks to overcome social issues and vulnerabilities of the citizens.

Hence, different approaches are required to shift from just theoretical knowledge gain and servicing to experimental approaches with the residents to create community development and resilience.

Cities are considered as complex adaptive system, wherein, the communities and the environment are interdependent along with other subsystems, and the concept of resilience is seen as a help to face the challenges of a complex socio-ecological system (Frantzeskaki 2016).

Traditionally, urban resilience is closely linked to "engineering resilience" where the urban system is seen to have a stable state, and disturbances are attended one by one without considering the interconnectedness of social and economic factors to a city's system. However, the static nature does not capture the dynamism of cities. According to UN-Habitat (2017), it is necessary that the entire system needs to be viewed as a whole for achieving urban resilience rather than having delivered programs that address urban challenges separately.

Figure 4.3 makes it clear that engineering resilience may be problematic as it is only focusing on a single state and does not consider non-physical factors. While the ecological approach still shows some flexibility and recognizes the characteristic of adaptability in a system.

Implementation of resilience in a city is not merely about responding to shocks or stressors but also the way it is perceived and understood as a concept (Shamsuddin 2020). Greater and efficient multilevel stakeholder engagement at vertical and horizontal levels will enable interconnecting resilience in policies, economic development, and in reducing climate-induced impacts. The idea of urban resilience is multifaceted and highly challenging. Therefore, there is a need to bridge the gap between theory and implementation of urban resilience (Coaffee et al. 2018).

Factors such as availability of resources, coordination between government departments, documentation procedures, and willingness from officials can highly influence the implementation process. To restructure the functions and activities of

	NUMBER OF EQUILIBRIUMS	MEASURE FOR RESILIENCE	NATURE OF DISTURBANCES	EMPHASIS
Engineering resilience	one	speed of return to the single equilibrium	- predictable - external - shocks	- resistance and recovery - efficiency, predictability
Ecological resilience	multiple	magnitude of shocks that can be absorbed, before the threshold to enter a new equilibrium is crossed, as well as degree of self-organisation and capacity for learning	- predictable and unpredictable - external - shocks	- persistence - adaptability, flexibility - resourcefulness, efficiency, diversity
Social-ecological resilience	none, continuously changing	magnitude of shocks and stresses that are continuously absorbed, as well as advanced degree of self-organisation and capacity for learning by social-ecological systems	- predictable and unpredictable - internal and external - shocks and stresses	- persistence - adaptability, flexibility - human potential to transform its surroundings (human agency)

Fig. 4.3 Analysis of resilience thinking (*Source* UN-Habitat 2017)

government departments in a traditional bureaucratic setup and shit toward horizontal integration is a major task for developing resilience (Coaffee et al. 2018).

Collaboration with private sector for services such as IT, investments, consultancies, and think tanks will further enhance the process of building resilience in a city. The Resilient Cities Congress held in 2018 had promoted the inclusion of private sector. Along with government and private sectors, communities play an important role in building resilience. Inclusion of communities in building resilience will help in better understanding of the social fabric of a city which will further lead to better decision-making processes (Frantzeskaki 2016).

4.2.3 Advances in Urban Resilience

An increasing number of international organizations has developed frameworks and formed alliances addressing their concerns related to the concept of resilience. The C40 Cities Climate Leadership group established in 2005 aims to make the registered 95 cities climate resilient. Another major project dedicated to help cities become resilient in social, economic, and physical aspects is the “100 Resilient Cities” pioneered by the Rockfeller Foundation. It helps formulate each city a resilient strategy that shall further inspire other cities of all sizes to adopt resilience.

The 2030 agenda for sustainable development comprises of 17 goals out of which the Sustainable Development Goal 11 looks specifically into making cities resilient, inclusive and sustainable. A non-governmental organization, Cities Alliance works for eradicating poverty in cities through programs that support strengthening policy frameworks, building local skills and capacity, implementing strategic urban planning, and facilitating investments. The Ecological Sequestration Trust in 2011 was formed to improve energy, water, and food security at city-regional scale.

"Resilient Cities" Congress in 2018 organized by ICLEI is considered to be one of the major events on urban resilience that brought various policy makers and a few researchers together to focus on natural disasters specifically, and discuss policies and mechanisms required to mitigate them in cities. According to a research conducted by Rogov and Rozenbalt (2018), the term "urban resilience" is more linked with to ecological concepts such as "natural disasters," "ecosystem services," and "climate resilience" while it is very rarely linked with social and economic domains.

Urban living labs are considered to be a place-explicit experimentation in neighborhoods and cities which promote the involvement of various stakeholders like residents, municipal, and local authorities, knowledge partners and other private actors to find solutions to urban challenges in the concerned area. This helps in generating new knowledge on urban issues that are under-researched and will effectively help in implementing solutions. Voytenko et al. (2016) have presented five characteristics of an urban living lab: (1) Geographical embeddedness, (2) experimentation and learning, (3) participation and user involvement, (4) leadership and ownership, and (5) evaluation and refinement.

As cities continue to grow, they need to deal with uncertainties and challenges like climate change, food insecurity, lack of essential services to the poor, urban waste and water mismanagement, and lack of effective land use planning that could result in construction on risk prone areas. Such unfavorable circumstances lead to inequalities in cities which further create social unrest and increase the chances of riots. Thus, urban resilience is turning out to be a favorable concept that can help cities derive maximum benefits from agglomeration of economies and help in minimizing externalities (Ribeiro and Gonçalves 2019).

4.3 Urban Resilience Concepts and Definitions

From the vast body of research available on the concept of urban resilience, in this section, 30 definitions are reviewed and presented from the most influential publications. Out of the mentioned definitions, some of are similar yet unclear due to the divergent contexts of urban. The definition of urban resilience presented in Table 4.1 are sorted according to the study domain they have been mentioned in, such as agricultural and biologicals sciences, engineering, environmental sciences, social science, management and accounting, and energy.

As seen from Table 4.1, most of the definitions are either based on the context of threat to a system or on urban sustainability. However, the main aim of all studies and definitions is to add quality to the living conditions of the urban residents. The area of study which highlighted the most about resilience is climate studies or climate change—since it is considered to be one of the major challenges along with urbanization and population growth in cities (Ribeiro and Gonçalves 2019).

Table 4.1 Definitions of urban resilience

Subject domain	Definition of urban resilience	Author(s)
Agricultural and biological sciences	Resilience is the degree to which cities tolerate the change before reorganizing around a new set of structures and processes and depends on the ability of cities to maintain their eco-systemic and human functions simultaneously	Alberti et al. (2003)
	It is recommended that resilience only be used in a restricted sense to describe specific system attributes relating to: (i) the amount of disturbances a system can absorb and remain within the same state or domain of attraction and (ii) that the system is able to self-organize	Klein et al. (2003)
	Resilience is the ability of a city to absorb disturbances while maintaining its functions and structures	Lu and Stead (2013)
	Urban resilience can be defined in evolutionary terms as a proactive vision for planning, policy formulation, and strategic direction in which communities play a vital role in resilient place modeling through their active learning ability, robustness, capacity for innovation, and adaptability	Mehmood (2016)
Engineering	A resilient city is a sustainable network of physical systems and human communities	Godschalk (2003)
	Resilience is the ability of individuals, communities, institutions, companies, and systems in a city to survive, adapt, and grow regardless of the type of chronic stress and acute shocks to which they are subject	Spaans and Waterhout (2017)

(continued)

Table 4.1 (continued)

Subject domain	Definition of urban resilience	Author(s)
Engineering; social sciences	Urban resilience to flooding is a city's ability to tolerate flooding and reorganize if physical damage and socio-economic disturbances occur to prevent death and injury and maintain current socio-economic identity	Liao (2012)
	A more comprehensive definition of a resilient city emphasizes a community's overall ability and ability to withstand stress, survive, adapt, and recover from a crisis or disaster, and move forward quickly	Wagner and Breil (2013)
Energy; engineering; social sciences	Resilience in cities usually refers to the ability to absorb, adapt, and respond to changes in an urban system	Desouza and Flanery (2013)
Social sciences	Urban resilience is the ability of a city to recover from destruction	Campanella (2006)
	Resilience is the ability of an urban asset, location, and / or system to provide predictable performance	Brugmann (2012)
	Resilience is the ability to resist and recover from disruptive challenges	Coaffee (2013)
	Urban resilience refers to the ability of an urban system and all its socio-ecological and sociotechnical networks to maintain or rapidly return to the desired functions in the face of a disturbance and adapt to change, and to rapidly transform systems that limit capacity adaptive current or future	Meerow et al. (2016)

(continued)

Table 4.1 (continued)

Subject domain	Definition of urban resilience	Author(s)
	Urban resilience is the capacity of a city and its urban systems (social, economic, natural, human, technical, physical) to absorb the first damage, to reduce the impacts (changes, tensions, destruction, or uncertainty) from a disturbance (shock, natural disaster, changing weather, disasters, crises, or disruptive events), to adapt to change, and to systems that limit current or future adaptive capacity	Ribeiro and Gonçalves (2019)
Environmental sciences	Resilience is the ability of a socio-ecological system to sustain a given set of ecosystem services in the face of uncertainty and change for a community	Ernstson (2008)
	A resilient system is a system that can tolerate disturbances by means of characteristics or measures that limit its impacts, reducing or neutralizing damages and disturbances, and allowing the system to respond, recover, and adapt quickly to such disturbances	Wardekker et al. (2010)
	Resilience refers to a set of urban ecosystems that provide benefits to livelihoods and urban well-being	McPhearson et al. (2015)
Environmental science; social sciences	Resilience is the ability of a system to adjust to changing conditions	Pickett et al. (2004)
	Urban resilience means extending the concept of resilience from technical systems to social systems, particularly to cities, and their ability to recover and continue to provide their main functions of life, commerce, industry, government, and social gathering in the face of calamities and other hazards	Hamilton (2009)

(continued)

Table 4.1 (continued)

Subject domain	Definition of urban resilience	Author(s)
	From a resilient perspective, governance can be thought as a propositional collective action to sustain and improve a regime, or to trigger a transition from the system to a preferable regime	Ernstson et al. (2010)
	Resilience is the ability of systems to organize and recover from changes and disruptions without change to other states-that is, systems that are "safe to fail"	Ahern (2011)
	Although urban resilience usually refers only to the ability to maintain functions and structures, it must be framed in the visions of resilience (system persistence), transition (incremental system change), and transformation (system reconfiguration)	Chelleri (2012)
	A city resilient to disasters can be understood as a city that can successfully support measures to strengthen individuals, communities, and institutions to: (a) reduce or avoid and future risks; (b) reduce current and future susceptibility to resist risks; (c) establish mechanisms and functional structures for disaster response; and (d) establish functional mechanisms and structures for disaster recovery	Wamsler et al. (2013)
Business management and accounting; social sciences	A climate-resilient city is one that can resist climate stress, respond effectively to climate- related risks, and quickly recover from residual negative impacts	Henstra (2012)
	Resilience is the ability of an individual, community, or institution to respond dynamically and effectively to changing climatic conditions, continuing to function at an acceptable level	Brown et al. (2012)

(continued)

Table 4.1 (continued)

Subject domain	Definition of urban resilience	Author(s)
	Resilience is the ability of populations and urban systems to resist a wide range of hazards and stresses	Romero-Lankao and Gnatz (2013)
Business management and accounting; psychology;	Urban resilience refers both to design changes (structural, architectural, spatial planning) and to management and governance measures that aim to prevent or mitigate the physical and social vulnerability of urban areas, to protect life, property, and the economic activity of the city	Coaffee and O'Hare (2008)
Business management and accounting; energy; engineering; social science;	Urban resilience usually refers to the ability of a city or urban system to resist a wide range of shocks and tensions	Leichenko (2011)
Agricultural and biological sciences; environmental science; social sciences;	Urban resilience is the ability of a city to persist without qualitative changes in its structure and function, despite the disturbances	Wu and Wu (2013)
Earth and planetary sciences; social sciences	In the case of urban adaptation to climate, a resilience-based approach encourages practitioners to consider innovation and change to help recover from tensions and shocks that may or may not be predictable	Tyler and Moench (2012)

Source Ribeiro and Gonçalves (2019)

According to Meerow et al. (2016), certain number of definitions are underdeveloped and have not addressed the conceptual tensions that exist in accordance to urban resilience. Some of the key conceptual tensions are:

1. Characterization of urban: Definitions of urban resilience are mostly considered as vague as they do not define the meaning of "urban,"
2. Notion of equilibrium: While certain definitions of resilience depict single-state equilibrium others adopt the notion of multi-state equilibrium,
3. Resilience as a positive concept: The definitions mentioned in the tables embrace urban resilience as a positive and a desirable characteristic,
4. Pathway to resilience: The definitions specify three pathways to resilience, that are, persistence, transition, and transformation,
5. Understanding of adaptation: The next tension builds around the concept of adaptation which means that certain definitions are built on such specific context which is undermining the system's flexibility and ability to cope during sudden and unexpected threats. There are many definitions in the literature which focus on general adaptability and flexibility to prepare the situation under any sort of threat,
6. Timescale of action: Only a certain number of definitions reflect upon the time scale of the system after the disturbance under the fields of natural disasters and climate change. None of the other definitions' emphasis on the speed of recovery.

The definition developed by Meerow et al. (2016) specifically addresses these six tensions which is flexible to be used by researchers from any field. The definition also offers multiple ways to be resilient and most importantly recognizes the timescale in terms of taking an action. While Ribeiro and Gonçalves (2019) build on the table by Meerow et al. (2016); yet, their definition focuses on the four basic pillar of resilience, that are, resistance, recovery, adaptation, and transformation. However, the main difference that exist in all the definitions presented above lies in the way they have conceived a system to absorb, tolerate, adjust, reorganize, support, resist, respond, recover, and transform from the disturbances.

4.4 Dimensions and Characteristics of Urban Resilience

4.4.1 Dimensions of Urban Resilience

Urban resilience can be divided into dimensions that are important for characterizing and evaluating resilience. Ostadtaghizadeh et al. (2015) conducted a study to examine the tools and models available for community disaster resilience assessment. The research summarized various dimensions, mainly divided into five parts as shown in Table 4.2: (1) Natural (2) Infrastructural (3) Economic (4) Institutional, and (5) Social.

Table 4.2 Dimensions and sub-dimension

Domain	Synonyms or sub-categories
Social	Human Capital, Lifestyle and Community Competence, Society and Economy, Community Capital, Social and Cultural Capital, Population and Demographics Environmental, Risk Knowledge
Economic	Economic Development, Society and Economy
Institutional	Governance, Organized Governmental Services, Coastal Resource Management, Warning and Evacuation, Emergency Response, Disaster Recovery
Physical	Physical Infrastructure, Infrastructural, Land Use and Structural Design
Natural	Ecosystem

Source Ostadtaghizadeh et al. (2015)

While Ostadtaghizadeh et al. (2015) summarized the indicators broadly into these five dimensions, the availability of definitions and concepts regarding these domains makes it difficult to form a common framework. This calls for an international attention for developing a framework to be used as a foundation for the development of resilience plans and programs in cities. Another important aspect, while developing measures that are valid and reliable is to first establish the "cross cultural utility of the variables" (Ostadtaghizadeh et al. 2015).

Similarly, Sharifi (2016) also carried out an analysis from various studies based on community resilient assessment tools to develop a framework that could be used to evaluate the performance of the resilient assessment tools by addressing to multiple dimensions and their sub-dimensions along with their resilience criteria. The dimensions he focused on are: environmental, social, economic, built environment and infrastructure, and institutional. Built environment and infrastructure dimension addressed by Sharifi (2016) is what Ostadtaghizadeh et al. (2015) has described as "physical" dimension.

Ribeiro and Gonçalves (2019) have defined the five dimensions in their research based on the same dimensions developed by Ostadtaghizadeh et al. (2015):

1. Physical Dimension: "…includes the assignment of the resilience in infrastructures"
2. Natural Dimension: "…includes the ecological and environmental resilience"
3. Economic Dimension: "…includes the development of societies and economies"
4. Institutional Dimension: "… includes the governance and mitigation policies"
5. Social Dimension: "…resilience of communities and people in general"

After careful examination and analysis, Ribeiro and Gonçalves (2019) and Sharifi (2016) observed that institutional and social dimensions were discussed most often in the literature. These two dimensions show a strong link between themselves which means that measures such as leadership and strategic structure are critical for building resilience in an urban area (Ribeiro and Gonçalves 2019). On the contrary, the environmental dimension receives very limited attention despite having importance in building resilience of a city (Sharifi 2016). Being a multifaceted concept, it is best to achieve resilience by looking into all the five dimensions. Hence, Sharifi (2016) emphasize paying more attention to the environmental dimension along with other dimensions.

Cutter et al. (2010) in their research described the dimensions with respect to disaster resilience. While they considered the social, infrastructural, institutional, and cultural components or dimensions, they did not consider the fifth dimension of environment separately but rather included it in the institutional dimension. This replaced the fifth element to community dimension. The dimensions are described as below:

1. Social: This dimension aimed at addressing the differences that persist between and among the communities by creating between the demographic segments to social capacity.
2. Economic: This dimension measures the economic attributes of citizens such as employment type and amount, homeownership, division of incomes, and business size.
3. Institutional: It aims at addressing sub-components of mitigation, planning, and earlier disaster events and experiences.
4. Infrastructural: It is based on community's capacity to respond and recover. It includes assessment of aspects, such as the number of private owned property that is vulnerable to the risks and economic loses.
5. Community: In broader terms, this dimension falls under social dimension; however, it assesses an individual's relations between their neighborhood and community. Sense of community, public participation, and place attachment are three elements that are required to be understood.

Hence, we observe that the dimensions of resilience might differ if put under the context of a specific types of resilience such as disaster resilience (Cutter et al. 2010). This makes the point of having a basic framework that can be modified according to the field, more affirmative.

4.4.2 Characteristics of Urban Resilience

After discussing the definitions and dimensions of urban resilience, there is a need to identify the characteristics for the evaluation of resilient frameworks of urban systems which in their origin are complex since the technical components come together with the social ones. The available literature discusses resilience characteristics in different contexts such as community-based resilience, urban energy system, resource efficiency, and urban climate change. However, the mentioned characteristics in these papers are not only specific to their mentioned concept but rather, urban resilience as a whole.

Ribeiro and Gonçalves (2019) identified eleven relevant characteristics that would make an urban system resilient enough, they are related to redundancy, diversity, efficiency, robustness, connectivity, adaptation, resources, independence, innovation, inclusion, and integration. The aim to identify these characteristics was to make it easier and efficient for the local stakeholders to implement their resilience frameworks. They argue that there is a need for the local stakeholders to frame urban resilience in a way that is flexible and adaptable in relation to everchanging global environment and to the needs and specific situation of a locality.

Sharifi and Yamagata (2016) present resilience characteristics as "principles" necessary for the urban energy framework developed for assessing energy resilience in an urban system. Further, Tyler and Moench (2012) and Spaans and Waterhout (2017) present resilience characteristics in their respective conceptual frameworks for assessing urban climate adaptation in cities. Building on the characteristics described by various researchers, 14 characteristics, that are, *robustness, diversity, redundancy, connectivity, flexibility, resourcefulness, agility, efficiency, adaptive learning, modularity, creativity, equity, inclusive, and foresight capacity* have been identified and described for making a city resilient.

Robustness: The ability of a system to withstand a shock or stresses without the major functions and roles of a system undergoing a dynamic shift (Sharifi and Yamagata 2016). A system needs to be well-conceived, designed, and managed to be able to withstand the external forces (Spaans and Waterhout 2017). With the anticipation ability to predict potential failures and ensure safety, a system ensures that it is robust.

Redundancy: This characteristic means having extra components as spare capacity just in case. Having redundant capacities and functions is essential as they can compensate for the failed components in case of major disruptive events (Tyler and Moench 2012; Sharifi and Yamagata 2016). This ensures that the entire system would not stop functioning due to failure of one component. The substitutable component should be a high-level priority at a city level and cost-effective (Spaans and Waterhout 2017). Tyler and Moench (2012) term it as safe failure where the interdependence of functions or the system supports each other which avoids the entire system to fail. In other words, the major services can still be delivered despite a system failure.

Flexibility: A flexible urban system means that it has the capacity and ability to adapt itself to changing conditions or if a system goes through disturbances and disruptions. Such a system can immediately detect threats and failures and evolve and adapt accordingly (Sharifi and Yamagata 2016; Spaans and Waterhout 2017). According to Roggema (2014), in a flexible urban system both physical and functional elements should be designed and developed in a way that would allow disassembling and rearrangement. This would ensure multifunctionality of the system to cope with disturbances in a shorter time frame. This calls for the system to execute necessary tasks under different types of conditions and modifying assets in regard to new conditions (Tyler and Moench 2012). To achieve flexibility, introduction of new technologies and/or integrating indigenous knowledge is a useful approach (Spaans and Waterhout 2017).

Agility: It is imperative for a system to return to its normal or better functionality by mobilizing its resources in a given time period (Sharifi and Yamagata 2016). According to Sharifi and Yamagata (2016), this process should be exhibited in a timely manner to achieve resilience.

Adaptive Capacity: This characteristic would help in determining the vulnerabilities to future shocks and stresses enhancing the adaptation process under changing conditions (Sharifi and Yamagata 2016). The adaptation process also includes having the appropriate knowledge and ability to respond rapidly to shocks and stresses in an urban system (Ribeiro and Gonçalves 2019). The ability to learn from past experiences and events would help in building capacity of a system to respond accordingly and improve its performance (Tyler and Moench 2012). It is necessary for the actors and institutes to keep gathering new knowledge for enhancing resilience in their city (Tyler and Moench 2012; Sharifi and Yamagata 2016).

Modularity: Another way to strengthen an urban system's resilience is to develop self-organization in a system which doesn't allow centralization of power and resources or relying on outside physical intervention (Ribeiro and Gonçalves 2019; Sharifi and Yamagata 2016). This characteristic centers around strengthening local community engagement, cross sectoral partnerships, and vertical and horizontal institutional setup for informed decision-making process (Sharifi and Yamagata 2016). The system should be independent and self-reliant to be able to function the bare necessity when going through a disturbance.

Resourcefulness: The various actors responsible for building a city's resilience should always ensure the availability of sufficient resources to identify, plan, and respond to risks and disruptions (Sharifi and Yamagata 2016). This means the citizens should have the option of meeting their basic requirements readily available. This also includes access to financial assets and other assets of the actors who are collaborating in making an urban system resilient (Tyler and Moench 2012). Hence, heavy investments are required in anticipating events and setting priorities (Spaans and Waterhout 2017).

Creativity: This characteristic emphasizes importance of integrating technological and non-technological innovation into management and planning aspects of an urban system (Sharifi and Yamagata 2016). It is necessary for strengthening the ability of restoring a system's functioning under limited conditions (Ribeiro and Gonçalves 2019).

Equity: It is considered to be one of the most essential characteristics for achieving resilience. For an energy system, for instance, it is based on fairly distributing the energy resources across the city and ensuring accessibility. Also, just options should be provided to the ones bearing the brunt of the consequences during production, transmission, and distribution of energy (Sharifi and Yamagata 2016).

Foresight Capacity: A resilient system should be able to anticipate future uncertainties and situations along with having the ability to visualize possible outcomes and consequences (Sharifi and Yamagata 2016). This characteristic is considered to be necessary for disaster management and absorbing an initial shock. Lack of this characteristic might exacerbate the risk when exposed to hazards and disasters (Sharifi and Yamagata 2016).

Diversity: It works on the principle that there should be various options available to deal with disturbances and disruptions in a city—with more options, the ability to adapt to a diverse set of circumstances will increase (Ribeiro and Gonçalves 2019).

According to Tyler and Moench (2012), a resilient system has two types of diversity—spatial and functional diversity. Spatial diversity would mean that all the functions of a system are not affected at one particular time, while, functional diversity would make a resilient system have more than one way to meet a given need (Tyler and Moench 2012).

Inclusiveness: Day by day it is being considered necessary to facilitate the participation and involvement of institutes, increasing community involved practices, mobilization of resources during recovery period, and smooth flow of information among the stakeholders through horizontal and vertical integration (Sharifi and Yamagata 2016). Transparency, accountability, and responsiveness are the widely acknowledged principles of good governance (UNDP 1997) needed for decision-making process especially including the vulnerable. These offer a sense of shared ownership for building resilience of city.

Connectivity: It is important to have a system where there is efficient coordination of various preparatory and recovery actions among various stakeholders from different sectors. Without such a characteristic, the system would fail to recover after going through a shock (Sharifi and Yamagata 2016; Riberiro and Gonçalves 2019).

Efficiency: This characteristic aims at building a positive relation with the functioning of a system with a static nature along with the systems that operate dynamically (Ribeiro and Gonçalves 2019). In terms of resource consumption, it is necessary for a city to have a resource efficiency agenda that will help a city become resilient by reducing the risk of shortfalls in essential resources, such as water, food, energy, and materials. Resources help in functioning of a city and being resilient would mean more efficient resource use and management (UNDP 1997).

The resilience of a system is exhibited only while going through shocks and stresses. However, these characteristics of the system exist irrespective of the exposure to disturbances (Tyler and Moench 2012). The characteristics of an urban system should be seen as functioning in a complex and interconnected network that do not function in isolation. Also, they should not only be viewed from a technical perspective. Every urban system would be different, and so it is impossible to provide specific solutions for all conditions and disturbances. However, the growing literature can help in providing a fundamental framework for these characteristics by establishing a fundamental or baseline framework. This way, it becomes easier to monitor changes is resilience which further helps in comparing one place to another (Cutter et al. 2010).

4.5 Conclusions

We are now living in the era of increasing risks and uncertainties. Annually, cities are hit with different types of adverse events ranging from natural disasters to political conflicts and public health crises. Indeed, many cities are already vulnerable to adverse events and, in many parts of the world, conditions may become more sever considering the projected trends of urban population growth. Climate change is expected to further increase the frequency and intensity of adverse events, and this is likely to further increase the pressure on cities and their already constrained resources. Additionally, the recent pandemic demonstrated that cities need to also plan and prepare for major public health crises in future (Sharifi and Khavarian-Garmsir 2020).

It is due to increasing awareness about such threats, risks, and uncertainties that many cities around the world are increasingly building on their efforts to enhance urban resilience. This is manifested in the increasing number of plans, programs, policies, and frameworks that are developed annually in many cities around the world. Through investment in such initiatives, cities expect to minimize potential losses in future and ensure building capacities to better respond to, recover from, and adapt to adverse events. One potential impediment to the proper design and implementation of resilience plans, programs, policies, and frameworks is the incomplete understanding of the resilience concept itself. This issue becomes even more complicated when considering the fact that resilience is a contested notion and various definitions exist for it depending on the background, field, context, and objectives of the stakeholders.

In an effort to better understand different conceptualizations of resilience in the context of urban planning, this chapter elaborated on the genealogy of the resilience concept and its underlying principles and characteristics. It was argued that resilience as a concept has an old history in fields such as physics and psychology. In such fields, resilience has mainly focused on abilities to absorb shocks and return to pre-shock equilibrium conditions in a timely manner. Unlike other fields, resilience has been introduced to and used in urban studies only since a few decades ago. In their conceptualizations of resilience urban scholars have been particularly influenced by ecologists. In the ecological domain, it is recognized that return to pre-disaster equilibrium state may be neither possible nor desirable. Instead, depending on the severity of the shock and the conditions of the system, it may transform into a new state(s). This is well-aligned with urban issues and phenomena, as cities are often considered as socio-ecological systems. In addition to ecology, urban scholars and practitioners have relied on the vast body of literature from other fields to conceptualize resilience depending on their specific purposes. In fact, the wealth of literature published in other fields have provided an opportunity to adopt the resilience concept flexibly in urban research and practice. Generally, the dominant approaches that guide such conceptualizations of urban resilience are categorized as engineering, socio-ecological, and adaptive. The latter one has gained more momentum in the recent years considering the increasing recognition of the concept of living with risk and the need for continuous improvement and evolvement.

This chapter is concluded by elaborating on various underlying resilience characteristics such as robustness, redundancy, flexibility, agility, adaptive capacity, modularity, resourcefulness, creativity, equity, foresight capacity, diversity, inclusiveness, connectivity, and efficiency. These characteristics are essential for developing more objective resilience plans, programs, policies, and frameworks. They could also contribute to making the resilience concept more tangible to various stakeholders. These characteristics have also guided design and implementation of case studies that have been reported in the latter part of this volume.

References

Ahern J (2011) From fail-safe to safe-to-fail: sustainability and resilience in the new urban world. Landsc Urban Plan 100(4):341–343. https://doi.org/10.1016/J.LANDURBPLAN.2011.02.021

Alberti M, Marzluff JM, Shulenberger E, Bradley G, Ryan C, Zumbrunnen C (2003) Integrating humans into ecology: opportunities and challenges for studying urban ecosystems. Bioscience 53(12):1169–1179. https://doi.org/10.1641/0006-3568(2003)053[1169:IHIEOA]2.0.CO

Berkes F, Ross H (2013) Community resilience: toward an integrated approach. Soc Nat Resour 26(1):5–20. https://doi.org/10.1080/08941920.2012.736605

Brown A, Dayal A, Rio CRD (2012) From practice to theory: emerging lessons from Asia for building urban climate change resilience. Environ Urban 24(2):531–556. https://doi.org/10.1177/0956247812456490

Brugmann J (2012) Financing the resilient city. Environ Urban 24(1):215–232. https://doi.org/10.1177/0956247812437130

Campanella TJ (2006) Urban resilience and the recovery of New Orleans. J Am Plann Assoc 72(2):141–146. https://doi.org/10.1080/01944360608976734

Carpenter S, Walker B, Anderies JM, Abel N (2001) From metaphor to measurement: resilience of what to what? Ecosystems 4:765–781. https://doi.org/10.1007/s10021-001-0045-9

Chelleri L (2012) From the "resilient city" to urban resilience. A review essay on understanding and integrating the resilience perspective for urban systems. Doc d'Anàlisi Geogràfica 58(2):287–306. http://dag.revista.uab.es/article/view/v58-n2-chelleri/pdf-en#.WrkF1_Oo-Ug.mendeley

Coaffee J (2013) Towards next-generation urban resilience in planning practice: from securitization to integrated place making. Plan Pract Res 28(3):323–339. https://doi.org/10.1080/02697459.2013.787693

Coaffee J, Therrein MC, Chelleri L, Henstra D (2018) Urban resilience implementation: a policy challenge and research agenda for the 21st century. J Contingencies Cris Manag 26(13):403–410. https://www.researchgate.net/publication/327210845_Urban_resilience_implementation_A_policy_challenge_and_research_agenda_for_the_21st_century

Cutter SL, Burton CG, Emrich CT (2010) Disaster resilience indicators for benchmarking baseline conditions. J Homel Secur Emerg Manag 7(1). https://doi.org/10.2202/1547-7355.1732

Desouza KC, Flanery TH (2013) Designing, planning, and managing resilient cities: a conceptual framework. Cities 35:89–99. https://doi.org/10.1016/J.CITIES.2013.06.003

Emilsson T, Sang OA (2017) Impacts of climate change on urban areas and nature-based solutions for adaptation. In: Kabisch N, Korn H, Stadler J, Bonn A (eds) Nature-based solutions to climate change adaptation in urban areas. Theory and practice of urban sustainability transitions. Springer, Cham. https://doi.org/10.1007/978-3-319-56091-5_2

Ernstson H (2008) The social production of ecosystem services: lessons from urban resilience research. In: Ernston H (ed) Rhizomia: actors, networks and resilience in urban landscapes. PhD Thesis, Stockholm University

Ernstson H, van der Leeuw SE, Redman CL, Meffert DJ, Davis G, Alfsen C, Elmqvist T (2010) Urban transitions: on urban resilience and human-dominated ecosystems. Ambio 39(8):531–545. https://doi.org/10.1007/s13280-010-0081-9

Falk DA, Watts AC, Thode AE (2019) Scaling ecological resilience. Front Ecol Evol. https://doi.org/10.3389/fevo.2019.00275

Filho WL, Balogun AL, Olayide OE, Azeiteiro UM, Ayal DY, Muñoz P, Nagy GZ, Bynoe P, Oguge O, Toamukum NY, Saroar M, Li C (2019) Assessing the impacts of climate change in cities and their adaptive capacity: towards transformative approaches to climate change adaptation and poverty reduction in urban areas in a set of developing countries. Sci Total Environ 692:1175–1190. https://doi.org/10.1016/j.scitotenv.2019.07.227

Folke C (2006) Resilience: the emergence of a perspective for social–ecological systems analyses. Glob Environ Chang 16:253–267. https://doi.org/10.1016/j.gloenvcha.2006.04.002

Frantzeskaki N (2016) Urban resilience: a concept for co-creating cities of the future. https://urbact.eu/sites/default/files/resilient_europe_baseline_study.pdf. Accessed May 5, 2021.

Garmestani AS, Allen CR, Gunderson L (2009) Panarchy: discontinuities reveal similarities in the dynamic system structure of ecological and social systems. Ecol Soc 14(1):15 [online]. http://www.ecologyandsociety.org/vol14/iss1/art15/

Garmezy N (1971) Vulnerability research and the issue of primary prevention. Am J Orthopsychiatry 41(1):101–116. https://doi.org/10.1111/j.1939-0025.1971.tb01111.x

Gößling-Reisemann S, Hellige HD, Thier P (2018) The resilience concept: from its historical roots to theoretical framework for critical infrastructure design (artec-paper, 217). Universität Bremen, Forschungszentrum Nachhaltigkeit (artec), Bremen. https://nbn-resolving.org/urn:nbn:de:0168-ssoar-59351-6

Godschalk DR (2003) Urban hazard mitigation: creating resilient cities. Nat Hazard Rev 4(3):136–143. https://doi.org/10.1061/(ASCE)1527-6988

Gonçalves A (2018) World cities day—why do we need resilient cities? You Matter, October 31. https://youmatter.world/en/world-cities-day-2018-why-do-we-need-resilient-cities/

Gunderson LH (2000) Ecological resilience—in theory and application. Annu Rev Ecol Syst 31:425–439. https://www.jstor.org/stable/221739?seq=1#metadata_info_tab_contents

Gunderson LH, Holling CS (2002) Panarchy: understanding transformations in human and natural systems. Island Press, Washington, DC

Hamilton WAH (2009) Resilience and the city: the water sector. Proc Inst Civ Eng-Urban Des Plan 162(3):109–121. https://doi.org/10.1680/udap.2009.162.3.109

Henstra D (2012) Toward the climate-resilient city: extreme weather and urban climate adaptation policies in two Canadian provinces. J Comp Policy Anal: Res Pract 14(2):175–194. https://doi.org/10.1080/13876988.2012.665215

Huq S, Kovats S, Reid H, Satterthwaite D (2007) Reducing risks to cities from disasters and climate change. Environ Urban 19(1):3–15. https://doi.org/10.1177/0956247807078058

Iliuk O, Teperik D (2018) The universe of resilience: from physics of materials through psychology to national security. https://icds.ee/en/the-universe-of-resilience-from-physics-of-materials-through-psychology-to-national-security/

Kim D, Lim U (2016) Urban resilience in climate change adaptation: a conceptual framework. Sustain 8(5):1–17. https://doi.org/10.3390/su8040405

Klein RJT, Nicholls RJ, Thomalla F (2003) Resilience to natural hazards: how useful is this concept? Glob Environ Chang Part B: Environ Hazards 5(1):35–45. https://doi.org/10.1016/j.hazards.2004.02.001

Lehmann AP (2015) Sprawling cities, growing risks? World Economic Forum, January 15. https://www.weforum.org/agenda/2015/01/sprawling-cities-growing-risks/

Leichenko R (2011) Climate change and urban resilience. Curr Opin Environ Sustain 3(3):164–168. https://doi.org/10.1016/J.COSUST.2010.12.014

Liao KH (2012) A theory on urban resilience to floods—a basis for alternative planning practices. Ecol Soc 17(4). https://doi.org/10.5751/ES-05231-170448

Lu P, Stead D (2013) Understanding the notion of resilience in spatial planning: a case study of Rotterdam, The Netherlands. Cities 35:200–212. https://doi.org/10.1016/J.CITIES.2013.06.001

Meerow S, Newell JP, Stults M (2016) Defining urban resilience: a review. Landsc Urban Plan 147:38–49. https://doi.org/10.1016/j.landurbplan.2015.11.011

Mehmood A (2016) Of resilient places: planning for urban resilience. Eur Plan Stud 24(2):407–419. https://doi.org/10.1080/09654313.2015.1082980

Olsson L, Jerneck A, Thoren H, Persson J, O'Byrne D (2015) Why resilience is unappealing to social science: theoretical and empirical investigations of the scientific use of resilience. Sci Adv 1(4):1–11. https://doi.org/10.1126/sciadv.1400217

Ostadtaghizadeh A, Ardalan A, Paton D, Jabbari H, Khankeh HR (2015) Community disaster resilience: a systematic review on assessment models and tools. PLOS Curr Disasters 1. https://doi.org/10.1371/currents.dis.f224ef8efbdfcf1d508dd0de4d8210ed

Pickett STA, Cadenasso ML, Grove JM (2004) Resilient cities: meaning, models, and metaphor for integrating the ecological, socio-economic, and planning realms. Landsc Urban Plan 69(4):369–384. https://doi.org/10.1016/j.landurbplan.2003.10.035

Reckien D, Creutzig F, Fernandez B, Lwasa S, Tovar-Restrepo M, Mcevoy D, Satterthwaite D (2017) Climate change, equity and the Sustainable Development Goals: an urban perspective. Environ Urban 29(1):159–182. https://doi.org/10.1177/0956247816677778

Ribeiro P, Gonçalves L (2019) Urban resilience: a conceptual framework. Sustain Cities Soc 50. https://doi.org/10.1016/j.scs.2019.101625

Roggema R (2014) Towards enhanced resilience in city design: a proposition. Land 3(2):460–481. https://doi.org/10.3390/land3020460

Rogov M, Rozenbalt C (2018) Urban resilience discourse analysis: towards a multi-level approach to cities. Sustain 10(12). https://doi.org/10.3390/su10124431

Romero-Lankao P, Gnatz DM (2013) Exploring urban transformations in Latin America. Curr Opin Environ Sustain 5(3–4):358–367. https://doi.org/10.1016/J.COSUST.2013.07.008

Shamsuddin S (2020) Resilience resistance: the challenges and implications of urban resilience implementation. Cities 103. https://doi.org/10.1016/j.cities.2020.102763

Sharifi A (2016) A critical review of selected tools for assessing community resilience. Ecol Ind 69:629–647. https://doi.org/10.1016/j.ecolind.2016.05.023

Sharifi A, Chelleri L, Fox-Lent C, Grafakos S, Pathak M, Olazabal M, Moloney S, Yumagulova L, Yamagata Y (2017) Conceptualizing dimensions and characteristics of urban resilience: insights from a co-design process. Sustain 9(6):1032. https://www.mdpi.com/2071-1050/9/6/1032

Sharifi A, Khavarian-Garmsir AR (2020) The COVID-19 pandemic: impacts on cities and major lessons for urban planning, design, and management. Sci Total Environ 749. https://doi.org/10.1016/j.scitotenv.2020.142391

Sharifi A, Pathak M, Joshi C, He B-J (2021) A systematic review of the health co-benefits of urban climate change adaptation. Sustain Cities Soc 74:103190. https://doi.org/10.1016/j.scs.2021.103190

Sharifi A, Yamagata Y (2016) Principles and criteria for assessing urban energy resilience: a literature review. Renew Sustain Energy Rev 60:1654–1677. https://doi.org/10.1016/j.rser.2016.03.028

Shean M (2015) Current theories relation to resilience and young people. A literature review. https://evidenceforlearning.org.au/assets/Grant-Round-II-Resilience/Current-theories-relating-to-resilience-and-young-people.pdf

Spaans M, Waterhout M (2017) Building up resilience in cities worldwide—Rotterdam as participant in the 100 Resilient Cities Programme. Cities 61:109–116. https://doi.org/10.1016/j.cities.2016.05.011

Sundstorm SM, Allen CR (2019) The adaptive cycle: more than a metaphor. Ecol Complex 39. https://doi.org/10.1016/j.ecocom.2019.100767

Thorén H (2014) Resilience as a unifying concept. Int Stud Philos Sci 28(3):303–324. https://doi.org/10.1080/02698595.2014.953343

Tyler S, Moench M (2012) A framework for urban climate resilience. Climate Dev 4(4):311–326. https://doi.org/10.1080/17565529.2012.745389

UN-Habitat (2017) Trends in urban resilience. United Nations Human Settlements Programme

United Nations (2018) 2018 revision of world urbanization prospects. https://www.un.org/development/desa/publications/2018-revision-of-world-urbanization-prospects.html

United Nations Development Programme (1997) Governance for sustainable human development: a UNDP policy document. http://mirror.undp.org/magnet/policy/

Voytenko Y, McCormick K, Evans J, Schliwa G (2016) Urban living labs for sustainability and low carbon cities in Europe: towards a research agenda. J Clean Prod 123:45–54. https://doi.org/10.1016/j.jclepro.2015.08.053

Wagner I, Breil P (2013) The role of ecohydrology in creating more resilient cities. Ecohydrol Hydrobiol 13(2):113–134. https://doi.org/10.1016/J.ECOHYD.2013.06.002

Wamsler C, Brink E, Rivera C (2013) Planning for climate change in urban areas: from theory to practice. J Clean Prod 50:68–81. https://doi.org/10.1016/J.JCLEPRO.2012.12.008

Wardekker JA, de Jong A, Knoop JM, van der Sluijs JP (2010) Operationalising a resilience approach to adapting an urban delta to uncertain climate changes. Technol Forecast Soc Chang 77(6):987–998. https://doi.org/10.1016/J.TECHFORE.2009.11.005

Wu J, Wu T (2013) Resilience in ecology and urban design: linking theory and practice for sustainable cities. In: Pickett STA, Cadenasso ML, McGrath B (eds) Ecological resilience as a foundation for urban design and sustainability. Springer, Dordrecht, pp 211–229. https://doi.org/10.1007/978-94-007-5341-9_10

Chapter 5
Resilient-Smart Cities: Theoretical Insights

Ke Xiong, Ayyoob Sharifi, and Bao-Jie He

Abstract Cities, the main settlements of human beings, are facing mega challenges of climate change, urbanization, population increase, economic growth, and environmental deterioration. To address such challenges, the goal of sustainable cities and communities has been advocated by the United Nations. In particular, smart city has been applied to integrate digital technologies and sensors to improve the efficiency of assets, resources, and services in urban operations. In comparison, the resilient city is expected to improve urban resilience (e.g., prevention, impact reduction, recovery, adaptation) to disasters and emergencies. However, limited studies have analyzed how to ensure a normal condition for smart city under extreme conditions and how to ensure a resilient city can efficiently respond to disasters and extreme events. Therefore, this chapter aims to address such research gaps for the integration of smart city and resilient city, namely resilient-smart city, in order to better ensure sustainable urban development under various mega challenges. This chapter discusses how six components of smart city (i.e., governance, people, life, mobility, economy, and environment) contribute the resilient city in four aspects of health and well-being, economy and society, urban systems and services, and leadership and strategy to indicate the possibilities of the integration of smart city and resilient city. Moreover, this chapter points out challenges hindering resilient-smart city development and provides corresponding suggestions to overcome such challenges. Overall,

K. Xiong · B.-J. He
School of Architecture and Urban Planning, Chongqing University, Chongqing 400045, China
e-mail: 20141513095@cqu.edu.cn

B.-J. He
e-mail: baojie.he@cqu.edu.cn

Key Laboratory of New Technology for Construction of Cities in Mountain Area, Ministry of Education, Chongqing University, Chongqing 400045, China

A. Sharifi (✉)
Graduate School of Humanities and Social Sciences, Hiroshima University, Higashihiroshima, Japan
e-mail: sharifi@hiroshima-u.ac.jp

Graduate School of Advanced Science and Engineering, Hiroshima University, Hiroshima 739-8511, Japan

A. Sharifi and P. Salehi (eds.), *Resilient Smart Cities*, The Urban Book Series,
https://doi.org/10.1007/978-3-030-95037-8_5

this chapter is expected to open a vision for further development of resilient-smart cities that can contribute to the achievement of sustainable urban development goals.

Keywords Smart city · Resilient city · Information and communications technologies · Internet of things · Assessment indicator system

5.1 Introduction

The Earth and human beings have been facing several mega challenges such as climate change, environmental deterioration, rapid urbanization, and unbalanced and unbridled economic development in recent years. In particular, following global urbanization, one of the most transformative trends (Korhonen and Snakin 2015), such challenges in cities are more prominent. For instance, while cities are the home to only 55% of the world population, they consume about 67% of global primary energy and emit more than 70% of the greenhouse gases (IPCC 2014). By 2030 when the urbanization rate reaches 60%, nearly 75% of carbon emissions and energy consumption will be concentrated in cities (Kennedy et al. 2014). Moreover, while cities only account for 4% of the land on the Earth (Pouffary and Rogers 2014), they will accommodate nearly five billion people by 2030. Accordingly, various problems, such as traffic congestion, environmental pollution, urban climate change, suitable housing, and safety are also critical (Gasco-Hernandez 2018).

There is an urgent need to control, alleviate, and address such challenges to provide citizens with basic needs of urban infrastructure and ensure the sustainability of society, economy, and environment (Eremia et al. 2017; Macke et al. 2018). In the last decade, development of smart cities has been widely recognized as an important approach for enhancing urban management capacity to deal with such challenges (Albino et al. 2015; Bansal et al. 2015), and many cities have practically implemented smart city projects. For instance, the city of Dubuque, Iowa, United States, digitized and connected all resources of the city, including water, electricity, oil, and natural gas, for intelligently reducing urban energy consumption and costs, while meeting citizens' needs through monitoring, analysis, and integration of various data (use 2009). The 'Your accessible transport network' plan, by Transport for London, is a project for the improvement of transport capacity and services across London, England, through the collection of real-time information from subway cards, mobile phones, and social networks and prompt and effective responses (TfL 2012). Partly due to the effective operation of this system, during the 2012 Olympic Games when users increased by 25%, the transport system in London still worked well (TfL 2013).

Cities, the main human settlements, are becoming increasingly important to secure citizens with sound and holistic functions. Under many extreme conditions (e.g., extreme weather, natural disasters, pandemic, and other emergencies), cities are required to be resilient to reduce associated losses and recover from potential impacts. To achieve so, it is necessary to adopt measures to improve a city's capability and capacity in dealing with disasters while ensuring that there are sufficient resources to reduce or avoid the harm caused by sudden disasters to cities and their inhabitants (Sharifi and Khavarian-Garmsir 2020). Developing resilient cities has been suggested as one of the solutions to such extreme conditions. Resilient city initiatives aim to strengthen the planning, coping, recovery, and adaptability of cities to a wide gamut of adverse events, including climate-induced disasters and natural disasters. For instance, after the Sandy storm, the United States developed disaster management and response systems under long-term climate change risks, in order to quickly recover facilities and services in cities. Gorakhpur, India, a city under the severe influence of urban flooding and the associated impacts of dengue fever, malaria, and Japanese encephalitis, mainstreamed resilient solid waste management systems and upgraded the drainage and sewerage system to improve urban resilience to urban flooding (Du 2019). Overall, to overcome urban challenges and secure safe, comfortable, and livable urban environments, both smart city and resilient city have attracted extensive attention of scholars and decision-makers in recent years (Sharifi et al. 2021; Zhu et al. 2019).

5.1.1 Concept and Development of Smart City

The smart city paradigm was first proposed at an international conference in San Francisco, California, in 1990, with the theme of 'smart cities, fast systems, and global networks', shedding a light on how technologies can reshape cities (Gibson et al. 1992). In the following years, its connotation gradually expanded (Fig. 5.1). Until 2008, the 'Smart Earth' proposed by IBM (full name: International Business Machines Corporation) was aimed at applying smart technologies to all aspects of life (e.g., medical care, transportation, currency, and infrastructure) and thereby making the planet increasingly intelligent. Afterward, there has been a clear vision of smart city among the public (Paroutis et al. 2014). In particular, a smart city covers a relatively broad range of views (Dong et al. 2020), while the conceptual understanding of smart cities is mostly relevant to technology, economy, management, etc., judging

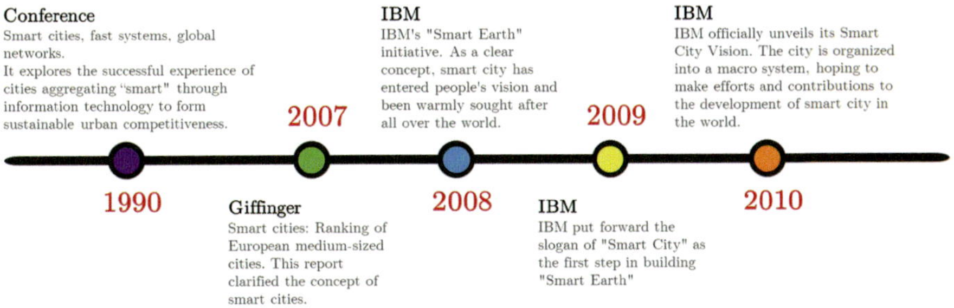

Fig. 5.1 Development of the smart city concept. 1990s (Gibson et al. 1992); 2007 (Giffinger et al. 2007); 2008–2010 (Paroutis et al. 2014)

from existing studies and findings (Zhang 2019). In addition, more recently, there has been an increasing recognition of non-technological and soft dimensions of smart city. For instance, the role of people as the end users of smart cities is increasingly emphasized (Sharifi 2019, 2020b).

At the technical level, a smart city captures the information of an entire city and conducts data analysis to achieve real-time monitoring of urban resources. Accordingly, a smart city is defined as an urban area that integrates information and communication technologies with traditional infrastructure and employs new digital technologies for coordination and integration of urban functions and operations. A smart city, therefore, is a tool or platform, capable of increasing competitiveness and thereby improving communities and quality of life (Batty et al. 2012). For example, using video surveillance and big data analytics can contribute to ensuring urban safety supervision and early warning more quickly and intelligently.

At the economic level, smart city approach can contribute to creating urban areas where the economy has achieved smart growth, and smart industry accounts for a large proportion of urban industry with a high growth rate (Davis and Weinstein 2003). For instance, emerging industries of smart motor systems, smart logistics, smart buildings, and smart grids have been the economic engine of many cities. Because of the significant reduction of carbon emissions, more importantly, such industries are important drivers to low-carbon economy (The Climate Group and Global eSustainability Initiative 2008). Smart economy also entails further promotion of start-ups and better training of people to enable them to be active contributors to the digitalized economic system (Sharifi 2019).

At the management level: A smart city refers to the utilization of modern technologies such as Internet of Things (IoT) that are being explored and tested, big data, and cloud computing, to enable urban infrastructure and services (e.g., urban management, education, health, public safety, transportation, power grid, and water resources) to be intelligent, interconnected, and efficient. Such an approach makes cities manageable, practically promoting sustainable urban development and improving people's quality of life (Han and Hawken 2018; Lytras and Visvizi 2018; Maye et al. 2016).

5.1.2 Concept and Development of Resilient City

The resilient city generally indicates a city should be able to face either internal or external shocks and still maintain its main functions during disasters or crisis events (Sharifi 2016; Sharifi and Yamagata 2018). Such an expectation enables resilient city to receive wide attention and efforts across various disciplines, such as ecology, urban planning, economics, and sociology. Resilience idea was originated in disciplines such as physics and psychology and then applied to other fields such as ecology. In ecology, it aims to uncover the fluctuations and interactions between two competitor communities in the biosphere. Its scope gradually expanded from natural ecology (i.e., the relationship between stability and diversity in the ecosystem) to human ecology (i.e., the ability of cities to respond to climate change by changing the political structure) (Bahadur and Tanner 2014).

In summary, research efforts on resilience have gone through three stages (Fig. 5.2), including engineering resilience (Alexander 2013), ecological resilience (Holling 1973), and social-ecological resilience (Berkes and Folke 1998). Engineering resilience is to describe the capability of a single and static system in restoring the original balance and maintaining system stability after being subjected to external shocks, showing an orderly and linear characteristic. Ecological resilience emphasizes the ability to adapt to external shocks and to characterize the robustness of a complex system with complex and nonlinear changes (multi-equilibria system). Social-ecological resilience emphasizes that humans and nature are regarded as a whole, and different levels of systems interact with each other to form a dynamic equilibrium process. Social-ecological resilience has a strong ability of learning and adapting with a certain degree of variability and innovation (Holling 2001).

While there have been several decades since the resilience was proposed, the definition of resilience is still not universal given its broad application in many disciplines (Sharifi 2020c). Among various definitions, Holling first put forward the theory of 'hierarchical structure, chaos, adaptive cycle' (Holling 1973), which explored the connotation of sustainable development and laid the ideological foundation for the formation of urban resilience theory (Zhao et al. 2020). Professor Mileti at the University of Colorado in the United States who introduced the resilience concept into urban planning pointed out that *a resilient city mainly responds to climate change and resists disasters through the construction and improvement of urban physical environment and infrastructure* (Mileti 1999). UN-Habitat pointed out that *a resilient city assesses, plans and acts to prepare for and respond to all hazards – sudden and slow-onset, expected and unexpected* (UN-Habitat).

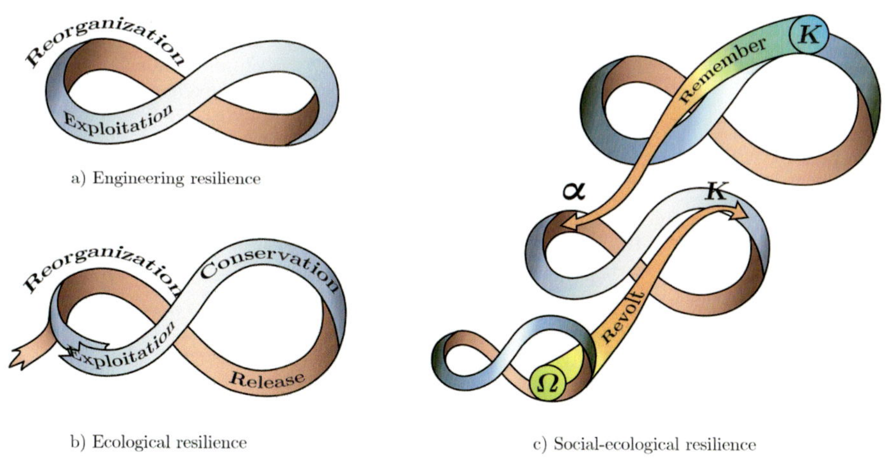

Fig. 5.2 Characteristics of resilience concept in different stages (*Source* Gunderson and Holling 2002)

Meerow and others provide a more systematic and comprehensive interpretation of resilient cities (Yan and Tang 2020), by defining a resilient city as '*the urban system and its social-ecological-technical networks at all time and space scales maintain/recover in time to the required functions when disturbed and adapt to changes and rapidly transform systems that limit current or future adaptive capacity*' (Meerow et al. 2016). Furthermore, the International Council for Regional Sustainable Development (ICLEI) mainstreamed the topic of 'Resilient City' and introduced it into studies on cities and disaster prevention. Since then, urban resilience has received extensive attention and (Motesharrei et al. 2016) and is now a critical and hot topic in urban planning and urban geography. Moreover, many countries and organizations including the United States, Britain, Australia, Japan, etc., have formulated development strategies, policies, and projects based on the 'resilience' theory (Table 5.1).

Table 5.1 Strategies, policies, and projects for the development of global representative resilient cities

Country/organization	Strategy/policy/project	Source	Time
Hyogo Prefecture, Japan	Hyogo Declaration	https://undocs.org/en/A/CONF.206/6	2005
London, England	Managing Risks and Increasing Resilience	https://www.london.gov.uk/WHAT-WE-DO/environment/environment-publications/managing-risks-and-increasing-resilience-our	2011
UNDRR	Making Cities Resilient	https://www.undrr.org/publication/making-cities-resilient-report-2012	2012
New York, USA	A Stronger, More Resilient New York	https://www1.nyc.gov/site/sirr/report/report.page	2013
Toronto, Canada	Resilient City—Preparing for a Changing Climate	https://www.toronto.ca/legdocs/mmis/2016/pe/bgrd/backgroundfile-98049.pdf	2013
Rockefeller Foundation	100 Resilient Cities	https://resilient-cities.sphaera.world	2013
USAID	The Building Community Resilience in Timor-Leste	https://www.usaid.gov/timor-leste/project-descriptions/building-community-resilience-timor-leste	2014
Earthquake Emergency Initiative	Urban Resilience Master Planning	https://emi-megacities.org/urban-resilience-master-planning/	2015
The Commonwealth of Australia	Building a more secure and resilient Australia	https://budget.gov.au/2021-22/content/download/glossy_resilient.pdf	2021

In summary, existing understandings indicate that smart city and resilient city concepts present their consistent expectations in *making cities and human settlements inclusive, safe, resilient and sustainable* (Goal11, UN SDGs). A smart city concentrates on improving efficiency, capacity and performance of assets, resources, and services (e.g., transportation, utilities, and many public spaces such as libraries, hospitals, museums, and exhibition halls), along with the growth in societal needs and increasing urbanization. A resilient city is expected to improve urban safety and resilience during urban crises and extreme conditions, protecting people from direct economic losses and deaths, and quickly recovering urban systems to normal.

Both smart city and resilient city could generate complementary benefits in improving urban sustainability. On the one hand, a smart city technically integrates intelligent methods and sensors for real-time monitoring urban situations and generating immediate and prompt responses. This is vital for the development of resilient city to obtain useful and critical information that is unavailable with only physical approaches under crises and extreme events, to support effective decisions and strategies for overcoming extreme events. On the other hand, a resilient city aims to secure human beings, ecosystem, assets, resources, and services, which is a critical premise of a smart city for higher efficiency, larger capacity, and better performance. In addition, the operation of digital methods and sensors in providing real-time urban services requires a safe and resilient environment, particularly under crises and extreme events. Integrating resilient city and smart city by developing resilient city smartly and enhancing smart city resilience should be an important issue for the achievement of *Sustainable Cities and Communities* (Goal11, UN SDGs) under mega challenges of climate change, environmental deterioration, urbanization, and economic development. However, existing studies have considered smart city and resilient city separately and only a few studies have explored the integration of smart city and resilient city (Hassankhani et al. 2021; Sharifi et al. 2021). To fill the gap, this chapter aims to explore the potential of integrating smart city and resilient city concepts for the development of resilient-smart cities, particularly to enhance the robustness and quality of both resilient city and smart city with the inclusion of information technologies and smart solutions such as big data and Internet of Things.

The remainder of this chapter is structured as follows. Section 5.2 describes why the integration of smart city and resilient city is needed. Section 5.3 presents the contribution of smart city integration for urban resilience. Section 5.4 discusses the challenges and opportunities for integrating smart city and resilient city, and proposes relevant suggestions. Finally, Sect. 5.5 summarizes and concludes this paper.

5.2 The Need for Resilient-Smart Cities

Both smart city and resilient city provide a path for city optimization and upgrading under various challenges. As shown in Fig. 5.3, smart city is capable of addressing a broad range of urban problems, like resource shortages, traffic congestion, environmental pollution, etc. In comparison, resilient city is expected to address severe

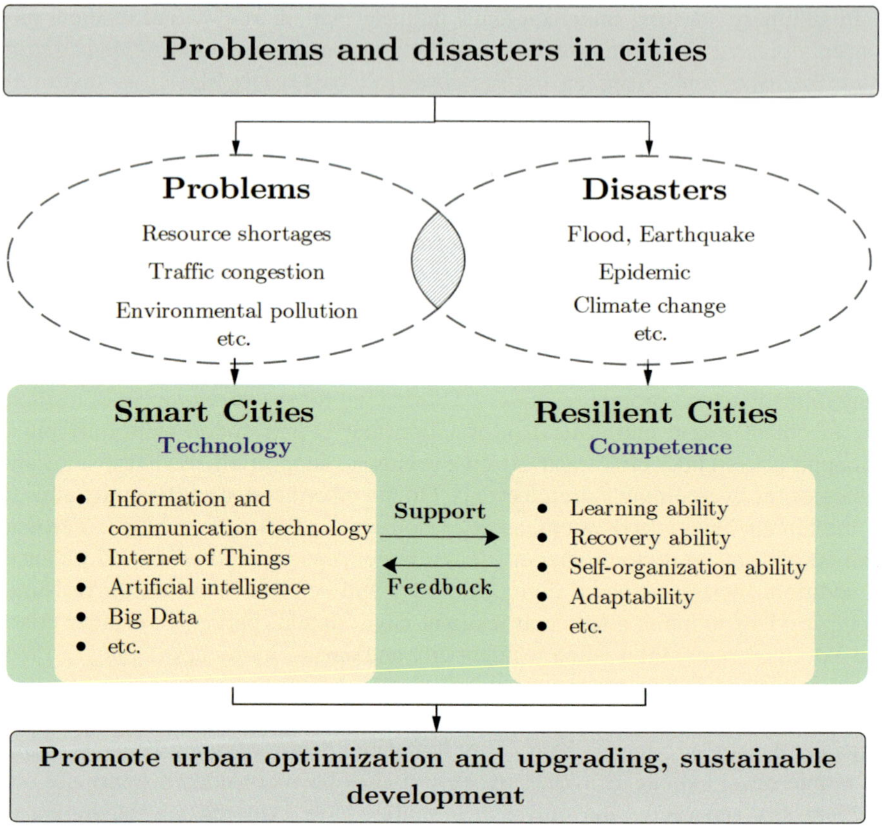

Fig. 5.3 Ways to promote urban development of SC and RC

problems caused by major disasters with strong consequences on environments, lives, economy, etc. While smart city presents a technical means of urban development by integrating information and communication technologies for real-time sensing, analysis, and integration of various key information of the core system for urban operation, smart city is compromised to quickly cope with disasters and reduce associated impacts. Therefore, there is a need to improve the capability of smart city in dealing with urban disasters rather than only day-to-day problems.

Resilient city can respond to and resist adverse situations of urban flooding, extreme heat, emergencies, pandemics, etc., but it is difficult to quickly identify disasters, promptly predict risks, and effectively generate decisions to resist and respond to disasters using traditional methods. Accordingly, it becomes important to improve the efficiency of addressing disasters by shortening response time and avoiding problems evolving into disasters. Such a consideration is consistent with the current trend of resilient city studies, where the main themes have been shifting to prevention before light disasters from overemphasizing recovery after severe disasters (Meerow et al. 2016). This should be supported by the inclusion of new measures and technologies and explore novel paths and models.

To enhance capacity and performance of smart city and resilient city, accordingly, this section suggests the integration of resilient city and smart city, namely resilient-smart city, for an innovative model of urban development and support for a city to conduct sustainable construction in an all-round way, thereby approaching the goals of inclusive, safe, resilient, and sustainable cities and communities. As shown in Fig. 5.3, smart city and resilient city complement each other for achieving sustainable urban development. On the one hand, the resilient city cannot be separated from the management technology of smart city, where smart city provides technical support. On the other hand, resilience provides positive feedback by enhancing urban resilience to ensure the healthy development of smart city. The needs for resilient-smart city are fourfold as follows.

First, the smart city provides a technical support for the development of resilient city in terms of information and communications technology. Moreover, the technology-based response features of smart city can be a management plug-in module for resilient city, making up for shortcomings (e.g., unavailability of real-time information and data, lack of evidence-based and effective decisions) of resilient city. Modern information and communications technologies such as the Internet of Things and big data analysis technologies provide open opportunities for the development of resilient city (Shah et al. 2019). The integration of smart city solutions in resilient city initiatives covers a wide scope of practices, such as real-time crime monitoring maps, predictive policing and security staff, natural disaster (climate risk) monitoring and assessment, urban information management, and urban disaster relief and disaster prevention and mitigation systems (Desouza and Flanery 2013; Kontokosta and Malik 2018; Takewaki et al. 2011; Zhang et al. 2019). During the COVID-19 pandemic, for instance, information and communications technology and big data combined relevant data with geospatial and temporal information to track data and build a complete information monitoring and dissemination platform. It allowed people to grasp essential data timely, thereby hindering the spread of the pandemic and minimizing adverse impacts (Sharifi and Khavarian-Garmsir 2020).

Second, smart city supports disaster prevention in the development of resilient city given its capability of monitoring and managing infrastructures such as transportation, buildings, electricity, and water in achieving energy and resource efficiency. For instance, the regular detection of conditions of underground water pipelines, tracking and investigation of the cleanliness of pipelines in a real-time manner by the municipal government of Paris, France, using geographic information systems, provides a database to implement intelligent management of underground drainage system and improves water efficiency on the one hand. On the other hand, it offers urban hydrological systems and stormwater infrastructure management required to improve resilience to flooding (esri 2019; WRT 2010).

Third, sound development of smart city requires positive feedback from resilient city, because smart city, an important factor for achieving urban sustainable development, has an endogenous relationship with the improvement of urban resilience

(Song 2020). However, the pure embedding of information and communications technologies into urban infrastructure and operation management systems does not necessarily promote the healthy and safe development of cities (Boulos et al. 2015). Resilient city acts as a foundation for smart city planning (Moraci et al. 2018). For instance, the increasing high temperature in cities is a challenge to the normal operation of sensors and transmission systems, where development of heat-resilient cities for cooling cities and communities is an important prerequisite.

Fourth, the information security is also an important issue in building resilient cities to determine the performance and outcomes of smart city. Therefore, resilient city, a positive feedback to emergencies, should quickly adapt to current or future systems and overcome the negative impact of smart city on social development (Hiller and Blanke 2017). For example, the SusCity project in Lisbon, Portugal, adopts a new architecture of the Internet of Things, using protection and recovery modules to analyze faults, take measures to solve them, and provide positive feedback on the Internet of Things infrastructure in the smart city environment to reduce the adverse effects of technology (Abreu et al. 2017).

5.3 Smart Cities for Urban Resilience

Conceptually, the application of smart city technologies helps improve urban resilience and the improvement of urban resilience generates positive feedback to the quality and performance of smart city. Nevertheless, it is worth discussing how smart city technologies can contribute to urban resilience, in order to identify possibilities of resilient-smart city development, given the fact that application of smart city technologies can potentially impede urban resilience. A typical case is that smart city is vulnerable to cybersecurity attacks that may detract from urban resilience (Beck 2017). This section starts with the analysis of the assessment indicator systems of smart city and resilient city, following which the possible contributions of smart city to urban resilience are analyzed in four aspects including health and well-being, economy and society, infrastructure and ecosystems, and leadership and strategy.

5.3.1 Assessment Indicator System for Smart City and Resilient City

As aforementioned, scholars have made various attempts to define the smart city concept in different disciplines such as technology, economy, and management. The frameworks which they have presented to assess city smartness in these different disciplines also vary (Sharifi 2020a). The most typical frameworks include the smart city model for European cities developed by Giffinger and Gudrun (2010), the basic component of smart city model by Nam and Pardo (2011), the smart city initiatives framework by Chourabi et al. (2012), and the Smart City Wheel (SCW) by Cohen (2013). Among these frameworks, the Smart City Wheel (Fig. 5.4) represents one of the most prominent concepts and characterizations of smart city. It is constructed based on six critical dimensions, including governance, people, life, mobility, economy, and environment proposed by (Giffinger et al. 2007). These dimensions are also widely accepted and used by other scholars, enterprises, and governments for smart city research, assessment, and building (Dong et al. 2020).

The framework of the City Resilience Index (CRI) (Fig. 5.5), proposed by the Rockefeller Foundation, is one of the most holistic ones for resilient city assessment (Wei et al. 2017). This framework presents a comprehensive set of indicators, variables, and metrics that are related to urban resilience (ARUP 2014), in four dimensions including health and well-being, economy and society, infrastructure and ecosystems, and leadership and strategy. Each of these dimensions consists of three

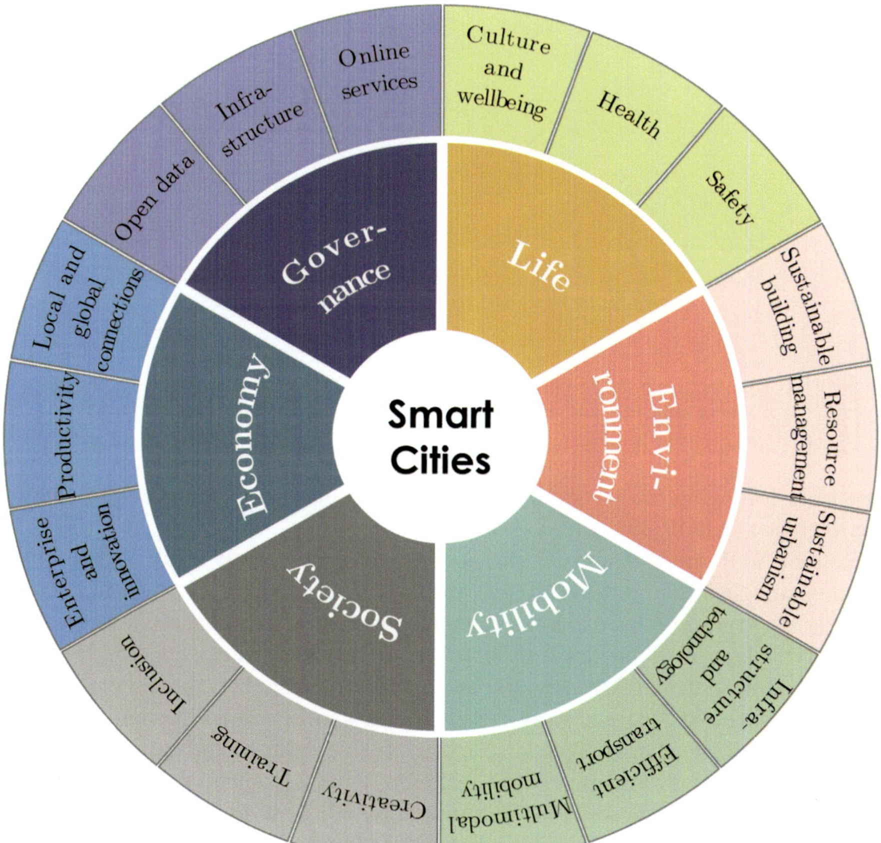

Fig. 5.4 Assessment indicator system of smart cities (*Source* TUWIEN 2015)

Fig. 5.5 Assessment indicator system of resilient cities (*Source* ARUP 2014)

indicators following the principles of feedback, robustness, redundancy, richness, and tolerance (ARUP 2014).

Both SCW framework and CRI framework comprehensively cover several urban components. Such components could overlap and supplement to a certain content, offering the opportunities to identify connections between smart city and resilient city. For instance, resilient city is expected to provide adequate safeguards to human life and health in the CRI framework, and smart city should ensure people's health and safety in the SCW framework. The urban systems and services defined in the CRI framework are enabled by diverse and affordable multi-modal transport systems and information and communications technology networks, and contingency planning. Such a vision indicates that the integration of smart city technology into the construction of resilient city has initially been considered in resilient city framework. In the following parts, the connections between smart city and resilient city, namely the contribution of smart city to urban resilience, will be detailed. In particular, six

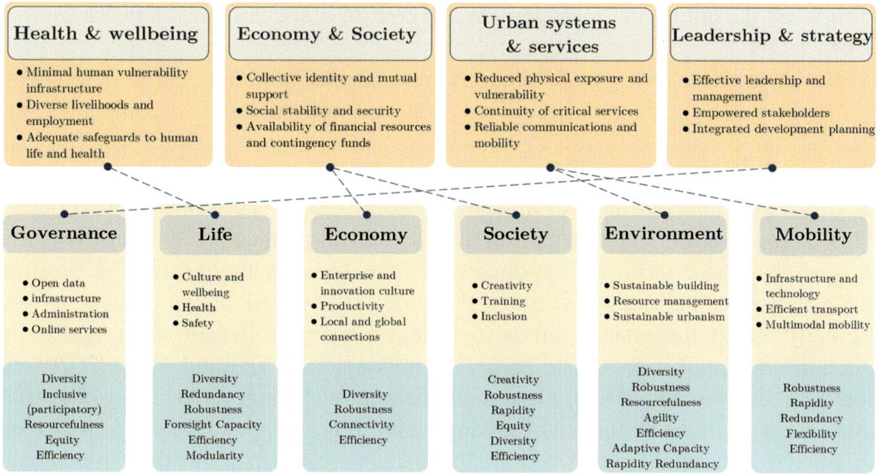

Fig. 5.6 Connections between smart city and resilient city

scopes of indicators in the smart city assessment indicator system are linked with the four scopes in the resilient city assessment framework (Fig. 5.6).

5.3.2 *Health and Well-Being*

The category of health and well-being indicates to what extent a city meets the basic needs of citizens. It is characterized by three indicators including minimal human vulnerability, diverse livelihoods and employment, and adequate safeguards to human life and health. Using information and communications technology or other modern technologies, smart life contributes to these three indicators, reflecting the characteristics of urban resilience such as integration, systems-based operation, anti-disturbance, redundancy, robustness, and rapidity.

First, smart life ensures the basic needs such as water sanitation, energy, and food. In terms of water, for instance, smart life is engaged to assess and manage water resources through monitoring and dispatching systems, recycling and reuse systems, and ensuring water quality and safety so as to efficiently and reasonably meet the needs of stakeholders and realize the smart governance of urban domestic sewage. A case of New York, United States, is that smart city technology was applied in the planning process by deploying sensors in the urban underground pipe network for monitoring the operation of water resources. Another case is the installation of 24 weather monitoring stations of Vienna Pipe Network Company, through the close cooperation with the National Meteorological Administration, for timely forecasting of the rainstorm direction and precipitation and sending feedback to the control center, to follow up the intelligent management and control of the drainage network (Marks 2014).

Second, new technology applications (new media, communication equipment, semiconductors, wired and wireless services, and other diversified companies) for smart life can support economic development, thereby providing more opportunities of access to finance, skills training, and business support. For example, Toronto, a city with a high concentration of information service enterprises, has approximately 148,000 employees in the information service industry in the metropolitan area. The information service industry contributes an annual sale of more than $32.5 billion and its annual export value exceeds $6.2 billion (TOP 2018).

Finally, smart life enables cities to collect relevant information through early detection and rapid response to crises or emergencies (Yao and Wang 2020), to coordinate disaster relief work, assess the degree of damage to urban system, and further strengthen the resilience of urban system. It is an approach to improve urban resilience to disasters (Fujinawa et al. 2015) and ensure urban dwellers' health and safety. There have been many practical cases of crises detection and response in smart life. The integrated sensors in Buffalo Bay Park in Houston, United States, provide disaster prevention information by detecting behaviors and degree of hurricane and flood danger for reducing associated losses to a certain extent (Jason Iken and Brown 2009). During the COVID-19 pandemic, information and communication technologies and big data analytics were utilized to combine relevant data with geospatial and temporal information to enable people to grasp relevant data as promptly as possible, thereby controlling the spread of the epidemic, minimizing its impact, and enhancing the resilience to the pandemic (Sharifi and Khavarian-Garmsir 2020). Singapore's urban public safety supervision system plan facilitates rapid detection, real-time response, and coordination of events that affect urban public safety. This is achieved through unified supervision, information integration, efficient coordination, network integration, information interaction, and data sharing between different urban public safety services and monitoring systems.

Smart life can also adopt information and communications technology that integrates health facilities and services, and responsive emergency services, providing faster and more convenient services. For example, many megacities such as Tokyo and New York adopt electronic medical record systems to integrate various clinical information systems and knowledge bases, set up sensors and wireless networks, and adopt a series of technical measures such as cloud computing. These technologies greatly enhance diagnosis and treatment activities of the medical staff and improve the accuracy of medical procedures. In addition, online medical information exchange system and mobile medical applications have been developed to provide residents with medical and health services anytime and anywhere, realizing the dynamic transmission of the physiological conditions of family patients and ensuring resident's health and safety. Such services are also expected to provide other resilience co-benefits. For instance, telemedicine minimizes the need to visit hospitals and medical centers, thereby allowing remote access in case of disruptions in the transportation network or when social distancing is needed as was the case during the COVID-19 pandemic (Sharifi et al. 2021). Additionally, through reducing travel demand, telework can contribute to climate change mitigation efforts as transportation sector is one of the major contributors to greenhouse gas emissions (GHGs).

5.3.3 Economy and Society

Economy and society need a collective identity and mutual support, social stability and security, and availability of financial resources and contingency funds. A smart economy is characterized by an innovative spirit, flexibility of labor market, international embeddedness, ability to transform, and so on (TUWIEN 2015). It can be diverse, robust, connected and efficient. Among other things, the characteristics of smart society include affinity to lifelong learning, flexibility, creativity, and people's participation in public life (azbil 2019). Smart society is capable of promoting the flexibility, creativity, robustness, rapidity, equity, diversity, and efficiency in cities.

Smart economy provides sound management of city finances and diverse revenue streams. In particular, a wide application of information and communication technologies (i.e., electronic medical treatment, electronic technology products, etc.) provides advantages to attract enterprises' investment and talents. Data mining and visualization of smart economic decision-making and operations, such as the sharing economy, improve economic diversification and capital flow allocation efficiency and make urban economy management more comprehensive. In addition, the economic dimensions of a smart city can provide strong support for disaster prediction and preparation, disaster reduction, and recovery. It is also the basis for the development of smart infrastructure and government management (Oliva and Lazzeretti 2018). Toronto has attracted many world-leading high-tech companies by its advanced information and communication services. Built upon the information service industry and cluster development strategy, Toronto has become one of the most innovative research and investment hubs around the globe. Through attracting private sector developers to provide start-up capital, Toronto's Lakeside Community has built a new cutting-edge network facility, which in turn, creates new opportunities for the community to attract more digital media and other innovative companies (TOP 2018).

In addition, the application of blockchain improves the reliability and transparency of transactions to ensure the security of financial and other transactions (Desroches and Taylor 2018) and provide support for sound management of city finances, which is of significance for a well-functioning city. As a supplement to the traditional economy, 'sharing economy' guarantees the normal establishment of a part of the supply–demand relationship. While striving to maintain normal operations, different social forces share the risks associated with shared resources, shared services, and shared technologies (Long 2020), greatly improving the allocation and utilization efficiency of various resources in a city.

Smart society development not only guarantees a safe and stable operation of medical care, office, teaching, etc., but also promotes social relationships and networks, integrated communities. First, it provides a cross-domain linkage for disease prevention and control (e.g., 'Healthcare' in Hong Kong), which improves the accessibility of medical information. Second, with the employment of information and communication technology, smart society promotes closer and more effective connections within and between enterprises or schools, improves the fairness of educational resources, and enhances the effectiveness of social management. Given the needs of COVID-19 prevention and control, applications of remote collaborative office and distance education are also emerging rapidly (Wen 2020). In particular, the remote collaboration and cloud office systems have been developed and optimized to provide technical support for employees to work at home during such special periods. The adoption of DingTalk and other network platforms has been prevalent for large-scale online teaching to minimize the impact of the epidemic on normal teaching. Overall, these applications are typical cases that greatly improve the efficiency of remote information collection and communication and facilitate better work-life balance (Li and Long 2020). Such approach to smart society development provides support for the speed, fairness, and diversity of resilient cities. Third, the use of technology and new media in smart society improves public awareness of environmental protection and disaster self-rescue, thereby enhancing risk reduction, recovery, and learning and innovation capacities. Hangzhou city in China, for instance, has developed a comprehensive security experience hall through video surveillance, simulation operations, game interaction, and Virtual Reality (VR) virtual experience. This allows residents to learn various hazard safety measures and prevention skills that are closely related to life and strengthen residents' awareness of safety precautions.

5.3.4 Urban Systems and Services

Resilient urban systems and services could be assessed by three indicators of reduced physical exposure and vulnerability, continuity of critical services, and reliable communications and mobility, which have a close connection with the smart environment and smart mobility. For the smart environment, one of the assessment indicators is sustainable resource management, in order to make environment management sounder and critical infrastructure safer. This, at the same time, indicates smart environment embodies the diversity, robustness, resourcefulness, agility, efficiency, adaptive capacity, rapidity, and redundancy of a resilient city. Smart mobility is consistent with the goal of reliable communications and mobility in resilient cities, in terms of integrated transport networks, information and communications technology, and emergency communication services. This makes it related to resilient city principles of robustness, rapidity, redundancy, flexibility, and efficiency.

Sustainable management of smart environment facilitates sustainable building and resource management and contributes to meeting demands of critical infrastructure, optimizing resource allocation in energy supply, and strengthening urban monitoring systems to alleviate deterioration and improve restoration of ecological systems. Many countries and regions have implemented smart environment measures to address a variety of problems emerged in urban development (e.g., increased demand on critical infrastructure, ecosystem degradation, etc.). A case of the critical infrastructure is the Huoshenshan and Leishenshan Hospitals in Wuhan China that were built with high quality standards in a short timeframe in the early stage of the COVID-19 outbreak, using prefabricated rapid construction and Building Information Modelling technologies. This provides efficient and accurate support for building and infrastructure construction in response to emergencies (Li and Long 2020). With the extensive use of renewable solar energy resources, Barcelona has developed a smart energy management (e.g., energy-saving management mode) system for buildings, water supply systems, information billboards, charging infrastructure, etc., and the city is among those with the highest density of solar panels in Europe (ZIGURAT 2019). Different types of electronic trash bins, where trash cans are connected to underground pipelines through their respective valves, have been built in Stockholm to classify and sort trash (Fourneris 2020). Moreover, waste treatment efficiency has been also improved by automatic control system, high-speed transportation system, and automatic separation and transportation systems.

Smart environment systems can automatically collect and monitor various resources related to the human living environment such as water, electricity, and atmosphere in real time. The systems detect and deal with various adverse events in time and continuously carry out plans to establish more comprehensive ecosystem management and flood risk management systems to ensure cities operate resiliently. For example, cities can build smart grids to promote on-demand mutual conversion and distribution of energy on a unified platform and encourage innovative methods for increased uptake of clean and renewable energy sources and technologies (Gargiulo and Zucaro 2015; Moraci et al. 2018; Zach et al. 2019). This can also provide flexible power and information and communications technologies to support for the operation of critical infrastructure in the post-disaster stage (Alqahtani et al. 2018). The Green University of Tokyo Project promoted by Tokyo, Japan, uses information technology to reduce electricity consumption and carbon emissions, improving urban environment in a smart and intelligent way. The plan connects the air-conditioning, lighting, power supply, safety facilities, and other subsystems in the buildings to form a compatible integrated system and conduct intelligent data analysis to realize intelligent, dynamic, and effective configuration and management of electric energy supply and consumption (Esaki 2021; GUTP 2008).

Sustainable, innovative, and safe transport systems of smart mobility provide infrastructure systems for urban transportation, facilitate integrated transportation networks, help maintain smooth connectivity of roads and communication networks, and restore water or power supply in time (Rus et al. 2018). These series of methods improve the efficiency of transportation system, reduce energy consumption, and improve quality of life. The Urban Transportation Master Plan and Electric Transportation Plan in Vienna, Austria, for example, have addressed problems of traffic congestion and exhaust pollution in urban construction on the one hand and have increased the utilization rate of environmentally friendly transportation and public transportation on the other. The Smart Commuter Initiative proposed by Toronto, Canada, was the first one to integrate advanced technologies such as expressway non-stop electronic toll collection and road traffic information collection to optimize expressway operations and improve transportation efficiency. Implementation of this initiative also improves the operation of traffic management system. Similarly, the highways in Tokyo, Japan, are controlled and monitored by information technology, and restructured information services are provided at any time to avoid various natural disasters and ensure safe operation.

Smart mobility can also adopt information and communications technology infrastructure to provide a basic guarantee for efficient logistics transportation and real-time full-process supervision of commodities and disaster relief medical supplies. The use of diversified contactless takeaway delivery services also provides basic living supplies for people staying at home while ensuring safety (Wen 2020). For example, smart logistics uses the Internet of Things technology to realize the exchange and sharing of information between things in the logistics process. This way, all links of logistics can be tracked and monitored in real time to realize the digitization and information of the entire logistics process, thereby upgrading logistics industry and reducing logistics costs. Moreover, the use of logistics robots can quickly work out the optimal path to increase logistics rate. Further, using blockchain can reduce logistics costs, production, and traceability of goods in transit and improve the efficiency of supply chain management.

5.3.5 Leadership and Strategy

The leadership and strategy of resilient city are assessed by effective leadership and management, empowered stakeholders, and integrated development planning. The smart governance prioritizes effective measures, such as participation in decision-making, transparent governance, and public and social services, to improve leadership and management and facilitate access to up-to-date information and knowledge to enable people and organizations to take appropriate actions. The smart governance dimension is related to the principles of diversity, inclusive (participatory), resourcefulness, equity, and efficiency of cities.

Smart governance employs information and communications technologies to provide government services and exchange information and integrate various independent systems in the face of disasters (Ruhlandt 2018). The intelligent decision-making system, supervision system, and others improve the ‘fineness’ and ‘precision’ of decision-making (Zhou and Fu 2020; Zhu et al. 2019). Through e-government, public government service platforms (e.g., transportation, energy, and water) and government websites have become more service-oriented to provide citizens with consulting information and services to improve the quality of life and satisfaction of citizens. The Wellbeing Toronto Website helps residents to better understand their communities and city government and strengthens the connection and communication between the public and the government. The first-hand public information published on this webpage provides the municipal government with references for relevant resolutions in order to provide public services more in line with the needs of citizens. Smart governance can also assist government managers and medical workers in dynamic management and monitoring through the Internet of Things. For example, the Disease Control AI Analysis Platform (WDCIP) developed by Weizhi Technology is based on related technologies to detect people in close contact, predict the high-risk transmission area of the epidemic, and assist in the analysis of the dynamics of the pandemic's spread. It also provides decision support for relevant government departments and health systems (Yan and Han 2020).

5.4 Challenges and Suggestions for Integrating Smart City and Resilience

Overall, smart city and resilient city have strong connections according to the analysis of the relationships in their assessment indicator systems. Such connections also indicate the potential to integrate smart city and resilient city for sustainable urban development in various aspects such as health and well-being, economy and society, urban system and services, and leadership and strategy. Moreover, the potential integration may promote the efficient and comprehensive operation of society, economy, infrastructure, and management and can enhance the city's redundancy, diversity, and anti-disturbance characteristics.

Within the context of sustainable urban development, there are many opportunities for resilient-smart city development. This has received the support from emerging studies. For instance, the use of emerging big data, cloud computing, and information and communications technology solutions to help cities survive and operate under extreme pressure has been advocated (Palmieri et al. 2016; Soyata et al. 2019; Yang et al. 2017). Urban structures should integrate smart technologies and systems to improve robustness, redundancy, resource richness, and rapidity (Desroches and Taylor 2018). Nevertheless, these studies are only preliminary, and in-depth research, exploration, and testing are needed.

While there are opportunities and possibilities as analyzed in Sect. 5.3, there are a variety of challenges for resilient-smart city development. First, the original goals of smart city and resilient city are different, implying the different requirements in development so that they cannot completely overlap. The elementary focus of a resilient city lies in the urban safety, reflected by the reduction of the casualties and economic losses caused to the city by natural disasters and emergencies and the capabilities of quickly recovering to a normal state after the disaster. In comparison, the focus of a smart city is on the efficient operation of the urban system (Zhu et al. 2020). There could be contradictions between them.

Smart city technology is not a panacea (Boulos et al. 2015), and the introduction of smart technology and thoughts may have negative impacts on resilient city, thereby detracting from urban resilience. First, many defects and shortcomings have been exposed in the application process of smart technologies (Li and Long 2020). If the big data system in smart city fails to collect and analyze key data at high frequency during the emergency period of disasters and make sensitive predictions accurately, it may lead to the collapse of the system. Second, if a city relies too much on smart city technology, in the process of smart city planning, the services are prone to be highly dependent on data and are more susceptible to potential data flow interruptions (Kotevska et al. 2017), also reducing the resilience of infrastructure given security threats (Beck 2017). Third, since the data among different departments cannot be always shared and connected in a timely and effective manner, data conflicts and miscommunications may emerge, making it impossible to efficiently trace the source. For example, for flood prevention and disaster relief, the use of drones and various high-precision sensors has greatly improved the predictability of disaster losses. However, once the data center is paralyzed, it will result in a huge obstacle to continue the work (Beck 2017). Fourth, less consideration has been given to the environment, and social equity and justice (Kaika 2017; Viitanen and Kingston 2014). This is contrary to one of the goals of resilient city (social stability and security) (ARUP 2014). Overall, while the use of smart technology can alleviate environmental problems and emergencies to a certain extent, it is prone to negative effects in terms of data privacy and uncontrolled power relations (Sharifi and Khavarian-Garmsir 2020). For example, for COVID-19 prevention and control, the application of various health codes has greatly improved the efficiency of monitoring and early warning. However, its effectiveness requires a large amount of public identities and travel information gathered by the platforms, which are relevant to public privacy and may lead to multiple risks such as external attacks and internal information leakages (Zhang 2020).

Nevertheless, it is necessary to overcome potential technical and social risks brought by smart city and take the improvement of urban resilience as positive feedback in the smart city development, and build a robust resilient-smart city, so as to manage or solve urban problems and disasters more intelligently, effectively, and accurately. To support this, several suggestions have been provided: (1) development of a comprehensive and systematic resilient-smart city assessment system that incorporates the resilience elements into the smart city assessment system (Song 2020) and strengthen the emergency prediction and sensitive analysis capabilities through big data analytics (Wang and Zhao 2020); (2) improvement of urban resilience as the main direction of smart city development in practice. This requires guiding smart city planning with resilient thinking and promoting resilient-smart city construction with a diversified governance structure. Platforms for urban big data collection and analytics or information centers can ensure seamless data sharing, thereby, offering advantages to technology and data resource organization and providing urban decision-makers with more efficient and accurate support means; (3) enhancement of data privacy and security based on the principle of social fairness. This will require optimizing institutional mechanisms for data development and utilization, strengthening privacy and confidentiality, combining technology-driven and human-driven methods, and enhancing civic awareness to enhance adaptability to future events (Sharifi and Khavarian-Garmsir 2020).

5.5 Conclusions

Combination of the concepts of smart city and resilient city to develop resilient-smart cities is drawing increasing attention for and accelerating sustainable urban development under a variety of mega challenges of climate change, urbanization, environmental deterioration, unbridled economic growth, and population growth. This chapter advances the theoretical understanding of the concept of resilient-smart city through exploring the possibilities of the integration of smart city and resilient city concepts. In particular, the possibilities were evidenced through analyzing the connections between the six components of smart city, namely governance, people, life, mobility, economy, and environment and the four components of resilient city, namely health and well-being, economy and society, infrastructure and ecosystems, and leadership and strategy. In the resilient-smart city framework, smart city solutions and technologies provide technical support to ensure a resilient city can deal with disasters and emergencies in an efficient manner, while the resilient city provides positive feedback for smart city in resisting external interferences and disturbances. Nevertheless, information accuracy, data security, data sharing, and associated social equity and justice issues are challenges a smart city can generate when being integrated into resilient city systems. To overcome such challenges, this chapter also presented suggestions to be considered in the development of resilient-smart city assessment system, the improvement of resilience of smart city, and the enhancement of social issues in smart technology utilization. Overall, this chapter is expected to provide scholars and practitioners with a point of reference regarding the principles and characteristics of resilient-smart city. This, in turn, is expected to lead to better recognitions of resilient-smart city in academia and practice.

Acknowledgements This chapter was partially funded by Asia-Pacific Network for Global Change Research (Funder ID: https://doi.org/10.13039/100005536).

References

Abreu DP, Velasquez K, Curado M, Monteiro E (2017) A resilient Internet of Things architecture for smart cities. Ann Telecommun 72(1–2):19–30. https://doi.org/10.1007/s12243-016-0530-y

Albino V, Berardi U, Dangelico RM (2015) Smart cities: definitions, dimensions, performance, and initiatives. J Urban Technol 22(1):3–21. https://doi.org/10.1080/10630732.2014.942092

Alexander DE (2013) Resilience and disaster risk reduction: an etymological journey. Nat Hazard 13(11):2707–2716. https://doi.org/10.5194/nhess-13-2707-2013

Alqahtani A, Tipper D, Kelly-Pitou K (2018) Locating microgrids to improve smart city resilience. Paper presented at the 2018 Resilience Week (RWS)

ARUP (2014) City Resilience Index. https://www.arup.com/perspectives/publications/research/section/city-resilience-index

azbil (2019) Smart society: building a better future for citizens with smart technologies. https://www.azbil.com/top/pickup/whitepaper/pdf/Azbil_Smart_Society_WP_12022020.pdf

Bahadur A, Tanner T (2014) Transformational resilience thinking: putting people, power and politics at the heart of urban climate resilience. Environ Urban 26(1):200–214. https://doi.org/10.1177/0956247814522154
Bansal N, Mukherjee M, Gairola A (2015) From poverty, inequality to smart city. Springer Transactions in Civil and Environmental Engineering. Springer, Singapore
Batty M, Axhausen KW, Giannotti F, Pozdnoukhov A, Bazzani A, Wachowicz M, Ouzounis G, Portugali Y (2012) Smart cities of the future. Eur Phys J-Spec Top 214(1):481–518. https://doi.org/10.1140/epjst/e2012-01703-3
Beck K (2017) Smart security? Evaluating security resiliency in the U.S. Department of Transportation's Smart City Challenge. Transp Res Rec 2604(1):37–43. https://doi.org/10.3141/2604-05.2021/02/18
Berkes F, Folke C (1998) Linking social and ecological systems: management practices and social mechanisms for building resilience. Cambridge University Press, Cambridge
Boulos MNK, Tsouros AD, Holopainen A (2015) 'Social, innovative and smart cities are happy and resilient': insights from the WHO EURO 2014 International Healthy Cities Conference. Int J Health Geogr 14(3). https://doi.org/10.1186/1476-072x-14-3
Chourabi H, Nam T, Walker S, Gil-Garcia JR, Mellouli S, Nahon K, Pardon TA, Scholl HJ (2012) Understanding smart cities: an integrative framework. In: 2012 45th Hawaii international conference on system sciences, January, pp 2289–2297. IEEE
Cohen B (2013) The smart city wheel. Smart Circle
Davis DR, Weinstein DE (2003) Market access, economic geography and comparative advantage: an empirical test. J Int Econ 59(1):1–23. https://doi.org/10.1016/S0022-1996(02)00088-0
Desouza KC, Flanery TH (2013) Designing, planning, and managing resilient cities: a conceptual framework. Cities 35(SI):89–99. https://doi.org/10.1016/j.cities.2013.06.003
Desroches R, Taylor JE (2018) The promise of smart and resilient cities. Bridge 48(2):13–20
Dong XJ, Shi T, Zhang W, Zhou Q (2020) Temporal and spatial differences in the resilience of smart cities and their influencing factors: evidence from non-provincial cities in China. Sustain 12(4). https://doi.org/10.3390/su12041321
Du J (2019) In Gorakhpur, India, citizens use nature to prevent floods. https://www.wri.org/insights/gorakhpur-india-citizens-use-nature-prevent-floods
Eremia M, Toma L, Sanduleac M (2017) The smart city concept in the 21st century. Procedia Eng 181:12–19. https://doi.org/10.1016/j.proeng.2017.02.357
Esaki H (2021) Open and smart data sharing platform for smart city. Special Issue of New Technologies for Smart City, Proc Inst Electr Install Eng Jpn (Japanese Edition) 41(4):191–194
esri (2019) Explore Paris with ArcGIS online. https://www.arcgis.com/apps/MapJournal/index.html?appid=d04845880c0844708f550f2173deb042
Fourneris C (2020) Climate control: Stockholm named world's smartest city as it aims for climate positive footprint. https://www.euronews.com/next/2020/01/27/climate-control-stockholm-named-world-s-smartest-city-as-it-aims-for-carbon-positive-footp
Fujinawa Y, Kouda R, Noda Y (2015) The resilient smart city (an proposal). J Disaster Res 10(2):319–325. https://doi.org/10.20965/jdr.2015.p0319
Gargiulo C, Zucaro F (2015) Smartness and urban resilience. A model of energy saving. Tema J Land Use, Mobil Environ (SI):81–102. http://gateway.isiknowledge.com/gateway/Gateway.cgi?GWVersion=2&SrcAuth=AegeanSoftware&SrcApp=NoteExpress&DestLinkType=FullRecord&DestApp=WOS&KeyUT=000384506000006
Gasco-Hernandez M (2018) Building a smart city: lessons from Barcelona. Commun ACM 61(4):50–57. https://doi.org/10.1145/3117800
Gibson DV, Kozmetsky G, Smilor RW (1992) Technopolis phenomenon: smart cities. Rowman & Littlefield, Lanham, MD
Giffinger R, Fertner C, Kramar H, Meijers E (2007) City-ranking of European medium-sized cities. http://www.smart-cities.eu/download/city_ranking_final.pdf
Giffinger R, Gudrun H (2010) Smart cities ranking: an effective instrument for the positioning of the cities? ACE: Arch City Environ 4(12):7–26

Goal11 (UN SDGs) Goal 11: sustainable cities and communities. https://www.globalgoals.org/11-sustainable-cities-and-communities
Gunderson LH, Holling CS (2002) Panarchy: understanding transformations in human and natural systems. Island Press, Washington, DC
GUTP (2008) Green University of Tokyo Project. https://www.gutp.jp/
Han H, Hawken S (2018) Introduction: innovation and identity in next-generation smart cities. City Cult Soc 12(1):1–4. https://doi.org/10.1016/j.ccs.2017.12.003
Hassankhani M, Alidadi M, Sharifi A, Azhdari A (2021) Smart city and crisis management: lessons for the COVID-19 pandemic. Int J Environ Res Public Health 18(15):7736. https://www.mdpi.com/1660-4601/18/15/7736
Hiller JS, Blanke JM (2017) Smart cities, big data, and the resilience of privacy. Hast LJ 68:309. https://repository.uchastings.edu/cgi/viewcontent.cgi?article=1007&context=hastings_law_journal
Holling CS (1973) Resilience and stability of ecological systems. Annu Rev Ecol Syst 4(1):1–23. https://doi.org/10.1146/annurev.es.04.110173.000245
Holling CS (2001) Understanding the complexity of economic, ecological, and social systems. Ecosystems 4(5):390–405. https://doi.org/10.1007/s10021-001-0101-5
IPCC (2014) Climate change 2014—mitigation of climate change. United Kingdom and New York. https://www.ipcc.ch/site/assets/uploads/2018/02/ipcc_wg3_ar5_full.pdf
Jason Iken PE, Brown W (2009) Use of multi-sensor inspection in Houston to minimize pipeline rupture risk. Trenchless Technol (Chinese Edition) (5):45–46. http://qikan.cqvip.com/Qikan/Article/Detail?id=32238965
Kaika M (2017) 'Don't call me resilient again!': the new urban agenda as immunology … or … what happens when communities refuse to be vaccinated with 'smart cities' and indicators. Environ Urban 29(1):89–102. https://doi.org/10.1177/0956247816684763
Kennedy CA, Ibrahim N, Hoornweg D (2014) Low-carbon infrastructure strategies for cities. Nat Clim Chang 4(5):343–346. https://doi.org/10.1038/Nclimate2160
Kontokosta CE, Malik A (2018) The Resilience to Emergencies and Disasters Index: applying big data to benchmark and validate neighborhood resilience capacity. Sustain Cities Soc 36:272–285. https://doi.org/10.1016/j.scs.2017.10.025
Korhonen J, Snakin JP (2015) Quantifying the relationship of resilience and eco-efficiency in complex adaptive energy systems. Ecol Econ 120:83–92. https://doi.org/10.1016/j.ecolecon.2015.09.006
Kotevska O, Kusne AG, Samarov DV, Lbath A, Battou A (2017) Dynamic network model for smart city data-loss resilience case study: city-to-city network for crime analytics. IEEE Access 5:20524–20535. https://doi.org/10.1109/ACCESS.2017.2757841
Li W, Long Y (2020) Technology and city: pan-smart city technology enhances urban resilience. Shanghai Urban Plan (Chinese Edition) (02):64–71. https://kns.cnki.net/KCMS/detail/detail.aspx?dbcode=CJFD&filename=HCSG202002013
Long Y (2020) Pan-intelligent city technology to improve urban resilience-response to the 2020 novel coronavirus pneumonia incident written talk. Urban Plan (Chinese Edition) 2:115–125
Lytras MD, Visvizi A (2018) Who uses smart city services and what to make of it: toward interdisciplinary smart cities research. Sustain 10(6). https://www.mdpi.com/2071-1050/10/6/1998
Macke J, Casagrande RM, Sarate JAR, Silva KA (2018) Smart city and quality of life: citizens' perception in a Brazilian case study. J Clean Prod 182:717–726. https://doi.org/10.1016/j.jclepro.2018.02.078
Marks P (2014) Legume with a view. New Sci 221(2952):17–18. https://doi.org/10.1016/S0262-4079(14)60124-X
Maye A, Peter M, Feng Y, Wang XH, Chen YS (2016) Managing smart cities: a literature review on Smart City Governance. Int Rev Adm Sci (Chinese Edition) 82(04):150–166
Meerow S, Newell JP, Stults M (2016) Defining urban resilience: a review. Landsc Urban Plan 147:38–49. https://doi.org/10.1016/j.landurbplan.2015.11.011

Mileti D (1999) Disasters by design: a reassessment of natural hazards in the United States. Joseph Henry Press, Washington, DC

Moraci F, Fazia C, Errigo MF (2018) Smart tools for energy resilient city. Annales de Chimie-Science des Materiaux 42(4):459–470. https://doi.org/10.3166/Acsm.42.459-470

Motesharrei S, Rivas J, Kalnay E, Asrar GR, Busalacchi AJ, Cahalan RF, Cane MA, Colwell RR, Feng K, Franklin RS, Hubacek K, Miralles-Wilhelm F, Miyoshi T, Ruth M, Sagdeev S, Shirmohammadi A, Shukla J, Srebric J, Yakovenko V, Zeng N (2016) Modeling sustainability: population, inequality, consumption, and bidirectional coupling of the earth and human systems. Natl Sci Rev 3(4):470–494. https://doi.org/10.1093/nsr/nww081

Nam T, Pardo TA (2011) Conceptualizing smart city with dimensions of technology, people, and institutions. In: Proceedings of the 12th annual international digital government research conference: digital government innovation in challenging times, June, pp 282–291

Oliva S, Lazzeretti L (2018) Measuring the economic resilience of natural disasters: an analysis of major earthquakes in Japan. City Cult Soc 15:53–59. https://doi.org/10.1016/j.ccs.2018.05.005

Palmieri F, Ficco M, Pardi S, Castiglione A (2016) A cloud-based architecture for emergency management and first responders localization in smart city environments. Comput Electr Eng 56:810–830. https://doi.org/10.1016/j.compeleceng.2016.02.012

Paroutis S, Bennett M, Heracleous L (2014) A strategic view on smart city technology: the case of IBM Smarter Cities during a recession. Technol Forecast Soc Chang 89:262–272. https://doi.org/10.1016/j.techfore.2013.08.041

Pouffary S, Rogers H (2014) Climate finance for cities and buildings—a handbook for local governments. https://www.eldis.org/document/A69822

Ruhlandt RWS (2018) The governance of smart cities: a systematic literature review. Cities 81:1–23. https://doi.org/10.1016/j.cities.2018.02.014

Rus K, Kilar V, Koren D (2018) Resilience assessment of complex urban systems to natural disasters: a new literature review. Int J Disaster Risk Reduct 31:311–330. https://doi.org/10.1016/j.ijdrr.2018.05.015

Shah SA, Seker DZ, Rathore MM, Hameed S, Yahia SB, Draheim D (2019) Towards disaster resilient smart cities: can internet of things and big data analytics be the game changers? IEEE Access 7:91885–91903. https://doi.org/10.1109/ACCESS.2019.2928233

Sharifi A (2016) A critical review of selected tools for assessing community resilience. Ecol Ind 69:629–647. https://doi.org/10.1016/j.ecolind.2016.05.023

Sharifi A (2019) A critical review of selected smart city assessment tools and indicator sets. J Clean Prod 233:1269–1283. https://doi.org/10.1016/j.jclepro.2019.06.172

Sharifi A (2020a) A global dataset on tools, frameworks, and indicator sets for smart city assessment. Data Brief 29:105364. https://doi.org/10.1016/j.dib.2020.105364

Sharifi A (2020b) A typology of smart city assessment tools and indicator sets. Sustain Cities Soc 53:101936. https://doi.org/10.1016/j.scs.2019.101936

Sharifi A (2020c) Urban resilience assessment: mapping knowledge structure and trends. Sustain (Switzerland) 12(15). https://doi.org/10.3390/SU12155918

Sharifi A, Khavarian-Garmsir AR (2020) The COVID-19 pandemic: impacts on cities and major lessons for urban planning, design, and management. Sci Total Environ 749:142391. https://doi.org/10.1016/j.scitotenv.2020.142391

Sharifi A, Khavarian-Garmsir AR, Kummitha RKR (2021) Contributions of smart city solutions and technologies to resilience against the COVID-19 pandemic: a literature review. Sustain 13(14):8018. https://www.mdpi.com/2071-1050/13/14/8018

Sharifi A, Yamagata Y (2018) Resilience-oriented urban planning. In: Yamagata Y, Sharifi A (eds) Resilience-oriented urban planning: theoretical and empirical insights. Springer, Cham, pp 3–27

Song L (2020) Is intelligence compatible with resilience? Resilience assessment and development path of smart city construction. Soc Sci (Chinese Edition) (03):21–32. https://doi.org/10.13644/j.cnki.cn31-1112.2020.03.004

Soyata T, Habibzadeh H, Ekenna C, Nussbaum B, Lozano J (2019) Smart city in crisis: technology and policy concerns. Sustain Cities Soc 50:101566. https://doi.org/10.1016/j.scs.2019.101566

Takewaki I, Fujita K, Yamamoto K, Takabatake H (2011) Smart passive damper control for greater building earthquake resilience in sustainable cities. Sustain Cities Soc 1(1):3–15. https://doi.org/10.1016/j.scs.2010.08.002
TfL (2012) Your accessible transport network. https://content.tfl.gov.uk/your-accessible-transport-network.pdf
TfL (2013) The London 2012 Games transport legacy: one year on. https://tfl.gov.uk/info-for/media/press-releases/2013/july/the-london-2012-games-transport-legacy-one-year-on
The Climate Group and Global eSustainability Initiative (2008) Smart 2020: enabling the low carbon economy in the information age. https://www.compromisorse.com/upload/estudios/000/36/smart2020.pdf
TOP (2018) There are 12 cases of smart cities around the world, we release them all at once! You're welcome to take it! https://www.sohu.com/a/218151840_472773
TUWIEN (2015) The smart city model. http://www.smart-cities.eu/?cid=2&ver=4
UN-Habitat. Urban resilience hub. https://urbanresiliencehub.org/what-is-urban-resilience/
use (2009) Smarter Sustainable Dubuque. https://use.metropolis.org/case-studies/smarter-sustainable-dubuque
Viitanen J, Kingston R (2014) Smart cities and green growth: outsourcing democratic and environmental resilience to the global technology sector. Environ Plan A-Econ Space 46(4):803–819. https://doi.org/10.1068/a46242
Wang P, Zhao L (2020) The future of smart city from the stress test of new crown disease. https://www.thepaper.cn/newsDetail_forward_5980938
Wei D, Yimin S, Al E (2017) 戴伟,孙一民,韩·迈尔, 等.气候变化下的三角洲城市韧性规划研究[J]. 城市规划 41(12):26–34 (in Vol. 2021).
Wen F (2020) Working from home: the new normal of publishing work in the anti-epidemic period. International Weekly Publication (Chinese Edition), p 5.
WRT (2010) Paris—Water Remunicipalisation Tracker. http://www.remunicipalisation.org/#case_Paris
Yang C, Su G, Chen J (2017) Using big data to enhance crisis response and disaster resilience for a smart city. Paper presented at the 2017 IEEE 2nd International Conference on Big Data Analysis (ICBDA)
Yan S, Tang J (2020) Progress on the theory and practice of resilient city. West J Hum Settl (Chinese Edition) 82(04):150–166. CNKI:SUN:SNSH.0.2020-02-017
Yan Y, Han Q (2020) Weizhi technology (WAYZ): disease control AI analysis platform. http://sh.people.com.cn/n2/2020/0225/c396182-33828003.html
Yao F, Wang Y (2020) Towards resilient and smart cities: a real-time urban analytical and geo-visual system for social media streaming data. Sustain Cities Soc 63(102448). https://doi.org/10.1016/j.scs.2020.102448
Zach F, Kretschmer F, Stoeglehner G (2019) Integrating energy demand and local renewable energy sources in smart urban development zones: new options for climate-friendly resilient urban planning. Energies 12(19). https://doi.org/10.3390/en12193672
Zhang F, Li Z, Li N, Fang D (2019) Assessment of urban human mobility perturbation under extreme weather events: a case study in Nanjing, China. Sustain Cities Soc 50:101671. https://doi.org/10.1016/j.scs.2019.101671
Zhang N (2020) Strengthen toughness as the main axis of smart city development. China Emergency Management News (Chinese Edition), p 1
Zhang T (2019) Smart city studies in Huai'an (Master). Southeast University, China
Zhao R-D, Fang C, Liu H (2020) Progress and prospect of urban resilience research. Prog Geogr Sci (Chinese Edition) 39(10):1717–1731. https://doi.org/10.18306/dlkxjz.2020.10.011
Zhou P, Fu R (2020) Smart city: how Smart Cities can help improve urban resilience. http://www.c114.com.cn/news/118/a1129125.html
Zhu S, Li D, Feng H (2019) Is smart city resilient? Evidence from China. Sustain Cities Soc 50:101636. https://doi.org/10.1016/j.scs.2019.101636

Zhu SY, Li DZ, Feng HB, Gu TT, Hewage K, Sadiq R (2020) Smart city and resilient city: differences and connections. Wiley Interdiscip Rev-Data Min Knowl Discov 10(6). https://doi.org/10.1002/widm.1388

ZIGURAT (2019) Smart city series: the Barcelona experience. https://www.e-zigurat.com/blog/en/smart-city-barcelona-experience/

Chapter 6
Smart Cities and Urban Resilience: Insights from a Delphi Survey

Nae-Wen Kuo, Ayyoob Sharifi, and Chong-En Li

Abstract Cities worldwide are exposed to an expansive range of climate-related disasters, and thus, enhancing urban resilience is increasingly critical and has become a major goal of city authorities. With the rapid development of technology, the concept of a "smart city" is also becoming popular. A vast body of research has been published on urban resilience as well as smart city. There are also many tools and indicator sets for their assessment. However, there have been limited efforts to synchronously study these two concepts. Urban resilience and smart city have the potential to be merged, which is what this research calls "smart city resilience" and implies deploying "smart solutions" for urban resilience and sustainable city management. However, this trend is still in its infancy worldwide, and further exploration is needed. Additionally, assessment methods and approaches, such as a toolkit for assessing the current situation and making cross-city comparisons, also need to be developed. Hence, the purpose of this research was to investigate the indicators that should be included in an assessment toolkit. A panel of 13 experts participated in the Delphi survey, and the analytic hierarchy process was used to find the relative weight of each indicator. Finally, the opinions toward the assessment toolkit from the experts were discussed further. Results can inform future efforts toward developing toolkits for assessing smart city resilience.

Keywords Smart city · Urban resilience · Smart city resilience · Indicators · Delphi survey

---

N.-W. Kuo (✉) · C.-E. Li
Department of Geography, National Taiwan Normal University, Taipei City 106308, Taiwan
e-mail: niven@ntnu.edu.tw

C.-E. Li
e-mail: chongen.li@ntnu.edu.tw

A. Sharifi
Graduate School of Humanities and Social Sciences, Hiroshima University, Higashihiroshima, Japan
e-mail: sharifi@hiroshima-u.ac.jp

Graduate School of Advanced Science and Engineering, Hiroshima University, Hiroshima 739-8511, Japan

A. Sharifi and P. Salehi (eds.), *Resilient Smart Cities*, The Urban Book Series,
https://doi.org/10.1007/978-3-030-95037-8_6

6.1 Introduction

Cities are places in which majority of people and economic activities are concentrated. Over the past few decades, many people have migrated to metropolitan areas from rural areas and continue to do so. According to the UN (2018), since 2007 more than half of the population worldwide has lived in cities, and the degree of urbanization is poised to increase to 68% by 2050. In addition to socio-economic challenges, the predominance of urban dwelling creates significant environmental concerns, including resource consumption, production of solid waste, air pollution, and water pollution. Additionally, many urban issues evolve into climate-related disasters and many cities are highly vulnerable to extreme climatic events. According to basic disaster theory, risk is equal to the multiplication of hazard, vulnerability, and exposure. Cities with high population density are likely to have higher exposure to and also more significant risk of disasters. To protect these vulnerable places, scientists from different fields have been engaged in relevant research and have adopted some special terms to communicate with each other. For example, determining how to increase cities' resilience has become a popular research purpose, and the word "resilience" is now generally accepted since the United Nations used it to frame the Goal 11 of Sustainable Development Goals (SDGs). Meanwhile, with the flourishing development of information and communication technologies (ICTs), city governance has become smarter. Leveraging ICTs for governance is a hallmark of "smart cities" which are aimed at developing capabilities to efficiently manage many urban challenges. The ultimate goal of developing a smart city is resolve various urban problems through ICT-based technologies connected up as an urban infrastructure (Lee et al. 2014).

Initiatives enabled by ICTs are deemed essential for helping cities develop transformative solutions to address challenges related to global climate changes and create equitable, sustainable, and resilient communities. Essentially, the concepts of urban resilience and smart city have the potential to be merged (Baron 2012; Zhou et al. 2021). Many cities worldwide are vulnerable to numerous types of climate-related disasters, and thus, enhancing resilience is expected to be at the core of smart city development efforts. This could lead to a new concept that can be referred to as "smart city resilience."

Although a vast body of knowledge exists on smart city projects' contributions to sustainability and quality of life, little is known about their contributions to climate resilience and climate action planning. In the other word, smart city projects based on ICTs have been proposed and implemented with the aim to enhance climate resilience. However, evaluating the effectiveness of such projects and the value of implementation experiences is challenging. To evaluate the effectiveness of smart city projects and to compare them across cities and areas, a set of assessment toolkits should be developed. Such toolkits can build upon the large number of tools and indicators sets that already exist for assessing city smartness (Sharifi 2019) and/or resilience (Sharifi 2016, 2020c). In fact, developing tools that can simultaneously take both resilience and smartness into account can help planners and policy makers save efforts and resources and also consider potential synergies and trade-offs between smartness and resilience. For this purpose, we conducted a Delphi survey to collect the opinions of experts on indicators that can be used for developing tools for assessing smart city resilience. The analytic hierarchy process (AHP) also was used to determine the weights of each assessment indicator in the smart city resilience-assessment toolkit. The toolkit can be used to examine the actual and/or potential performance of cases worldwide in further research. The results and its findings can provide helpful scientific and policy knowledge on integrating resilience thinking into smart city development and can enable smart city developers to measure their contributions to disaster resilience.

The remainder of this chapter is structured as follows. The methods used for Delphi and AHP surveys are explained in the next section. Results are presented and discussed in Sect. 6.3. Finally, Sect. 6.4 concludes the research by providing some final remarks and recommendations.

6.2 Methods

6.2.1 Delphi Method and AHP Survey

The Delphi technique is a procedure to systematically collect experts' opinions. In the process of Delphi survey, all participants are anonymous to the others. It is an excellent method to avoid some of the disadvantages of roundtable meetings, including the bandwagon effect, the spiral of silence, and groupthink (Linstone and Turoff 1975). AHP is a method to address complex decisions. It is especially useful in group decision-making; it condenses the decision problem into intelligible categories, and the categories into many related indicators to construct a structural hierarchy model (R. W. Saaty 1987; T. L. Saaty 1980, 1982). The decision-maker creates a matrix for pairwise comparison of each category by using a 9-point scale. This research adopted the AHP approach to determine the weights of the indicators to be used to assess smart city resilience.

6.2.2 Questionnaire Survey Procedures

This study invited 13 experts from different countries to participate in the questionnaire survey. The experts specialize in the research fields of urban resilience, smart cities, sustainable cities, and ICT applications. In the Round 1, summarized results of a literature review on the topic were shared with participants to familiarize them with the work's scope and structure. This also included a list of potentially relevant indicators for assessing smart city resilience. Further details on the contents of the first round of the Delphi process are shown in Table 6.1. All experts were asked to

Table 6.1 Questions in round 1

Serial number		Context
Part 1: Meanings of urban resilience, smart city, and smart solution	Q1	What urban resilience and smart city mean to you?
	Q2	What do you think smart city solutions, enabled by Information and Communication Technologies add to the existing discourses of urban resilience?
	Q3	Resilient urban systems feature multiple characteristics such as robustness, diversity, redundancy, connectivity, complexity, flexibility, resourcefulness, learning capacity, agility, efficiency, adaptive capacity, modularity (independence and self-organization), creativity, and equity. Are there any characteristics you think should be added or removed? Please justify your response in few words
	Q4	Do you think smart solutions can enable or constrain these characteristics? If so, in what way? Please provide examples if possible (see the Glossary at the end of this document for definitions)
Part 2: Establishing toolkits for assessing smart solutions	Q5	Do you think a toolkit for assessing contributions of smart city initiatives to resilience is useful? If so, what do you think the potential contributions of such a toolkit will be?
	Q6	We have identified a set of potentially measurable indicator to be used for assessing contributions of smart cities to disaster resilience. These indicators are categorized into 7 themes, namely, Environment, Economy, People, Governance, Living, Mobility, and Data (see the Excel sheet). This list is not comprehensive, and we hope you would be able to help us improve it. Are there potentially measurable indicator in the list you think should be removed and/or are there additional measurable indicator you think need to be considered?
Part 3: Experts' opinions regarding assessment toolkits	Q7	What stakeholders do you think should be involved in the processes aimed at assessing contributions of smart city solutions to disaster resilience?
	Q8	Smart city initiatives are practiced at different scales, ranging from small scale projects to city-wide initiatives. In your view should assessment indicators for different scales be different? If so, why and how should they be different?
	Q9	In your view, what challenges and barriers should be addressed when assessing contributions of smart city solutions to disaster resilience?

write their opinions in the initial questionnaire survey. Next, all participants' survey responses were anonymously summarized and fed back to each participant as part of Round 2 of questionnaire survey for participants' input. The participants' opinions were expected to converge. If views would have still been divergent, third or possibly fourth stages of the Delphi survey may have been needed.

Two rounds of the questionnaire survey were conducted in this research; 13 experts participated in the first-round questionnaire survey and 9 experts participated in both the first and second rounds. The first-round questionnaire was primarily organized into three parts: (a) the meanings of urban resilience, smart city, and smart city resilience, (b) establishing the toolkits for assessing smart solution, and (c) giving opinions about the suggested potential indicators for assessing smart city resilience. The first-round questionnaire is shown in Table 6.1 and was sent to participants by e-mail. Since the questionnaire survey design was open-ended, the results were analyzed through text analysis in this round.

The second-round questionnaire was designed on the basis of the outcomes of the first round. The opinions on three questions (Q1, Q4, and Q8) were divergent, and the opinions on other questions (Q2, Q3, Q5, Q6, Q7, and Q9) were converged. Hence, we designed three new questions (Q10, Q11, and Q12) to further the discussion prompted by the original three questions (Q1, Q4, and Q8) and modified specific topics according to all participants' comments and suggestions (see the next section for more details). In addition, two new questions (Q13 and Q14) were added in the second-round questionnaire to further discuss the outcome of Q6. Also, Q13 was used to select the potentially measurable indicators in each category, and Q14 was used to weigh the seven assessment categories by using the AHP method. Finally, there were a total of five questions in the second-round questionnaire.

6.3 Results and Discussion

6.3.1 *Part 1: Meanings of Urban Resilience, Smart City, and Smart City Resilience*

Because the key terms in this study lack robust and coherent meanings (Cavada et al. 2014), we needed to reach a consensus on these concepts' definitions before discussing the topics. In Question 1 of the first round, each participant offered many different meanings of urban resilience and smart city because the question was open-ended. Various experts repeatedly mentioned several concepts in this process. We collected those high-frequency words to design the closed-end Question 10 in the second round, with the aim to make the opinions more convergent (Tables 6.2 and 6.3). Finally, there was consensus on definition of these two concepts, specifically there was agreement that urban resilience is the networked capability of the urban systems to recover from unpredicted shocks (e.g., natural disasters, climate change, and terrorism) and long-term disturbance; and smart city means that by relying on ICT and emerging technologies the city becomes more effective and efficient in terms of operations and urban management and is capable of improving quality of life. Other proposed definitions should not be considered per se *wrong* but rather broad definitions of these two terms.

After the meanings of urban resilience and smart city were determined, we requested participants' opinions on the term smart city resilience in Question 2. Most respondents agreed that smart city can promote urban resilience (85%) and that ICTs

Table 6.2 Closed-end question on the meaning of urban resilience in round 2

Urban resilience is the	☐ ability ☐ characteristic ☐ networked capability	of the	☐ urban systems ☐ city ☐ society ☐ government ☐ metropolitan region	to	☐ recover from ☐ respond to ☐ resist to ☐ react to ☐ prepare for ☐ adapt to ☐ withstand ☐ recuperate from

☐ unpredicted ☐ sudden ☐ unprecedented	☐ shocks ☐ events ☐ change ☐ disastrous conditions ☐ threats	(e.g., natural disasters, climate change and terrorism) and long-term	☐ stresses ☐ disturbance

Table 6.3 Closed-end question on the meaning of "smart city" in round 2

Smart city means the city is	☐ smarter ☐ better	at	☐ urban planning ☐ resources use ☐ citizen daily-life convenience ☐ urban management	and to

☐ solve urban problems ☐ improve urban livings ☐ optimize the efficiency of operations	, which relies on ICTs and other emerging technologies

are an indispensable component of urban resilience (15%). Smart solutions have the potential to reduce operations and management costs of urban areas while extending the reach and impact of city services. Specifically, discussions of smart solutions have contributed to two critical ideas to the discourse on ICT-enabled resilience. First, when understanding ICT-enabled resilience as a preventative and preemptive line of defense, the ICT-enabled resilience discourse can highlight systemic inadequacies (presenting in the form of risks and vulnerabilities) that can potentially benefit from ICT interventions. Second, when understanding ICT-enabled resilience as a response mechanism, ICT-enabled resilience discourse can aid in examining how ICTs can perform the role of facilitating community-led and context-specific responses to risks and vulnerabilities as opposed to a role that overrides grounded reactions to threats. At the individual level, ICT-enabled resilience can help in disseminating essential government policies and hazard warnings, as well as promoting general services and initiatives like bike sharing and using greener technologies. At the institutional level, ICTs can help various government departments (both vertically and horizontally) cooperate, integrate their planning and policies into one comprehensive plan, and ease practical communication barriers.

We examined the literature to identify 14 characteristics of resilient urban systems that are widely recognized. These are, namely, robustness, diversity, redundancy, connectivity, flexibility, resourcefulness, agility (rapid response), efficiency, adaptive/learning capacity, modularity (independence and self-organization), creativity (innovation), equity, inclusiveness (participatory), and foresight capacity (Sharifi and Yamagata 2016; Tyler and Moench 2012). We asked the participants to modify these characteristics' definitions in Question 3. The respondents confirmed these meanings as shown in Table 6.4. It is worth noting that the participants also proposed characteristics other than those mentioned above, including rapidity, knowledge, resistance, transformability, reliability, humanity, sensitivity (or self-recognition), social and ecological memory, horizontal and vertical cooperation and integration, socio-ecological-technological systems, nature-based solutions, and livability (listed by frequency of mention). These additional characteristics can also be helpful in understanding urban resilience under specific situations.

After discussing urban resilience characteristics, we attempted to identify, by posing Questions 4 and 11, smart solutions that can enable or constrain the characteristics. The results of two rounds of surveys are shown in Fig. 6.1. All experts agreed that the smart solutions can enable urban resilience by strengthening robustness, flexibility, and resourcefulness characteristics. There was, however, limited agreement on the characteristic of equity. In an opinion, one expert was unsure whether smart solutions can improve cities' equity because such improvement depends on citizens' access to smart techniques. Indeed, this is related to the issue of digital divide that has always been a concern in the context of efforts aimed at smart city development.

6.3.2 Part 2: Establishing Toolkits for Assessing Smart City Resilience

To assess the current situation of city's resilience and make cross-city comparisons of smart solutions' contributions to resilience were two key purposes for developing a set of toolkits. However, Question 5 of the first round revealed that participants' opinions on this issue were diverse. Most respondents (77%) hold a positive attitude to the usefulness of a toolkit for assessing the contribution of smart city programs to resilience, and one of the experts specifically believes that the toolkit can be used to discover the strengths and weaknesses of development policies for urban planners. However, some respondents (23%) thought that a toolkit would not be an appropriate approach because it is not a flexible mechanism; hence, it would not be helpful in creating a comprehensive tool for all cities. Further, because a smart city is not a crystalized concept, its contours are rather vague and subject to interpretation, with such an amorphous concept apparently confounding the understanding of our expert panel. Therefore, we designed more specific questions as follows.

We listed several potentially measurable indicators in Question 6 of the first round to assess smart cities' contributions to disaster resilience and categorized them into

Table 6.4 Recognized characteristics of resilient urban systems in previous studies

Characteristics	Meaning
Robustness	"Robustness refers to a system's strength to withstand short-term(sudden), acute internal and external shocks without suffering from major degradation of the main functions. To achieve this and enhance system security, the system needs to have the ability to counteract and/or absorb the disturbance" (Sharifi and Yamagata 2016)
Diversity	"Refers to the degree to which multiple distinct functions, that can be used simultaneously, are included in the system. The aim of this principle is to hedge against supply disruptions and ensure that a variety of options (resources, instruments, etc.) for dealing with disturbances and ensuring functionality exist in an urban system" (Sharifi and Yamagata 2016)
Redundancy	"Redundant capacity refers to the availability of (substitutable) components with similar (even overlapping) functions in the urban system to enhance its adaptive capacity and ability to absorb shocks, give it reserve capacity for problem solving, and ensure that uncertain events causing the failure or displacement of one component would not result in the failure of the whole system. In a system featuring redundant capacity, exclusion of an element should not result in significant loss of functioning" (Sharifi and Yamagata 2016)
Connectivity	An urban system includes multiple sub-systems. A resilient system should be capable of establishing connections between those sub-systems and coordinate their activities in order to enhance effectiveness and efficiency of operations. "Without this capacity the existing resources would not be effectively utilized to prepare for the disaster, the system will not be able to achieve its full absorptive capacity, and consequently there would be procrastination in the recovery efforts" (Sharifi and Yamagata 2016)
Flexibility	"Flexibility means that a system should have the ability to 'adapt to changing conditions' and undergo a safe failure by changing its configuration. A flexible system is capable of sensing threats, immediately detecting the failure and making prompt changes at smaller scales of its subsystems and thereby maintain overall performance during disaster. In the context of energy systems, this could (e.g.,) refer to the ability to shift between different energy configurations or adjust regulations or prices according to changing conditions" (Sharifi and Yamagata 2016)
Resourcefulness	"Relates to the adequacy of resources at the disposal of urban planners and decision makers to appropriately identify, prepare for, respond, and recover from potential disruptions. This includes having appropriate capacity to understand status quo, and identify patterns, potential threats, and contingencies" (Sharifi and Yamagata 2016)
Agility (rapid response)	"Represents the system's capacity to mobilize the resources necessary for recovery and return to normal functioning within an acceptable time frame. Agility is essential for avoiding cascading failures that can result in the disruption of other functions in the system" (Sharifi and Yamagata 2016)
Efficiency	"Means that the proportion of energy and resources provided by an urban system to the resources given to it as input, should be positive to improve resource use productivity and avoid waste" (Sharifi and Yamagata 2016)

(continued)

Table 6.4 (continued)

Characteristics	Meaning
Adaptive/learning capacity	"Refers to an urban system's capacity to learn from the disaster to reduce its pre-disturbance vulnerabilities and enhance its capacity to adapt to the changing conditions. Adaptability implies recognition of the inherent vulnerability of the system components, availability of appropriate knowledge and assignment of authority to prioritize tasks at the time of crisis, and ability to respond with rapidity in order to facilitate a 'safe-to-fail' (or at least 'soft-fail') urban system. A resilient urban system should entail 'adaptive' cycles that 'alternate between long periods of aggregation and transformation of resources and shorter periods that create opportunities for innovation', thereby ensuring survivability of the system" (Sharifi and Yamagata 2016)
Modularity (independence and self-organization)	"A resilient system should possess a 'certain degree of self-reliance that gives it the ability to maintain a minimum acceptable level of functioning (without external support) when influenced by disturbance'. A self-organized system discourages centralization of resources and authorities and should involve community-based management characterized by strengthened local communities capable of independently responding to disaster, cross-scale partnerships, and 'horizontal' and 'vertical' institutional connections that provide direct feedback to the system and enable better informed decision making. Furthermore, it should entail the ability to build upon and strengthen networks established to respond to an earlier disturbance" (Sharifi and Yamagata 2016)
Creativity (innovation)	"This principle represents the 'urban system's ability to use the disruption as an opportunity to attain a more advanced state'. This requires utilizing innovation (both technological and non-technological) in management, planning, and design of urban system. Innovation is essential to enhance various resilience abilities and avoid being overwhelmed by the constantly changing nature of risks" (Sharifi and Yamagata 2016)
Equity	"Equity plays an essential role for achievement of resilience. This is to ensure that all urban citizens have the ability to utilize services to prepare/plan for, cope with and recover from disruptions. Also, justice is needed in terms of exposure to adverse impacts. This is to ensure that marginalized and poor people do not bear the brunt of those impacts" (Sharifi and Yamagata 2016)
Inclusive (participatory)	Engagement of various stakeholders in planning and decision-making processes enhances social capital and improves planning, absorption, recovery, and adaptation capacities (Sharifi and Yamagata 2016)
Foresight capacity	"Any resilient system must be able to face the uncertainty and relativity of the future conditions. The characteristic of disaster is entangled with uncertainty and nonlinearity of the impacts and behaviors of a portfolio of endogenous and exogenous forces that can potentially become sources of disturbance in the system. This principle is essential for disaster preparation and also absorption of initial shocks. It implies that only preparation based on shortcomings exposed by past events is not enough and forecasting methods should also be applied in preparation to respond to newer risks that may unfold in the future" (Sharifi and Yamagata 2016)

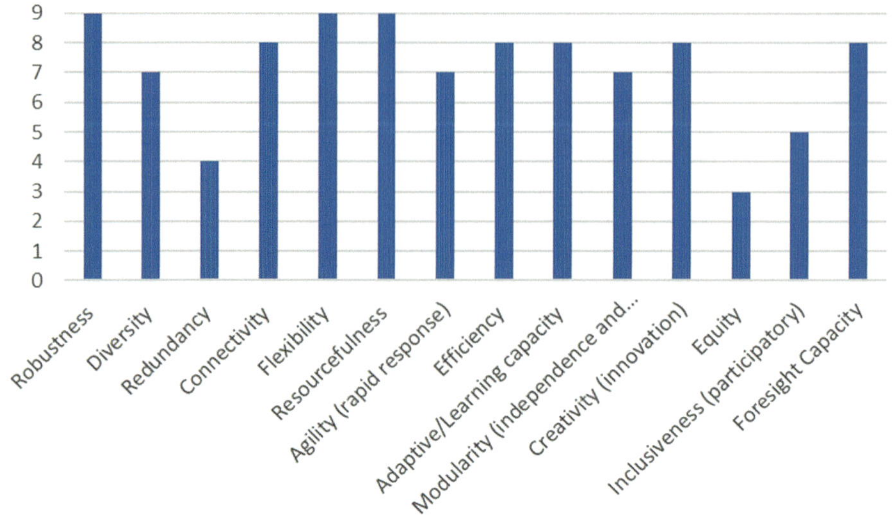

Fig. 6.1 Relationship between smart solutions and the 14 characteristics

seven themes: environment, economy, people, governance, living, mobility, and data. These indicators have been selected based on previously published comprehensive studies on smart city assessment (Sharifi 2019, 2020a, 2020b). In the survey, some participants opined that specific indicators in the list could not be easily quantified. Undoubtedly, these indicators are so numerous that not all of them are suitable for assessment. Thus, for the second round, closed-end Question 13 was added to prompt participants to choose the most essential indicators in each category. Each participant could choose the top one-third of all indicators. All candidate indicators and the frequency selected by respondents are shown in Table 6.5. The high-ranking potentially measurable indicators are marked with an asterisk. Based on the results, we determined that those selected indicators are likely to be more suitable for measuring the associated categories.

After the discussion following two rounds of surveys, participants were asked to complete an AHP survey to ascertain how each category might be more important than others to determine the weights in the second round. An empty pairwise comparison is given in Question 14, and all experts gave a weighting for seven assessment "categories" of the smart city assessment framework based on the AHP method. Generally, a score of 1 means two categories have equal importance, and scores of 3, 5, 7, and 9, respectively, mean that one category is moderately, strongly, very strongly, or extremely more important than another category (i.e., the number closer to 9 means that the difference between the two categories is more significant). The AHP results (Table 6.6) indicated that the "people" category received the highest score (0.207), followed by, in descending order, environment (0.173), governance (0.157), living (0.153), data (0.110), economy (0.106), and mobility (0.094). The survey revealed that regarding smart cities, all participants believed that the "people" aspect could

Table 6.5 Potentially measurable indicators

Category	Potentially measurable indicator	Frequency (%)
Category 1: Environment	Availability of water quality monitoring sensors (ICT enabled)	67**
	Availability of air quality monitoring sensors (ICT enabled)	67**
	Availability of smart metering systems (water/energy)	11
	Prediction of water, energy, and food requirements using consumption data (for resource demand management)	33
	Energy/water demand management through home energy/water management systems that provide real-time information to residents for behavior change	44
	Decentralized and modular infrastructure systems that function inter-operably	0
	Penetration of clean and renewable energy sources	56*
	Penetration of smart grids	44
	Availability of energy saving infrastructure to enable smart energy control (e.g., large-scale deployment of solar PVs, Vehicle to grid, etc.)	22
	Integration of smart solutions in waste collection, disposal, and treatment	56*
	Resilient urban development (mixed use, compact, job-housing proximity, etc.)	56*
	Universal design of ICT tools and services to ensure service accessibility for people with disabilities and special needs	11
	Affordability and accessibility of smart devices and ICT services	67**
	Take measures for automatic isolation of disrupted components (of energy/water systems) to avoid domino effects	0
	Use ICT and take measures for real-time detection of failures (e.g., leakage, water contamination, sewage discharge, energy theft, pressure anomality, etc.)	56*
	Implement measures to facilitate (real-time) communication between consumers and utilities during emergencies to improve resource conservation	11

(continued)

Table 6.5 (continued)

Category	Potentially measurable indicator	Frequency (%)
	Take measure for real-time communication of air/water quality to residents during emergency conditions	33
	Use smart location finding technologies such as RFID for locating infrastructure buried under debris and for expediting the clean-up process	22
	Utilize decentralized energy/water systems enabled by ICT to maintain resource supply during the recovery process (e.g., integration of PV, EV, and Home Energy Management Systems)	22
	Use event-detection software such as CANARY to monitor quality of resources during the recovery stage	11
	Promoting environmental-friendly behavior through illustrative methods for communication of consumption patterns using smart devices	44
	Use ICT to Restore/reconstruct the energy/water infrastructure in a way that less resources will be consumed in future	22
	Seizing the disaster as an opportunity for technological upgrade	11
	Long-term contributions of smart solutions to resource conservation	33
Category 2: Economy	Share of e-business and e-commerce transactions	33
	Contribution of knowledge economy and ICT initiatives to GDP (%)	67**
	Number of start-ups	22
	Funding for smart city projects (public/private finance, crowdsourced, etc.)	56*
	Online and ICT-enabled tourism promotion	11
	Smart pricing (e.g., apply time of use tariffs for energy/water conservation, especially under emergency conditions; demand-based pricing; congestion pricing)	44
	Avoided damage attributable to smart solutions	22
	Home-based work and workspace flexibilization (teleworking)	22
	Timetable flexibilization, working hour flexibility	0
	Use of smart solutions to reduce recovery costs	22

(continued)

Table 6.5 (continued)

Category	Potentially measurable indicator	Frequency (%)
	Long-term contribution of smart solutions to household and municipal cost saving	56*
	ICT-enabled flexibility (e.g., online job advertisement and application, etc.) and improvement of traditional industry and job market	44
Category 3: People	Level of digital and ICT literacy and technical capability	78**
	Educational programs to improve digital skills	33
	Internet penetration rate	33
	Social networking penetration rate	11
	Implementation of e-learning programs (reducing the need for physical presence under undesirable conditions)	44
Category 4: Living	Availability of digital health portals and hospitals archiving/using electronic health records	33
	Availability of telemedicine infrastructure (for disaster response, assisting aging population, etc.)	44
	Installing surveillance systems for crime protection	22
	Use of ICT and smart technologies for monitoring and control of diseases, and for Improving diagnostic, operation, and treatment methods	78**
	Use of telemedicine solutions to provide healthcare services to disaster affected areas	11
	Long-term contribution of smart solutions to demand reductions in the society	33
	Long-term contributions of smart solutions to safety and security	44
	Effectiveness of ICT solutions in improving general well-being	33
Category 5: Mobility	Availability of infrastructure to collect real-time traffic flow information (intelligent traffic management)	67**
	Availability of wireless infrastructure (hotspots, etc.)	33
	Integrate Web 3.0 into disaster management	22
	Car and bike sharing services, ridesharing etc.	33
	Autonomous Vehicle (AV) testing and deployment	22

(continued)

Table 6.5 (continued)

Category	Potentially measurable indicator	Frequency (%)
	Penetration level of green transportation modes and infrastructure (e.g., EVs, EV charging stations, etc.)	44
	Availability of real-time information about transit services and parking	33
	Penetration level of broadband internet connection	33
	Quality of internet service (rate of coverage by mobile broadband)	33
	ICT accessibility (e.g., smartphone penetration, PC ownership rate, internet penetration, etc.)	33
	Availability of ad hoc solutions to maintain network connectivity in case the conventional cellular network is disrupted (e.g., Drone empowered small cellular networks)	11
	Multi-modal public transportation	67**
	Maintenance and regular revision of the ICT infrastructure	33
	Use of IOT devices to improve situational awareness and facilitate collaboration and resource sharing between different groups during the post-disaster recovery process (use of IOT-based communication)	44
	Use of smart solutions such as Vehicular Ad-hoc Networks (VANETs) (V2V, V2I, V2X) to facilitate emergency communication between vehicles in order to reduce evacuation chaos and optimize access of rescue teams	11
	Use the disruptive event to upgrade the transportation system to become more climate resilient and user friendly	33
	Update configuration and structure of ICT devices to improve communication performance in future	11
	Long-term transportation demand trends, effectiveness of smart solutions in reducing vehicle ownership, and transportation demand/ travel distance	11
	ICT-enabled transportation damage and fatalities reduction	22
Category 6: Governance	Availability of city resilience plans and strategies	67**
	Availability of smart city vision and roadmap	56*
	Plans and strategies for performance monitoring and assessment of smart cities	44
	Availability of risk governance plans and integration of smart solutions in them	11

(continued)

Table 6.5 (continued)

Category	Potentially measurable indicator	Frequency (%)
	Laws and regulatory frameworks for smart city planning that include strategies to protect consumer privacy and regulate data ownership and access	67**
	Availability of a one-stop platform for data integration and for online accessibility and coordination of city services	44
	Availability of online civic engagement and feedback systems	56*
	Data sharing policies and the state of data/information sharing among various institutions	33
	Shared architecture for multi-level governance and inter-agency collaboration	56*
	Implementation of ICT-enabled scenario making	0
	Trained personnel to respond to cyber security threats	0
	Trained personnel to operate ICT-enabled systems	33
	Install early warning systems for disaster risk management	56*
	Implementation of early warning systems to communicate emergency information to stakeholders	0
	Use real-time data obtained during emergency conditions to update scenarios and estimate impacts	0
	Use platforms/ applications for communication between authorities and citizens during the recovery process (feedback mechanisms)	33
	Enable real-time tracking of situations by authorities during the recovery process (e.g., data on supply and demand of resources, monitoring progress, etc.)	22
	Prepare and disseminate information related to disruptive event and its impacts to raise awareness and ensure enhancing coping capacity in future	11
	Use lessons learned from the event to improve governance procedures	11
Category 7: Data	Infrastructure for data collection and storage	56*
	Measures for crowdsourced data collection	0
	Open data platforms for making information (governmental, etc.) open to the public	89**

(continued)

Table 6.5 (continued)

Category	Potentially measurable indicator	Frequency (%)
	Implementation of cyber security measures (backup systems, virus and threat protection, firewall and network protection systems)	33
	Big data analytics capabilities for data quality control, data classification, and data processing	78**
	Apply measures to isolate contaminated data to avert system-wide losses	0
	Use of cloud computing services for collection and sharing of data related to the disaster	22
	Link data platforms to each other (to promote interoperability)	33
	Use data collected during the disruptive event to improve future projections	22
	Evidenced-based planning and decision-making informed by big data analytics	67**
	Big data analytics' contribution to upgrading on modes of operation and planning process	0

Note "*" indicates more than half of the experts selected this indicator; "**" indicates more than two-thirds of the experts selected this indicator

Table 6.6 Weighting score from AHP survey

Category	Weighting score
Environment	0.173
Economy	0.106
People	0.207
Living	0.153
Mobility	0.094
Governance	0.157
Data	0.110

most contribute to disaster resilience. Indeed, as the end users stakeholders of smart cities, people and their capacities are essential for determining the extent to which a city is resilient to adverse events.

6.3.3 Part 3: Experts' Opinions Regarding Assessment Toolkits

For Part 3, we designed three primary questions. First, we inquired which stakeholders should be involved in the process of assessing the contributions of smart city solutions to disaster resilience. Most respondents of Question 7 opined that the (national and local) government, citizen/neighborhood groups, experts, companies, local media, and civil society representatives should be involved. This is a clear indication of the significance of getting multiple stakeholders engaged in the process. We next asked participants whether assessment indicators for different scales should be modified. In other words, whether customized indicators would be needed in different contexts and scales. According to the responses for Question 8, participants believed that the toolkit should be tailored to some specific city or country first and then extended globally. Moreover, they believed that the purpose of the toolkits should be clear and specific. Two respondents opined that assessment at a city-wide scale is more meaningful and that different levels of stakeholders should be consulted for different scales. However, one respondent contended that the assessment indicators do not need to be different because this would lead to too many variables. This can, in turn, cause challenges in terms of implementing the toolkit as more efforts and resources would be needed. Since the opinions are still not convergent enough, we asked participants in Question 12 in the second round whether the smart cities assessment framework is more meaningful in certain scales. According to their responses, most participants believed that the meaningfulness of the framework depends on the smart city services or projects under examination. Some participants argued that the preferred scale might be at the city level for smart cities but that the communities should be targeted units for urban resilience; thus, the framework should combine the two. However, when asked to choose between them, two-thirds of participants chose the city-wide scale and did not agree on the desirability of the community scale. The reason was that developing a smart city entails work on all aspects of the city (e.g., transportation, medical treatment, and government affairs) and considering all those aspects may not be feasible at a sub-city scale. A city-wide scale relates to the efforts of various planning bodies and governance units; many social and political groups are also within the scope of a city-wide scale. Much construction work is proposed and approved at the city level. Thus, only through a city-wide assessment can researchers acquire comprehensive data. Further, smart solutions at a community level are sometimes mixed with metropolitan or national responses and policies. In other words, they are highly entangled with and dependent on activities at higher scales of action.

Finally, in response to Question 9, the participants indicated some challenges for and barriers to assessing smart city solutions' contributions to disaster resilience. For example, it is difficult to collect real-time, accurate, available data (since government agencies tend to keep data within their organizations) on how smart city infrastructure contributes to disaster resilience, especially because many smart cities are still under construction. Moreover, privacy (surveillance capitalism), human rights, IT security threats (e.g., those linked with the deployment of 5G), issues of social bias in AI algorithms, technological limitations, and nature-based solutions (and how to integrate these in smart solutions) should be considered. Also, local norms and values should be considered when adopting new technologies. Some respondents urged that evaluation of different smart city initiatives must occur at different scales in different contexts with the same measures and indicators. For example, indicators such as crime rates and the number of surveillance cameras have been misused to target racial minorities.

6.4 Conclusion

Smart city initiatives enabled by ICTs are deemed essential for helping cities to develop transformative solutions to address the challenges of global climate change and to create equitable, sustainable, and resilient communities. An extensive body of knowledge has been developed on the contributions of such projects to sustainability and quality of life, but little is known about their contributions to climate resilience and climate action planning. Because many cities around the world are vulnerable to a wide array of climate-related disasters, enhancing resilience is expected to be at the center of smart city development efforts. Accordingly, this research project is focused on investigating the contributions of selected smart city projects to climate resilience. For this purpose, our research group first conducted a Delphi survey involving a panel of experts to develop a smart city resilience-assessment toolkit.

In the first stage of this research, the Delphi method was implemented to clarify three primary problems: (1) the meanings of urban resilience, smart city, and smart city resilience, (2) the framework for toolkits for assessing smart solution, and (3) the experts' opinions of the constituent indicators of assessment toolkits.

In the first part, the experts reached a consensus on the meanings of urban resilience, smart city, and smart solution. Urban resilience is the networked capability of the urban systems to recover from unpredicted shocks (e.g., natural disasters, climate change, and terrorism) and long-term disturbances; and smart city means that by relying on ICT and emerging technologies the city becomes more effective and efficient in terms of operations and urban management and is capable of improving quality of life. These previously vaguely defined terms were clarified through scholars' viewpoints. The Delphi study results also demonstrated that most participants agreed that smart city can help improve urban resilience (85%) and that ICT is an indispensable part of urban resilience. This could be the essence of the smart city resilience concept. Additionally, this research further explored the comprehensive characteristics of "urban resilience," which are robustness, diversity, redundancy, connectivity, flexibility, resourcefulness, agility (rapid response), efficiency, adaptive/learning capacity, modularity (independence and self-organization), creativity (innovation), equity, inclusive (participatory), and foresight capacity. A noteworthy finding was that all experts agreed that smart solutions can promote urban resilience through increasing robustness, flexibility, and resourcefulness; the characteristic of "equity" received the lowest agreement. This finding seems to indicate that equity cannot necessarily be solved through smart solutions based on ICTs.

The framework of assessment toolkits that can be used to evaluate smart city projects (smart solutions) was also explored in this research. A total number of 78 candidate indicators in seven categories were designed in the questionnaire, and the top 24 high-ranking indicators were chosen by the experts. Moreover, the weights of the seven categories were determined through the AHP method. The final results affirm that the relevance of these seven categories is sequenced as people, environment, governance, living, data, economy, and mobility. The assessment toolkit is planned to be tested in some smart city projects in Japanese and Taiwanese cities, and they may be used to compare various projects in other cities.

The third part of this research investigated the experts' opinions of the assessment toolkits. Most experts suggested that government (central and local), citizens, experts, companies, local media, and civil society representatives should be involved in the process to assess the contributions of smart city solutions to disaster resilience. In most cases, the toolkits seemed to be more meaningful and useful at the city level than at the community level. Some critical challenges for and barriers to assessing smart city solutions' contributions to disaster resilience were found in this research. For example, collecting real-time, accurate, and available data in most cities is difficult; in particular, collecting such data in cities in developing countries could be difficult. Data accessibility also may be a critical problem in older cities in well-developed countries. Additionally, privacy and human rights issues are often proposed as being critical by sociologists. Other challenges, such as IT security threats, the issues of social bias in AI algorithms, technological limitations, and considering local norms and values, should also be considered when implementing new technologies (smart city solutions) in urban governance. Despite these challenges, smart solutions have many potentials to contribute to resilience as was demonstrated during the recent pandemic (Hassankhani et al. 2021; Sharifi et al. 2021), and it is expected that research and practice on "smart city resilience" will receive more attention in future.

Acknowledgements This chapter was partially funded by Asia-Pacific Network for Global Change Research (Funder ID: https://doi.org/10.13039/100005536). We would also like to sincerely thank the experts who took time and contributed to the survey. Any errors of omission or commission remain the responsibility of the authors.

References

Baron M (2012) Do we need smart cities for resilience? [In:] Urban economic resilience-new concept for post-industrial city transition. J Econ Manag 10:32–46

Cavada M, Hunt D, Rogers C (2014) Smart cities: contradicting definitions and unclear measures. Paper presented at the World Sustainability Forum

Hassankhani M, Alidadi M, Sharifi A, Azhdari A (2021) Smart city and crisis management: lessons for the COVID-19 pandemic. Int J Environ Res Public Health 18(15):7736. https://www.mdpi.com/1660-4601/18/15/7736

Lee JH, Hancock MG, Hu M-C (2014) Towards an effective framework for building smart cities: lessons from Seoul and San Francisco. Technol Forecast Soc Chang 89:80–99. https://doi.org/10.1016/j.techfore.2013.08.033

Linstone HA, Turoff M (1975) The Delphi method: techniques and applications. Addison-Wesley Pub. Co., Advanced Book Program, Reading, MA

Saaty RW (1987) The analytic hierarchy process—what it is and how it is used. Math Model 9(3):161–176. https://doi.org/10.1016/0270-0255(87)90473-8

Saaty TL (1980) The analytic hierarchy process: planning, priority setting, resource allocation. McGraw-Hill International Book Co., New York and London

Saaty TL (1982) The analytic hierarchy process: a new approach to deal with fuzziness in architecture. Arch Sci Rev 25(3):64–69. https://doi.org/10.1080/00038628.1982.9696499

Sharifi A (2016) A critical review of selected tools for assessing community resilience. Ecol Indic 69:629–647. https://doi.org/10.1016/j.ecolind.2016.05.023

Sharifi A (2019) A critical review of selected smart city assessment tools and indicator sets. J Clean Prod 233:1269–1283. https://doi.org/10.1016/j.jclepro.2019.06.172

Sharifi A (2020a) A global dataset on tools, frameworks, and indicator sets for smart city assessment. Data in Brief 29:105364. https://doi.org/10.1016/j.dib.2020.105364

Sharifi A (2020b) A typology of smart city assessment tools and indicator sets. Sustain Cities Soc 53:101936. https://doi.org/10.1016/j.scs.2019.101936

Sharifi A (2020c) Urban resilience assessment: mapping knowledge structure and trends. Sustain (Switzerland) 12(15). https://doi.org/10.3390/SU12155918

Sharifi A, Khavarian-Garmsir AR, Kummitha RKR (2021) Contributions of smart city solutions and technologies to resilience against the COVID-19 pandemic: a literature review. Sustain 13(14):8018. https://www.mdpi.com/2071-1050/13/14/8018

Sharifi A, Yamagata Y (2016) Principles and criteria for assessing urban energy resilience: a literature review. Renew Sustain Energy Rev 60:1654–1677. https://doi.org/10.1016/j.rser.2016.03.028

Tyler S, Moench M (2012) A framework for urban climate resilience. Clim Dev 4(4):311–326. https://doi.org/10.1080/17565529.2012.745389

UN (2018) World urbanization prospects: the 2018 revision, Online edn. Department of Economic and Social Affairs PD, New York, NY

Zhou Q, Zhu M, Qiao Y, Zhang X, Chen J (2021) Achieving resilience through smart cities? Evidence from China. Habitat Int 111:102348. https://doi.org/10.1016/j.habitatint.2021.102348

Part II
Empirical Insights from Case Studies

Chapter 7
Resilient Smart Cities: Contributions to Pandemic Control and Other Co-benefits

Maria Rebecca Quintero and Ayyoob Sharifi

Abstract The COVID-19 pandemic disrupted daily lives and operations in many parts of the world. Being home to more than half of the world's population, cities were particularly hit hard by the pandemic. Different socioeconomic, institutional, and technological measures and policies have been adopted by cities in their efforts to control the pandemic. This chapter is focused on those measures and policies enabled by smart technologies and solutions. COVID-19 was the first global pandemic that occurred after digital revolution. It was, therefore, no surprise that smart technologies and solutions have been deployed at a large scale to deal with it. It is argued that this has even accelerated adoption of such technologies and solutions. By focusing on the planning, absorption, recovery, and adaptation capacities, this chapter discusses how smart solutions and technologies have contributed to resilience against the pandemic. In terms of planning, it is discussed that planning and existence of smart city infrastructure have enhanced different resilience characteristics such as connectivity, innovation, and resourcefulness that have helped some cities be less affected by the pandemic. These characteristics and availability and deployment of smart infrastructure have also enabled cities to absorb the initial shocks through, among other things, better tracing and tracking. Smart solutions and technologies have also enhanced resilience characteristics such as connectivity, creativity, agility, flexibility, and inclusion, thereby helping cities to resume their functionalities in a more timely manner. This, for instance, has been achieved through teleworking, telemedicine, automatic operations, etc. Lastly, contributions to adaptation had fostered connectivity, learning capacity, and flexibility. It is expected that the use of technology will lead to positive behavioral changes that may last even after the pandemic. Despite all these positive contributions, there are concerns about privacy and digital divide that need to be duly considered and addressed for more effective uptake and implementation of smart city solutions and technologies.

M. R. Quintero (✉)
School of Architecture and Built Environment, University of Newcastle, Newcastle, NSW, Australia

A. Sharifi
Graduate School of Humanities and Social Sciences, Hiroshima University, Higashihiroshima, Japan
e-mail: sharifi@hiroshima-u.ac.jp

A. Sharifi and P. Salehi (eds.), *Resilient Smart Cities*, The Urban Book Series,
https://doi.org/10.1007/978-3-030-95037-8_7

Keywords Smart city · Urban resilience · COVID-19 · Pandemic · Recovery · Adaptation

7.1 Introduction

7.1.1 COVID-19 Pandemic and Its Global Impacts

A novel coronavirus, identified as SARS-Cov-2, brought about the Severe Acute Respiratory Syndrome—Coronavirus Disease 2019 (COVID-19) pandemic (Lal et al. 2020). The earliest documented case of COVID-19 was from Wuhan, Hubei province, China on the 31st of December 2019 (Shi et al. 2020), initially identified by the Chinese Center for Disease Control and Prevention (Li et al. 2020). The symptoms varied from mild to severe such as, but not limited to, signs comparable to the common cold or flu, interminable chest pains, respiratory problems, confusion, and pale or bluish lips, nail beds, or skin (CDC 2020). It is a highly infectious disease in which lack of intervention had spread the SARS-Cov-2 from a patient to a median of three people, higher than the infectivity rate of two for Ebola and one for the common flu (Liu et al. 2020). Although the numbers of infected by SARS-Cov-2 were higher, its mortality rate was 3.44%, which was lower than other coronaviruses such as, the 9.19% rate of SARS-Cov and the 34.4% rate of MERS-Cov (Khanna et al. 2020). However, the upsurge in the number of infected cases and deaths globally, and the absence of human's pre-existing immunity to the virus, expeditiously turned this outbreak into a pandemic (Thomson 2020; World Health Organization 2020). On the 30th of January 2020, the World Health Organization (WHO) announced that COVID-19 is a Public Health Emergency of International Concern (PHEIC) (World Health Organization 2020) and was declared a pandemic by the 11th of March 2020, as it had proliferated to 113 countries (Sułkowski 2020; Khanna et al. 2020).

Moreover, COVID-19 had adverse economic, health, and social repercussions globally (Oldekop et al. 2020; Sułkowski 2020). Consequently, the outbreak had severely disrupted systems and processes such as, but not limited to, public health, the economy, global supply chains, demand for goods and services, transportation, virtual communication, and international travel and tourism (World Trade Organization 2020; Khan et al. 2020; Sułkowski 2020). Among the courses of action for the risk reduction of the COVID-19 pandemic were strict lock downs, isolation, social distancing, case detection, human contact tracing, quarantine, and widespread travel ban (Long and Feng 2020; Khanna et al. 2020). Worldwide, the drastic measures that countries and regions have adapted to limit the effect of the outbreak were diverse and have absolutely altered the way people live within a short length of time (Thomson 2020; Sułkowski 2020). Furthermore, the challenges faced due to COVID-19 have emphasized the importance of global strategy for development in which the focus deals with problems collectively to achieve equitable and sustainable solutions (Oldekop et al. 2020).

As of 8th of September 2021, WHO reported 221,134,742 COVID-19 confirmed cases and 4,574,089 confirmed deaths (World Health Organisation 2021). To date, COVID-19 vaccines have been already rolled out in which 5,352,927,296 doses were administered (World Health Organisation 2021). However, it has been argued that the response to the pandemic would be much improved by comprehensive partnership and collaboration in prioritizing the necessity of strengthening measures in the containment of the infection in communities, expeditious vaccination, and the call for the International Monetary Fund (IMF) and multilateral development banks to provide more emergency funding, with the agreement of the G20 member countries (Sachs et al. 2021).

7.1.2 Impacts of the Pandemic on Cities

Undoubtedly, the spread of COVID-19 in our globalized society meant that the novel disease not only affected countries in general, but devastatingly impacted people's lives in cities as well (UN Habitat 2020). As social beings, humans benefitted from cities through interactions, knowledge and information exchange, and cooperation (Kang et al. 2020). However, this pandemic provided a different challenge compared to diseases caused by other coronaviruses, such as the SARS-Cov and MERS-Cov, because these viruses were transmitted to humans through certain animals (Abe and Ye 2013), while SARS-Cov-2 was extremely infectious and transmissible between people, that everyday social interactions posed a high risk of further spread (Kang et al. 2020).

Aside from the medical response to this biological disaster, the protective social approaches such as, but not limited to, contact tracing, social distancing, travel bans, and lockdowns produced secondary problems, namely, the reorientation of communities' social structure in terms of work and everyday life, unstable housing costs, drastic economic changes, and concerns about privacy (Kang et al. 2020; Honey-Rosés et al. 2020). In fact, recent studies have identified socioeconomic impact as one of the categorical issues from this pandemic. While other impacts were determined from categories such as environmental quality, management and governance, and transportation and urban design (Sharifi and Khavarian-Garmsir 2020).

Moreover, cities have the highest number of confirmed COVID-19 cases, comprised of more than 90% (Pomeroy 2020). However, it was observed that the high density of the population of an urban area does not directly correlate to the increase in the number of cases. In fact, overcrowding and the management of urbanization were identified as factors which increased the likelihood of risk of infection in major urban areas such as Lagos, Lima, London, Madrid, Milan, New York, Paris, Sao Paolo, and Tokyo (UN Habitat 2020). Furthermore, the economic recession and the predicted decline of local government revenue by 15 to 20% meant less financial resources for urban development projects and services such as, but not limited to, public transport, poverty reduction, public healthcare, affordable housing, water, and sanitation

(Wahba et al. 2020). In addition, people who lived in overcrowded informal settlements and slum areas experienced further vulnerability due to inadequate, sometimes even non-existent, basic services and support that made it impossible to implement social distancing and follow recommended hygienic practices to counter the spread of the infection (Wahba et al. 2020).

Also, lockdown measures drove people to increase the use of technology and the internet for online schooling and working remotely. However, the pandemic also highlighted existing inequalities. For instance, the work from home approach succeeded with white-collar employees but, not for the workers in the informal sector who relied on their day-to-day income to survive (de Waal 2020). Additionally, students from poor families lack the technological amenities that were necessary for their online lessons and essential learning (UN Habitat 2020).

7.1.3 Urban Planning and Management Gaps and Difficulties

There are currently more people living in urban areas compared to rural areas (Carmichael et al. 2019; Lerner et al. 2018). The increasing global population is predicted to be absorbed by cities in the coming decades. To elaborate, in 2015 urban areas accommodated 54% of the total human population, and by 2036 it is estimated that it would rise up to 62% (United Nations 2018; McDonald et al. 2020). Studies have shown that the upward trend in urban population had critical consequences on land use (Angel et al. 2012), air quality (Cole and Neumayer 2004), water quality and security (Flörke et al. 2018), climate change and energy use (Rodríguez 2007; Güneralp et al. 2017), food security (Regmi 2001), and health and well-being (Carmichael et al. 2019). Furthermore, the strengthening recognition of the connection between these multidisciplinary factors demonstrated the value and the crucial role that cities and urbanization play in addressing international, national, and local development issues, resilience, and sustainable development. Also, as rapid urbanization and urban population continue to proliferate, the need for effective planning and management for cities become more significant (UN Habitat 2020).

However, urban planning and management is a multidisciplinary field, which deals with different sectors to implement extensive tasks and objectives, that would aid the community in attaining their development goals; so, there are numerous and complex challenges in it (Boarnet and Takahashi 2005). First, the knowledge and tools for planning and managing urban areas remains fragmented, which makes the approaches and solutions unreliable and confusing for cities (Polidoro et al. 2012). Second, research focus, data, and analysis about the urban environment, which urban planners could use for knowledge source, policy decision-making, and governance remain segregated (Duhr et al. 2020; Shwartz et al. 2014; Huang et al. 2018). In fact, it was argued that the incompatibility of data format, timescale, and resolution is more the reason for the disengagement between researchers and planners, and not data unavailability (Duhr et al. 2020). Third, the development policies from the national level down to the local level of governing bodies are disassociated, and there is a lack of essential key stakeholder consultation and participation (Huang et al. 2018). Fourth, the need for more comprehensive and effective master plans with strong spatial analysis that consider factors such as, but not limited to, urban sprawl, rapid unplanned urban growth (Polidoro et al. 2012), vital infrastructure aside from roads, appropriate locations for commercial and industrial sectors, and identification of unplanned areas (Huang et al. 2018). Additionally, political intervention and corruption in cities are concerns because it had an impact on the planning, implementation, and level of autonomy for planners in urban development projects (Polidoro et al. 2012). Moreover, the limited understanding and consideration for climate change and various disaster hazards and risks, such as this pandemic, are also a challenge that needs to be grappled with (Huang et al. 2018).

Against this background, the main objective of this chapter is to examine how smart city solutions and technologies have contributed to resilience against the COVID-19 pandemic. Before elaborating on this issue, we discuss some theoretical concepts related to resilience and smartness in the next section.

7.2 Theoretical Concepts

7.2.1 Resilience

Resilience is a concept that is extensively used in current literature, policy documents, reports, and decision-making, as complex interactions between and within various systems further acknowledge the relevance and necessity of maintaining their functions despite of internal and external shocks (Connelly et al. 2017). The term resilience first came from physics, which denoted an object's reaction from the impact of an extrinsic force (Peng et al. 2017). Then, studies in ecology from the 1970s used the term to pertain to a system's capacity to anticipate and recover from any external disturbances (Holling 1973; Cai et al. 2012). Moreover, in the context of urban areas, resilience was referred to as the capability of a city to absorb disruptions from its usual operations and still be able to revive and resume its operations regardless of the changes and disturbances (Lhomme et al. 2011; Alliance 2007; Folke et al. 2010; Wikström 2013; Sharifi et al. 2017).

Furthermore, there are differing ideas and definitions about resilience. Indeed, there is a lack of a unifying concept partly due to various technical languages used in different disciplines, which impede the effective sharing of ideas (Connelly et al. 2017). Even in the context of urban areas, contradictory definitions of resilience exist. This was derived from factors such as, but not limited to, various perspectives from social and ecological research, methodologies, dissimilarity on base ideas, and the ways of thinking and reasoning about it (Rezaei and Bastaminia 2016). Table 7.1 shows the different definition and concepts of resilience of cities, collated by Rezaei

Table 7.1 Definitions and concepts of resilience of cities (Rezaei and Bastaminia 2016)

Author	Definition
Holling (1996)	Urban resilience is further capacity or the ability of a city to digest disturbance or is the amount of disturbance that a system can digest it before the system structure is changed by changing variables
Buckle et al. (2000)	Quality of people, communities, agencies, and infrastructure reduces vulnerability. Not only lack of vulnerability but the capacity to prevent and reduce damages and then at the next stage maintenance of ideal conditions in cities as much as possible in case of incidence of harms, and then in third stage to recover from the effects
Gunderson et al. (2002)	The severity of disturbance that a city can absorb it before structure of cities to be converted to a different structure through change in variables and processes that are controlled by behavior
Cardona et al. (2003)	Capacity of damaged cities or ecosystems to digest the negative effects and rehabilitate them
Manyena (2006)	Capacity of a city at risks to become adjusted, resist, or change in order to reach an acceptable level of operations and structure and its continuation. This is determined by a degree that the social system is capable of organizing and increasing capacity, learning from past disasters and improving assessments of reducing the possibility of its own risk
Cutter et al. (2010)	Urban resilience is called to absorption capacity and basic and special performances, as well as capacity of recovery, "Return to Balance", after disaster
Pelling (2003)	Urban resilience is the ability of biological systems and organisms to resist or return to normal status against shocks, disasters, diseases, and other changes
Birkmann et al. (2013)	Urban resilience is considered capacity of damaged communities or ecosystems to digest the negative effects and rehabilitate them
Turner (2014)	Change process of strengthening the capacity of population, communities, organizations, and forecasting, prevention, recovery and change of cities after the occurrence of shock, stress, and changes
Kärrholm et al. (2014)	The severity of disturbances that a city can absorb it before structure of the system to be converted to a different structure through change in variables and processes that are controlled by behavior
Hodson and Marvin (2014)	Resilience is called the capacity of cities to absorb disturbances and to keep necessary and inherent feedbacks, processes, and structures of the city

and Bastaminia (2016). Despite the definitions from various authors, the study had identified two common properties from their analysis, which were: first, resilience was regarded more as a means rather than an end goal. Second, urban resilience was more effective when it was flexible to change and less resistant.

Moreover, as different definitions of resilience were explored, there were prevailing features that were present. These features were critical functions or services, thresholds, time and scale, and memory and adaptive management (Connelly et al. 2017). In addition, published literature have identified capacities of resilience which were absorptive, adaptive, and restorative (Lhomme et al. 2013; Proag 2014; Fiksel 2003; Rose 2007; Tongyue et al. 2015; Sharifi 2016). Also, in urban and disaster studies, the planning and preparation capacity was deemed crucial in decreasing the impact of disasters and was linked closely to adaptive capacity (Rose and Liao 2005). More to the point, the features that were analyzed from the definitions of resilience by Connelly et al. (2017) and the resilience capacities discerned by various studies were presented in Table 7.2, which shows the relationship of how resilience was perceived and its application in different disciplines.

To elaborate, the ability to plan and prepare is strongly related to the critical functions of a system (Connelly et al. 2017). In fact, the key stakeholders determine the scale, connectivity, and the level of priority of these critical functions when considering the vulnerabilities, hazards, and risks that would possibly disrupt the system (Cumming et al. 2006). Then, the absorption capacity is linked to thresholds due to the robustness of the critical functions and the system (Connelly et al. 2017) and has significant influence on its adaptive capacity (Fath et al. 2015). The threshold serves as the system's limitation from disruptions and once this is exceeded, the unrecovered changes in the system are adapted and different function emerges (Dakos et al. 2008). Additionally, the recovery ability of resilience correlates to the amount of time it takes to recover the critical functions and the scale of the impact of the disruption (Allen et al. 2014). Ideally, agility in recovery is deemed important to restore the system's services with no change or as little change as possible (Connelly et al. 2017). Also, the adaptive capacity places emphasis on the flexibility of the system, not just on the known external shocks but, on the uncertain possibilities of other disturbances as well (Fath et al. 2015). Moreover, foresight is also an essential resilience characteristic in envisioning alternative solutions and investment on planning and preparing for future disruptive events (Park et al. 2013; Sharifi and Yamagata 2016).

7.2.2 Common Smart City Solutions

As the urban population grows and cities are perceived as essential for the economy, there is also an increase in the demand for basic public services, healthcare, education, water, energy, and sanitation (Rao and Prasad 2018; Ramaprasad et al. 2017; Neirotti et al. 2014). The aim to improve the population's quality of life is a race with the rapid urbanization and complex system of cities, and one of the ways to handle such

Table 7.2 Resilience features common to socio-ecology, psychology, organizations, and engineering and infrastructure, which are related to the temporal phases from the National Academy of Sciences (NAS) definition of resilience (Connelly et al. 2017)

NAS phase of resilience	Resilience feature	Description by application domain			
		Socio-ecological	Psychological	Organizational	Engineering and infrastructure
Plan	**Critical functions (services)**	A system function identified by stakeholders as an important dimension by which to assess system performance			
		Ecosystem services provided to society	Human psychological well-being	Goods and services provided to society	Services provided by physical and technical engineered system
Absorb	**Thresholds**	Intrinsic tolerance to stress or changes in conditions where exceeding a threshold perpetuates a regime shift			
		Used to identify natural breaks in scale	Based on sense of community and personal attributes	Linked to organizational adaptive capacity and to brittleness when close to threshold	Based on sensitivity of system functioning to changes in input variables
Recover	**Time (and scale)**	Duration of degraded system performance			
		Emphasis on dynamics over time	Emphasis on time of disruption (i.e., developmental stage: childhood vs adulthood)	Emphasis on time until recovery	Emphasis on time until recovery
Adapt	**Memory/adaptive management**	Change in management approach or other responses in anticipation of or enabled by learning from previous disruptions, events, or experiences			
		Ecological memory guides how ecosystem reorganizes after a disruption, which is maintained if the system has high modularity	Human and social memory, can enhance (through learning) or diminish (e.g., post-traumatic stress) psychological resilience	Corporate memory of challenges posed to the organization and management that enable modification and building of responsiveness to events	Re-designing of engineering systems designs based on past and potential future stressors

Table 7.3 Comparison of key attributes in the first- and second-generation smart city paradigms (Trencher, 2019)

	Smart city 1.0	Smart city 2.0
Focus of vision	Technology and economy	People, governance, and policy
Role of citizens	Passive role as sensors, end-users, or consumers	Active role as co-creators or contributors to innovation, problem-solving, and planning
Objective of technology and experimentation	• Optimize infrastructures and services • Serve demand-side interests and spur new business opportunities • Address universal technical agendas (energy, transport, economy)	• Mitigate or solve social challenges • Enhance citizen wellbeing and public services • Address specific endogenous problems and citizen needs
Approach	• Centralized (privileged actors) • Exogenous development	• Decentralized (diverse actors) • Endogenous development

challenges is to make the city smarter and efficient (Rao and Prasad 2018; Sharifi 2019, 2020).

There is a lack of common definition for a smart city (Rao and Prasad 2018; Ramaprasad et al. 2017). When its concept started in the last part of the twentieth century, the focal point was on the development of infrastructure and technology (Woetzel and Kuznetsova 2018; Ramaprasad et al. 2017; Neirotti et al. 2014; Trencher 2019). However, the current evolution of the idea also focuses on its effects, such as, quality of services, sustainable development, and improvement on the lives of the residents (Anthopoulos 2015; Woetzel and Kuznetsova 2018; David and Koch 2019; Papa et al. 2013; Trencher 2019). In Table 7.3, the development of the concept of smart cities shows through the differences between Smart city 1.0 and Smart city 2.0 in terms of vision, citizen role, objectives, and approach (Trencher 2019).

Additionally, it is argued that for a lot of cities, being a smart city is a complex challenge and an aspiration (Ramaprasad et al. 2017). Aside from being an aspiration, the interpretation of the concept varies per city or region and is contingent on a city's resources, development, and preparedness for the transformation of some of its systems and processes (Rao and Prasad 2018; Neirotti et al. 2014).

Primarily, one of the basic requirements for a smart city is the information and communication technology (ICT), as it integrates and manages all the factors in the system. In fact, ICT enables connectivity, which is fundamental to utilize and coordinate multiple resources within the urban system (Rao and Prasad 2018; Westraadt and Calitz 2018; Papa et al. 2013). However, it is argued that investment in ICT alone could not transform an area into a smart city but, it would also need its citizen's learning capacity and innovation to promote change (Neirotti et al. 2014). In addition, the internet of things (IoT) is another solution which makes it possible to link together control and monitoring devices such as, but not limited to, actuators, sensors,

RFID tags, mobile devices, and cameras, which are used to collect real-time data to be utilized and actioned appropriately (Rao and Prasad 2018; Westraadt and Calitz 2018). For instance, the smart home is a popular example of IoT, in which an infrared module is applied to control home machines and appliances (Jeong and Park 2019). Moreover, big data analytics is also a smart city solution, consequently from off-line and online datafication, and could even be obtained from cloud computing (Westraadt and Calitz 2018). Big data refers to the huge amount of information and details gathered about the citizens, potentially utilized for the management of urban systems and processes (Bibri and Krogstie 2020; David and Koch 2019). Additionally, the synthesis of data analytics, real-time monitoring, event management, and participatory technology aims for effective and well-organized delivery of public services and governance (Westraadt and Calitz 2018).

Moreover, it is argued that solutions for a smart city are not found only in the tangible elements or hard domains, and that the intangible elements or soft domains are also essential (Papa et al. 2015; Neirotti et al. 2014). Table 7.4 summarizes the hard domains, which are the infrastructure- and technology-focused solutions, while soft domains refer to the public services necessary to improve institutional position and social environment (Neirotti et al. 2014).

7.2.3 Framework for Enhancing Urban Resilience Through Smart Solutions

As urban areas are rapidly changing and expanding, the risks from disasters in cities have increased as well. Cities have experienced, and continues to experience, the consequences of climate change, unsustainable practices, and numerous hazards that hinder their development. In addition, existing socioeconomic problems have added to the burden of finding solutions to urban issues. With the consensus that cities play a critical role in global development (Rao and Prasad 2018), there is an urgent necessity for enhancing their resilience against any disruptions.

While cities have made efforts to enhance resilience, there has also been growing interest in smart city initiatives. The smart city concept was developed with the goal of improving the quality of life of citizens (Woetzel and Kuznetsova 2018; David and Koch 2019). Smart city solutions have a wide range of scope, which aim to enhance systems and services. Through the years, these types of solution have evolved and have focused on addressing different urban problems.

In relation to the points discussed, the authors developed a framework for enhancing urban resilience with the utilization of smart city solutions. The framework is illustrated in Fig. 7.1. The framework revolves around four major resilience abilities/stages, namely, planning, absorption, recovery, and adaptation (Sharifi et al. 2021).

To elaborate, the framework builds on the concept of urban resilience as a process, instead of a goal, and emphasizes adaptation to changes (Rezaei and Bastaminia

Table 7.4 Classified literature on the domains of a Smart City (Neirotti et al. 2014)

Prevalence of investment in:	Domain	Main objectives	References
"Hard" domains	Energy grids	Automated grids that employ ICT to deliver energy and enable information exchange about consumption between providers and users, with the aim of reducing costs and increasing reliability and transparency of energy supply systems	Chourabi et al. (2012), Correia and Wünstel (2011), Mahizhnan (1999) and Steria-Smart City (2011)
	Public lightning, natural resources, and water management	Managing public lighting and natural resources. Exploiting renewable resources such as heat, solar, cooling, water, and wind power	Accenture (2011), Correia and Wünstel (2011), Dirks et al. (2009), Hughes et al. (2013), Nam and Pardo (2011), The Climate Group et al. (2011), Think (2011) and Toppeta (2010)
	Waste management	Applying innovations in order to effectively manage the waste generated by people, business and city services. It includes waste collection, disposal, recycling, and recovery	Accenture (2011) and The Climate Group et al. (2011)
	Environment	Using technology to protect and better manage environmental resources and related infrastructure, with the ultimate goal of increasing sustainability. It includes pollution control	Atzori et al. (2010), Caragliu et al. (2009), Chourabi et al. (2012), Inayatullah (2011), Nam and Pardo (2011), Tiwari et al. (2011)

(continued)

Table 7.4 (continued)

Prevalence of investment in:	Domain	Main objectives	References
	Transport, mobility, and logistics	Optimizing logistics and transportation in urban areas by taking into account traffic conditions and energy consumption. Providing users with dynamic and multi-modal information for traffic and transport efficiency. Assuring sustainable public transportation by means of environmental friendly fuels and innovative propulsion systems	Atzori et al. (2010), Caragliu et al. (2009), Correia and Wünstel (2011), Dirks et al. (2009), Giffinger et al. (2007), La Greca et al. (2011), Munuzuri et al. (2005), Nam and pardo (2011), Steria-Smart City (2011), The Climate Group et al. (2011), Think (2011), Toppeta (2010) and Washburn et al. (2010)
	Office and residential buildings	Adopting sustainable building technologies to create living and working environments with reduced resources. Adapting or retrofitting existing structures to gain energy and water efficiency	Accenture (2011), Steria-Smart City (2011), The Climate Group et al. (2011), Think (2011), Washburn et al. (2010)
	Healthcare	Using ICT and remote assistance to prevent and diagnose diseases, and deliver the healthcare service. Providing all citizens with access to an efficient healthcare system characterized by adequate facilities and services	Accenture (2011), Atzori et al. (2010), Correia and Wünstel (2011), Dirks et al. (2009), Nam and Pardo (2011), The Climate Group et al. (2011) and Washburn et al. (2010)
	Public security	Helping public organizations to protect citizens' integrity and their goods. It includes the use of ICTs to feed real-time information to fire and police departments	Accenture (2011), Dirks et al. (2009), Nam and Pardo (2011) and Washburn et al. (2010)

(continued)

Table 7.4 (continued)

Prevalence of investment in:	Domain	Main objectives	References
Soft domains	Education and culture	Capitalizing system education policy, creating more opportunities for students and teachers using ICT tools. Promoting cultural events and motivating people participation. Managing entertainment, tourism and hospitality	Accenture (2011), Dirks et al. (2009), Mahizhnan (1999), Nam and Pardo (2011) and Washburn et al. (2010)
	Social inclusion and welfare	Making tools available to reduce barriers in social learning and participation, improving the quality of life, especially for the elder and disabled. Implementing social policies to attract and retain talented people	Atzori et al. (2010), Bakıcı et al. (2013), Caragliu et al. (2009), Chourabi et al. (2012), Correia and Wünstel (2011), Giffinger et al. (2007), Mahizhnan (1999) and Toppeta (2010)
	Public administration and (e-) government	Promoting digitized public administration, e-ballots and ICT-based transparency of government activities in order to enhance citizens empowerment and involvement in public management	Accenture (2011), Bakıcı et al. (2013), Caragliu et al. (2009), Chourabi et al. (2012), Correia and Wünstel (2011), Dirks et al. (2009), Giffinger et al. (2007), Odendaal (2003), Steria-Smart City (2011), Think (2011), Toppeta (2010) and Washburn et al. (2010)
	Economy	Facilitating innovation, entrepreneurship and integrating the city in national and global markets	Bakıcı et al. (2013), Caragliu et al. (2009), Chourabi et al. (2012), Correia and Wünstel (2011), Giffinger et al. (2007), Mahizhnan (1999) and Toppeta (2010)

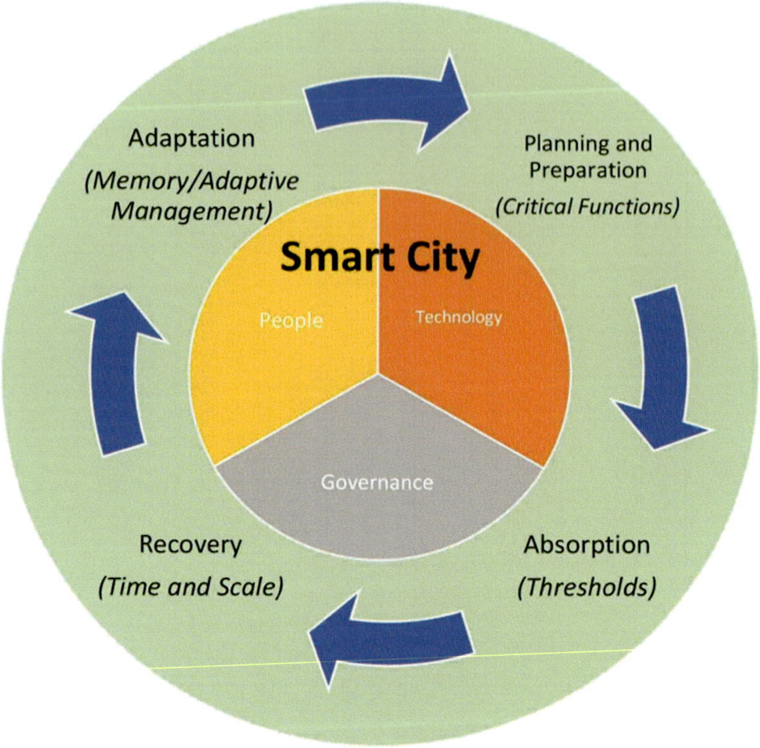

Fig. 7.1 Smart city solutions enhancing urban resilience (by author)

2016). The idea of resilience as a process and its flexibility to uncertainties are shown through the cycle of arrows. This is a nonlinear process, and each phase is dependent on the features of resilience that were impacted by disturbances on the system.

In planning and preparation, the critical functions are determined by key stakeholders. These stakeholders decide which services will be prioritized in case of disruptions. In the event of a disaster, absorption is the phase in which thresholds determine the limit on how much a city could withstand disturbances. Then, once the threshold is exceeded, recovery would be the next process. In this situation, the scale of the damage and the time it would take the city to resume its impacted critical functions would determine its recovery. However, there are cases in which recovery is impossible and adaptation would be the ideal process instead. In adaptation, it would be based on how the city learns from previous disasters and implement better processes for the future. Additionally, due to the cyclical process of resilience, adaptation links back to planning and preparation. This reflects that the knowledge gained from adaptation is strongly considered in planning for future uncertainties of disaster risks.

Moreover, the framework shows that the smart city is composed of different factors. These factors are technology, people, and governance (Skouby and Lynggaard 2014; Woetzel and Kuznetsova 2018; United Nations 2018). This builds on the concepts of previous studies that smart cities are not only about advancements in technology but, comprised of the people and effective governance as well. For smart city solutions to be beneficial, it is necessary to apply a comprehensive understanding of the system instead of addressing just individual urban problems and disruptions.

Furthermore, the framework demonstrates that at the core of the urban resilience as a process is the smart city and its solutions that enhance it. By adopting the nonlinear process of resilience and placing emphasis on the different factors of what makes a city smart, it strengthens urban resilience and fosters sustainable solutions for the future.

7.3 Contributions of Smart Cities to Resilience

7.3.1 Contributions to Planning and Preparation

Unexpected disasters and other catastrophic events, such as the COVID-19 pandemic, have drastically impacted cities (Yao and Wang 2020). Numerous endeavors and research were made to prevent, mitigate, respond, and recover from disasters. However, there is still a lot of work to be done to be able to successfully manage such difficulties (Shahat et al. 2020). Moreover, the increasing popularity of smart cities brought specific consideration on the function and effects of information and technology (ICT) in planning and preparation for dealing with disaster risks (Papa et al. 2013). Additionally, big data and ICTs are proving to be essential in disaster mitigation and support timely recovery (Yao and Wang 2020). Also, the population's capacity to learn from previous disasters not only resulted in further innovation of technological pursuits, but also strengthened the key role of human capital in a connected and inclusive city (Papa et al. 2015).

In the context of urban resilience, connectivity and innovation are major characteristics that have developed extensively through smart city solutions. Innovation brought about not only new technologies, products, and tools but, new courses of actions as well (UN Habitat 2020). Connection and exchange of information and data between citizens and governing bodies have aided decisions on what to plan and prepare for in times of crisis (Johnson et al. 2020; Yao and Wang 2020). In fact, cities that have effective systems for infectious disease surveillance through data analytics could reduce mortality rate by 5% (McKinsey Global Institute 2018). Predominantly, the cities that were a step ahead in preparation for the COVID-19 pandemic were those that had better connectivity, adoption, and inclusive system (Costa and Peixoto 2020).

Moreover, the adaptive learning applied from previous disease outbreaks due to experience has exhibited the resourcefulness of some cities in their preparation for another crisis like the COVID-19 pandemic, and the use of IoTs was in the forefront. For instance, the South Korea Digital Health was one of the initiatives from the country's adoption of the smart city concept years ago (Costa and Peixoto 2020; Whitelaw et al. 2020). Smart city solutions, such as Artificial Intelligence (AI), consumer health electronics, big data, block chain, and telemedicine were utilized in withstanding the COVID-19 outbreak (Whitelaw et al. 2020). In addition, a website with a Corona map gave information to people on which areas had confirmed cases. Also, apps such as Corona 100 m and a self-quarantine app were used to let citizens be aware of their proximity from infected areas and make sure they were following quarantine protocols, respectively (Wray 2020; UN Habitat 2020; Söderström 2020).

Similarly, Singapore's preparation for this pandemic was due to the SARS outbreak they had more than a decade ago (Costa and Peixoto 2020). The country had the TraceTogether app in place, which was part of their Smart Nation program. This program utilized various new technologies to provide smart solutions and improve the residents' quality of life. For instance, with the TraceTogether app installed in mobile phones, it helped detect people who were in close contact with confirmed cases and shared that data with the Ministry of Health for easy identification (Tandoc Jr and Lee 2020; Söderström 2020). To elaborate, the Bluetooth signal exchange in residents' mobile phones, when they were in close proximity from each other, would be data that was registered and kept for 21 days, and the public health department had access to it for their tracking of cases (Whitelaw et al. 2020). In addition, websites and messaging apps were used by the government to educate and inform the people in managing COVID-19 (Costa and Peixoto 2020). Similarly, a company in Romania developed an app called CovTrack, which utilized the same Bluetooth technology to track and trace people that have been infected (Tešić et al. 2020).

Additionally, Germany had used big data collated from a smartwatch application they developed to track people's health condition digitally, as one of their pandemic responses. This smart technology made it possible for the government to record an individual's sleep pattern, pulse, and temperature, and applied that data in analyzing the country's collective risk and status of COVID-19 infection situation, and planning for the best appropriate approach (Busvine 2020).

Another case is Taiwan, in which their well-prepared and efficient action to control the outbreak resulted in low number of infected cases and deaths (Wang et al. 2020). Big data was used ubiquitously as immigration and health insurance records were integrated, checked, and analyzed, which helped in contact tracing and identified people who needed to be tested (Whitelaw et al. 2020).

The application of several smart city solutions for the planning and preparation of urban areas to deal with disease outbreaks has aided in fighting the spread of COVID-19. However, the advantages from smart solutions in functions such as tracking, screening, and contact tracing to manage the infection have also presented some disadvantages. It has been argued that privacy, data security, lack of management and regulation, high production and maintenance costs, limited information on asymptomatic carriers, and unequal opportunity for people that have no mobile device or internet connection are some potential trade-offs that need to be taken into account (Whitelaw et al. 2020).

7.3.2 Contributions to Absorption

Aside from connectivity, smart city solutions have enhanced the robustness, flexibility, and resourcefulness of cities during the pandemic. Additionally, smart city solutions were crucial in maintaining connections and kept people productive while in lockdown or physically apart (UN Habitat 2020). The COVID-19 outbreak has facilitated the advancement in technology and the population's adoption of it (Xiao and Fan 2020; Chagoury 2020; Berg 2020). In fact, the drastic effect on communities, economy, and healthcare system has made the people and local governments more accepting of digital and technological tools to contain the spread of the disease and continue their city's critical functions as much as possible (Chagoury 2020). The smart city solutions that rapidly developed during the pandemic lockdown and physical distancing were in ICT and IoT systems such as digital and contactless payments, telehealth, remote work, online learning, robot deliveries, online shopping, and online entertainment (Xiao and Fan 2020; UN Habitat 2020).

Due to the infectious nature of COVID-19, along with the strict lockdown and social distancing measures, flexibility and resourcefulness of some regions were clearly demonstrated in the use of smart technology for telemedicine, with the aim of reducing the burden on clinics, hospitals, and health professionals in dealing with the pandemic (Wang et al. 2020; Whitelaw et al. 2020). It was reported that in the United States, almost half of the number of patients have utilized communication services such as Skype and Facetime to consult with a medical professional (Stokel-Walker 2020; McKinsey Global Institute 2018). In another case, Canada has recorded that from February to May 2020, the number of patients who consulted doctors through video conference calls increased from 1000 to 14,000 visits per day (Whitelaw et al. 2020).

Moreover, remote work and online learning have shown that resilient cities have robustness due to the aim of maintaining critical functions for the benefit of the economy and education. In this context, smart solutions such as, but not limited to, virtual meetings, virtual reality, AI-enabled robot teachers, virtual private networks, cloud technology, work collaboration tools, 3D printing, and facial recognition tools were enabled to mitigate the impact of the lockdowns as much as possible (Xiao and Fan 2020). However, the smart city technologies also reinforced the existing inequalities within cities as considerable number of the population lacked access to such innovations and faced difficulties with remote work and online learning (UN Habitat 2020). In fact, prior to the pandemic, it was recorded that high-income regions still had better conditions in terms of internet access, mobile phones, and wearable smart devices compared to both middle- and low-income regions (Bahia and Suardi 2019).

Additionally, the use of smart city solutions was not only identified through government initiatives. As a response to the lockdown, a case study has identified that some people from the tourism sector have utilized the digital sharing of copyrighted photos and videos, along with Google Maps, to continue their work through virtual tours despite the situation (Costa and Peixoto 2020). Furthermore, heritage sites and museums provided virtual tours as well, as just some of the numerous entertainments available online (Xiao and Fan 2020).

Moreover, in the case of Brazil, the lack of coordination of the national and local governments, along with overcrowded population and poor living conditions, has drastically increased the infected cases and exacerbated the impacts of COVID-19 in the country (Costa and Peixoto 2020). Due to the inaccurate reports of the public sector about the numbers affected by the pandemic, with the support of academic and scientific institutions, some groups utilize big data and IoTs for gaining better knowledge of the conditions and providing some solutions for the citizens (Urban and Nakada 2021). Smart solutions such as real-time maps of reported COVID-19 infection and open data source were readily accessed by the population to provide information of high-risk areas. In addition, several universities in the country developed a system that integrated all the pandemic data and executed statistical analysis of the population and the impact of the disease in the context of unemployment, education, violence, and poverty (Costa and Peixoto 2020). Such situations exhibited the modularity or self-organization characteristic that was decentralized and initiated locally.

7.3.3 Contributions to Recovery

In urban resilience, recovery is a capacity that is determined by the time it takes to return to normal conditions. This is influenced by the state of preparation and also the scale of a disruption (Connelly et al. 2017). Big data analytics and real-time information about a disaster could enhance the agility, connectivity, creativity, flexibility, and participatory characteristics of resilience. For example, the use of early warning systems provides cities with relevant data about the disaster and enables them to conduct more comprehensive analysis for better decision-making, damage assessment, relief efforts, and re-establishment of critical functions (Zhu et al. 2019; Yao and Wang 2020; Woetzel and Kuznetsova 2018).

The advancement of AI and its application in scientific research tools, robots, smart homes, toys, and medical diagnosis has been crucial in the development of services that affect our daily lives (Skouby and Lynggaard 2014). During the pandemic, robot-based delivery was one of the smart city solutions implemented as an answer to contactless delivery. In China, the increase in online shopping has encouraged major companies to deliver their service by robots (Xiao and Fan 2020). Additionally, drones were used to deliver essential supplies, provide alerts and information during lockdown and quarantine, and sanitize areas. Similarly, in Hong Kong (Costa and Peixoto 2020) and Nigeria (Chagoury 2020), the use of robots was considered safer in disinfecting public buses and trains and presented no risk in spreading the infection (Costa and Peixoto 2020; Scott and Coiera 2020; Chagoury 2020). These types of approaches showed that implementing smart technologies enhances creativity and agility and facilitates rapid management of the pandemic impact. Additionally, in Denmark, robots that used UV light to sanitize were positioned strategically in airports, hotels, offices, universities, government buildings, and hospitals to provide less burden to frontline workers (Bedi 2020).

Smart city solutions also had a positive contribution to the COVID-19 vaccine development. The efficient collation and sharing of data about every new disease provided more critical information and knowledge that was made possible by ICT-enabled systems (Rabi et al. 2020). Traditionally, the development of a drug for the treatment of a new disease could take years, or even decades, due to the overwhelming amount of data and necessary experiments to be done. In the case of the COVID-19 vaccines, AI had already sorted out the essential data and conducted calculations at a rapid pace (Smart City 2020; Ibrahim 2020). In fact, deep-learning based algorithms facilitated the recognition of potential SARS-CoV-2 protein structures just in several weeks (Malone et al. 2020). Also, an open data source, known as the COVID-19 Open Research Dataset (CORD-19), provided access to scientific research groups to a large number of documents on coronaviruses, including SARS-CoV-2 (Scott and Coiera 2020).

7.3.4 Contributions to Adaptation

The uncertainty and complexity of the impact of the COVID-19 pandemic worldwide reinforced the importance of a multidisciplinary approach to finding sustainable solutions (Pan and Zhang 2020). The capacity of cities previous to learn from previous disasters, and this novel coronavirus outbreak, had accelerated the trends toward utilization of smart city services (Chagoury 2020). Indeed, this crisis had demonstrated the rapid response and behavioral change of the urban population in terms of economic and business model, security, data privacy, and management of information (Davison 2020).

In the case of the Pimpri-Chinchwad Smart City, located at the region of Pune, the Pimpri-Chinchwad Municipal Corporation (PCMC) officials had adapted the smart technology and facilities into an Integrated Command and Control Centre (ICCC) to manage the COVID-19 outbreak in their locality (Costa and Peixoto 2020; Guha et al. 2020). The application of IoTs such as CCTV, real-time dashboard, GIS mapping, monitoring, and big data analytics had helped in dealing with the impact of the pandemic. Aside from the COVID-19 tracking and tracing functions, PCMC also implemented the SARATHI (System of Assisting Residents and Tourist through Helpline Information) which was an assistance platform to receive complaints or assistance requests from the population and was responded accordingly through smart analytics and categorization. In addition, the flexibility and learning capacity of other cities were demonstrated in the establishment of 45 more ICCCs across India. The crucial role that the ICCCs played in urban governance and management of the current disaster were recognized, which made the adaptation of such approach an essential consideration for future disasters (Guha et al. 2020).

Additionally, the public transport and freight sectors had experienced massive economic losses during the pandemic. The severe restrictions and lockdowns have suspended all transportation that posed a high risk for the spread of infection (Amir et al. 2020). Despite the drastic disruption, smart technology had fostered innovation, foresight capacity, and flexibility in developing solutions that adapted to the changes and uncertainty that this biological disaster had caused. For instance, the reduction in air pollution and improvement of air quality due to the halt in travel during the lockdown have strengthened the interest and investment of some cities in the UK for smart public transport. There is a massive plan to launch 3000 hydrogen buses in the cities of London, Aberdeen, Bimingham, Brighton, Liverpool, Manchester, Belfast, Glasgow, and Edinburgh, in partnership with the companies, Wrightbus and Ryse Hydrogen. This is in line with the goal to maintain the improvement of air quality even when the pandemic is over (Barrett 2020).

In India, the ride-hailing app Efleet had partnered with auto-rickshaws in the city of Jammu. The safety precautions such as installed protective screen between the driver and passenger areas, daily temperature checks for drivers, and provision of essential safety supplies such as face masks, sanitizers, and gloves were adapted to ensure hygienic transport (Amir et al. 2020). The use of the app provided an ease of transaction between consumers and the company and assured proper hygiene and safety measures for everyone. A case such as this reflects the prioritization of health approaches and related functions as one of the guiding principles in the adaptation and resilience of cities from future uncertainties (UN Habitat 2020).

Moreover, the connectivity between citizens and government was reinforced through smart technologies and has redefined the population's role as direct or indirect contributors to the urban resilience (Johnson et al. 2020). In fact, the important roles of citizen participation and the bottom-up approaches in smart cities are increasingly recognized. For instance, urban labs were established as collaborative efforts between cities, universities, tech companies, and local communities that aim to address urban issues. Some of the urban labs worldwide are MIT Senseable City Lab, University of Chicago's Urban Labs, Hyderabad Urban Lab, C4D Lab (University of Nairobi) and iHub in Nairobi, Laboratorio de Innovación Quito (LINQ), LABcapital in Bogotá, and the Laboratorio para la Ciudad in Mexico City (UN Habitat 2020). Furthermore, open data portals and open calls for innovative solutions could aid in improving public services and foster urban development. However, such approaches have also raised concerns about the standard, structure, and trustworthiness of the open data, which are challenges that need to be examined (McKinsey Global Institute 2018).

7.4 Discussions

The infectious nature of SARS-Cov-2 and the rapid spread of the COVID-19 disease worldwide had escalated it into a pandemic in a short amount of time. Its global impact had drastically affected systems and processes such as, but not limited to, the economy, global supply chains, transportation, and public health. Moreover, as centers of economic activities and other processes, cities were also severely impacted by the disease outbreak. Additionally, this biological disaster had highlighted existing urban planning issues and even exacerbated these issues in some instances. Undoubtedly, the necessity for effective response and management of this disaster and future disasters is crucial for the resilience of cities. In fact, it is argued that the pandemic has offered an unprecedented opportunity to learn lessons for improving urban resilience (Sharifi 2021).

In this chapter, the concepts of resilience and smart cities were explored. The definition of resilience, in the context of urban areas, was discussed with its phases of planning and preparation, absorption, recovery, and adaptation. Also, the definition of the smart city, along with the common smart city solutions, was reviewed. The smart solutions initially discussed were Big Data Analytics, Information and Communication Technology (ICT), and Internet of Things (IoTs). However, it was identified that smart city is not just all about advancements in technology and data. Furthermore, a framework for strengthening the resilience of urban areas due to smart city solutions was developed and analyzed.

Moreover, the contributions of smart city solutions to urban resilience were explored. These discussions were categorized according to its contribution to the phases of resilience. Also, some actual cases of smart city solutions were reviewed, and these are in the context of various cities response to the COVID-19 pandemic.

There is no general definition for resilience. In the urban context, resilience is understood as the ability and process of a city in dealing with external and internal shocks, and its flexibility to unknown disturbances. Resilience was analyzed in this study as a process in terms of its capacity to plan and prepare, absorb, recover, and adapt. Each phase had features which helped define it. Planning and preparation were based on the system's critical functions. Absorption depended on the threshold present in absorbing shocks. Recovery was determined through time and scale of the disturbance. Lastly, adaptation was based on memory, learning from, and management of functions that could not be recovered.

Like resilience, there was also no common definition for smart cities. Primarily, the goal of a smart city is to improve the quality of life of its population and improve operational efficiency. Initially, the concept of smart cities was focused on technology and infrastructure. This concept evolved as numerous studies on smart cities were conducted. The current idea is that although smart cities are heavily influenced by the advancement and utilization of technology, as demonstrated by the ICTs, IoTs, big data, etc., there is a growing recognition of the importance of the people and effective governance in smart solutions. The bottom-up approach and participatory measures are now deemed essential in the application of smart solutions to urban issues.

Additionally, a framework about the concepts of resilience and smart city was developed. This illustrated the essential components of smart city and its solutions. Also, it showed the role of these solutions to urban resiliency and how it related to planning and preparation, absorption, recovery, and adaptation.

Moreover, the contributions of smart city solutions to urban resilience were explored through its different phases, and in the context and real-world cases of

COVID-19 pandemic control. First, its contributions to planning and preparation have enhanced the resilience characteristics of connectivity, innovation, and resourcefulness. The utilization of big data and ICTs by different countries for functions such as contact tracing, tracking, and testing proved effective in early infection detection and management of the spread. However, concerns on data privacy, security, lack of regulation, and costs were also observed from the case studies.

Second, smart city solutions strengthened resilience characteristics such as connectivity, robustness, flexibility, and resourcefulness, when contributions to absorption were analyzed. The strict lockdown and physical distancing approaches brought about online learning and remote work. These initiatives were focused on maintaining critical functions in economy and education as much as possible. Smart solutions such as telemedicine, virtual meetings, work collaboration tools, social media, and other communication services were necessary to absorb the shock from the sudden distance and isolation. Additionally, government initiatives were not the only ones that applied smart solutions. Resilience was identified when bottom-up approaches by the people were initiated as they saw a gap in the situation. For instance, the case of Brazil in which smart technology was utilized by academic and scientific institutions to provide accurate, real-time updates on the COVID-19 pandemic to local communities.

Third, smart city solutions contributions to recovery have enhanced the resilience characteristics of connectivity, creativity, agility, flexibility, and inclusion. The cases in which the extensive use of artificial intelligence (AI) have brought about the essential application of drones and robots in certain situations. These smart technologies were used in contactless deliveries, provide alerts during lockdowns, and sanitize public areas; functions that prevented further spread of the disease. Another case was crucial role that AI played in the rapid development of COVID-19 vaccine. An endeavor that would usually take years to complete.

Lastly, contributions to adaptation had fostered connectivity, learning capacity, and flexibility in resilience. The response and behavioral change of the population in terms of information management, data privacy, and economic and business models were observed. The case of the establishment of Integrated Command and Control Center (ICCC) in Pimpri-Chinchwad Smart City, which was essential in fighting the COVID-19 pandemic, brought about 45 more ICCCs across India. This has also resulted in the government's investment on maintaining and establishing more ICCCs in planning for future disaster risks. Also, in the case of the public transport sector, due to the initiative to keep air pollution levels down after COVID, some cities in the UK are investing in smart buses. Furthermore, the widespread use of smart technology was observed to have changed how government bodies and citizens interact. The increasing interest in participatory approaches, encouraged by smart city solutions, has positioned citizens as direct or indirect contributors to decision-making for their city's development.

This article discussed the contributions of smart city solutions to urban resilience broadly. Further study is recommended to explore the specific resilience characteristics and how it was enhanced by smart city solutions. It is also suggested that the contributions of smart city solutions that were not heavily dependent on smart technologies, in the context of urban resilience and pandemics, be examined for future research.

7.5 Conclusions

Undoubtedly, the increasing importance of urban resilience requires a more comprehensive approach in addressing existing urban issues and the increasing number of disaster risks. The experience of the global population in the COVID-19 pandemic has reinforced the necessity for resilience characteristics such as, but not limited to, connectivity, innovation, resourcefulness, inclusion, and flexibility; so that communities would be able to plan and prepare, absorb, recover, and adapt from future disruptions. Additionally, as much as smart technologies and other smart solutions develop, it is imperative that the roles of the people and good governance do not fall behind. Also, it is essential that the advantages and disadvantages of smart city solutions be analyzed to prevent exacerbating existing socioeconomic urban issues such as inequality and poverty. Furthermore, the COVID-19 pandemic is far from over. As much as recent studies have tried to explore its impacts on the global population, there is still much to be observed and analyzed on its real long-lasting impacts and implications for urban resilience.

Acknowledgements This chapter was partially funded by Asia-Pacific Network for Global Change Research (Funder ID: https://doi.org/10.13039/100005536).

References

Abe M, Ye L (2013) Building resilient supply chains against natural disasters: the cases of Japan and Thailand. Glob Bus Rev 14:567–586

Accenture (2011) Building and managing an intelligent city. http://www.accenture.com/SiteCollectionDocuments/PDF/Accenture-Building-Managing-Intelligent-City.pdf

Allen CR, Angeler DG, Garmestani AS, Gunderson LH, Holling CS (2014) Panarchy: theory and application. Ecosystems 17:578–589

Alliance R (2007) Urban resilience research prospectus

Amir Y, Deb S, Alam MS, Rafat Y, Hameed S (2020) Real world solutions for smart cities transportation to be pandemic ready. In: 2020 Fifth international conference on research in computational intelligence and communication networks (ICRCICN), pp 159–164. IEEE

Angel S, Blei AM, Civco DL, Parent J (2012) Atlas of urban expansion. Lincoln Institute of Land Policy, Cambridge, MA

Anthopoulos L (2015) Defining smart city architecture for sustainability. In: *Proceedings of 14th electronic government and 7th electronic participation conference (IFIP2015)*, pp 140–147

Atzori L, Iera A, Morabito G (2010) The internet of things: A survey. Comput Netw 54(15):2787–2805

Bahia K, Suardi S (2019) The state of mobile internet connectivity 2019. GSMA Connected Society, London
Bakıcı T, Almirall E, Wareham J (2013) A smart city initiative: the case of Barcelona. J Knowl Econ 4(2):135–148
Barrett T (2020) Wrightbus announce plans for 3000 hydrogen buses [Online]. Air Quality News. https://airqualitynews.com/2020/05/01/wrightbus-announce-plans-for-3000-hydrogen-buses/ [Accessed].
Bedi I (2020) How smart cities are fighting the COVID-19 pandemic [Online]. Geospatial World. https://www.geospatialworld.net/blogs/how-smart-cities-are-fighting-the-covid-19-pandemic/ [Accessed].
Berg N (2020) COVID-19 has opened the floodgates for smart cities—whether we like it or not [Online]. Fast Company. https://www.fastcompany.com/90530580/covid-19-has-opened-the-floodgates-for-smart-cities-whether-we-like-it-or-not [Accessed].
Bibri SE, Krogstie J (2020) The emerging data–driven smart city and its innovative applied solutions for sustainability: the cases of London and Barcelona. Energy Inform 3:1–42
Birkmann J, Cardona OD, Carreño ML, Barbat AH, Pelling M, Schneiderbauer S, Kienberger S, Keiler M, Alexander D, Zeil P (2013) Framing vulnerability, risk and societal responses: the MOVE framework. Nat Hazards 67:193–211
Boarnet MG, Takahashi LM (2005) Bridging the gap between urban health and urban planning. Springer, Boston
Buckle P, Mars G, Smale S (2000) New approaches to assessing vulnerability and resilience. Aust J Emerg Manag 15:8–14
Busvine D (2020) Germany launches new smartwatch application to monitor coronavirus spread. Reuters; 7 April 2020
Cai J, Guo H, Wang D (2012) Review on the resilient city research overseas. Prog Geogr 31:1245–1255
Caragliu A, Del Bo C, Nijkamp P (7–9, October 2009) Smart cities in Europe. In: 3rd Central European conference in regional science—CERS. Slovak Republic
Cardona OD, Hurtado JE, Chardon AC (2003) Indicators for disaster risk management. In: First expert meeting on disaster risk conceptualization and indicator modelling. Manizales
Carmichael L, Townshend TG, Fischer TB, Lock K, Petrokofsky C, Sheppard A, Sweeting D, Ogilvie F (2019) Urban planning as an enabler of urban health: challenges and good practice in England following the 2012 planning and public health reforms. Land Use Policy 84:154–162
CDC (2020) Coronavirus disease 2019 (COVID-19)—symptoms [Online]. Centers for Disease Control and Prevention. [Accessed]
Chagoury R (2020) The pandemic is accelerating smart city tech [Online]. Smart Cities Dive. https://www.smartcitiesdive.com/news/the-pandemic-is-accelerating-smart-city-tech/587388/ [Accessed]
Chourabi H, Nam T, Walker S, Gil-Garcia JR, Mellouli S, Nahon K et al (4–7, January 2012) Understanding smart city initiatives: an integrative framework. In: 45th Hawaii international conference on system sciences, pp 2289–2297
Cole MA, Neumayer E (2004) Examining the impact of demographic factors on air pollution. Population and Environment 26:5–21
Connelly EB, Allen CR, Hatfield K, Palma-Oliveira JM, Woods DD, Linkov I (2017) Features of resilience. Environ Syst Decis 37:46–50
Correia LM, Wünstel K (2011) Smart Cities applications and requirements. White Paper of the Experts Working Group, Net!Works European Technology Platform. http://www.scribd.com/doc/87944173/White-Paper-Smart-Cities-Applications
Costa DG, Peixoto JPJ (2020) COVID-19 pandemic: a review of smart cities initiatives to face new outbreaks. IET Smart Cities 2:64–73
Cumming GS, Cumming DH, Redman CL (2006) Scale mismatches in social-ecological systems: causes, consequences, and solutions. Ecol Soc 11

Cutter SL, Burton CG, Emrich CT (2010) Disaster resilience indicators for benchmarking baseline conditions. J Homel Secur Emerg Manag 7

Dakos V, Scheffer M, Van Nes EH, Brovkin V, Petoukhov V, Held H (2008) Slowing down as an early warning signal for abrupt climate change. Proc Natl Acad Sci 105:14308–14312

David M, Koch F (2019) "Smart is not smart enough!" Anticipating critical raw material use in smart city concepts: the example of smart grids. Sustainability 11:4422

Davison RM (2020) The transformative potential of disruptions: a viewpoint. Int J Infor Manag 55:102149

De Waal A (2020) Coronavirus: why lockdowns may not be the answer in Africa. In: News B (ed) BBC News Africa. BBC

Dirks S, Keeling M, Dencik J (2009) How smart is your city. IBM Institute for Business Value. http://www-03.ibm.com/press/attachments/IBV_Smarter_Cities_-_Final.pdf

Duhr S, Gilbert H, Peters S (2020) 'Evidence--informed' metropolitan planning in Australia? Investigating spatial data availability, integration and governance. In: Australia UOS (ed) Adelaide: University of South Australia

Fath BD, Dean CA, Katzmair H (2015) Navigating the adaptive cycle: an approach to managing the resilience of social systems. Ecol Soc 20

Fiksel J (2003) Designing resilient, sustainable systems. Environ Sci Technol 37:5330–5339

Flörke M, Schneider C, Mcdonald RI (2018) Water competition between cities and agriculture driven by climate change and urban growth. Nat Sustain 1:51–58

Folke C, Carpenter SR., Walker B, Scheffer M, Chapin T, Rockström J (2010) Resilience thinking: integrating resilience, adaptability and transformability. Ecol Soc 15

Giffinger R, Fertner C, Kramar H, Kalasek R, Pichler-Milanovic N, Meijers E (2007) Smart cities: ranking of European medium-sized cities. Centre of Regional Science (SRF). http://www.smart-cities.eu/download/smart_cities_final_report.pdf

Guha A, Mallick A, Biswas D (2020) India smart cities COVID-19 response. In: India D (ed) Public sector. Deloitte, UK

Gunderson LH, Holling C, Pritchard L, Peterson GD (2002) Resilience of large-scale resource systems. Scope-Sci Comm Probl Environ Int Counc Sci Unions 60:3–20

Güneralp B, Zhou Y, Ürge-Vorsatz D, Gupta M, Yu S, Patel PL, Fragkias M, Li X, Seto KC (2017) Global scenarios of urban density and its impacts on building energy use through 2050. Proc Natl Acad Sci 114:8945–8950

Hodson M, Marvin S (2014) After sustainable cities? Routledge

Holling CS (1973) Resilience and stability of ecological systems. Annu Rev Ecol Syst 4:1–23

Holling CS (1996) Engineering resilience versus ecological resilience. Eng Ecol Constraints 31:32

Honey-Rosés J, Anguelovski I, Chireh VK, Daher C, Konijnendijk Van Den Bosch C, Litt JS, Mawani V, Mccall MK, Orellana A, Oscilowicz E (2020) The impact of COVID-19 on public space: an early review of the emerging questions–design, perceptions and inequities. Cities Health 1–17

Huang C-Y, Namangaya A, Lugakingira MW, Cantada ID (2018) Impact and effectiveness of urban planning in Tanzania secondary cities Transla

Hughes S, Pincetl S, Boone C (2013) Triple exposure: regulatory, climatic, and political drivers of water management in Los Angeles. Cities 32:51–69

Ibrahim M (2020) The fourth industrial revolution combatting COVID-19: the role of smart and sustainable cities.

Inayatullah S (2011) City futures in transformation: emerging issues and case studies. Futures 43(7):654–661

Jeong Y-S, Park JH (2019) IoT and smart city technology: challenges, opportunities, and solutions. J Inf Process Syst 15:233–238

Johnson PA, Robinson PJ, Philpot S (2020) Type, tweet, tap, and pass: how smart city technology is creating a transactional citizen. GovMent Inf Q 37:101414

Kang M, Choi Y, Kim J, Lee KO, Lee S, Park IK, Park J, Seo I (2020) COVID-19 impact on city and region: what's next after lockdown? Int J Urban Sci 24:297–315

Kärrholm M, Nylund K, De La Fuente PP (2014) Spatial resilience and urban planning: addressing the interdependence of urban retail areas. Cities 36:121–130
Khan N, Fahad S, Faisal S, Naushad M (2020) Quarantine role in the control of corona virus in the world and its impact on the world economy. Available at SSRN 3556940
Khanna RC, Cicinelli MV, Gilbert SS, Honavar SG, Murthy GV (2020) COVID-19 pandemic: lessons learned and future directions. Indian J Ophthalmol 68:703
La Greca P, Barbarossa L, Ignaccolo M, Inturri G, Martinico F (2011) The density dilemma: a proposal for introducing smart growth principles in a sprawling settlement within Catania Metropolitan Area. Cities 28(6):527–535
Lal P, Kumar A, Kumar S, Kumari S, Saikia P, Dayanandan A, Adhikari D, Khan M (2020) The dark cloud with a silver lining: assessing the impact of the SARS COVID-19 pandemic on the global environment. Sci Total Environ 732:139297
Lerner AM, Eakin HC, Tellman E, Bausch JC, Aguilar BH (2018) Governing the gaps in water governance and land-use planning in a megacity: the example of hydrological risk in Mexico City. Cities 83:61–70
Lhomme S, Serre D, Diab Y, Laganier R (2011) A methodology to produce interdependent networks disturbance scenarios. In: Vulnerability, uncertainty, and risk: analysis, modeling, and management
Lhomme S, Serre D, Diab Y, Laganier R (2013) Analyzing resilience of urban networks: a preliminary step towards more flood resilient cities. Nat Hazards Earth Syst Sci 13:221–230
Li Q, Guan X, Wu P, Wang X, Zhou L, Tong Y, Ren R, Leung KS, Lau EH, Wong JY (2020) Early transmission dynamics in Wuhan, China, of novel coronavirus–infected pneumonia. N Engl J Med
Liu Y, Gayle AA, Wilder-Smith A, Rocklöv J (2020) The reproductive number of COVID-19 is higher compared to SARS coronavirus. J Travel Med
Long HW, Feng WJ (2020) Research report on companies' survival and development strategy during a novel coronavirus epidemic. UIBE Press, Beijing
Mahizhnan A (1999) Smart cities: the Singapore case. Cities 16(1):13–18
Malone B, Simovski B, Moliné C, Cheng J, Gheorghe M, Fontenelle H, Vardaxis I, Tennøe S, Malmberg J-A, Stratford R (2020) Artificial intelligence predicts the immunogenic landscape of SARS-CoV-2 leading to universal blueprints for vaccine designs. Sci Rep 10:1–14
Manyena SB (2006) The concept of resilience revisited. Disasters 30:434–450
Mcdonald RI, Mansur AV, Ascensão F, Crossman K, Elmqvist T, Gonzalez A, Güneralp B, Haase D, Hamann M, Hillel O (2020) Research gaps in knowledge of the impact of urban growth on biodiversity. Nat Sustain 3:16–24
Mckinsey Global Institute (2018) Smart cities: digital solutions for a more livable future. In: Company M (ed) USA
Munuzuri J, Larraneta J, Onieva L, Cortes P (2005) Solutions applicable by local administrations for urban logistics improvement. Cities 22(1):15–28
Nam T, Pardo TA (12–15, June 2011) Conceptualizing smart city with dimensions of technology, people, and institutions. In: 12th annual international conference on digital government research
Neirotti P, De Marco A, Cagliano AC, Mangano G, Scorrano F (2014) Current trends in Smart City initiatives: some stylised facts. Cities 38:25–36
Odendaal N (2003) Information and communication technology and local governance: understanding the difference between cities in developed and emerging economies. Comput Environ Urban Syst 27(6):585–607
Oldekop JA, Horner R, Hulme D, Adhikari R, Agarwal B, Alford M, Bakewell O, Banks N, Barrientos S, Bastia T (2020) COVID-19 and the case for global development. World Dev 134:105044
Pan SL, Zhang S (2020) From fighting COVID-19 pandemic to tackling sustainable development goals: an opportunity for responsible information systems research. Int J Inf Manag 55:102196

Papa R, Galderisi A, Vigo Majello MC, Saretta E (2015) Smart and resilient cities: a systemic approach for developing cross-sectoral strategies in the face of climate change. TeMA J Land Use Mobil Environ 8:19–49
Papa R, Gargiulo C, Galderisi A (2013) Towards an urban planners' perspective on smart city. TeMA J Land Use Mobil Environ 6:5–17
Park J, Seager TP, Rao PSC, Convertino M, Linkov I (2013) Integrating risk and resilience approaches to catastrophe management in engineering systems. Risk Anal 33:356–367
Pelling M (2003) The vulnerability of cities: natural disasters and social resilience. Earthscan
Peng C, Yuan M, Gu C, Peng Z, Ming T (2017) A review of the theory and practice of regional resilience. Sustain Cities Soc 29:86–96
Polidoro M, De Lollo JA, Barros MVF (2012) Urban sprawl and the challenges for urban planning. J Environ Prot 3: 1010–1019
Pomeroy J (2020) Cities of opportunities: connecting culture and innovation. Routledge
Proag V (2014) Assessing and measuring resilience. Procedia Econ Financ 18:222–229
Rabi FA, Al Zoubi MS, Kasasbeh GA, Salameh DM, Al-Nasser AD (2020) SARS-CoV-2 and coronavirus disease 2019: What we know so far. Pathogens 9:231
Ramaprasad A, Sánchez-Ortiz A, Syn T (2017) A unified definition of a smart city. In: International conference on electronic government. Springer, pp 13–24
Rao SK, Prasad R (2018) Impact of 5G technologies on smart city implementation. Wirel Pers Commun 100:161–176
Regmi A (2001) Changing structure of global food consumption and trade: an introduction. Changing structure of global food consumption and trade. Anita Regmi 1:1–3
Rezaei M, Bastaminia A (2016) Evaluation of dimensions, approaches and concepts of resilience in urban societies with an emphasis on natural disasters. J Fundam Appl Sci 8:1630–1649
Rodríguez J (2007) United Nations expert group meeting on population distribution, urbanization, internal migration and development. U N Secr N Y 21–23
Rose A (2007) Economic resilience to natural and man-made disasters: multidisciplinary origins and contextual dimensions. Environ Hazards 7:383–398
Rose A, Liao SY (2005) Modeling regional economic resilience to disasters: a computable general equilibrium analysis of water service disruptions. J RegNal Sci 45:75–112
Sachs JD, Karim SA, Aknin L, Allen J, Brosbøl K, Barron GC, Daszak P, Espinosa MF, Gaspar V, Gaviria A (2021) Priorities for the COVID-19 pandemic at the start of 2021: statement of the Lancet COVID-19 Commission The Lancet 397(10278): 947–950
Scott IA, Coiera EW (2020) Can AI help in the fight against COVID-19? Med J Aust 213:439–441.e2
Shahat E, Hyun CT, Yeom C (2020) Conceptualizing smart disaster governance: an integrative conceptual framework. Sustainability 12:9536
Sharifi A (2016) A critical review of selected tools for assessing community resilience. Ecol Indic 69:629–647
Sharifi A (2019) A critical review of selected smart city assessment tools and indicator sets. J Clean Prod 233:1269–1283
Sharifi A (2020) A typology of smart city assessment tools and indicator sets. Sustain Cities Soc 53:101936
Sharifi A (2021) The COVID-19 pandemic: lessons for urban resilience. In: Linkov I, Keenan JM, Trump BD (eds) COVID-19: systemic risk and resilience. Springer International Publishing, Cham
Sharifi A, Chelleri L, Fox-Lent C, Grafakos S, Pathak M, Olazabal M, Moloney S, Yumagulova L, Yamagata Y (2017) Conceptualizing dimensions and characteristics of urban resilience: insights from a co-design process. Sustainability 9:1032
Sharifi A, Khavarian-Garmsir AR (2020) The COVID-19 pandemic: impacts on cities and major lessons for urban planning, design, and management. Sci Total Environ 142391
Sharifi A, Khavarian-Garmsir AR, Kummitha RKR (2021) Contributions of smart city solutions and technologies to resilience against the COVID-19 pandemic: a literature review. Sustainability 13:8018

Sharifi A, Yamagata Y (2016) Principles and criteria for assessing urban energy resilience: a literature review. Renew Sustain Energy Rev 60:1654–1677
Shi H, Han X, Jiang N, Cao Y, Alwalid O, Gu J, Fan Y, Zheng C (2020) Radiological findings from 81 patients with COVID-19 pneumonia in Wuhan, China: a descriptive study. Lancet Infect Dis 20:425–434
Shwartz A, Turbé A, Julliard R, Simon L, Prévot A-C (2014) Outstanding challenges for urban conservation research and action. Glob Environ Chang 28:39–49
Skouby KE, Lynggaard P (2014) Smart home and smart city solutions enabled by 5G, IoT, AAI and CoT services. In: 2014 international conference on contemporary computing and informatics (IC3I), pp 874–878. IEEE
Smart City (2020) Can AI help the world find the medicine to treat COVID-19? [Online]. Smart City Press, Australia. https://www.smartcity.press/ai-accelerating-covid-vaccine/ [Accessed]
Söderström O (2020) The three modes of existence of the pandemic smart city. Urban Geogr 1–9
Steria-Smart City (2011) Smart cities will be enabled by smart IT. http://www.steria.com/uk/fileadmin/assets/media/STE3899Smart_Cities_brochure_08_APP.PDF
Stokel-Walker C (2020) Why telemedicine is here to stay. bmj 371
Sułkowski Ł (2020) Covid-19 pandemic; recession, virtual revolution leading to de-globalization? J Intercult Manag 12:1–11
Tandoc JR, EC, Lee JCB (2020) When viruses and misinformation spread: How young Singaporeans navigated uncertainty in the early stages of the COVID-19 outbreak. New Media Soc 1461444820968212
Tešić D, Blagojević D, Lukić A (2020) Bringing'smart'into cities to fight pandemics: with the reference to the COVID-19. Zb Rad Departmana Za Geogr, Turiz I Hotel, 99–112
The Climate Group, ARUP, Accenture, The University of Nottingham (2011) Information marketplaces the new economics of cities. http://www.theclimategroup.org/_assets/files/information_marketplaces_05_12_11.pdf
Think (2011) Smart cities initiative: How to foster a quick transition towards local sustainable energy systems. http://www.symple.tm.fr/uploaded/pdf/THINK_smart_cities.pdf
Thomson B (2020) The COVID-19 pandemic: a global natural experiment. Circulation 142:14–16
Tiwari R, Cervero R, Schipper L (2011) Driving CO2 reduction by integrating transport and urban design strategy. Cities 28(5):394–405
Tongyue L, Pinyi N, Chaolin G (2015) A review on research frameworks of resilient cities. Urban Plan Forum 218:23–31
Toppeta D (2010) The smart city vision: How innovation and ICT can build smart, "liveable", sustainable cities. The Innovation Knowledge Foundation. Think!Report, 005/2010
Trencher G (2019) Towards the smart city 2.0: empirical evidence of using smartness as a tool for tackling social challenges. Technol Forecast Soc Chang 142:117–128
Turner MD (2014) Political ecology I: An alliance with resilience? Prog Hum Geogr 38:616–623
Un Habitat (2020) World Cities Report 2020: the value of sustainable urbanization. In: Program UNHS (ed) World cities report. UN Habitat, Nairobi
United Nations (2018) 2018 revision of world urbanization prospects. May, 16, 2018
Urban RC, Nakada LYK (2021) COVID-19 pandemic: solid waste and environmental impacts in Brazil. Sci Total Environ 755:142471
Wahba S, Sharif MM, Mizutori M, Sorkin L (2020) Cities are on the front lines of COVID-19 [Online]. World Bank, Washington, DC. https://www.worldbank.org/en/about/contacts [Accessed]
Wang CJ, Ng CY, Brook RH (2020) Response to COVID-19 in Taiwan: big data analytics, new technology, and proactive testing. JAMA 323:1341–1342
Washburn D, Sindhu U, Balaouras S, Dines RA, Hayes NM, Nelson LE (2010) Helping CIOs understand "smart city" initiatives: defining the smart city, its drivers, and the role of the CIO. Forrester Research, Inc. http://public.dhe.ibm.com/partnerworld/pub/smb/smarterplanet/forr_help_cios_und_smart_city_initiatives.pdf

Westraadt L, Calitz A (2018) A gap analysis of new smart city solutions for integrated city planning and management. In: Proceedings of the Annual Conference of the South African Institute of Computer Scientists and Information Technologists, 2018, pp 145–153
Whitelaw S, Mamas MA, Topol E, Van Spall HG (2020) Applications of digital technology in COVID-19 pandemic planning and response. The Lancet Digital Health
Wikström A (2013) The challenge of change: planning for social urban resilience: an analysis of contemporary planning aims and practices
Woetzel J, Kuznetsova E (2018) Smart city solutions: What drives citizen adoption around the globe. McKinsey Center for Government, McKinsey&Company
World Health Organization (2020) Report of the WHO-China joint mission on coronavirus disease 2019 (COVID-19). WHO, Geneva
World Health Organisation (2021) Weekly operational update on COVID-19—16 March 2021. In: WHO (ed) Situation reports. WHO, Geneva
World Trade Organization (2020) Trade set to plunge as COVID-19 pandemic upends global economy. World Trade Organization, Geneva
Wray S (2020) South Korea to step-up online coronavirus tracking. Smart Cities World
Xiao Y, Fan Z (2020) Technology trends to watch in the COVID-19 pandemic. In: World Economic Forum
Yao F, Wang Y (2020) Towards resilient and smart cities: a real-time urban analytical and geo-visual system for social media streaming data. Sustain Cities Soc 63:102448
Zhu S, Li D, Feng H (2019) Is smart city resilient? Evidence from China. Sustain Cities Soc 50:101636

Chapter 8
Contributions of Smart City Projects to Resilience: Lessons Learned from Case Studies

Hasan Masrur and Ayyoob Sharifi

Abstract Smart cities are often characterized by using ICT-enabled solutions in various socio-economic, institutional, and environmental fields to enhance quality of life, sustainability, and resilience and to preserve the competitive potential of cities in an increasingly interconnected network of cities. While the concept of "smart city" has been around for a while, recently there is a growing interest in using smart city solutions and technologies for enhancing resilience worldwide. It is vital to recognize the effect of smart cities on improving urban resilience, especially with regard to climate adaptation and mitigation. As a preliminary step toward this goal, we have created a database of smart city projects and initiatives with actual and/or potential contributions to resilience. Our database of approximately 300 case studies tries to investigate the resilience steps and smart solutions taken by smart cities around the world under categorized indicator sets. Results show that most of the smart city projects are mainly aimed at the reduction of CO_2 emission. Regarding the resilience stage, we considered four stages, namely, planning, absorption, recovery, and adaptation. It was found that the projects are related to different stages, particularly, adaptation and absorption. In terms of sectoral focus, energy sector has received the most attention by the smart city planners and policymakers. Concerning smart city dimensions, "living" has received the utmost attention, followed by "mobility" and "data" that have also received considerable attention. Much of the projects are owned by the government and are participatory in terms of governance. It is important to note that most of the projects have paid attention to multiple smart city "dimensions" and can contribute to different "resilience characteristics." This evidence-based quantitative analysis of global smart city projects could be used to highlight the success

H. Masrur
Faculty of Engineering, University of the Ryukyus, 1 Senbaru, Nishihara, Okinawa 903-0213, Japan

A. Sharifi (✉)
Graduate School of Humanities and Social Sciences, Hiroshima University, Higashihiroshima, Japan
e-mail: sharifi@hiroshima-u.ac.jp

Graduate School of Advanced Science and Engineering, Hiroshima University, Higashihiroshima, Hiroshima, Japan

A. Sharifi and P. Salehi (eds.), *Resilient Smart Cities*, The Urban Book Series,
https://doi.org/10.1007/978-3-030-95037-8_8

factors, trends, and gaps. The results can be used to develop more effective future pathways that could contribute to achieving sustainable development goals.

Keywords Smart city · Resilience · Case study · Information and communication technologies · Climate change adaptation and mitigation

8.1 Introduction

Cities have always been experiencing transitions in terms of innovation, economic development, culture, population, and so on. At the same time, cities remain one of the most surviving entities, as history has proven (Vale and Campanella 2005). However, since developments in global urbanization has intensified, urban authorities face increasing difficulties in satisfying the dynamically shifting needs of people while resolving the challenges and uncertainties of global sustainability (Sharifi 2019, 2020b; Stratigea et al. 2015). As the world is in the verge of climate change induced disasters, cities are in the forefront to embrace the grave repercussions that are specially driven by the increasing trends in the emission of greenhouse gases (mainly CO_2) (Perboli and Rosano 2020).

To tackle these threats, the idea of "resilience" has attracted enormous interest on the national agenda and is also increasingly gaining support by the urban studies literature although at present, there seems to be no universally accepted definition of urban resilience (Shafiei Dastjerdi et al. 2021). The term resilience can be applied as a guide to developing planning, absorption, recovery, and adaptation capacities in various urban multi-scale structures to tackle a broad spectrum of challenges concerning social, economic, environmental, and physical systems of cities (Shafiei Dastjerdi et al. 2021; Sharifi 2020c). Planning refers to actions taken before the occurrence of the adverse event to ensure risk mitigation and prepare resources necessary for dealing with these incidents. Absorption entails the capacity to go minimize the potential losses when a disastrous event hits the cities. Recovery refers to the ability to return to normal conditions in a timely manner. Finally, adaptation entails taking the adverse event as an opportunity to learn lessons to ensure better performance under future adverse events that may be even more severe. The vast literature on urban resilience demonstrates that various efforts across different sectors need to be taken to enhance resilience of cities. These include efforts related to social, economic, environmental, physical, institutional, and technological dimensions (Sharifi 2016). The later dimension has a specific focus on smart solutions that are enabled by advances in information and communication technologies and big data analytics. Acting in collaboration with technology and infrastructure providers, regional governments, and community officials now understand the opportunities that smart city projects (SCPs) would offer in terms of resiliency, economic opportunity, sustainability, and quality of life. Cities are seeking higher levels and a greater variety of convergence between technology-driven public services and solutions, as the number of smart city initiatives grows globally. However, there is still limited knowledge in the literature on the nexus between smart city and resilience. Identifying the basic features and patterns of smart cities is critical in achieving sustainable and equitable urban growth that also contributes to better resilience against the future adverse events that are expected to be more frequent and severe (Perboli and Rosano 2020).

SCPs and urban case studies will contribute to our knowledge of responses to the possible threats and damages. Therefore, an overview of the SCPs and their potential and actual contributions to resilience is desperately needed. In response to this need, in this chapter, we collected information of SCPs around the world, and we will deal with several questions, for example, what are the scale of these projects? Are these projects specifically focused on cities, buildings, neighborhoods, or regions? and who are the project owners? Government, private entities or are there are multiple stakeholders like public private partnerships? We will also look into the geographical locations of these projects, the specific types of smart solutions, the major dimensions that have received more attention in the SCPs, sectors that are mostly focused on. Additionally, and most importantly we will investigate the resilience characteristics and the stages of these projects and will also try to find out what are issues and problems that these projects are resilient against. In other words, they intend to enhance resilience against what types of threats and stressors.

8.2 Methodology

Using a search string that included several terms relevant to SCPs, data for analysis was collected from the Scopus database as seen in Table 8.1. The search phrase was created to include as many research as possible that focused on evaluating smart city initiatives and their potential and/or actual contributions to resilience. Here, resilience was considered in a broad sense. In other words, any contributions to planning, absorption, recovery, and adaptation to any types of adverse events were considered as relevant. Cases are particularly based on geographical locations i.e., regions, cities, buildings, and neighborhood were considered. In terms of article sources, we considered peer-reviewed sources such as journals and conference proceedings. We, however, acknowledge that in not all projects, the stakeholders and concerned individuals publish their project outputs in journals and conference proceedings. We followed the procedure described in A. Sharifi (2020c) for screening peer-reviewed papers. On June 13, 2020, a literature search was done for an indefinite period, yielding 1622 papers. All of the recovered documents' titles and abstracts were scrutinized to rule out those that were not on focused SCPs. The conclusions of several

Table 8.1 Keywords for the search string

Database	Search string
Scopus	TITLE-ABS-KEY (("smart city" OR "smart cities") AND ("case stud*"))

publications were also examined in case it was not possible to determine relevance based on titles and abstracts only. During the screening process, two significant categories of publications were eliminated. We excluded most of the returned papers that are focused on generic smart city concepts which does not reflect the any specific case studies. We also didn't include any that were not focused on a specific case study area. Only 124 papers remained in the database after the filtering procedure was completed. As for non-peer-reviewed sources, we considered information reported in websites, white papers, and reports. Among the non-peer-reviewed sources, SmartCitiesWorld (SCW 2020) has been instrumental for collecting the updates and information of the smart city projects around the world. This website shares news and views related to SCPs to solve urban challenges and to make cities more resilient and sustainable. A total of 168 cases are included from the non-peer-reviewed sources. We prepared two separate excel files—one for peer-reviewed sources and another one is from non-peer-reviewed sources, and finally they were merged to perform the final analysis.

The criteria and principle of assessing the case studies are depicted in Table 8.2. Around 22 parameters are explicitly introduced to analyze the state of potential and/or actual contributions of smart city projects around the world to resilience. It is important to remember that resilience is a multi-faceted notion with no uniform definition. However, three ways of defining resilience are frequently used: engineering, ecological, and adaptive. Engineering resilience is primarily concerned with physical properties like robustness, which allow for shock prevention. Ecological resilience stresses qualities such as system flexibility and the system's ability to handle shocks and quickly recover to a steady state (or states). Details related to each parameter are provided in Table 8.2. As can be seen, we have made efforts to collect data related to various aspects of smart city projects. In terms of stages of resilience, as mentioned earlier, we have considered four stages, namely, planning, absorption, recovery, and adaptation. As for resilience characteristics, multiple characteristics have been considered based on the existing literature. In particular, we have considered the following characteristics: robustness, diversity, redundancy, connectivity, flexibility, resourcefulness, agility, efficiency, adaptive capacity, modularity, creativity, equity, inclusiveness, foresight capacity, and low-carbon development. For further details about resilience dimensions, abilities, and characteristics, interested readers are referred to (Sharifi and Yamagata 2016a, b). These characteristics have also been elaborated on in other chapters of this volume.

8.3 Results

8.3.1 Overall Focus

As shown in Fig. 8.1, in terms of geographical scale, most of the projects are based on the city scale, followed by region, building, and neighborhood. This is an important finding, indicating that the significance of implementing city-wide initiatives

Table 8.2 Parameters, criteria, and principle used for collecting data to assess the studies

Parameters	Definition
Project name	Refers to the name of the project given by the project owner or initiator. If there is none, we titled the case study using the related keyword(s)
Paper title	Title of the paper, normally project report published as white paper or in the journals and conference proceedings
Publication	Refers to the date of publication
Source	Refers to the name of journals or other sources from which the information was extracted
Timeline	start-finish date of the project
Scale of project	Building /neighborhood/ city/ region
Project owner	Public/private/public and private (mixed)
Geographic focus	Refers to the geographic location of the case study project
Size (area) of the project	Refers to the size of the project in hectares
Project budget	Total budget used for developing the project
Type of smart solution	Specific type of smart technology/solution (e.g., smart metering, smart surveillance, early warning system, etc.)
Related smart city dimension	Seven dimensions are considered—Data, people, mobility, environment, living, governance, and economy
Energy sector	The energy generators involved in SCPs, for example, renewable energy sources
Transportation sector	Public transportation/ autonomous vehicles/others
Building sector	Heating demand, cooling demand, water demand, and other usage
Waste management sector	Whether the smart technology/solution has focused on waste management
Water sector	Whether the smart technology/solution has focused on water management
Governance	The types of governance best fit to the case study: monitoring and valuation/planning and management/ Supervisory basis and law enforcement/pricing/ Local government support and incentives
Mitigation contribution	SCPs contribution to mitigate CO_2 emission, overall cost, energy consumption, and so on
Adaptation contribution	SCPs' contribution to adjust or adapt in response to the different shocks, particularly those related to climate change
Equity benefits?	Any indications of equity benefits of the project
Health benefits?	Any indications of health benefits of the project
Other benefits	Any indications of other benefits of the project
Any trade-offs?	Whether implementation of the smart city project has resulted in unwanted negative impacts

(continued)

Table 8.2 (continued)

Parameters	Definition
Resilience characteristic	The resilience characteristics (robustness, diversity, redundancy, connectivity, flexibility, resourcefulness, agility, efficiency, adaptive capacity, modularity, creativity, equity, inclusiveness, foresight capacity, and low-carbon) of the SCP
Resilience stage	The resilience stage involved with the SCP (i.e., planning, absorption, recovery, or adaptation)
Resilience to what?	To find out resilience to what type of stressor or adverse event is discussed (e.g., flooding, heat wave, earthquake, etc.)

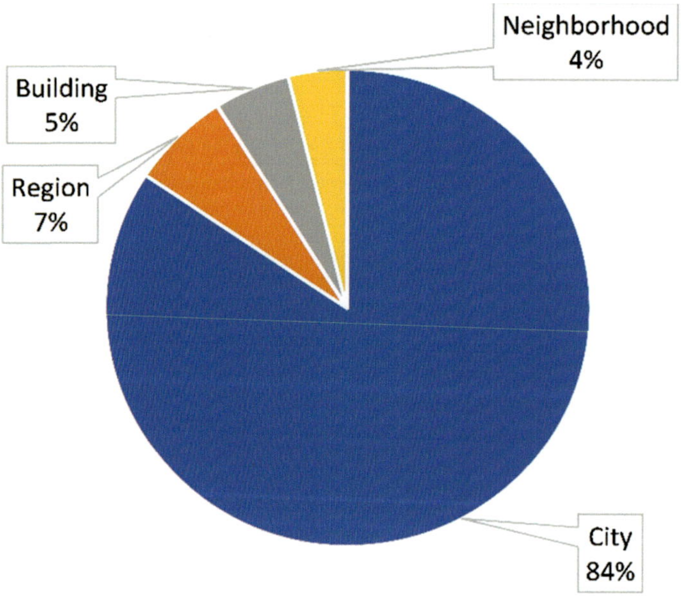

Fig. 8.1 Share of smart city projects from the studied data

is well-recognized. While initiative focused on building and neighborhood scales are also important for improving urban efficiency and enhancing resilience, city-wide initiatives are more likely to facilitate holistic and systemic approaches, engage different stakeholders in the processes, and also provide benefits to a larger number of urban residents. Adopting systemic approaches can, in turn, facilitate increasing co-benefits and synergies and minimize trade-offs and conflicts between different planning and design related measures and actions (Sharifi 2020a, 2021a). Most of the studied projects are in Europe, followed by North America and Asia. A very few projects are found for African countries. European Union (EU) initiatives to invest in and promote smart city projects (such as H2020 program) is likely to be the main driver for this advancement in Europe. As part of these initiatives, capital and main cities in Europe are also considered to implement the SCPs. Cities in North America and Asia (particularly East Asian countries) have also made many efforts to develop and implement smart city projects. The fact that other regions of the world are lagging behind, however, is something that raises concerns about equitable access to digital resources in a world economy that is increasingly dependent on digital networks and smart solutions and technologies. This may have significant implications for achieving the Sustainable Development Goals. Global partnership and support from the leading countries in the area of information and communication technologies may be needed to fill this gap.

8.3.2 Sector-wise Focus

Energy sector received the most attention as pointed out in Fig. 8.2. This is not surprising given the fact that many cities have developed targets related to climate change mitigation that are directly or indirectly related to the energy sector. In addition, enhancing energy efficiency to ensure continued availability, accessibility, and affordability of energy resources has always been high on the agenda of planners and policy makers (Sharifi and Yamagata 2016a). There have been major advances in terms of deployment and availability of renewable energy sources and technologies, rapid cost declination of Photovoltaic (PV) modules and other technologies, storage devices, home energy management systems (HEMS), smart metering devices, etc. Introduction of smart grids, microgrids, and smart houses have indeed played

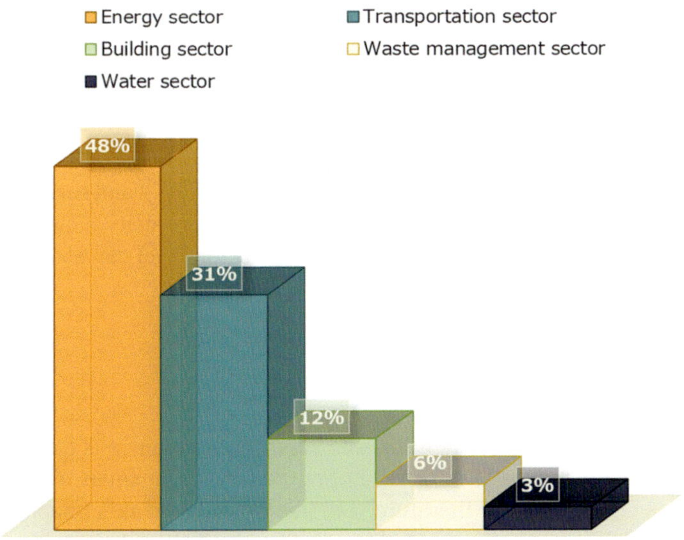

Fig. 8.2 Focused sector of smart city projects

key role in this regard. While energy sustainability has not always been the main goal of smart city projects, the smartness agenda pushes forward to sustainability measures as suggested by the authors in Haarstad and Wathne (2019). Gathering insights from the "Lighthouse" city projects sponsored by H2020 program, they claimed that advanced technology seldom governs sustainability-related activities of smart city projects (SCPs).

Water and waste management sector have been found to be trailing behind because of no proper planning and high costing. This is a sector that certainly needs more attention. Climate change is expected to increase water stress in many parts of the world, making it necessary to enhance efficiency of water resource management in cities. Use of smart technologies such as smart meters or smart monitoring systems may be effective as such technologies can reduce water waste through leaking and also promote a more environmental conscious behavior that can contribute to water resource conservation. Climate change may also result in increased precipitation in some other parts of the world, making it necessary to use smart technologies for better stormwater management and also for facilitating better response to urban flooding risks.

Transportation sector has been moderately considered in the projects. Mainly electric vehicles (EV) that include electric cars, buses, and bikes are employed as a smart measure. EV charging stations, intelligent traffic system, bike sharing, and smart car parking are also some smart technologies related to transportation. With more advances in autonomous vehicles, it is expected that more attention to smart technologies will be paid in the coming decades. Also, there have been advances in other fronts such as vehicle to vehicle communication systems, vehicle to grid, and ride sourcing services that are expected to further grow in the coming decades (Khavarian-Garmsir et al. 2021). If designed and implemented appropriately, such services and technologies are expected to also provide multiple benefits for environmental, social, and economic sustainability.

Building sector includes manly initiative related to retrofitting, adaptation of ZEB's, reductions in demands for heating, cooling, water, and so on. Further, energy management systems such as home energy management system (HEMS), building energy management system (BEMS), internet of things (IOT), and sensors have been widely implemented to make the buildings "smart." Given its manageable scale, further penetration of smart technologies at the building level can also be expected.

8.3.3 Smart City Dimensions

Figure 8.3 shows the level of attention that smart city dimensions have received in our examined case studies. It is worth noting that multiple smart city dimensions have been mentioned in the literature. For the purpose of this chapter, we have used the classification introduced in Ayyoob Sharifi (2019). As can be seen, there are seven dimensions, namely, economy, governance, environment, living, mobility, people, and data. As mentioned by Ayyoob Sharifi (2019), economy is related to

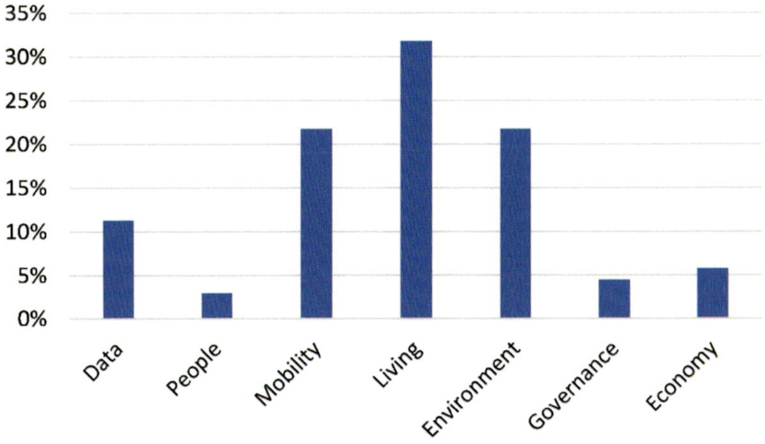

Fig. 8.3 Frequency of smart city dimensions considered in the studied smart city projects

issues such as "innovation/innovation culture, knowledge economy, entrepreneurship, finance, tourism, employment, local and global interconnectedness (international embeddedness), productivity and efficiency, flexibility of the labor market, and economic impacts of smart city projects;" the people dimension is concerned with "education/ lifelong learning, level of qualification/ ICT skills, and cosmopolitanism culture/open mindedness;" governance deals with "visioning and leadership, legal and regulatory frameworks, citizen participation, transparency, public and social services, and efficient and integrated urban management;" environment is focused on "environmental monitoring and management, general infrastructure, built environment/planning and design, materials, energy resources, water resources, waste (solid waste, waste water, sewage), and environmental quality/pollution;" living is related to issues such as "social cohesion/inclusion, equity and justice, cultural development, housing/livelihood quality, healthcare, safety and security, and convenience and satisfaction/ subjective well-being;" mobility has a broad meaning and deals with various issues such as "transport infrastructure, transportation management, ICT infrastructure, ICT management, and ICT accessibility;" and finally the data dimension is concerned data-related matters such as data openness, sensing and collecting data, judging (analytics) data, reacting based on the analyzed data, and using results of data analytics to learn lessons and improve future performance (Sharifi 2019). Regarding dimension-based focus, it is worth noting that on many occasions, projects considered multiple dimensions simultaneously. When counted separately, most of the projects focus on improving "living" (~32%) conditions of the people. This is not surprising as enhancing quality of life is one of the main objectives of smart city projects. Mobility and environment have been emphasized equally (~22%). The reasonably high attention to environment is encouraging. However, given the urgency of taking climate actions, more attention to environment is needed. "Data" has also been taken account in 12% of smart city cases. People, governance, and economy are less considered (less than 10%). These dimensions obviously require more attention. Overall, a more balanced attention to multiple dimensions would be desirable.

8.3.4 SCPs Challenges

Cities are now facing multiple grave challenges at the same time, with environmental threats being the most serious. Urbanization process, which stresses basic services, in combination with more frequent and intense weather events connected to climate change, is worsening the effect of environmental hazards. Flooding, tropical cyclones, heat waves, and diseases are all common natural hazards that are on the rise. The studied SCPs also testify this fact as seen Figs. 8.4 and 8.5. More than 70% of the issues experienced by the SCPs are related to environmental concerns due to

Fig. 8.4 Targeted urban challenges or shocks involved in the SCPs

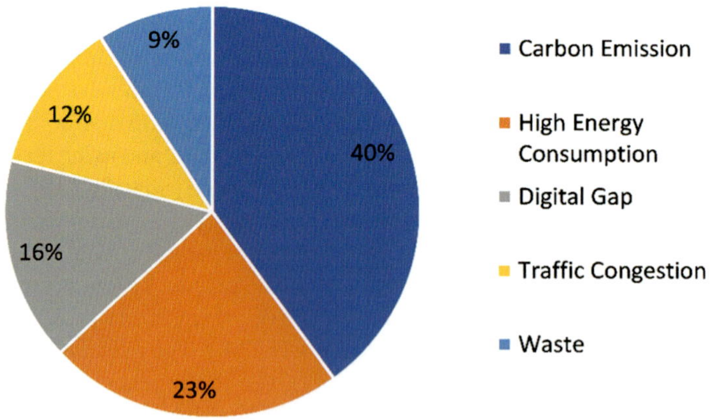

Fig. 8.5 Most frequent issues encountered by the SCPs

high energy consumption and carbon emission. Urban development diminishes water availability and often negatively impacts catchment areas and agricultural fields while increasing energy consumption. Then, there is the continuous depletion of natural resources such as fossil fuels and water. Hence, SCPs need to ensure meeting the basic human needs with limited resources, without worsening earth's current condition. Crowding and traffic jams continue to be major issues in urban areas. Dealing with digital gaps and cyber security is also critical. Interestingly, aging and a declining birthrate have emerged as new challenges in this century. Waste management is another big issue that cities face every day. In addition to environmental issues, our studied SCPs identified that the digital gap, traffic congestion and waste are the most frequent issues by the percentages of 16, 12, and 9, respectively. Recently, curing pandemics and outbreaks like COVID-19 has become a huge issue for SCPs (Sharifi and Khavarian-Garmsir 2020; Sharifi et al. 2021).

8.3.5 *Dominant Smart City Solutions and Technologies*

Most of the projects have adopted smart solutions related to energy systems as seen in Fig. 8.6. This is because climate change, as well as a rising list of additional risks such as cyber assaults, terrorism, technological flaws, and market volatility, all influence the reliability of energy supply in cities. To minimize the negative repercussions of interruptions in energy supply, concerted measures are required that recognize the interconnections and interdependencies between energy and other sectors. Distributed energy resources like solar and wind, as well as related energy technologies like PV panels and wind turbines, are being increasingly implemented in order to achieve a carbon-free energy system. In most cases, smart solutions and technologies have contributed to better deployment of these energy technologies. Other renewable technologies based on hydropower, biomass, and geothermal sources, on the other hand, are not widely used. This is due to the scarcity of these resources and

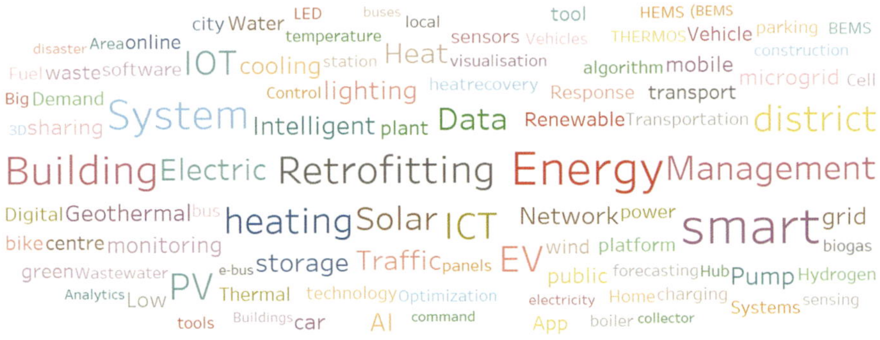

Fig. 8.6 Smart solutions adopted by the projects

the high cost of related technologies when implemented in smart city projects. Wind and solar, on the other hand, are readily available, and their upfront technology costs are also decreasing as a result of government policies. It is worth noting that many projects already implemented hydrogen fueled energy system.

Many of the smart solutions emphasize on building retrofitting and improved energy management system while adopting demand response, energy forecasting, and other AI-driven processes. District heating and cooling are one of the frequently used smart solutions specially in Europe and North America owing to the weather conditions of those regions. However, implementation of smart grid systems and renewable energy based microgrids are now widely being embraced, according to recent findings (Masrur et al. 2021a, b). These are basically hybrid systems backed up by fossil fuel although some target for a 100% renewable system. Some cases which are still in the planning phase use multi objective optimization algorithms, for example, maximizing thermal comfort while minimizing energy usage in a smart apartment/building reducing the electric bills of the users while using strategies for increasing the profit of the utilities. Technologies related to wastewater treatment and general waste management are also smart solutions that are on the top priority list of the city planners and policy makers.

There are many ICT related technologies that are used as smart solutions such as IOT, sensors, 3D mapping, data visualization tools, different user-friendly applications, and mobile and home applications. It seems that AI and machine learning has been widely deployed in the SCPs as they can integrate and operate different types of smart technologies, ensuring the success of any smart projects.

In the transportation sector, EV, sensors, and street lighting using LED are being adopted as smart solutions. Intelligent traffic management systems based on big data analytics and data-driven systems are becoming increasingly popular and are being extensively utilized for practically all mobility-related SCPs.

8.3.6 Resiliency Characteristics

A world map is created (Fig. 8.7) using the information from our database of SCPs by Tableau software. City names and resiliency stages of each smart city project are given as input. Hence, along with the locations of the SCPs, it shows most of the SCPs are related to the adaptation stage followed by recovery, absorption, and planning. This indicates that over three decades of smart city planning has been effective in contributing to better penetration of smart solutions and lessons learned from these activities have contributed to better adaptation abilities (Sharifi et al. 2021a). It also should be noted that adaptation is more dominant in European, while a large number of Asian projects can mainly be attributed to the absorption process. In the U.S., SCPs are not dominated by a single resilience condition, but they belong to different stages. In India, many SCPs are under development and newly initiated. These could be mainly considered as related to the adaptation stage. Overall, more time is needed to better judge the relevance of the projects to the four different phases.

Fig. 8.7 Frequency of different resilience stages of the projects in different cities

All the projects demonstrate multiple resilience characteristics as shown in Fig. 8.8. Robustness is the ability to counteract and/or absorb disturbance. The completed projects and achieved significant reduction in either CO_2 emission, traffic congestion, or energy saving contribute to better robustness and stability against shocks. Such achievements could also contribute to enhancing other resilience characteristics such as efficiency. Many projects feature all different resilience characteristics simultaneously. According to our findings, more than 20% of the SCPs are

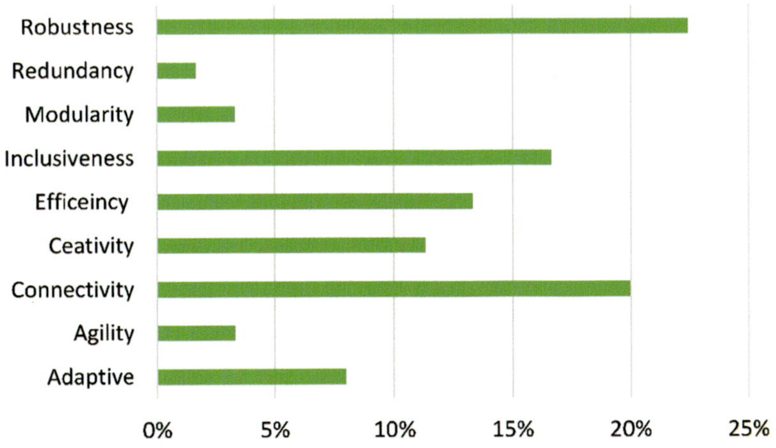

Fig. 8.8 Resiliency characteristics and trends of smart city projects

robust in nature. These projects have the characteristics of connectivity and inclusiveness. Inclusiveness can be referred to the engagement of various stakeholders in planning and decision-making processes that enhances social capital and improves planning, absorption, recovery, and adaptation capacities of the projects. Connectivity, however, is a term used to describe a system that has multiple sub-systems that work well together. According to our research, approximately 17% of SCPs are inclusive, while 20% of projects have connectivity characteristics. The projects are also found to be efficient and creative, even though these characteristics are less emphasized (10–15%) as per our dataset. Finally, only 5–10% of projects are reported to demonstrate agility, redundancy, and modularity as resilience attributes. The state of attention to different characteristics across the studied projects is shown in Fig. 8.8. Given the significance of agility, redundancy, and modularity for disaster resilience, these characteristics deserve more attention.

8.3.7 Project Ownership and Governance

Most of the projects are obviously government owned, but there is also significant public private partnership and contribution, and often universities come into the picture. According to our research, 61% of SCPs are funded and run by the government, 31% are jointly owned by public and private stakeholders, and only 9% are controlled by private entities. This information is depicted on Fig. 8.9. However, recently there is an increasing trend of private companies contributing for the smart

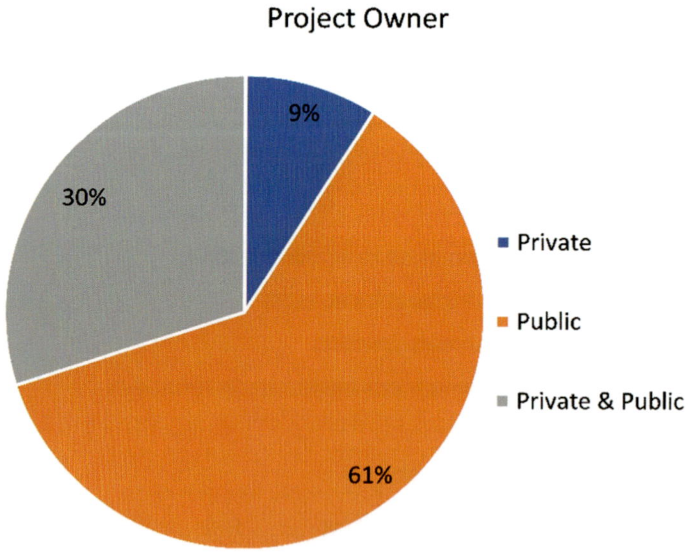

Fig. 8.9 Percentage of different project owners

cities in the form of public private partnership or even private alone. For example, in Japan, Toyota is building a smart city called Woven city. They plan to introduce a living laboratory, which is a fully connected ecosystem powered by hydrogen fuel cells and it is to be founded at the base of Mount Fuji. Another company Hitachi recently has joined "Kashiwa no ha" project, located at the city of Kashiwa, Chiba prefecture, and it aims to build a human friendly and environmentally sustainable city. It should also be noted that many private companies in other parts of the world have also initiated investing in smart city projects. The fact that only 9% of the projects in our study belong to the private sector is indeed surprising. This could be due to limited reporting of such projects in peer-reviewed articles and requires more research in future. It, particularly, requires doing a more systematic analysis of non-academic literature. Also, conducting surveys among private sector entities could be helpful for gaining more information in this regard.

As for the governance, most of the projects are participatory in nature, public service has always been a key governance factor while designing the smart city projects followed by the integrated management, visioning, and legal framework. Figure 8.10 indicates the percentage of these indicators. It is also noticed that visioning is always linked to participation, public service, or integrated management.

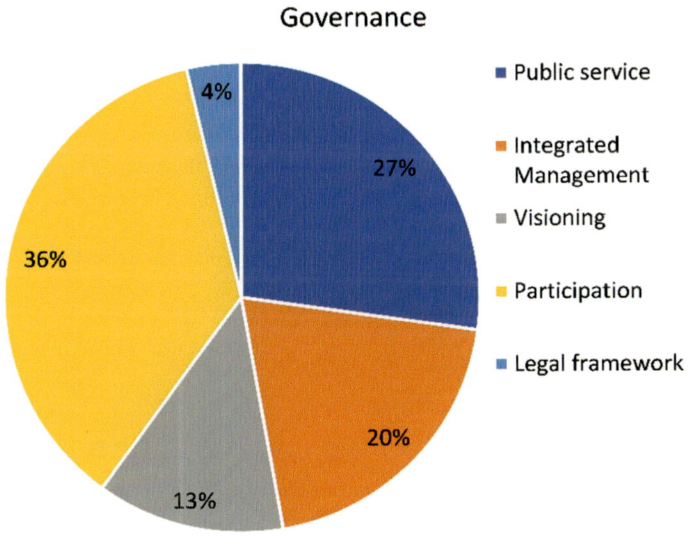

Fig. 8.10 Share of various indicators related to governance

8.4 Conclusions

The ever-increasing influx of population and economic activity in urban centers entails actions to address a wide range of climatic and non-climatic risks that jeopardize city functioning. This need is widely acknowledged in scientific and policy circles, and there are now numerous smart city projects focused on providing ways to improve urban resilience. To understand the impacts of these smart city projects, 292 cases have been studied. Data related to the case studies are collected from different publications, gray literature, and websites. These initiatives are primarily centered in developed countries, with just a few projects identified in developing countries, which might be due to a lack of a smart city plans/projects or information that is not made publicly available. It is found that majority of the projects are based on the city scale, which is very likely. Most of the projects are mainly aimed at the reduction of CO_2 emission. Regarding the resilience stage, the projects contribute to different stages and particularly, adaptation and absorption. Energy sector received the peak attention by the smart city planners and policymakers. Concerning smart city dimensions, living has received the utmost attention. Mobility and data are equally considered as other important smart city dimensions. Much of the projects are owned by the government and are participatory in terms of governance. It is important to note that most of the projects pay attention to multiple smart city "dimensions" and "resilience characteristics." However, the commerce and societal well-being are two crucial elements that have received little attention. As a result, SCPs should pay more attention to the social and economic dimensions of resilience. More focus on health-related solutions is likely to be adopted in the future years, since the recent COVID-19 pandemic has sparked fresh debates about the need of urban resilience (Sharifi 2021b).

Acknowledgements This chapter was partially funded by Asia-Pacific Network for Global Change Research (Funder ID: https://doi.org/10.13039/100005536).

References

Haarstad H, Wathne MW (2019) Are smart city projects catalyzing urban energy sustainability? Energy Policy 129:918–925. https://doi.org/10.1016/j.enpol.2019.03.001

Khavarian-Garmsir AR, Sharifi A, Hajian Hossein Abadi M (2021) The social, economic, and environmental impacts of ridesourcing services: a literature review. Futur Transp 1(2):268–289. https://www.mdpi.com/2673-7590/1/2/16

Masrur H, Senjyu T, Islam MR, Kouzani AZ, Mahmud MAP (2021) Resilience-oriented dispatch of microgrids considering grid interruptions. IEEE Trans Appl Supercond 31(8):1–5. https://doi.org/10.1109/TASC.2021.3094423

Masrur H, Sharifi A, Islam MR, Hossain MA, Senjyu T (2021) Optimal and economic operation of microgrids to leverage resilience benefits during grid outages. Int J Electr Power Energy Syst 132:107137. https://doi.org/10.1016/j.ijepes.2021.107137

Perboli G, Rosano M (2020) A taxonomic analysis of smart city projects in North America and Europe. Sustainability 12(18):7813. https://www.mdpi.com/2071-1050/12/18/7813
SCW (2020) Smart cities world. https://www.smartcitiesworld.net/home. Accessed 22 February 2021
Shafiei Dastjerdi M, Lak A, Ghaffari A, Sharifi A (2021) A conceptual framework for resilient place assessment based on spatial resilience approach: an integrative review. Urban Clim 36:100794. https://doi.org/10.1016/j.uclim.2021.100794
Sharifi A (2016) A critical review of selected tools for assessing community resilience. Ecol Indic 69:629–647. https://doi.org/10.1016/j.ecolind.2016.05.023
Sharifi A (2019) A critical review of selected smart city assessment tools and indicator sets. J Clean Prod 233:1269–1283. https://doi.org/10.1016/j.jclepro.2019.06.172
Sharifi A (2020a) Trade-offs and conflicts between urban climate change mitigation and adaptation measures: a literature review. J Clean Prod 276:122813. https://doi.org/10.1016/j.jclepro.2020.122813
Sharifi A (2020b) A typology of smart city assessment tools and indicator sets. Sustain Cities Soc 53:101936. https://doi.org/10.1016/j.scs.2019.101936
Sharifi A (2020c) Urban resilience assessment: mapping knowledge structure and trends. Sustainability (Switzerland), 12(15). https://doi.org/10.3390/SU12155918
Sharifi A (2021a) Co-benefits and synergies between urban climate change mitigation and adaptation measures: a literature review. Sci Total Environ 750:141642. https://doi.org/10.1016/j.scitotenv.2020.141642
Sharifi A (2021b) The COVID-19 pandemic: lessons for urban resilience. In: Linkov I, Keenan JM, Trump BD (eds) COVID-19: systemic risk and resilience. Springer International Publishing, pp 285–297
Sharifi A, Allam Z, Feizizadeh B, Ghamari H (2021a) Three decades of research on smart cities: mapping knowledge structure and trends. Sustainability 13(13):7140. https://www.mdpi.com/2071-1050/13/13/7140
Sharifi A, Khavarian-Garmsir AR (2020) The COVID-19 pandemic: impacts on cities and major lessons for urban planning, design, and management. Sci Total Environ 749. https://doi.org/10.1016/j.scitotenv.2020.142391
Sharifi A, Khavarian-Garmsir AR, Kummitha RKR (2021b) Contributions of smart city solutions and technologies to resilience against the COVID-19 pandemic: a literature review. Sustainability 13(14):8018. https://www.mdpi.com/2071-1050/13/14/8018
Sharifi A, Yamagata Y (2016a) Principles and criteria for assessing urban energy resilience: a literature review. Renew Sustain Energy Rev 60:1654–1677. https://doi.org/10.1016/j.rser.2016.03.028
Sharifi A, Yamagata Y (2016b) Urban resilience assessment: multiple dimensions, criteria, and indicators. In: Yamagata Y, Maruyama H (eds) Urban resilience: a transformative approach. Springer International Publishing, pp 259–276
Stratigea A, Papadopoulou C-A, Panagiotopoulou M (2015) Tools and technologies for planning the development of smart cities. J Urban Technol 22(2):43–62. https://doi.org/10.1080/10630732.2015.1018725
Vale LJ, Campanella TJ (2005) Introduction: the cities rise again. Oxford University Press

Chapter 9
Do Smart Cities Projects Contribute to Urban Resilience? A Case Study Based in Taipei City, Taiwan

Nae-Wen Kuo and Chong-En Li

Abstract The trend of urbanization has increased the urban density of cities, resulting in larger populations facing greater disaster-related threats and risks. Furthermore, climate change is expected to increase the frequency, intensity, and effects of specific types of extreme weather events, thus raising disaster risk. Increasing urban resilience is a key challenge of urban governance and is essential to the development of forward-looking plans for reducing vulnerability and disaster threats. Furthermore, the rapid advancement of information and communication technologies (ICTs) and the Internet of things (IoT) has led to the increasing application of technologies from these fields in urban governance. The implementation of ICTs- and IoT-based smart city projects is becoming increasingly common worldwide. Nevertheless, the following questions require exploration: Do those smart city projects contribute significantly to urban resilience significantly? How should the performance of such projects be evaluated with respect to the dimensions of urban resilience? Accordingly, the main purpose of the present study was to develop an assessment toolkit and evaluate the contribution of smart city projects implemented in Taipei City. Two evaluation systems were developed, and eight experts were invited to participate in the evaluations. The first evaluation system involved "design principles for creating more resilient cities" developed by ResilientCity.org. The second system involved the "City Resilience Index" established by Arup Group Limited. After a screening was conducted, 10 Taipei-based smart city projects were selected as study cases, and the evaluation results for these projects with respect to multiple dimensions were discussed. The assessment toolkit introduced in the present study can be used by the city government to formulate strategies for addressing multiple problems. For example, future smart city projects should emphasize the design principles of redundancy and diversity. Furthermore, current ICTs- or IoT-based smart city projects overemphasized the characteristics of new technology and sometimes neglected key

---

N.-W. Kuo (✉) · C.-E. Li
Department of Geography, National Taiwan Normal University, Taipei, Taiwan
e-mail: niven@ntnu.edu.tw

C.-E. Li
e-mail: chongen.li@ntnu.edu.tw

A. Sharifi and P. Salehi (eds.), *Resilient Smart Cities*, The Urban Book Series,
https://doi.org/10.1007/978-3-030-95037-8_9

issues such as empowering stakeholders, which may result in challenges related to urban resilience enhancement and sustainable urban management.

Keywords Urban resilience · Smart city · ICTs · And IoTs

9.1 Introduction

9.1.1 Urban Resilience

The ongoing digital revolution has increased people's reliance on staying connected to the world. Currently, cities provide suitable conditions such as faster Internet, higher education institutes, and more real-time information flow. Thus, people are drawn to live in metropolitan areas. According to the United Nations report, the degree of urbanization reached 55% in 2018 and is projected to increase to 68% in 2050 (United Nations et al. 2019). When population density increases, the risk of disaster increases. Any sudden shocks may cause the collapse of an urban system, and common threats include extreme weather, terrorist attacks, pandemics, social problems, and water source pollution. Therefore, cities must have the ability to resist those threats; in other words, they must exhibit urban resilience. However, this term is still not well defined. Studies conducted by researchers such as Meerow et al. (2016) have proposed some useful definitions (Table 9.1). Several international organizations and institutes are also engaged in the development of resilient cities, and all of them define the concept of resilience differently (Table 9.2). Among them, some are even working on developing resilient cities by using evaluation assessment indicators or system architecture. After analyzing their framework and standards, we discovered that the dimensions of the resilience framework designed and constructed by ResilientCity.org and the Arup Group Limited are more complete relative to the other analyzed frameworks.

Table 9.1 Definition of resilience (urban resilience) in previous studies

Author (year)	Definition
Alberti et al. (2003)	… the degree to which cities tolerate alteration before reorganizing around a new set of structures and processes (p. 1170)
Godschalk (2003)	… a sustainable network of physical systems and human communities (p. 137)
Pickett et al. (2004)	… the ability of a system to adjust in the face of changing conditions (p. 373)
Ernstson et al. (2010)	To sustain a certain dynamic regime, urban governance also needs to build transformative capacity to face uncertainty and change (p. 533)
Campanella (2006)	… the capacity of a city to rebound from destruction (p. 141)

Source Meerow et al. (2016)

Table 9.2 Definition of resilience (urban resilience) provided by major organizations and institutions

Organizations or institutions	Definition
United Nations Office for Disaster Risk Reduction (formerly UNISDR)	Resilience is the ability of a system, community or society exposed to hazards to resist, absorb, accommodate, adapt to, transform and recover from the effects of a hazard in a timely and efficient manner, including through the preservation and restoration of its essential basic structures and functions through risk management (United Nations Office for Disaster Risk Reduction 2017)
United Union-Habitat	Resilience refers to the ability of any urban system to maintain continuity through all shocks and stresses while positively adapting and transforming towards sustainability (United Union-Habitat 2021)
Local Governments for Suitability	A 'Resilient City' is prepared to absorb and recover from any shock or stress while maintaining its essential functions, structures, and identity as well as adapting and thriving in the face of continual change. Building resilience requires identifying and assessing hazard risks, reducing vulnerability and exposure, and lastly, increasing resistance, adaptive capacity, and emergency preparedness (Local Governments for Suitability 2021)
Department for International Development	Disaster Resilience is the ability of countries, communities and households to manage change, by maintaining or transforming living standards in the face of shocks or stresses—such as earthquakes, drought or violent conflict—without compromising their long-term prospects (Department for International Development 2011)
Lutheran World Relief	When we talk about resilience, we are describing the capacity of an individual, a household or a community to absorb, adapt and potentially transform in the face of shocks and stressors that are triggered by a disrupting event, like a natural disaster, an outbreak of disease or armed conflict (Lutheran World Relief, 2021)
Rockefeller Foundation	Resilience is the capacity of individuals, communities and systems to survive, adapt, and grow in the face of stress and shocks, and even transform when conditions require it (Rockefeller Foundation 2017)
ResilientCity.org	A Resilient City is one that has developed capacities to help absorb future shocks and stresses to its social, economic, and technical systems and infrastructures so as to still be able to maintain essentially the same functions, structures, systems, and identity (ResilientCity.org 2012)

(continued)

Table 9.2 (continued)

Organizations or institutions	Definition
United States Agency for International Development	Resilience—defined by USAID as the ability of people, households, communities, countries and systems to mitigate, adapt to and recover from shocks and stresses in a manner that reduces chronic vulnerability and facilitates inclusive growth (United States Agency for International Development 2013)
European Commission	Resilience is the ability of an individual, a community or a country to cope with, adapt and recover quickly from the impact of a disaster, violence or conflict. Resilience covers all stages of disaster, from prevention (when possible) to adaptation (when necessary), and includes positive transformation that strengthens the ability of current and future generations to meet their needs and withstand crises (European Commission 2016)
Organization for Economic Cooperation and Development	Resilience is the ability of households, communities, and nations to absorb and recover from shocks, whilst positively adapting and transforming their structures and means for living in the face of long-term stressors, change, and uncertainty (Organization for Economic Cooperation and Development 2014)
Arup Group Limited	City resilience describes the capacity of cities to function, so that the people living and working in cities–particularly the poor and vulnerable–survive and thrive no matter what stresses or shocks they encounter (Arup Group Limited 2018)

First, we observed that ResilientCity.org, which is an open Internet forum operated mainly by a group of architectural and urban planning researchers and practitioners in Canada (Baba et al., 2020), proposed several guidelines relating to future-proofing city toolkits, climate change, energy scarcity, resilient design principles, urban design principles, and building design principles. With respect to "design principles for creating more resilient cities," the forum argued the following:

> It will take a new set of planning and design principles to create more resilient cities, more resilient communities, and more resilient buildings. This will take a significant revision of current thinking. As a starting point, we propose the following as an overarching set of principles for creating greater urban resilience…. (ResilientCity.org, 2012)

The six principles are diversity, redundancy, modularity, independence of system components, feedback sensitivity, capacity for adaptation, environmental responsiveness, and integration. Their meanings are presented in Table 9.3. The guideline is valuable for designing a resilient city when we plan to advocate resilient city policy, project, and program; it is also useful for assessing existing projects and evaluating their contribution.

Moreover, Arup Group Limited developed the City Resilience Indicator System, which is supported by the Rockefeller Foundation. It provides a helpful assessment framework with four main categories and related indicators for measuring and monitoring contributions to cities' resilience. These two institutes cooperated to establish the 100 Resilient Cities organization to promote resilient city activities. The Rockefeller Foundation plans to select more than 100 cities worldwide as targets for active assistance and coaching, with the objective of aiding them in improving their resilience and further implementing the developed resilient city evaluation system. This framework allows cities worldwide to assess their vulnerabilities, weaknesses, and disaster threats and subsequently promote specific improvement strategies to enhance urban resilience. This framework and its related documents are continually reviewed by researchers, and it is regarded as the property assessment toolkit for achieving urban resilience in a practical manner.

Lu and Stead (2013) and Spaana and Waterhout (2017) have reviewed the development of Arup Group Limited's City Resilience Indicator System; they discussed its practicality and indicated its advantages. Spaana and Waterhout (2017) explained the success achieved by Rotterdam in the Netherlands with the assistance of the Resilient Cities 100 organization. The city self-examined and identified its vulnerabilities and then proposed relevant projects to improve its weak dimensions and sectors. Therrien et al. (2020) consolidated evidence and insights from previous studies on efforts to achieve urban resilience that were published from 2005 to 2017 and were based on the urban resilience framework.

Arup Group Limited's City resilience indicator system comprises the four dimensions of health and welfare, economy and society, infrastructure and environment, and leadership and strategy (Table 9.4). Each dimension has three core objectives that a resilient city must achieve; collectively, these 12 core objectives are the key evaluation indicators for checking or evaluating policies, projects, and programs aimed at

Table 9.3 Resilient design principles as proposed by ResilientCity.org

Principle	Meaning
Diversity	Increasing the diversity of the various systems that comprise our cities is important because with greater diversity comes an increased ability to thrive, survive and bounce back from external shocks and stresses. Diversity of systems reduces the potential negative impact to the whole city of the failure of any one particular system. Increasing the diversity of systems means that we will want to maximize the diversity of different business types, institutions, sources of food, and industries, etc.
Redundancy	The increasingly likelihood of more energetic weather events associated with global climate change means that cities and their communities will need to build the capacity for resilience to more frequent and powerful environmental shocks and stresses. An increased redundancy of key infrastructure systems—including electrical power, fuel supply, waste water processing and, most important, food and potable water supply—means that if one system is compromised, there is enough redundancy in the overall system to fill in for the compromised system until it can be replaced or repaired. Although redundancy reduces efficiency, its necessary increases resilience
Modularity and Independence of System Components	Resilience capacity will be increased when system components have enough independence that damage or failure of one part or component of a system is designed to have a low probability of inducing failure of other similar or related components in the system
Feedback Sensitivity	Feedback Sensitivity is a system's ability to detect and respond to changes in its constituent parts. The more quickly a system can detect and respond to changes throughout the system, the greater its potential for effectively coping with these changes, and thus for resilience. Social, economic, and technical systems designed with tight feedback loops will increase resilience. In our cities, urban density is one of the important foundations for loop tightness. Density provides for reduced time and costs for moving information and materials throughout the system in an efficient and effective manner
Capacity for Adaptation	Resilience capacity will be increased by the relative adaptability of the various systems that comprise a city. City systems and infrastructure that are designed to quickly adapt to changing conditions and requirements will increase the overall resilience capacity of a city
Environmental Responsiveness and Integration	The resilience capacity of a city is increased by how responsive and integrated its systems and functions are with its natural systems, services and resources. Environmental responsiveness and integration will not only reduce the cost of creating and maintaining technical infrastructure, but reduce the relative probability of infrastructure suffering significant negative impacts from the increasing environmental shocks and stresses associated with climate change

Table 9.4 Arup Group Limited's City Resilience Indicator System

Dimension	Core objectives
1. Health and welfare	1.1 Minimal Human Vulnerability
	1.2 Diverse Livelihood & Employment
	1.3 Effective Safeguards to Human Health & Life
2. Economy and society	2.1 Collective Identity & Mutual Support
	2.2 Comprehensive Security & Rules of Law
	Sustainable Economy
3. Infrastructure and environment	3.1 Reduced Exposure & Fragility
	3.2 Effective Provision of Critical Services
	Reliable Mobility & Communication
4. Leadership and strategy	4.1 Effective Leadership & Management
	4.2 Empowered Stakeholders
	4.3 Integrated Development Planning

enhancing urban resilience. The detailed definition for each indicator can be found in the original report (Arup Group Limited 2018).

9.1.2 Smart City

ICTs and IoT have developed rapidly in recent years. These advanced technologies help cities to manage their available resources more efficiently, which corresponds to the concept of smart city. Studies have proposed several definitions for the smart city concept; however, a standardized definition has not yet been established. Camero and Alba (2019) consolidated and reported on some commonly cited smart city definitions in Table 9.5.

Currently, the smart city concept is applied in several fields such as disaster prevention, transportation, governance, and environmental protection. Smart cities improve the living convenience of urban residents; however, not every smart city project necessarily enhance urban resilience. Some projects had this effect, it seems as a smart solution. Figure 9.1 illustrates a Venn diagram of the relationship between these terms. Projects that serve the objectives of both concepts should be promoted worldwide because they can further the goal of increasing urban resilience quickly and efficiently through new technologies. However, the literature related to smart solutions is not comprehensive, and a set of procedures for gradual evaluation must be developed.

9.2 Research Area: Taipei City

Taipei City is the capital city of Taiwan. It is a city with 2.6 million residents and a land area of 271.8 km^2. Taiwan is a high-income country with the central government managing an annual budget of more than NT$2 trillion (nearly US$72.4 billion) annual budget in 2019. Among the local governments, the Taipei City Government has the highest total and per capita general budget (Fig. 9.2) in the same year. (Directorate-General of Budget 2019) It is a strong base for developing a smart city.

Taipei City has been experimenting with the smart city concept for several years. However, the various departments and offices of the city government are required to independently implement their respective smart city projects without cooperation from each other in the past. Until 2016, the city government has been integrating

Table 9.5 Definitions of smart city

Author (year)	Definition
Hall et al. (2000)	A city that monitors and integrates conditions of all of its critical infrastructures, including roads, bridges, tunnels, rails/subways, airports, seaports, communications, water, power, even major buildings, can better optimize its resources, plan its preventive maintenance activities, and monitor security aspects while maximizing services to its citizens… systems and structures will monitor their own conditions and carry out self-repair, as 'needed'
Harrison et al. (2010)	Connecting the physical infrastructure, the IT infrastructure, the social infrastructure, and the business infrastructure to leverage the collective intelligence of the city
Caragliu et al. (2011)	We believe a city to be smart when investments in human and social capital and traditional (transport) and modern (Information and Communications Technology, ICTs) communication infrastructure fuel sustainable economic growth and a high quality of life, with a wise management of natural resources, through participatory governance
Manville et al. (2014)	A Smart City is a city seeking to address public issues via ICTs-based solutions on the basis of a multi-stakeholder, municipally based partnership
Giffinger and Pichler-Milanović (2007)	A Smart City is a city well performing in a forward-looking way in these six characteristics, built on the 'smart' combination of endowments and activities of self-decisive, independent and aware citizens
Dameri (2013)	A smart city is a well defined geographical area, in which high technologies such as ICTs, logistic, energy production, and so on, cooperate to create benefits for citizens in terms of well being, inclusion and participation, environmental quality, intelligent development; it is governed by a well defined pool of subjects, able to state the rules and policy for the city government and development

Source Camero and Alba (2019)

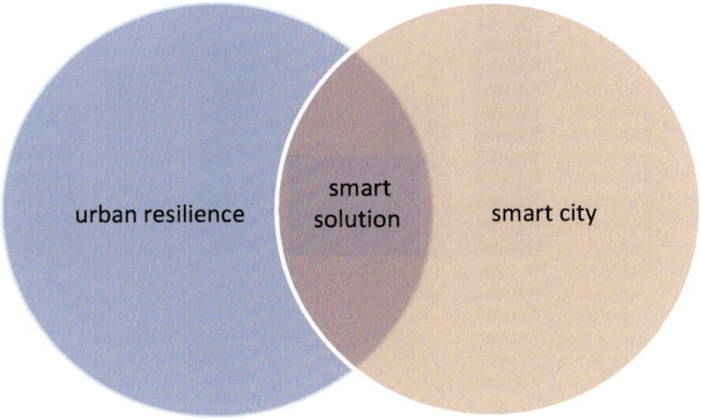

Fig. 9.1 Venn diagram of urban resilience and smart city

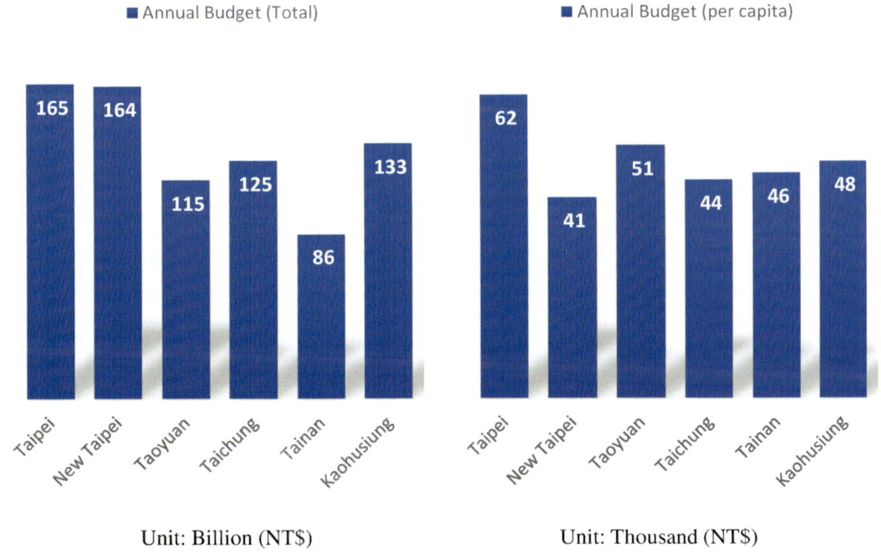

Fig. 9.2 Annual budget of Taipei City and other special municipalities in 2019

the various demands of every field and interacting with private sectors; specific moves include the promotion of the Smart Taipei concept and establishment of the Taipei Smart City Project Management Office under the Department of Information Technology. The primary purposes of the office are to classify the citizens' needs, identify local technology companies that can address these needs, and transform Taipei City into a "living lab" for the application of IoT-based smart solutions.

Many projects are currently being implemented in Taipei City in areas such as smart transportation, building infrastructure, education, environments, economy, security, health, and public administration. As of August 2021, a total of 236 projects are either ongoing or completed (Fig. 9.3). Several major projects are listed in Table 9.6.

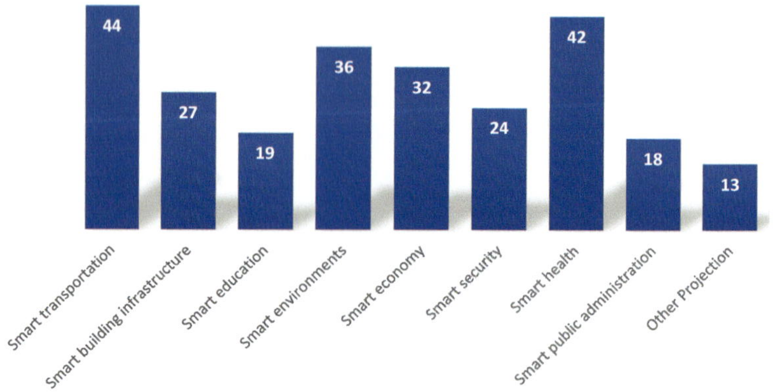

Fig. 9.3 Categories of smart city projects in Taipei

Table 9.6 Major smart city projects in Taipei

Categories	Projects
smart transportation	Smart transportation systems (shared bike, motorbike, electric vehicle, and smart parking systems), 3D AR indoor navigation in Taipei Main Station, parking grid automation, driverless buses/cars, traffic flow recognition system, car recorder value-added service, real-time bus/metro/traffic information…
Smart building infrastructure	Smart public housing, Intelligent structural monitoring system for Shezi Bridge, household water quality/usage monitoring, energy-saving air conditioning system…
Smart education	Online education, AR English learning amusement park…
Smart environments	Intelligent streetlamps, Microscale air quality sensor systems, Smart management system for Feitsui Reservoir, AIoT in forest farm, networked trash can…
Smart economy	Electronic payment, shopping zone's Kiosk multimedia information station…
Smart security	Smart 119 app, Liquefied petroleum gas safety monitoring and cloud management platform, intelligent police enforcement, disaster victims care services…
Smart health	New technology in municipal hospitals, wear-free elderly safety detection systems…
Smart public administration	Big database center, obsolete public property/second-hand goods shopping platform, tax calculation service, receipt cloudification…

Sources Websites of smarttaipei and smart city & IoT

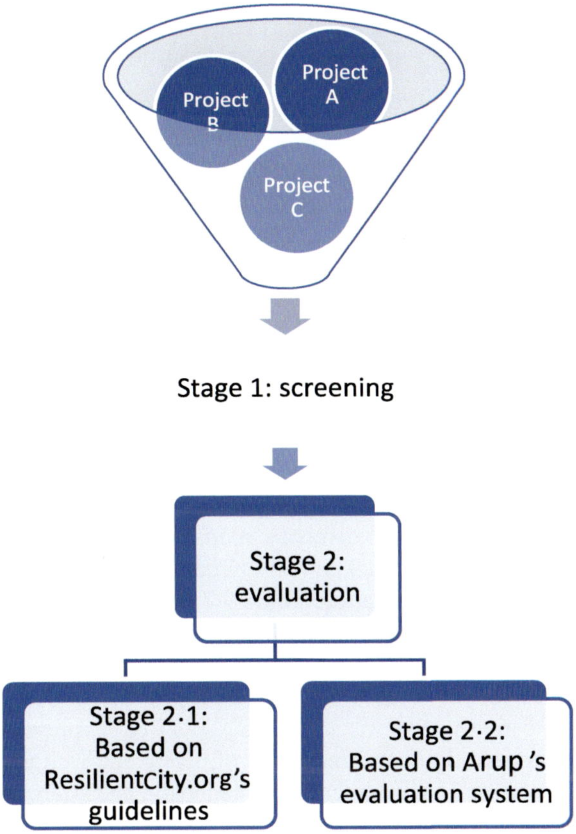

Fig. 9.4 Research procedure

9.3 Research Method and Procedure

The present study was conducted using an expert interview method, and experts from two universities in Taipei City[1] were invited to participate in the study. These experts specialized in the fields of urban resilience, smart city, and sustainable city research. The research process chart is presented in Fig. 9.4. First, we screened all smart city projects in Taipei and identified the high-potential cases that could contribute to urban resilience (Stage 1). Next, these high-potential cases were examined and evaluated by experts through face-to-face interviews. To ensure that the all experts shared the same cognition with respect to the meanings of urban resilience and smart city, we briefed them on the concepts of urban resilience and smart city and all of the smart city projects in Taipei City before allowing them to begin the screening and evaluation procedure.

In Stage 2, the contribution of each project was evaluated using two evaluation systems. The first evaluation system entailed the use of ResilientCity.org's "design

[1] National Taiwan Normal University and the University of Taipei.

Table 9.7 Example of an evaluation survey

Project	Diversity	Redundancy	Modularity and independence of system components	Feedback sensitivity	Capacity for adaptation	Environmental responsiveness and integration
Project A	○	×	×	○	○	×

principles for creating more resilient cities." The second evaluation system entailed the use of the "City Resilience Index" established by Arup Group Limited.

In Stage 2.1 of the present study, we explored the question of whether the smart city projects fit the design principles of a resilient city and asked the experts to determine whether each selected project satisfied each item related to design principles. The guidelines adopted in this research were based on the six design principles for a resilient city that were developed by ResilientCity.org (2012). The six principles are outlined as follows: diversity, redundancy, modularity and independence of system components, feedback sensitivity, capacity for adaptation, and environmental responsiveness and integration. Table 9.7 presents an example of how an evaluation was reported by an expert when they consider a given Project A to be compatible with the principles of diversity, feedback sensitivity, and capacity for adaptation.

To identify which dimensions a project contributed to, Arup Group Limited's evaluation system was used in Stage 2.2 of the present study. This evaluation system comprised the four dimensions of health and welfare, economy and society, infrastructure and environment, and leadership and strategy, which are collectively linked to 12 core objectives. We designed a checklist to allow the experts to evaluate each project easily. Specifically, they were only required to write "○" to indicate their opinion that a given project can achieve a particular assessment indicator.

We invited 10 experts to participate in our research survey, 8 of whom subsequently agreed to participate in our face-to-face interviews. When a participant encountered a problem, an assistant would act to ensure that all of the experts were providing responses on the basis of the same knowledge background and not being affected by interruptions or receiving guidance.

We collected experts' responses for 10 projects; hence, each assessment principle (indicator) was scored between 0 and 80, with the highest score of 80 indicating that all experts were of the opinion that all 10 projects were compatible with a particular assessment principle. For our statistical analysis, the cumulative total was used instead of an averaged figure such that the attributes of assessment principles (indicators) and projects could be compared easily.

9.4 Results and Discussion

9.4.1 *Stage 1: Screening Results*

After the screening procedure was conducted, 10 projects were selected as representative cases. Only 10 were selected because most of the other projects focused only on the applications of ICTs or IoT and were not directly linked to urban resilience; moreover, some projects were excessively small in scale and implemented citywide. A brief introduction of these 10 projects is presented in Table 9.8. These 10 projects including: smart transportation systems (shared bike, motorbike, electric vehicle, and smart parking systems), smart public housing, new technology in municipal hospitals, intelligent streetlamps, the smart 119 app, wear-free elderly safety detection systems, microscale air quality sensor systems, an intelligent structural monitoring system for Shezi Bridge, a smart management system for Feitsui Reservoir, and an liquefied petroleum gas safety monitoring and cloud management platform.

9.4.2 *Stage 2: Evaluation*

Eight experts participated in this study. They individually evaluated the 10 selected projects according to each project's compatibility with specific principles (indicators). Hence, the cumulative score for each principle ranged from 0 to 80 and measured the contribution of the selected smart city projects with respect to a given resilience principle.

The Stage 2.1 results are presented in Table 9.9. The experts believed that the 10 selected smart city projects helped enhance Taipei City's resilience with respect to the principle of feedback sensitivity, for which a 75% fit ratio was observed. The principle with the second highest score was the principle of capacity for adaptation (65%). However, fit ratios of 36 and 46% were reported for the principles of redundancy and diversity, respectively; in other words, the experts felt that the 10 projects do not provide sufficient contributions with respect to the principles of redundancy and diversity and that these two resilience principles should be emphasized in future projects.

A comparison of the 10 projects revealed that the smart transportation systems (shared bike, motorbike, electric vehicle, and smart parking systems) and the smart management system for Feitsui Reservoir scored the highest (31), suggesting that the experts believed that these two projects have the best potential to contribute to Taipei City's resilience. In addition, the smart transportation systems (shared bike, motorbike, electric vehicle, and smart parking systems) were considered to have performed well with respect to the principles of diversity ($S = 8$; $r = 100\%$), modularity and independence of system components ($S = 7$; $r = 87.5\%$), and capacity for adaptation ($S = 6$; $r = 75\%$).

Table 9.8 Brief introduction of selected projects

Selected projects	Brief introduction
Smart transportation systems (shared bike, motorbike, electric vehicle, and smart parking systems)	Improve the efficiency of using transportation equipment and parking spaces in a shared mode with smart way, reduce the number of private vehicles, and solve traffic problems
Smart public housing	Use ICTs to construct an intelligent community and reverse the negative stereotypes of public housing
New technology in municipal hospitals	Import information tools to improve medical efficiency and make caring for patients more intelligent
Intelligent streetlamps	Install sensors on street lights and using environmental sensing data and telecommunication users' traffic data to predict light intensity to save energy and improve public security issues
Smart 119 app	Use smartphone APP to achieve quick reports, location transmission, visualization, online first aid guidance, and other functions
Wear-free elderly safety detection systems	Use the sensors on the ceiling to grasp the breathing and movement trajectory of the elderly to predict or report the health status of the elderly, and achieve the optimal allocation of care human resources
Microscale air quality sensor systems	Distribute low-cost and small air quality monitoring devices throughout Taipei City to obtain rapidly changing air quality conditions, providing citizens a response
Intelligent structural monitoring system for Shezi Bridge	Strengthen the mastery of the structural status of Shezi Bridge to prevent damage and improve its maintenance quality
Smart management system for Feitsui Reservoir	Provide applications such as access control, pipelines, water level, weather monitoring, and intelligently manage reservoirs
Liquefied petroleum gas safety monitoring and cloud management platform	When the gas leaks, the gas supply is automatically blocked, and the relevant units are notified to deal with it to reduce disasters

However, the projects involving smart public housing, intelligent streetlamps, the smart 119 app, and wear-free elderly safety detection systems were considered to have contributed less to the resilience of the city relative to the other projects.

Table 9.10 presents the Stage 2.2 results, which were based on Arup Group Limited's evaluation system. The score analysis component of Stage 2.2 was identical to that of Stage 2.1. The fit ratio (r) of each indicator indicated the contribution made by the 10 selected smart city projects. The results suggested that the categories of

Table 9.9 Evaluation results based on ResilientCity.org's guidelines

Projects	Principles						
	Diversity	Redundancy	Modularity and independence of system components	Feedback sensitivity	Capacity for adaptation	Environmental responsiveness and integration	The sum score of project (P)
Smart transportation systems (shared bike, motorbike, electric vehicle, and smart parking systems)	8[a]	4	7	4	6	2	31
Smart public housing	4	3	1	5	6	4	23
New technology in municipal hospitals	5	4	6	5	7	2	29
Intelligent streetlamps	3	3	3	5	3	6	23
Smart 119 app	5	0[b]	4	6	4	4	27
Wear-free elderly safety detection systems	4	1	3	5	6	4	23
Microscale air quality sensor systems	4	2	5	8	5	5	29
Intelligent structural monitoring system for Shezi Bridge	1	4	5	8	5	5	28

(continued)

Table 9.9 (continued)

Projects	Principles						
	Diversity	Redundancy	Modularity and independence of system components	Feedback sensitivity	Capacity for adaptation	Environmental responsiveness and integration	The sum score of project (P)
Smart management system for Feitsui Reservoir	2	4	5	7	6	7	31
Liquefied petroleum gas safety monitoring and cloud management platform	1	4	4	7	4	5	25
The sum score of principle (*S*)	37	29	43	60	52	44	
The ratio of fit ($r = S/80$)	46%	36%	54%	75%	65%	55%	

[a]Indicates that eight experts agreed that this project conformed to the principle of diversity
[b]Indicates that all experts agreed that this project did not conform to the principle of redundancy.

Table 9.10 Evaluation results based on Arup Group Limited's Resilience City Indicator System

Categories	Health & well-being (total category score = 119)			Economy & society (total category score = 66)			Infratructure & ecosystems (total category score = 125)			Leadership & strategy (total category score = 79)			The sum scores of project (P)
Projects	Indicators												
	Minimal human vulnerability	Diverse livelihood & employment	Effective safeguards to human health & life	Collective identity & mutual Support	Comprehensive security & rules of law	Sustainable economy	Reduced exposure & fragility	Effective provision of critical services	Reliable mobility & communication	Effective leadership & management	Empowered stakeholders	Integrated development planning	
Smart transportation systems (shared bike, motorbike, electric vehicle, and smart parking systems)	3	5	3	1	2	1	5	7	7	2	2	6	44
Smart public housing	6	4	6	5	7	1	5	4	3	1	1	7	50
New technology in municipal hospitals	8	0	8	3	3	0	2	5	3	3	2	6	43
Intelligent streetlamps	4	0	2	0	4	0	5	5	3	4	0	3	30
Smart 119 app	8	1	7	1	5	2	3	6	5	3	2	2	45
Wear-free elderly safety detection systems	8	1	8	3	5	1	3	3	2	3	0	3	40

(continued)

Table 9.10 (continued)

Categories	Health & well-being (total category score = 119)			Economy & society (total category score = 66)			Infratructure & ecosystems (total category score = 125)			Leadership & strategy (total category score = 79)			The sum scores of project (P)
Projects	Indicators												
	Minimal human vulnerability	Diverse livelihood & employment	Effective safeguards to human health & life	Collective identity & mutual Support	Comprehensive security & rules of law	Sustainable economy	Reduced exposure & fragility	Effective provision of critical services	Reliable mobility & communication	Effective leadership & management	Empowered stakeholders	Integrated development planning	
Microscale air quality sensor systems	5	0	6	1	2	1	5	1	2	2	0	5	30
Intelligent structural monitoring system for Shezi Bridge	4	0	3	0	5	0	8	4	1	5	0	2	32
Smart management system for Feitsui Reservoir	5	0	1	0	5	1	8	5	2	4	0	5	36
Liquefied petroleum gas safety monitoring and cloud management platform	6	3	4	0	5	2	6	4	3	4	0	2	39
The sum score of indicators (*S*)	57	14	48	14	43	9	50	44	31	31	7	41	
The ratio of fit ($r = S/80$)	71.25%	17.50%	60.00%	17.50%	53.75%	11.25%	62.50%	55.00%	38.75%	38.75%	8.75%	51.25%	

infrastructure and ecosystems (total category score, TCS = 125) and health and well-being (TCS = 119) were the two dimensions to which the 10 projects contributed the most. By contrast, the 10 projects contributed the least to the categories of leadership and strategy (TCS = 79) and economy and society (TCS = 66).

Specifically, the indicators with the highest score were, in descending order, minimal human vulnerability, ($S = 57$) reduced exposure and fragility ($S = 50$), and effective safeguards to human health and life ($S = 48$); these three indicators belong to the two aforementioned categories with the highest scores (health and well-being and infrastructure and ecosystems). In summary, these 10 projects performed well with respect to minimizing human vulnerability, reducing exposure and fragility, and safeguarding human health and life. These items are always found among the core characteristics of most resilience city projects proposed and implemented in several countries, and our findings correspond to those reported by previous studies.

However, low scores were obtained for the following indicators: empowered stakeholders ($S = 7$), sustainable economy ($S = 9$), diverse livelihood and employment ($S = 14$), and collective identity and mutual support ($S = 14$). This indicates that the 10 smart city projects implemented in Taipei City performed poorly for these four indicators (relative to the other indicators); hence, improvements must be made. For example, the 10 smart city projects contributed little to the indicator empowered stakeholders, which is under the category of economy and society; this signifies that the 10 ICTs- and IoT-based projects neglected empowered stakeholders, which is a key topic in the political and social sectors for sustainable urban management.

For the performance of each smart city project, the smart public housing project ($S = 50$) received the highest score because public housing projects constitute a crucial topic that has attracted the attention of numerous scholars. Moreover, this project was determined to have performed well with respect to the following indicators: minimal human vulnerability ($S = 6$), effective safeguards to human health and life ($S = 6$), comprehensive security and rules of law ($S = 7$), and integrated development planning ($S = 7$). The project with the second and third highest scores were the smart 119 app ($S = 45$) and smart transportation systems (shared bike, motorbike, electric vehicle, and smart parking systems; $S = 44$). These three projects appeared to have made the most contributions to Taipei City's resilience and substantially improved the daily lives of most residents.

The evaluation results presented in Tables 9.8 and 9.9 can be challenging to read and interpret. Therefore, two infographics were designed to more clearly present the complex results (Figs. 9.5 and 9.6). The differences among six design principles are emphasized in Fig. 9.5, and the performance of the projects (a unique color is assigned for each project) can be compared easily.

Furthermore, the evaluation results obtained using Arup Group Limited's Resilience City Indicator System were more complex (relative to those obtained using the first system) because this system comprises 4 categories and 12 major evaluation indicators. Figure 9.6 presents the wind rose plot used to interpret the assessment results; all relevant information is included in this figure. The score for each indicator is indicated, the results are summarized into four categories, and the performance of the projects can be compared (a unique color is assigned for each

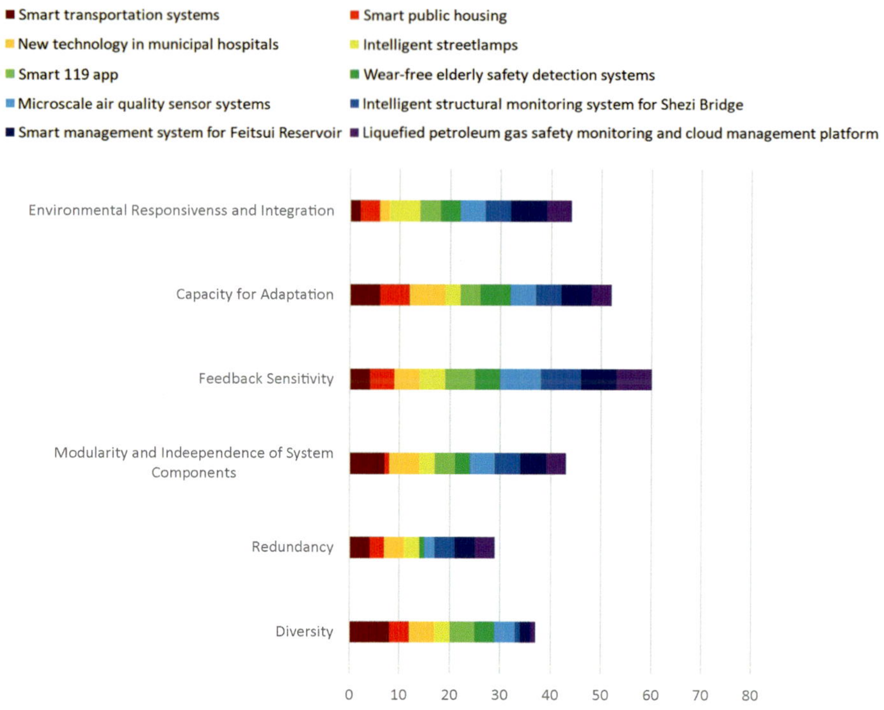

Fig. 9.5 Evaluation results based on ResilientCity.org's guidelines

project). These infographics are provided for interpreting the complex data because data visualization is also a key aspect of this type of research.

9.5 Conclusion

How urban resilience can be enhanced has become a key challenge in urban governance, and it is essential in the development of policies, strategies, and projects for enhancing resilience and reducing vulnerability and disaster threats. Furthermore, the rapid development of ICTs and IoT has increased the frequency at which these technologies are applied in urban governance. ICTs- and IoT-based smart city projects have become a trend in numerous cities, including Taipei City. Smart city projects are increasingly being implemented worldwide, and smart city solutions offer considerable potential as means of increasing urban resilience. Therefore, the practical effectiveness of these projects should be investigated. The following questions required exploration: Do smart city projects contribute substantially to the

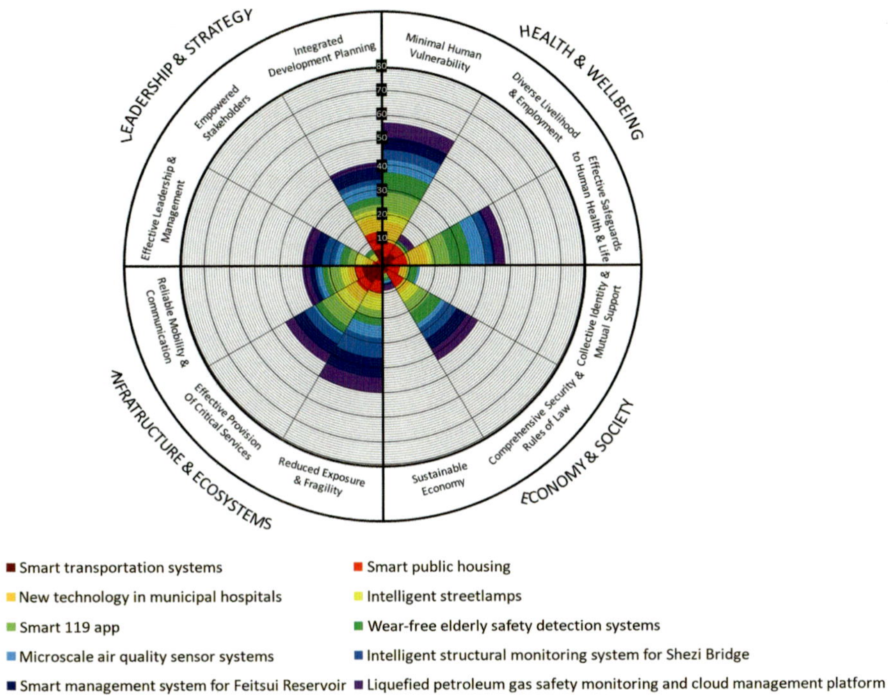

Fig. 9.6 Evaluation results based on Arup Group Limited's Resilience City Indicator System

enhancement of urban resilience? How should their performance be assessed with respect to the multiple dimensions of urban resilience?

Accordingly, the main purpose of the present study was to evaluate the contributions of Taipei City's smart city projects to its urban resilience. Two sets of evaluation systems were developed, and experts were invited to participate in the evaluation process. The first evaluation system was based on ResilientCity.org's design principles for creating more resilient cities. The second system was based on the Resilience City Indicator System established by Arup Group Limited. After a screening procedure was conducted, 10 Taipei City smart city projects were selected as the study cases, and the evaluation results for these projects with respect to multiple dimensions were discussed.

The Taipei City Government has developed numerous smart city projects to solve urban problems; however, only a small proportion (about 4.7%) of these smart city projects is considered to be able to contribute to the urban resilience of Taipei City. This is because most projects emphasize the only applications of new technology and follow differing objectives such that enhancing urban resilience is just one of many objectives; moreover, many projects were designed to solve a specific problem handled by a specific department. In addition, the city government is always focused on solving current problems, resulting in the development of short-sighted plans and projects.

The evaluation results obtained using the ResilientCity.org guidelines indicate that the smart city projects could improve the resilience of Taipei City with respect to the principles of feedback sensitivity (75%) and capacity for adaptation (65%). However, low fit ratios of 36 and 46% were reported for the principles of redundancy and diversity, respectively. Among the 6 design principles, the 10 projects could not contribute sufficiently with respect to the principles of redundancy and diversity. The weaknesses identified in the present study can be used by the city government to formulate strategies for improving these problems. For example, future smart projects must emphasize redundancy and diversity.

Furthermore, the evaluation results obtained using Arup Group Limited's Resilience City Indicator System indicate that the smart city projects contributed substantially to infrastructure and ecosystems (TCS = 125) and health and well-being (TCS = 119). However, the projects contributed less to the dimensions of leadership and strategy (TCS = 79) and economy and society (TCS = 66).

For individual indicators, the indicators with the highest score were minimal human vulnerability ($S = 57$), reduced exposure and fragility ($S = 50$), and effective safeguards to human health and life ($S = 48$). These items are always found among the core characteristics of most resilience city projects proposed and implemented in numerous countries, and our findings correspond to those of previous studies. However, low scores were reported for following indicators: empowered stakeholders ($S = 7$), sustainable economy ($S = 9$), diverse livelihood and employment ($S = 14$), and collective identity and mutual support ($S = 14$). In other words, the 10 smart city projects had little contributions to the empowered stakeholders subcategory (which is under the economy and society category), suggesting that the ICTs- and IoT-based smart city projects overemphasized the characteristic of new technology and neglected the empowerment of stakeholders, which is an essential challenge in the social component of sustainable urban management.

A comparison of the performance of the smart city projects revealed that the smart transportation systems (shared bike, motorbike, electric vehicle, and smart parking systems) and smart management of Feitsui Reservoir scored the highest (31) based on the ResilientCity.org guidelines. However, the smart public housing ($S = 50$), smart 119 app ($S = 45$), and smart transportation systems (shared bike, motorbike, electric vehicle, and smart parking systems) ($S = 44$) exhibited good performance when they were evaluated using Arup Group Limited's Resilience City Indicator System. This is a notable finding that should be investigated further in future studies; it suggests that the performance of one project cannot be judged comprehensively using only one evaluation system because different evaluation systems use different core assessment criteria. In addition, the characteristics of the multiple objectives of a smart city project may affect evaluation results.

Notably, the present study still has its limitations. First, due to the large number of smart city projects being implemented, establishing a standard screening procedure may be necessary to objectively identify representative cases. Second, different people may have a different understanding of a smart city and urban resilience. Therefore, all experts must have the same understanding regarding the definitions of principles and the selection of evaluation indicators for evaluating selected projects; otherwise, the results may be affected by bias.

Acknowledgements This chapter was partially funded by Asia-Pacific Network for Global Change Research (Funder ID: https://doi.org/10.13039/100005536).

References

Alberti M, Marzluff JM, Shulenberger E, Bradley G, Ryan C, Zumbrunnen C (2003) Integrating humans into ecology: opportunities and challenges for studying urban ecosystems. BioScience 53(12):1169–1179

Arup Group Limited (2018) City resilience index: understanding and measuring city resilience.

Baba K, Nagata Y, Kawakubo S, Tanaka M (2020) A framework and indicators of resilience. In Tanaka M, Baba K (eds) Resilient Policies in Asian Cities. Springer, pp 3–45. https://doi.org/10.1007/978-981-13-8600-8_1

Camero A, Alba E (2019) Smart city and information technology: a review. Cities 93:84–94

Campanella TJ (2006) Urban resilience and the recovery of New Orleans. J Am Plan Assoc 72(2):141–146

Caragliu A, Del Bo C, Nijkamp P (2011) Smart cities in Europe. J Urban Technol 18(2):65–82

Dameri RP (2013) Searching for smart city definition: a comprehensive proposal. Int J Comput Technol 11(5):2544–2551

Department for International Development (2011) Defining disaster resilience: a DFID approach paper. https://www.gov.uk/government/uploads/system/uploads/attachment_data/file/186874/defining-disaster-resilience-approach-paper.pdf

Directorate-General of Budget (2019) https://www.dgbas.gov.tw/np.asp?ctNode=41

Ernstson H, Van der Leeuw SE, Redman CL, Meffert DJ, Davis G, Alfsen C, Elmqvist T (2010) Urban transitions: on urban resilience and human-dominated ecosystems. Ambio 39(8): 531–545

European Commission (2016) Building resilience: the EU's approach. http://ec.europa.eu/echo/files/aid/countries/factsheets/thematic/resilience_en.pdf

Giffinger R, Pichler-Milanović N (2007) Smart cities: ranking of European medium-sized cities. Vienna University of Technology

Godschalk DR (2003) Urban hazard mitigation: creating resilient cities. Nat Hazards Rev 4(3):136–143.

Hall RE, Bowerman B, Braverman J, Taylor J, Todosow H, Von Wimmersperg U (2000) The vision of a smart city. Brookhaven National Lab, Ubtan

Harrison C, Eckman B, Hamilton R, Hartswick P, Kalagnanam J, Paraszczak J, Williams P (2010) Foundations for smarter cities. IBM J Res Dev 54(4):1–16

Local Governments for Suitability (2021) http://old.iclei.org/index.php?id=36

Lu P, Stead D (2013) Understanding the notion of resilience in spatial planning: a case study of Rotterdam, The Netherlands. Cities 35:200–212

Lutheran World Relief (2021) https://indepth.lwr.org/resilience

Manville C, Cochrane G, Cave J, Millard J, Pederson JK, Thaarup RK, Liebe A, Wissner M, Massink R, Kotterink B (2014) Mapping smart cities in the EU

Meerow S, Newell JP, Stults M (2016) Defining urban resilience: a review. Landsc Urban Plan 147:38–49

Organization for Economic Cooperation and Development (2014) Guidelines for resilience systems analysis: How to analyze risk and build a roadmap to resilience. https://www.oecd.org/dac/Resilience%20Systems%20Analysis%20FINAL.pdf

Pickett ST, Cadenasso ML, Grove JM (2004) Resilient cities: meaning, models, and metaphor for integrating the ecological, socio-economic, and planning realms. Landsc Urban Plan 69(4):369–384
ResilientCity.org (2012) https://www.resilientcity.org/index.cfm?pagepath=Resilience&id=11449
Rockefeller Foundation. https://www.rockefellerfoundation.org/our-work/topics/resilience/). Date Last Accessed 14 February 2017
Spaans M, Waterhout B (2017) Building up resilience in cities worldwide–Rotterdam as participant in the 100 Resilient Cities Programme. Cities 61:109–116
Therrien MC, Usher S, Matyas D (2020) Enabling strategies and impeding factors to urban resilience implementation: a scoping review. J Contingencies Cris Manag 28(1):83–102
United Nations, Department of Economic and Social Affairs, Population Division (2019) World urbanization prospects: the 2018 revision (ST/ESA/SER.A/420). United Nations
United Nations Office for Disaster Risk Reduction (2017) Disaster resilience scorecard for cities
United States Agency for International Development (2013) The resilience agenda: measuring resilience in USAID. https://www.usaid.gov/sites/default/files/documents/1866/Technical%20Note_Measuring%20Resilience%20in%20USAID_June%202013.pdf
United Union-Habitat (2021) https://unhabitat.org/resilience

Chapter 10
Envisioning Sustainable and Resilient Petaling Jaya Through Low-Carbon and Smart City Framework

Melasutra Md Dali, Ayyoob Sharifi, and Yasmin Mohd Adnan

Abstract The Malaysia Smart City Framework (MSCF) launched in September 2019 serves as a national guideline for cities and their local governments, as well as for other relevant agencies and stakeholders, in developing and implementing smart city initiatives. The involvement of national, state, and local authority agencies, along with the private sector, has coordinated and streamlined the efforts of developing smart cities across Malaysia. The United Nations Development Program (UNDP), the Sustainable Energy Development Authority (SEDA), the Ministry of Energy, Science, Technology, Environment and Climate Change (MESTECC), and the Global Environment Facility (GEF) (for financial support) have deployed a project that highlights cities with low-carbon entitled 'Green Technology Application for the Development of Low-Carbon Cities' (GTALCC) (Lee, N. (2019). Low-Carbon cities—Malaysia's response to global climate emergency. UNDP). This project is imminent to generate an ideal urban infrastructure that meets the demands of citizens seeking better quality of life. As such, this case study looked into the Petaling Jaya City Council (PJCC) Smart City initiatives based on various low-carbon projects for urban planning, as well as their attainment thus far. This study sheds light on the initiatives taken by PJCC that emphasise on low-carbon approach with a touch of modern intelligent technology to formulate a smart city that is resilient to climate

M. Md Dali
Centre for Sustainable Urban Planning and Real Estate (SUPRE), University of Malaya, 50603, Malaya, Malaysia
e-mail: melasutr@um.edu.my

Department of Urban and Regional Planning, Faculty of Built Environment, University of Malaya, 739-8530, Kuala Lumpur, Malaysia

A. Sharifi (✉)
Graduate School of Humanities and Social Sciences, Hiroshima University, Higashihiroshima, Japan
e-mail: sharifi@hiroshima-u.ac.jp

Y. Mohd Adnan
Department of Real Estate, Faculty of Built Environment, University of Malaya, Kuala Lumpur, Malaysia
e-mail: yasmin_alambina@um.edu.my

A. Sharifi and P. Salehi (eds.), *Resilient Smart Cities*, The Urban Book Series,
https://doi.org/10.1007/978-3-030-95037-8_10

change. The presented discussion is beneficial in terms of policy knowledge for other local authorities who wish to mimic such initiatives.

Keywords Smart city framework · Low-carbon resilient city · Resilience · Malaysia

10.1 Introduction

The share of global urban population is currently about 55% and this has been predicted to escalate up to 68% by 2050 (Ovington 2020). Cities are the engines for world economic activities that consume massive energy and emit undesired amounts of greenhouse gases (GHG). In order to reduce carbon emissions, urban centres must significantly reduce their energy consumption and start using alternative renewable energy sources. As numerous countries have begun setting their targets to decrease GHG emissions, it is integral for cities to significantly minimise their energy usage. Hence, this study assessed the initiatives taken by smart cities to decrease GHG emissions, as well as the role of internet of things (IoT) and telecommunication networks in smart cities. Ideal management of urbanisation, along with the related risks and opportunities, is crucial to attain higher economic progress, lower poverty rate, better human welfare, and greater environmental sustainability. This scenario highlights the pressing need to implement effective urban planning policies and strategies in addressing vital development primacies.

10.2 Sustainable Development and Climate Change Agenda: The Malaysian Perspective

Urbanisation is crucial in addressing the effects of climate change. Massive energy consumption by cities due to their role as economic and social hubs has led to uncontrollable GHG emissions. Hence, several aspects have been identified where energy conservation and emission reduction are viable in urban development, such as carbon-free transportation system, compact urban forms, better accessibility, and effective waste management, to name a few. These listed aspects also promote more job vacancies, better public health, lower pollution level, and greater economic progress. Economic and social urban development infused with low carbon goals catalyses new climate economy as part of the sustainable national growth agenda.

Numerous countries have prioritised policies that uphold low-carbon urban development. As such, this paper reviews the implementation of low-carbon urban policies in a city, their efficacy, and viable options towards green urban development. Additionally, this paper explored the role of policymakers in ascertaining the deployment of suitable smart infrastructure to meet the objective of smart city development. Based on the information gathered at the local-level initiatives and efforts, this present study unravels the mechanism appropriate for low-carbon urban development, while simultaneously identifying an appropriate smart city urban framework to address national objectives.

As a rapidly urbanising country in Southeast Asia, GHG emitted in Malaysia is higher than other developing countries within the region. In Malaysia, its carbon emissions are highly related to urban environment with the energy domain (transportation & electricity) contributing to 80% of total emission. Despite the absence of major calamities, except for flash floods in established cities (e.g., Kuala Lumpur), both climate change adaptation (CCA) and disaster risk reduction (DRR) were incorporated as part of the national agenda (Hashim 2018). This action substantially promotes emission reduction in the energy sector that saves cost and lowers carbon use. In fact, a number of cities across Malaysia have either devised an action plan or set a vision to minimise carbon use, while preparing GHG inventory to track actions that demand low carbon. The 'Green Technology Application for the Development of Low Carbon Cities' (GTALCC) project was initiated by United Nations Development Program (UNDP), Sustainable Energy Development Authority (SEDA), and Ministry of Energy, Science, Technology, Environment and Climate Change (MESTECC), while funded by Global Environment Facility (GEF) (Lee 2019).

In 2007, GHG emissions in Malaysia per capita were 10.8 tCO2e, while 292.9 million tCO2e were recorded at the national level. The fractions of the emissions were as follows: electricity emitted 26% of GHG emissions, while transportation contributed 16%, and solid waste recorded 12%. The transportation sector emitted the most GHG in Malaysia since 2009, and this was followed by the energy segment.

Malaysia has started addressing environmental problems since the 1972 United Nations Conference on Human Environment, such as the deployment of National Environmental Policy in 1992 Rio Summit. Initiatives taken by Malaysia to become resilient towards climate change started since 1976 as outlined in the Third Malaysia Plan. As shown in Table 10.1, the initiatives and views in climate resilience began

Table 10.1 Timeline of initiatives taken by Malaysia towards climate resilience

Year	Initiatives taken by Malaysia towards climate resilience
1976	Third Malaysia Plan
2005	National Physical Plan
2005	Land Use Planning Appraisal for Risk Areas
2005	National Security Council Directive no. 20
2009	National Climate Change Policy (NCCP)
2009	KL Action Plan (KLAP)
2009	National Green Technology Policy (NGTP)
2010	Low Carbon Cities Initiatives (LCCI)
2012	Green Technology Applications for the Development of Low Carbon Cities

Adopted and adapted from Jamaludin and Sulaiman (2018)

expanding over time with updates based on climate issues experienced in Malaysia (Jamaludin and Sulaiman 2018).

Green initiatives taken between 1992 and 2009 were amalgamated to devise sustainable policies that were incorporated into urban development plans, which were deployed by the government and local authorities. For instance, the National Physical Plan (NPP) initiated a spatial framework in 2005 for national physical development. The framework served as the basis for lower-tier development plans, including structural, special area, local, rural development, and transportation plans. It is the function of NPP to ascertain that the development plans adhere to the national policies and goals. By adhering to the spatial framework; more sustainable development is attained, national resources are optimally used, and duplicate infrastructure investment is hindered. Apart from being the reference to development, NPP serves as an integral framework to attain sustainable and integrated land use planning at both local and state levels throughout Malaysia.

Low carbon cities initiatives (LCCI) (later known as low carbon city framework (LCCF)) introduced in 2010 refer to a subset of sustainable cities agenda, which tracks carbon emissions at city level—a tool to deploy strategic policy development for sustainability goals in Malaysia. The LCCF updates improvement status of carbon emissions in sustainable cities for long run.

No single set of policies suits all contexts and countries. Nonetheless, several common aspects have been adopted in many countries to design national urban policy that emphasises on reducing carbon emissions. Similarly, several projects have been implemented in Malaysia in light of low-carbon effort.

10.2.1 Green Technology Applications for the Development of Low Carbon Cities (GTALCC) Project

Sustainable economic progress is strongly linked with a country's ability to pave the path for low carbon climate-resilient development. Hence, the five-year GTALCC project was initiated by the Malaysian government to track carbon reduction in cities, encourage green technology, and promote activities/plans to decrease carbon usage in light of national agenda. The GTALCC project specifies green growth as the strategic thrusts and as a game changer as well. Green strategies deployed in Malaysia can lower environmental and ecological issues, promote better quality of life and well-being, as well as strengthen energy, water, and food security. This decreases GHG emissions, enhances coastal and marine areas, conserves inland and terrestrial waters, as well as preserves the ecosystem.

The integrated solutions outlined in GTALCC project blankets these areas: (1) integrated planning for climate action, (2) crowding in private investment and supporting city-level climate finance instrument, as well as (3) driving zero-carbon transition in cities by integrating transportation, energy, building, and waste systems. The GTALCC project is implemented in five pilot cities; Hang Tuah Jaya, Putrajaya,

Cyberjaya, Iskandar Malaysia, and Petaling Jaya (PJ), to attain sustainability solutions. The project is related to the National Low Carbon Cities Masterplan initiated in 2020 that not only defines low-carbon cities (LCC), but also lists transition plans and key actions in moulding LCC across Malaysia (Rahman 2009). The GTALCC project is aligned with Sustainable Development Goals (SDGs) (Lee 2019).

The GTALCC project promotes integrated planning by moving away from the conventional 'grow first and clean-up later' framework to one that emphasises on sustainability for future generation gains by highlighting resilient, resource-efficient, low-carbon, and socially inclusive development. Evidently, specific targets have been outlined to lower GHG emissions in Malaysia by strengthening institutional and governance structure, expanding LCC ideology, formulating effective sustainable policies, as well as organising educational and awareness raising programmes.

10.2.2 Malaysia's Smart City, Green City, and Resilient Framework

A smart city uses Information and Communication Technology (ICT) to efficiently offer services, amenities, and infrastructure for its residents. In a smart city, its technocentric concept is extended to social and economic progress. According to Caragliu et al. (2011), a city becomes smart '*when investments in human and social capital and traditional and* modern *communication infrastructure fuel sustainable economic growth and a high quality life, with wise management of natural resources; through participatory governance*'. In Malaysian Smart City Framework (MSCF) (MHLG 2018, p. 7), 'smart city' denotes 'a city that uses technological advancement to address urban issues, improve quality of life, boost economic growth, develop a sustainable and safe environment, as well as encourage efficient urban management practices'. The MSCF is composed of Giffinger's (2007) six smart aspects: environment, smart economy, mobility, government, living, and people.

Innovative and transformative solutions can strengthen urban infrastructure in meeting the demands of citizens for better quality of life. The MSCF, introduced in September 2019 by the Ministry of Housing and Local Government, is a reference for local governments to turn cities into smart cities (Loo 2019; Adnan 2016). For smart cities to be more liveable, convenient, practical, and sustainable to their residents (MHLG 2018), MSCF deploys the use of ICT. Smart city initiatives overcome urban issues, including pollution, congestion, security, and misuse of natural resources. The Smart Cities Initiatives initiated during the 11th Malaysia Plan (2016–2020) embeds IoT and the following elements:

1. Smart governance via:
    - Public participation
    - Efficient public and social services
    - Private and public partnership
    - Transparent governance

2. Smart mobility via:

   - Efficient road accessibility
   - Efficient public transportation
   - Non-motorised accessibility
   - Availability of ICT infrastructure

3. Smart technology/economy via:

   - Economic growth and value creation
   - Innovative economic growth
   - Equitable wealth distribution
   - Entrepreneurship

4. Smart infrastructure via:

   - Clean Environmental protection
   - Green development
   - Green infrastructure
   - Green economy

5. Smart citizen via:

   - Caring community
   - High Human Development Index
   - Talented and skilled human capital
   - Racial harmony

The low-carbon approach implemented in Malaysia is elucidated from Green City ideas. Green City '*strives to lessen its environmental impact by reducing waste, tumbling emissions, fostering recycling, expediting the use of renewable energy, and boosting housing density; while expanding open space and encouraging the development of sustainable local business*' (Lomba-Fernández 2019). Similarly, Eco-City and Low-Carbon City concepts aim to lower carbon use and boost development potential.

During the United Nations Climate Change Conference held in Copenhagen in 2009, Malaysia highlighted that it is committed to decrease the intensity of carbon emission up to 40% by year 2020 with the aid of developed nations. Notably, great progress was recorded with 33% decrease in emission intensity for year 2015. To address the impacts of global climate change, Malaysia had given its pledge during the 2015 Paris Agreement to further lower its intensity of carbon emission by 45% in 2030.

The Making Cities Resilient Campaign (MCRC) launched in 2011 was the initial effort taken by the Malaysian government, along with the United Nations Office for Disaster Risk Reduction (UNISDR), to turn Kuala Lumpur, Putrajaya, and Melaka into resilient cities by raising awareness on resilience and reducing disaster risks amidst the public. In Japan, both MCRC and Hyogo Framework for Action (HFA) supported the Sendai Framework for Disaster Risk Reduction 2015–2030 (UNISDR 2005). City resilience enhances city performance in spite of hazards, instead of mitigating asset loss due to calamity.

10.2.3 Malaysian Smart City Framework at State Level

The MSCF reflects the top down–bottom up approach to realise the national agenda on climate change. Selangor is a progressive and dynamic state both economically and environmentally. The Selangor government aims to become a Smart State by 2025. The Selangor State Government has mandated its State Chief Minister and Selangor Incorporated to execute smart initiatives through the establishment of SSDU Innovations Sdn Bhd (Smart Selangor Delivery Unit) to become Smart Selangor. In adhering to three main objectives; Productivity, Liveability, and Sustainability, Smart Selangor is aimed at becoming a premier, regional Smart State in ASEAN by 2025. In realising the Smart Selangor vision to be the most liveable state in Malaysia; five enablers have been outlined to drive this initiative:

1. Smart People—yields ecosystem to generate connected, resilient, informative, and adaptable citizens
2. K-Economy—promotes business to apply innovative technology to scale-up business
3. Smart Environment—safeguards clean environment, water, and air
4. Smart Government—enables equitable, reliable, accountable, and accessible government services
5. Smart Infrastructure—improves digital and physical connectivity to lay the platform for smart state

10.3 Smart, Sustainable, Resilience Petaling Jaya Blueprint 2030

In 1951, Petaling Jaya (PJ) was a town meant to cater post-war housing demand and the population overflow of Kuala Lumpur (Thong 2006). It accommodated 70,000 residents in 10 km to the south-west of Kuala Lumpur. To date, good city planning has led the city to become an advance city in Malaysia (see Figs. 10.1, 10.2 and 10.3).

Until June 2006, the council was upgraded from PJ Municipal Council to Petaling Jaya City Council (PJCC) (Majlis Bandaraya Petaling Jaya) with 550,000 residents. At present, the administrative area of PJCC covers 97.2 square kilometres with 619,925 residents (PJCC 2020). Located in Greater Kuala Lumpur and close to Kuala Lumpur, it has the sprawl effect of population in terms of economy and connectivity (see Fig. 10.4).

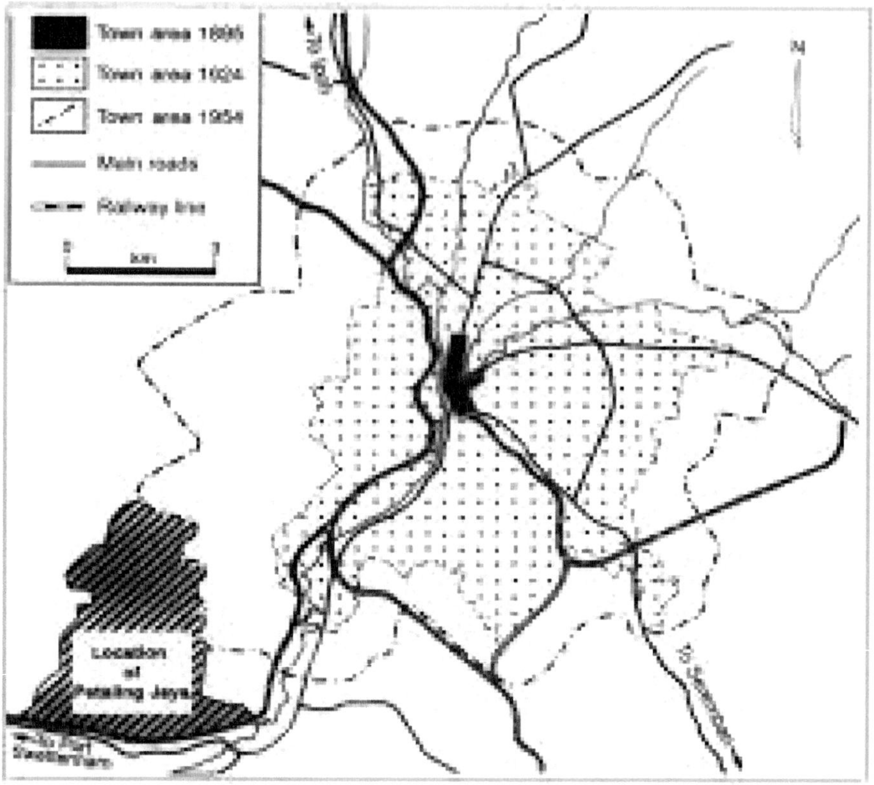

Fig. 10.1 Expansion of Kuala Lumpur and location of Petaling Jaya in 1948

With the council predicting a hike in its total population up to 1.8 million in 2050, PJCC has greater responsibilities for planning and managing the city. The city is composed of residential neighbourhood, a wide range of commercial and industrial activities, institutional facilities, financial and administrative points, as well as recreational attractions such as Bukit Gasing, Taman Bandaran Kelana Jaya, and Central Park Bandar Utama. The strategic location of PJ, which is adjacent with the capital city of Malaysia, Kuala Lumpur, has made the city one of the most visited and fast developing city in Malaysia. Covered with networks of highways, the city is designed with an excellent public transportation system, such as Light Rapid Transit (LRT), Commuter (KTM), and Mass Rapid Transit (MRT).

Rapid urbanisation in PJ has exposed the city to climate change phenomena, such as flooding, landslide, disease outbreak, and pollutions. On 16th April 2020, the city faced one of the massive flash floods throughout the year with the water height reaching 2 feet at Seksyen 51A (Noor 2020). Numerous factors have contributed to the climate change happening in PJ, mainly due to traffic congestion as a result of higher number of vehicles that emits greater carbon. Urban heat island is another

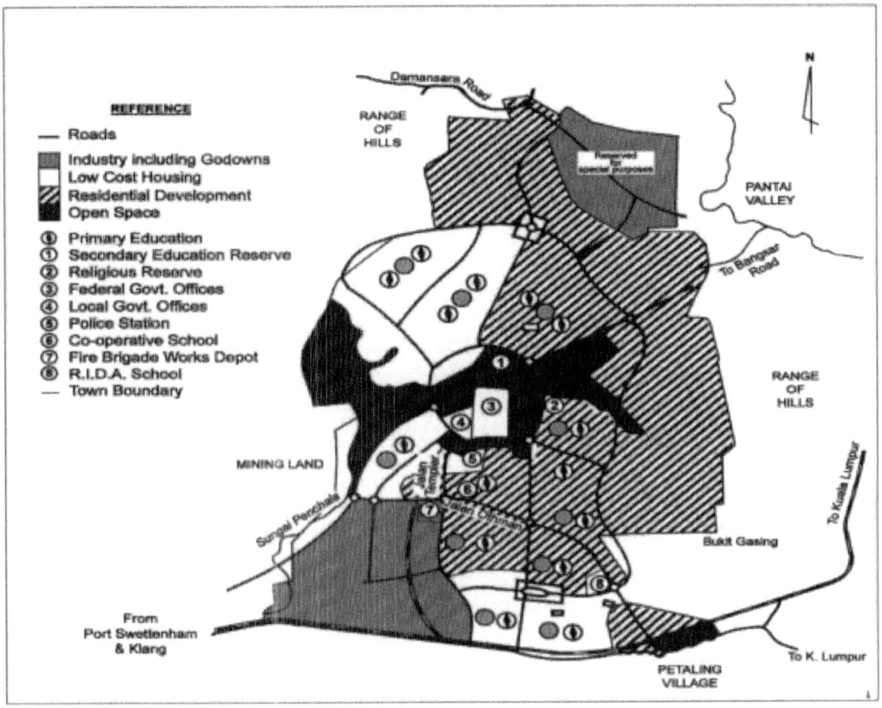

Fig. 10.2 Detailed Plan of Petaling Jaya in 1955 (*Source* Thong 2006)

scenario due to the rising amount of concrete buildings in the city. Hence, PJCC is devising a smart and sustainable plan to manage the city, particularly to lower GHG emission into the city atmosphere.

Being the first winner of Malaysia's National Earth Hour Capital, PJ stands out with its exceptional energy efficiency and low-carbon efforts. The leadership displayed by PJCC in addressing multiple issues and deploying green innovations has been applauded by international panel of juries.

By year 2030, PJ aims to be a sustainable and resilient city through the implementation of SMART PJ—'*a total solution of intelligent city planning and management*' (PJCC 2018). A framework has been built to achieve the following targets:

- 40% public transport usage
- 30% reduction of energy consumption
- 3% increase in the collection of 3R programme annually
- Universal and accessible design
- 2% reduction of carbon emission annually
- 10% reduction of crime cases
- Less than 10% flood-prone population

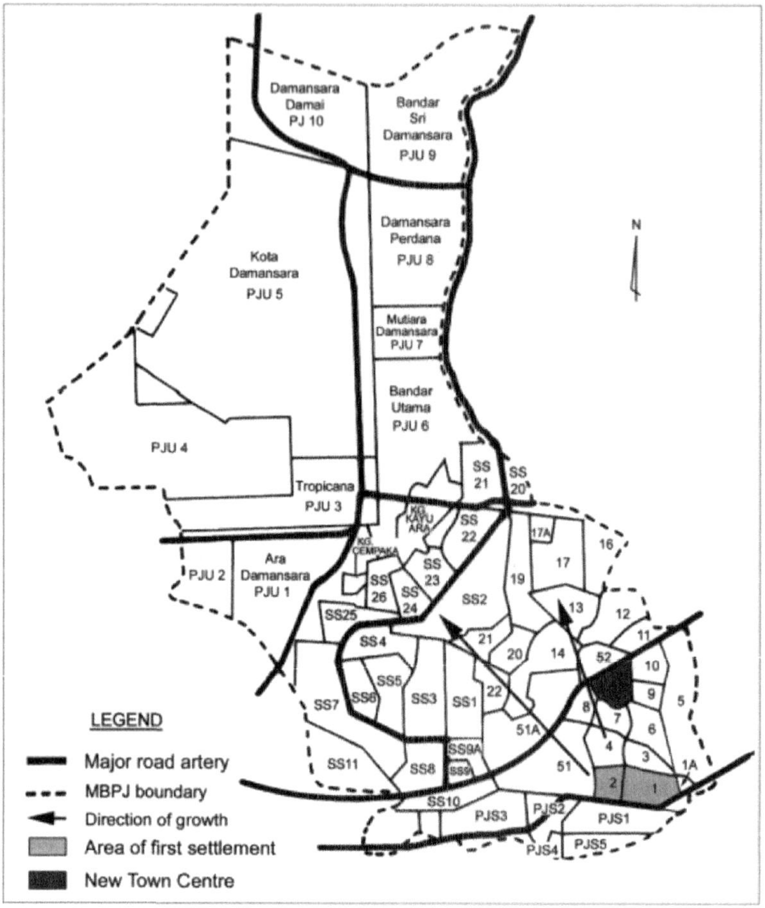

Fig. 10.3 Petaling Jaya City Boundary in 2006 (*Source* Thong 2006)

The SMART PJ applications as shown in Fig. 10.5, based on five main cores of Clean, Green, Safe, Connected, and Happy that act as dashboards, are developed in a mini command centre and delivered to PJ SMART Centre in report form. This guides the authority to make city planning decision for PJ based on actual city scenario.

Until year 2019, the dashboards had covered the following aspects:

1. City Planning and Development
2. PJ KITA Community, such as City Community Development
3. City Business
4. City Town Services
5. City Management & Enforcement

Smart PJ seeks to: (1) enhance climate resilience, (2) lower resources use & demand, (3) improve urban living quality, (4) deploy smart economy with low carbon, (5)

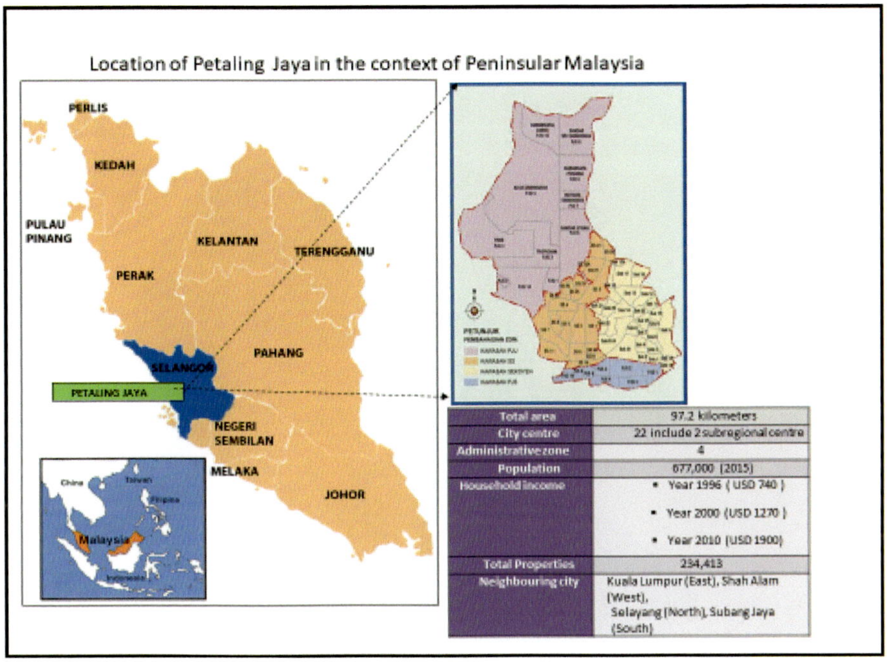

Fig. 10.4 Map of Petaling Jaya City Council

become an efficient city with prudent enforcement & integrated management, and (6) reduce CO_2 emissions. The PJCC Low Carbon City Action Plan was devised with government bodies and the Carbon Trust to calculate GHG inventory until 2030 with the goal of lowering CO_2 emissions by 30% (refer Table 10.2). The council had identified projects that could save 6 million tCO2 at RM 9 billion net; contributing to ~ 50% cost saving (CDP 2020). The 63% of GHG emission in PJ due to energy use for industrial and commercial buildings must be lowered to 25% in 2025 and further 25% in 2030. The Green Building Index initiated in 2008 has benefitted PJ. Data gathered from sensors, lights, and video cameras are integrated with information on resident behaviour to plan public infrastructure (e.g., retail zones, parking & road layouts, & healthcare centres).

The PJCC has faced many common challenges, while the solutions in terms of policy mix rely on some factors. The feasibility of deploying a policy must be weighed in, including local-national government coordination, budgets, social equity impact, burdensome administrative work, impact on local government capacity & resources, and technical expertise. Since varying policies impose varying opportunities and challenges, it is crucial to deploy the best mix of policies based on the capacity of the council. The following lists several significant aspects that should be weighed in:

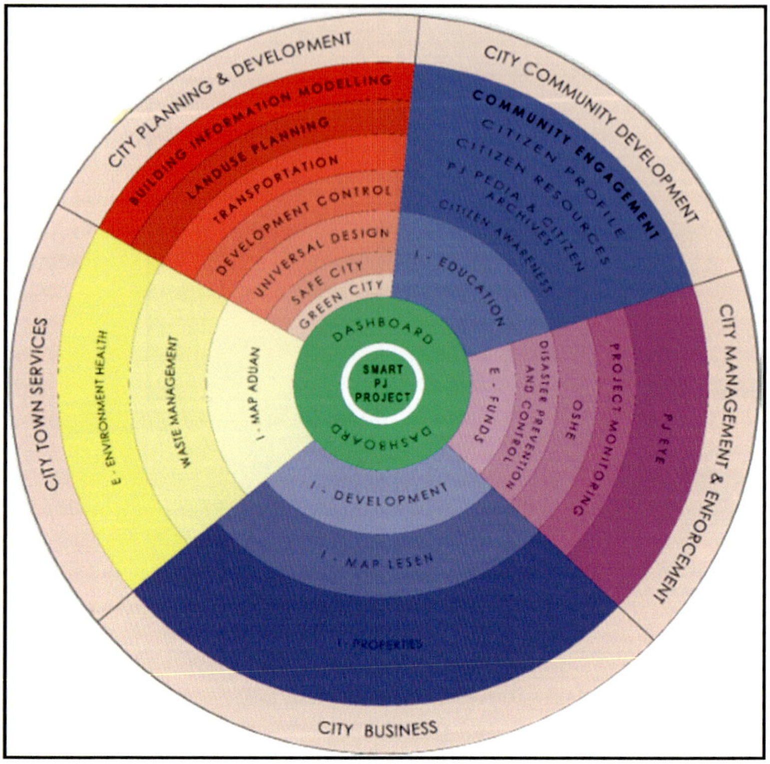

Fig. 10.5 PJ smart city framework (*Source* PJCC 2020)

- ***Governance and institutional arrangements, as well as legal authorities****:* Low carbon goals may be met with effective coordination at all government and non-governmental levels—among varied national bodies, high-level subnational agencies, and local government. The national governance structure can either ease or impede policy adoption. The selection of viable national policies also depends on the coordination of opportunities and challenges.
- ***Local government resources****:* Technical, financial, and administrative capacities are crucial for policy success. Notably, PJCC did not suffer from financial limitation as other cities, such as Putrajaya or Kuala Lumpur. Thus, burdensome policies should be hindered to enhance municipality capacity.
- ***Local opportunities*** *for urban economic growth, energy saving, and GHG reduction:* Many approaches boost economic growth, increase energy saving, and lower GHG emissions. Policy selection in line with low carbon aim is dictated by existing energy resources, rate of urbanisation & its infrastructure, as well as urban area type & distribution.
- ***Political and policy priorities****:* To enhance the political feasibility of policies related to low-carbon urban progress, the selected policies should reinforce

Table 10.2 Petaling Jaya City council mitigation plans and collaborators

Plan or activity	Key agencies or groups	Overview
Low Carbon Cities	PJCC, KeTTHA and	PJCC adopted the LCCF
Framework	Malaysian Institute	developed by KeTTHA and MIP, as
	of Planners	their guideline for its transition to LCC in 2010. The framework has four main areas; Urban Environment, Urban Transport, Urban Infrastructure and Buildings
		These main areas are divided into 35 sub-criteria for low-carbon performance
Green Technology Application for the Development of Low-Carbon Cities	PJCC, KeTTHA with funding from UNDP-GEF	The project facilitates low-carbon initiatives in participating cities through policy support, raising awareness and local capacities, as well as encouraging investments in low-carbon technology
Low-Carbon City Action Plan	PJCC and Carbon Trust	Aims to reduce city GHG emission by 10% in 2020, 20% in 2025, and 30% in 2030 using 2014 as the baseline energy consumption in residential, These are to be achieved by reducing commercial and industrial sectors, implementing more efficient new developments, morerenewable energy installation, and a shift away from fossil fuel-based vehicles
Carbon Management Plan	PJCC and Carbon Trust	The plan outlines a suite of specific measures, targeting a 25% reduction in PJCC's organisational GHG emissions compared to 2014. Some measures include LED street lights, upgraded cooling systems & building fabrics, and installation of photovoltaics in PJCC buildings
Staff energy saving	PJCC staff	Energy reducing behaviour including: switching office lights off during lunch break, setting office temperature to 24 °C,
		and setting computers to hibernation mode when not in use
		These changes are implemented based on

(continued)

Table 10.2 (continued)

Plan or activity	Key agencies or groups	Overview
		ISO 50001 (energy management) certification of municipal operations
		PJCC purchased 20 electric motorcycles
		for its community service staff
Urban greening	PJCC	A tree planting programme and forest
		conservation activities
GBI requirements for new	PJCC and	New bungalows and semi-detached housing must be fitted with
buildings	builders/developers	rainwater harvesting equipment. Commercial and mixed commercial
		developments must adhere to water and
		energy standards for GBI, use LED lighting
		live up to the requirements of Urban
		Stormwater Management Manual, and
		reserve 10–15% of the total development for green space
Homeowner energy	PJCC	A tax incentive scheme for homeowners to
efficiency rebate		retrofit their houses for energy & water
		efficiency and better waste management. The scheme offers up to 100% tax rebate of the total cost of retrofit or RM500

Adapted and modified from KeTTHA (2011), UNDP (2017), Carbon Trust (2015, 2015a), Carbon Climate Registry (2017), consultations with PJCC (2018)

ongoing policies that highlight poverty eradication, provision of job vacancies, enhanced urban sanitisation, lower road congestion, as well as better inclusion and accessibility. If new policies support energy efficiency, it is easier to attain distributed energy solutions and meet climate policy aims.

Essentially, the smarter the cities, the higher is their reliance on data-collecting systems—from updates on pollution based on environmental sensor to massive data retrieved from streams of video recording. Public safety agencies use mobile broadband and rapid data access to increase operational efficiency.

10.4 Research Significance on Sustainable, Smart, and Resilient City

This section discusses sustainable, smart, and resilient concepts based on initiatives and accepted benefits although they are not exhaustive. This paper highlights the achievement of PJ in becoming a sustainable, smart, and resilient city in terms of its sustainable environment, infrastructure, transportation system, and effective land development. The managers of such city should look into the needs and demands of both the residents and the environment. Multiple considerations must be weighed in at all development levels in the context of resilient and smart built environment.

This present research project is composed of two parts: (1) a combination of desk-based literature review of urban GHG studies, urban and disaster planning documents, and articles related to climate risks to urban infrastructure and green building design; and (2) stakeholder consultations with various PJCC departments, as well as their recent engagement and reporting to A LIST CITIES 2019 of Carbon Disclosure Project (CPD) (a non-profitable charity that runs the global disclosure system for investors, companies, cities, states, and regions to manage their environmental impacts).

This study applied the anticipated outcomes of the most impactful mitigation actions occurrence reported and presented in the CPD Cities 2019 document. The performance-based approach, which provides a holistic and comprehensive approach, denotes resilience as the ability of a city to sustain and meet several core functions. This performance-based approach can address issues related to scale, interdependency, and power dynamics; mainly because the functions of a city depend on the combination of systems, assets, actions, and practices deployed by numerous players (Refer to Table 10.3).

In this case study, the examination of city-wide GHG emissions date for LCC programmes was related to the resilience of individual sub-smart city sectors while considering city resilience as a system in itself. The purpose of close impactful mitigation analysis was to provide a measure of relative description based on the LCC initiatives implemented by PJCC. Cities implement indices to determine their targets towards resilience based on their actions taken to improve their performance (in each sub-indicator area). The examination involved the city envisaging its performance

Table 10.3 Priorities and low-carbon policy strategies, as well as their potential impact on resiliency

Smart city dimensions	Low-carbon development Initiatives	Resilience descriptions															Remarks
		1	2	3	4	5	6	7	8	9	10	11	12	13	14	15	
SMART Economy	Buildings: Carbon emissions reduction from industry																Mandatory Green Building Index (GBI) for all new development projects in Petaling Jaya
	Instruments to fund low-carbon projects																Property Tax rebates for environmentally friendly household owners to facilitate low-carbon residential units
	Developing the green economy																Petaling Jaya Smart City Concept
	Instruments to fund low-carbon projects																Selection & Continuance of assessment tax rebates as pilot project under the UNDP funded GTALCC project
	Community-Scale Development: Brownfield redevelopment programmes																PJ Sect. 13 urban renewal
	Developing the green economy																Uplifting living standards for Petaling Jaya poor through green economy

(continued)

Table 10.3 (continued)

Smart city dimensions	Low-carbon development Initiatives	Resilience descriptions															Remarks
		1	2	3	4	5	6	7	8	9	10	11	12	13	14	15	
	Material Recovery Facilities																MRF at sub-city scale is required so that resources from waste are recovered and volume of waste disposed is reduced
SMART People	Waste Recyclables and organics separation from other waste																SS20 Community Composting initiative
	Waste Recyclables and organics separa tion from other waste																Food Waste Composting at Desa Mentari
SMART Governance	Buildings: Energy efficiency/ retrofit measures																Petaling Jaya City Council Building Carbon Management Plan
	Waste prevention policies and programmes																Launching of MBPJ's Zero Single-Use Plastics campaign
	Finance and Economic Development: Instruments to fund low-carbon projects																Establishment of RM 3 million community grant programme by Petaling Jaya City Council

(continued)

Table 10.3 (continued)

Smart city dimensions	Low-carbon development Initiatives	Resilience descriptions															Remarks
		1	2	3	4	5	6	7	8	9	10	11	12	13	14	15	
	Low Energy Office of PJ																Retrofitting existing Council buildings to be green and energy efficient. Four buildings were identified and scope of work mainly involved cooling and lighting systems
SMART Mobility	Mass Transit: Improves bus infrastructure, services, and operations																Petaling Jaya City Free bus service
	Mass Transit: Improve rail, metro and tram infrastructure, services, and operations																Free bus service for School children and terminates at Taman Jaya LRT Station
SMART Environment	MSMA 2 (Urban Stormwater Management Manual for Malaysia 2nd edition)																Guideline for key players in managing stormwater from flood, erosion, minimise environmental impact runoff, and enhance urban landscape and ecology
	Rehabilitation and Development of Environmentally Sensitive Areas																Assist authority controlling any form of development thus safeguarding the Environmentally Sensitive Area

(continued)

Table 10.3 (continued)

Smart city dimensions	Low-carbon development Initiatives	Resilience descriptions															Remarks
		1	2	3	4	5	6	7	8	9	10	11	12	13	14	15	
	Flood Mapping																On-Site Detention Tanks were built and drainage systems were upgraded. This enhanced the capacity to detain water during heavy downpour or runoff
	Landslide Risk Mapping																Guidelines on planning and development of Gasing Hill area in Petaling Jaya based on the level of danger and risk
	Outdoor Lighting: LED/CFL/other luminaire technologies																Installation of LED Street Lighting in Petaling Jaya
	Water use efficiency projects																Rainwater Harvesting system implementation
SMART Living	Global Protocol for Community Greenhouse Gas Emissions Inventories (GPC)																Outlines the strategy for reducing the carbon intensity of the cities' buildings, transport, and waste infrastructure
	Community-Scale Development: Eco-district development strategy																Installation of more closed-circuit TV cameras

(continued)

Table 10.3 (continued)

Smart city dimensions	Low-carbon development Initiatives	Resilience descriptions															Remarks
		1	2	3	4	5	6	7	8	9	10	11	12	13	14	15	
	Community-Scale Development: Eco-district development strategy																Formation of Sustainable Community in I-Play Park in Damansara Damai
	Waste: Recycling or composting collections and/or facilities																PJ Eco Recycling Plaza
	Waste: Recyclables and organics separation from other waste 2																Smart PJ Waste Solution Lab in SS2, Petaling Jaya
	Community-Scale Development: Eco-district development strategy																The Learning Cities Award 2019 by The UNESCO Institute for Lifelong Learning (UIL)
	Food and Agriculture: Encourage sustainable food production and consumption																Petaling Jaya Agrobox Scheme

(continued)

Table 10.3 (continued)

Smart city dimensions	Low-carbon development Initiatives	Resilience descriptions															Remarks
		1	2	3	4	5	6	7	8	9	10	11	12	13	14	15	
	Food and Agriculture: Encourage sustainable food production and consumption																Petaling Jaya City Council Food Bank Scheme
	Upscaling of biogas facilities for storage and usage at various sites																Installation of various food digestors at many places and seeking funding to implement biogas storage and upkeep facilities to provide renewable energy for various facility usages
	Total count of characteristics	**28**	**23**	**29**	**23**	**28**	**21**	**29**	**7**	**28**	**17**	**14**	**22**	**26**	**16**	**20**	
	Resilient characteristics	Low-carbon	Robustness	Diversity	Redundancy	Connectivity	Flexibility	Resourcefulness	Agility (rapid response),	Efficiency	Adaptive/ Learning capacity	Modularity	Creativity	Equity	Inclusiveness	Foresight Capacity	

Adopted and modified from CDP (2020). Petaling Jaya City Council—Cities 2019

(undertaken actions) against each sub-initiative based on the following qualitative items:

1. The 15 descriptions of resiliency characteristics (as defined and discussed in the Asia–Pacific Network for Global Change Research Project)
2. The overall initiative was categorised into 5 Smart City themes to link smart city with resiliency

The implementation of LCC is an intricate process that entails governance, planning mechanisms, monitoring and evaluation, financing, as well as context and process. Petaling Jaya (PJ) is one of the pilot studies of LCCF that have implemented many planning mechanisms as strategic tools to enhance both the planning and deployment of LCC programmes. Based on the analysis of initiatives, strategies, and characteristics occurrence, several characteristics seemed to stand out, such as the following:

Diversity is the key characteristic that represents variety, multiplicity, and mixture that require PJ to energise its economy and urban activities, engage its residents, enhance its built environment, and promote green programmes. It implies the extent numerous distinct functions can be concurrently applied within the system. This principle is aimed at hedging against disruption in supply, ascertaining the availability of multiple options (e.g., resources & instruments) to overcome drawbacks, and acknowledging the functionality of the urban system.

Diverse support mechanisms and livelihood opportunities enable residents to respond proactively to the changing city conditions without compromising their quality of life. Accessibility to business support, finance, and training of skills leaves one with multiple options in securing the necessary assets for meeting the fundamental demands. Secure and long-term livelihoods enable individuals to accumulate savings so as to support them during crises. The dynamic PJCC has an advantage over diversity in terms of ethnicity and group diversity, thus enhancing economic growth, as well as communal and cultural activities. Petaling Jaya (PJ) enjoys inclusive labour policies and relevant skills for emerging employment with high residing population due its close location to Kuala Lumpur and foreign residents that create cultural diversity, thus generating high demand and expectations for diverse services.

The MSMA 2nd Edition—Urban Stormwater Management Manual for Malaysia—guides designers, regulators, and planners for managing storm water (Dept. of Irrigation and Drainage 2012; Nor Azazi Zakaria 2014). It is vital to ensure the consistency of planning, designing, and maintaining storm water management system across local, state, and federal authorities; landscape architecture; as well as environmental, urban development, and environmental engineers.

Resourcefulness refers to resources used by urban planners to determine, prepare for, respond, and recover from any disruption, if any. This is inclusive of a viable capacity to comprehend the status quo, apart from identifying contingencies, patterns, and lurking threats.

For instance, the residents in SS20 located in PJ have adopted modern machinery after years of practising conventional food composting procedure. They successfully began with the single-stream waste-recycling technique, followed by conventional composting method at their home and at the community park. To date, compost derived from modern technique is applied as fertilisers in their gardens for growing plants and vegetables. One benefit noted from this composting activity is reduction in leachate from garbage compactors. Worth RM30, 000; the 30-L composting machine was purchased using funds pooled by the residents, PJCC awards, as well as funding received from councillors, government, and assemblyman. Food waste collected by the residents was processed in the machine thrice a week. While the conventional method consumes a month, the machine only takes 20 h to complete the composting process. Nevertheless, the machine is unviable for composting stone seeds and bones.

In another example, a waste solution lab established in Jalan SS2/63 applied the circular economy technique to recycle waste into beneficial items for the society, which differs from linear economy that denotes disposal of garbage at landfill. The PJCC reported that 49% of the daily 600-tonne disposed waste was organic waste. Hence, waste in PJ city could be reduced by half if all organic waste is collected and turned into compost. After starting its operation in May 2017, the lab, which was a pilot project, used two compost machines to process 15 tonnes of food waste into detergent, liquid fertiliser, and biogas on monthly basis. The effort involved 30 hawkers from Ibu Mee stalls, SS2 morning market, and food court who provided organic waste. Additionally, PJCC gathered more organic waste from local small traders and residents from other PJ areas. One may also visit a gallery that was setup to view the anaerobic digester machine on display, besides learning about the advantages of recycling and adopting the green concept. A mini greenhouse filled with plants and vegetation was also available for viewing, which used bio-organic fertiliser generated from organic waste. Biogas is used to fuel the carbonisation process for converting coconut harsh to bio-charcoal. The organic fertiliser, along with other products yielded from the compost machine, may be purchased by visitors.

Low-carbon: The PJCC has initiated some steps to lower carbon in the city through its own natural resources. Such an example is the tree planting initiative under the adaptation goal. A total of 100,000 trees are to be planted by the city council to reduce urban heat island impact by year 2030. The council had set interim targets for 2020 and 2025, whereby 8% of the goals were met in 2019.

The PJCC has been the only Malaysian local council that offered assessment rebates, namely PJ Homeowners Low-Carbon and Green Initiative assessment rebate scheme, to its residents who embrace green concept. After its initiation in 2011, RM414, 380.48 of waived assessment was reported until 2018 involving 1,240 households in the PJ city. In fact, PJCC received the Green Apple Award in 2013 by The Green Organisation in London for this initiative. Points of energy saving were awarded for installing LED lighting (at least 70%), solar heaters and panels, roof insulation, rainwater harvesting system, and flush box with less than six litres of water. Homes that practice recycling (e.g., rainwater for cleaning); carry out composting; use bicycles, hybrid vehicles, and public transportation; and cover 50% of open space with plants and greenery are given points. The two least counted features in the analysis are modularity and agility elements.

Modularity (independence and self-organise) denotes loosely-coupled components. A resilient system can be-recombine in many ways to deal with environmental changes. The SMART centre in PJCC headquarters located at Jalan Yong Shook Lin offers information about its city planning to those interested. This SMART centre is part of the Smart PJ Project towards becoming a sustainable city by year 2030. Over RM15 mil was spent to boost the following 10 sectors: information and control planning, green and safe city, PJ Eye, community engagement, transportation, digital publication, management of waste and complaints. The Smart PJ Project involves erection of physical structures, including the integrated digitised systems and the SMART centre. This RM500, 000 worth structure provides imminent information about the projects undertaken by PJCC, pothole-ridden areas, and crime hotspots, as extracted from the city's geographical integrated system (GIS). For example, city planners can gather data about land use along with detailed mapping based on images captured by drones. Ultimately, PJCC can work better with its stakeholders with all information being readily available. Both private and public sectors can devise viable projects upon identifying the emerging needs. The PJCC can make better decisions based on real-time and integrated data using this SMART system.

A city with **agile characteristic** can respond more rapidly to initial crisis and better leverage the changing situation as a phase returning to normal. An instance of PJ agility can be observed from the disaster preparation systems called Flood Mapping and Landslide Risk Mapping to withstand short-term (sudden) shocks without major degradation. The systems identify hotspots for flash flood, while simultaneously providing relevant planning and development guidelines based on the level of the danger and risk. These provide benefits in disaster preparedness and risk reduction, besides enhancing resilience shifts to more sustainable behaviour. The agility in the PJ SMART centre is almost completely supported by digitalisation.

10.5 Concluding Remarks

City resiliency is a property of a complex adaptive system. It emerges as a result of many interactions within the system. As such, it is naturally associated at national, state, and local levels, but it does not equate to the sum of its parts. A collection of very resilient initiatives does not necessarily make a resilient city. In a nutshell, the discussion serves to paint an optimistic outlook of how PJCC has been and will continue to address urban challenges with smart systems. A fresh urban development strategy is sought to improve liveability, promote growth, and strengthen the overall competitiveness of PJ to be at par with other cities. The strategic planning undertaken by PJ must go beyond the realm of technological development due to the significance of technology and innovation in dictating the progress of a city. Digitalisation is crucial to enable smooth city operations, whereas real-time information offers consistent and effective services especially when devising crisis response plan. Information accessibility to stakeholders, citizens, and collaborators is critical for exceptional customer service. The deployment of a holistic strategy, which embeds software and hardware systems, demands constant updating and monitoring. In promoting technology and innovation in PJ, such holistic implementation must be executed at all built environment levels while facilitating urban innovations to tap into the full potential of the smart city by. The SMART Centre provides a good test-bed for PJCC to integrate the initiatives feasibly and comprehensively in the challenging urban setting, apart from nurturing the supportive culture and the institutional setup.

On top of the initiatives taken by PJCC, a comprehensive Resilient City Framework is sought to holistically integrate the urban system, inclusive of the environment, governance, infrastructure, health & well-being, social, and economy. Such a framework can facilitate the city PJ to evaluate the degree of their resilience by identifying limitations and devising effective strategies towards better resilience.

Acknowledgements This chapter was partially funded by Asia-Pacific Network for Global Change Research (Funder ID: https://doi.org/10.13039/100005536).

References

Adnan YM, Hamzah H, Dali MM, Daud MN, Alias A (2016) An initiatives-based assessment framework for smart city. Plan Malays J (5):13–22

Caragliu A, Del Bo C, Nijkamp P (2011) Smart cities in Europe. J Urban Technol 18(2):65–82

Carbon Climate Registry (2017, January 31) Petaling Jaya city council Malaysia report. Retrieved from http://carbonn.org/data/report/commitments/?tx_datareport_pi1%5Buid%5D=644

Carbon Trust (2015) MBPJ low carbon city action plan, Majlis Bandaraya Petaling Jaya 2015–2030. Majlis Bandaraya Petaling Jaya and Carbon Trust

CDP (2020) Petaling Jaya City Council—Cities 2019. CDP

DID, D. o (2012) Urban stormwater management manual for Malaysia (MSMA 2nd Edition). Government of Malaysia

Giffinger R (2007) Smart cities ranking: an effective instrument for the positioning of cities. ACE: Archit. City and Environment, pp 7–25

Hashim Ob (2018) Making cities resilient: The Malaysian experience. Putrajaya Corporation

Jamaludin IS, Sulaiman N (2018) Malaysia resilient initiatives: case study of Melaka. J Malays Inst PlanS, 19–20

KeTTHA (2011) Ministry of energy, green technology and water (Kementerian Tenaga, Teknologi Hijau dan Air)

Lee N (2019) Low Carbon cities—Malaysia's response to global climate emergency. UNDP

Lomba-Fernández C, Hernantes J, Labaka L (2019) Guide for climate-resilient cities: an urban critical infrastructures approach. Sustainability 11(17):4727

Loo. (2019, September 23) Ministry launches Malaysia smart city framework. theSundaily

Ministry of Housing and Local Government (2018) Malaysia smart Ccity framework. Kuala Lumpur

Noor HH (2020, April 16) Banjir Kilat di Kelana Jaya dan Seapark. myMetro

Nor Azazi Zakaria AA (2014) MSMA 2nd Edition—application of green infrastructures for solving. Persidangan Air Kebangasaan 2014. Universiti Sains Malaysia, pp 1–4

Petaling Jaya City Council Malaysia Report (2017, January 31) Retrieved from http://carbonn.org/data/report/commitments/?tx_datareport_pi1%5Buid%5D=644

PJCC PJ (2018) Smart, sustainbale, resiliene petaling jaya blueprint. Petaling jaya. Selangor, Malaysia

PJCC PJ (2020) Petaling Jaya city council official portal. Retrieved from Petaling Jaya City Council Official Portal http://www.mbpj.gov.my/en/mbpj/profile/background
Rahman HA (2009) Global climate change and its effects on human habitat and environment in Malaysia. Malays J Environ Manag, 17–32
Thong LB (2006) Petaling Jaya: the early development and growth of Malaysia's first new town. J Malays Branch R Asiat Soc 6(12):19
Tom Ovington GH (2020) How smart cities can help tackle climate change. Frontier Economics
UNISDR (2005, January) World conference on disaster eduction. ISDR, pp 18–22
United Nation Development Programme (2017) Low carbon cities framework and assessment system, version 1. 2. 10th annual performance report

Chapter 11
Digital Solutions for Resilient Cities: A Critical Assessment of Resilience in Smart City Initiatives in Melbourne, Victoria

Leila Irajifar and Khanh N. Vu

Abstract Urban resilience and smart cities have emerged as a critical agenda for urban development in the twenty-first century. The growing emphasis on smart and resilience concepts is mostly due to increasing shocks and stresses related to the environmental, economic, social, and technological pressures which is also exacerbated by the uncertainty associated with rapid urbanisation, climate change, and resource limitations. While digital smart solutions are becoming increasingly critical in addressing these challenges, it is essential to consider its broadest systemic impacts to ensure that new vulnerabilities are not created, and resilience compromised by adopting and using digital technologies in urban systems (Kupers, R., & Foden, M. (2017). Learning for resilience and complex systems thinking. Agenda setting scoping studies summary report. The Resilience Shift.) note "no complexity, no resilience" but conversely, it is also important to recognise that "systems can fail, even if everything works as it is supposed to". The aim of this paper is to investigate if the complexity drawn by introducing smart digital technologies in urban systems enhance resilience or create vulnerabilities? For this purpose, a deep case study analysis is conducted in Melbourne, Australia investigating the extent to which the current smart city initiatives contribute to the urban resilience attributes or have the potential to do so. Melbourne has dubbed as the most livable city in the world in several years and leads the nation as the most innovative city in Australia. Yet, despite its many efforts and relative wealth overall, the city faces risks and stresses that weaken the fabric of the society, which entrench disadvantage and may trigger the shocks of the future. Melbourne is exposed to natural disasters such as extreme

The original version of the chapter has been revised: Co-author name has been included. A correction to this chapter can be found at
https://doi.org/10.1007/978-3-030-95037-8_21

L. Irajifar (✉)
School of Architecture and Urban Design, RMIT University, Melbourne, Australia
e-mail: leila.irajifar@rmit.edu.au

K. N. Vu
Centre for Urban Transitions, Swinburne University of Technology, Melbourne, VIC, Australia

A. Sharifi and P. Salehi (eds.), *Resilient Smart Cities*, The Urban Book Series,
https://doi.org/10.1007/978-3-030-95037-8_11

heat, bushfires and floods, extremist acts, and of course the pandemics. There is an increasing number of smart city initiatives taking place in Melbourne, however, the extent to which these initiatives are aligned with the overall strategic plan of the city for resilience and sustainability objectives or just ad-hoc projects for testing new exciting technologies, is what we have investigated in this chapter.

The significance and innovation of this chapter lie in its systematic examination of the contradictory promises, perils, and tensions of smart city solutions. This will facilitate incorporating resilience thinking in the design of smart city projects -to be optimised alongside traditional criteria like quality, cost, and avoid the potential risks of the smart solutions on city resilience.

Keywords Digital solutions · Smart cities · Resilience · ICT technologies · Melbourne · Climate resilience

11.1 Introduction

Cities are increasingly using the power of emerging technologies, and the digital transformation is revolutionising different aspects of our societies. On the other hand, confronting with more and more shocks, stresses and challenges, makes building resilience in our cities a necessity. The digitalisation and the fast transformation of how cities function in future are inevitable and thus to make building resilience and sustainability keep pace with digital transformations, they need to be integrated for multiple dividend initiatives, by incorporating resilience thinking in smart city projects and digital solutions for building resilience.

Following on previous chapters on theoretical and conceptual discussions on this topic, in this chapter, we aim to investigate the contributions of smart city solutions to resilience through case study analysis in Melbourne, Australia. An action research study on implementation of Melbourne Resilience Strategy by (Fastenrath et al. 2019) showed that "a reconceptualisation and new analytical dimensions are needed to understand urban resilience as an urban innovation strategy". Below we first review the climate change activities and smart city initiatives in Australia. In following sections, we will focus on Melbourne initiatives in detail.

11.1.1 Climate Change Ada Activities in Australia

Historically, Indigenous Australians—Aboriginal and Torres Strait Islander people—had a strong background in dealing with and adapting to various unstable climate conditions due to their close connections with the local nature (Green et al. 2010; Prober et al. 2011). In 2005, adaptation to climate change played a prominent role in Australia's policies through the National Climate Change Adaptation Programme (NCCAP) (Allen Consulting Group 2005). The Council of Australian Government subsequently adopted the National Climate Change Adaptation Framework in 2007 in an attempt to evaluate the climate change impacts on the crucial sectors (Commonwealth of Australia 2007).

On the NCCAP, the National Climate Change Adaptation Research Facility (NCCARF) was founded in 2008 and then allocated AU$58.8 million in a ten-year funding package to various research projects and networks conducted from 2008 to 2017, with the focus on developing capacity and policymaking process (National Climate Change Adaptation Research Facility 2014). The Climate Adaptation Flagship launched by the Commonwealth Scientific Industry Research Organisation (CSIRO) in 2009 was another part of the NCCAP for the purpose of providing citizens, businesses, and policymakers with more effective and efficient manners of adaptation (Commonwealth Scientific Industry & Research Organisation 2009). Generally, Australian municipal councils are mainly responsible for local adaptation planning that somewhat illustrates the big picture of the national adaptation strategy (Pearce et al. 2018). However, successful climate change adaptation demands more effective collaborations among all government levels embracing a shared vision for the future (McEvoy et al. 2013; Nalau et al. 2015).

The Paris Agreement 2015 underlined the shared long-term goals and entailed all parties' more tremendous efforts to address climate change's grand challenges through their Intended Nationally Determined Contributions (INDCs) (United Nations 2015b). In this agreement, Australia as a party in the United Framework Convention on Climate Change (UNFCCC) affirmed the strong commitment to decreasing "greenhouse gas emissions by 26–28% below 2005 levels by 2030" (United Nations 2015a, p. 1). To assess the progress towards the committed target, the INFCCC required a regular submission of a National Communication and a Biennial Report every four years and two years, respectively. Moreover, the Australian government annually conducts National Inventory Reports regularly supplied with quarterly updates (Australian Government 2020b; United Nations 2020). Based on the UNFCCC's Database, the Australia's Seventh National Communication and the Fourth Biennial Report were, respectively, submitted in 2017 and 2019 as the latest versions.

These latest reports provided the up-to-date information regarding Australia's 2020 targets, adaptation and mitigation activities, economic and technical assistance for developing countries, and cooperation in developing capacity (Commonwealth of Australia 2017, 2019). More detailed, the level of Australia's emissions per capita in 2018 reached the lowest point since 1990, with 21.5 t CO2-e per person in 2018 compared with 33.5 t CO2-e in 1990 and 22.1 t CO2-e in 2016 (Commonwealth of Australia 2017, pp. 11–12; 2019, p. 10). Essentially, Australia's emissions per capita decreased 59.8% from 1990 to 2008 in five key sectors: energy, agriculture, industrial processes and product use, land use and change, and waste (Commonwealth of Australia, 2019, pp. 10–11). In general, Australia has somewhat demonstrated the commitment to the INDC in terms of strictly controlling the level of emissions and frequently submitting national reports to the UNFCCC (Commonwealth of Australia 2019, pp. 42–76; United Nations 2015a). As a result, the Australian government is progressing 2021–2030 towards Australia's 2030 targets of the INDC under the UNFCCC and the second commitment period (CP2) under the Kyoto Protocol.

11.1.2 Smart City Movements in Australia

In line with the 2015 Paris Agreement, the New Urban Agenda of the United Nations also highlighted a global commitment to the smart city vision that leverages innovative technologies to produce more inclusive, sustainable, and resilient cities (United Nations 2017). In Australia, the Smart Cities Plan was formally initiated in 2016, with three primary pillars: smart investment, smart policy, and smart technology (Commonwealth of Australia 2016). Accordingly, numerous smart city initiatives were launched across the country. However, a focus on the Australian smart city agenda is needed to be sharpened for two key reasons. First, Australian cities are quickly outdistanced by other competitors in the race for smart city development. Second, the Australian economy is rapidly transitioning towards positioning itself as a knowledge-based economy (Maalsen et al. 2018; Pettit et al. 2018). Thus, many Australian cities have varied smart city movements to formulate a sharpened focus for the shared smart city vision.

Since 2019, three types of smart city initiatives have claimed precedence over others in Australia. These initiatives aim to extract as much information acquired by the gathered data from the extensive networks as possible during the piloting phase. Specifically, platform projects, together with smart infrastructure and communication networks, have performed their leading roles in further facilitating Australia's smart city movements (KPMG 2019). The Australian Urban Research Infrastructure Network (AURIN) and its initial application—AURIN Portal—are the quintessence of e-infrastructure initiatives which aim to create a platform to support smart city research, development, and policymaking. The AURIN project was initiated in 2010 to support various organisations in accessing various inter-related datasets that range from social attributes to economic indicators (Sinnott and the AURIN Technical Team 2016).

Economic growth has been one of the main drivers for the global explosion of smart city agendas (Bakıcı et al., 2013). Nevertheless, the citizens must also be centred on these urban agendas (Hollands 2008; Maalsen et al. 2018). In an effort to foster smart city development in Australia, the government has executed the Smart Cities and Suburbs programme with three significant commitments: the investment of $50 million for infrastructure projects, the formation of a unit of infrastructure financing, and the cooperation between the three levels of government and the citizens (Commonwealth of Australia 2016). Australian cities have thus tended to adopt human-centric approaches to smart city strategies in preference to "urban labelling" processes (Hollands 2008, p. 303; KPMG 2019; Maalsen et al. 2018). Specifically, the local governments have leveraged open data and innovative technologies to increase the social engagement in co-designing and co-creating cities capable of providing higher living standards, stimulating economic growth, and making inclusive places (Burgoyne and Maalsen 2017).

While many Australian cities are gradually transitioning towards the knowledge-based economy, Melbourne's City has a particularly advantageous position on this transition pathway. Melbourne has explicitly formulated its economy based firmly on innovation and knowledge rather than mining and manufacturing (Burgoyne and Maalsen 2017). Therefore, the local government is taking advantage of the potentialities of smart technologies and smart city movements to establish its leading position in the digital transformation among other Australian cities (City of Melbourne 2020b; Maalsen et al. 2018; Sancino and Hudson 2020). In response to the impacts of climate change and the challenges in delivering liveability and sustainability, the City of Melbourne has implemented numerous smart city initiatives which have been mainly concerned with infrastructure improvement, community engagement, open data, and data platform (City of Melbourne 2020a, 2020b; State Government of Victoria 2016a, 2016b). Furthermore, the local government aimed at working closely with their local businesses and citizens to facilitate technological innovations "to design, develop and test the best ways to live, work, and play in Melbourne" (Anthopoulos 2017; City of Melbourne 2020b).

11.2 Materials and Methods for Case Study Analysis

The case study in this research is focused on Melbourne Greater Region. As a "city of cities", Melbourne is made up of 32 local government councils, comprising of many diverse local neighbourhoods with their own unique character, cultural mix, and set of advantages and challenges. Melbourne is a multicultural city of approximately 5 million residents, originating from more than 180 different countries (Melbourne 2016). It has been consistently considered as one of the most liveable cities in the world according to the Global Liveability Ranking annually published by the Economist Intelligence Unit. Melbourne has always been among top liveable cities in the world, but it has its own challenges in facing shocks and disruptions (Fig. 11.1).

In this section for materials and methods, we will briefly review Melbourne's Resilience Strategy and Melbourne as a smart city to clearly show the context of the case study, then the assessment method will be described later in this section.

EXAMPLES OF MELBOURNE'S CHRONIC STRESSES	EXAMPLES OF MELBOURNE'S ACUTE SHOCKS
• Rapid population growth	• Bushfires
• Increasing social inequality	• Floods
• Increasing pressure on natural assets	• Heatwaves
• Unemployment particularly among young people	• Pandemics
• Climate change	• Infrastructure- related emergencies
• Increasing rates of alcoholism & family violence	• Extremist acts, including cyber crime

Fig. 11.1 Melbourne's shocks and stresses identified in resilient Melbourne strategy

11.2.1 The Resilient Melbourne Strategy

The City of Melbourne was among the first 32 cities that joined 100RC Resilient Cities Initiative by Rockefeller foundation in 2013. For of its "city of cities" nature, the work was at a metropolitan scale, collaborating with all 32 councils of metropolitan Melbourne and was based on the Resilience City Framework developed by ARUP. Resilient Melbourne Framework (RMF) is comprised of four categories and 12 key indicators that describe the fundamental attributes of a resilient city (Fig. 11.3). In developing the RMS, the gaps were identified over consultations and collaborations between different actors including individuals, public, private, and academic sectors. The identified actions are mostly in these three categories: adaptation (reducing the exposure), surviving (enduring disruptions and bouncing back), thriving (improving people's quality of life), and embed (incorporating resilience thinking in our ways of living and working).

As shown in Fig. 11.1, the city faces a diverse range of "shocks" and "stresses" that increasingly pressure the city's infrastructure and population. For example, one of the main stressors is the rapid population growth the city has been facing for more than ten years, and similar growth rates are expected for the following decades, with the population projected to 8 million by 2050. This will have consequences for the population and of course for authorities that provide infrastructure and other public services and facilities (e.g. energy, transport, water, sewage, schools, health services).

11.2.2 Melbourne as a Smart City

As discussed in the above sections, the smart city initiatives in Australia have focussed primarily on developing platforms, constructing smart infrastructure, and improving communication networks based on national development strategies. Nevertheless, there have been many smart city initiatives at the local level that have lied beyond the national strategies, so-called piecemeals in the study of (Dowling et al. 2019). Therefore, to clarify the contributions of smart city initiatives in Victoria across various scales, we chose the Melbourne LGA and the Victoria State as our primary local and regional study areas, respectively. Analysing smart initiatives' characteristics, both strategic and piecemeal ones, the authors attempted to explore these initiatives' different contributions to the selected study area's resilience.

To analyse smart initiatives' characteristics against resilience, we employed an established dataset of natural disaster resilience score at the LGA level from the Australian Natural Disaster Resilience Index (ANDRI). The LGA level is determined as the third level of Australia's government and is allocated under the state and federal levels. The ANDRI is identified by the coping capacity and the adaptive capacity of the selected area. More specifically, the coping capacity is measured by six primary areas, consisting of social character, economic capital, infrastructure and planning, emergency services, community capital, and information and engagement.

Likewise, the adaptive capacity is assessed by governance, policy, leadership, social, and community engagement. The relevant indicators of these primary areas are shown in Table 11.1.

Smart city initiatives in Australia have a great deal of support and political will behind it—the Australian government allocated A$50 million just for the smarter cities and suburbs programme in 2015. The Smart Cities Plan and City Deals were formally initiated in 2016, with three primary pillars: smart investment, smart policy, and smart technology—and then, numerous smart city initiatives were launched across the country. Australian economy is rapidly transitioning towards positioning itself as a knowledge-based economy—Thus, there have been varied smart city movements in many Australian cities to formulate a sharpened focus for the shared smart city vision. And Melbourne is among the lead cities in this area with many different smart city initiatives running at different scales in different stages but Melbourne's vision as a smart city is pretty simple—it aims to enhance the aspects of the city that make it uniquely Melbourne, and intelligently prepare for the changing needs of the community, the environment, and the economy. This emphasis on the needs of community is an important aspect of this vision. These are all, some examples of existing programmes and initiatives in Melbourne—digital twin, 24-h pedestrian counting system, city labs, technology test beds, different open data platforms, etc.….

To categorise the collected set of smart initiatives in Victoria, the authors proposed two fundamental categorisation types: (1) thematic fields and (2) development strategies. Specifically, the thematic fields consist of seven essential overarching areas based on which each smart city initiative's particular function is identified and developed. The seven thematic fields include (1) Energy and environment, (2) Infrastructure and buildings, (3) Safety and security, (4) Governance, (5) Economy, (6) Mobility, and (7) Living and community engagement. Accordingly, one project can be categorised into one or more thematic fields based on its functions and purposes. Another type of categorisation is focussed primarily on the LGA's detailed development strategy. This type of categorisation was initially proposed in the study of Dowling et al. (2019) that divided smart city initiatives implemented in Sydney and Melbourne into strategic projects and piecemeal ones. According to Dowling et al. (2019), some projects were considered strategic projects, meaning that they have been developed based on the national development plan, while many other projects that were categorised as piecemeal projects have been developed beyond the national smart development plan.

As articulated above, Australia's Federal government has explicitly initiated the national agenda through the Smart City Plan since 2016 to stimulate national investments in infrastructure investment and leverage innovative technologies to foster the growth of Australia's digital economy (Commonwealth of Australia 2016). To execute this strategy, the Federal government has triggered two following programmes, which have worked as two primary delivery mechanisms, namely Smart Cities and Suburbs and City Deals Programmes (Australian Government 2020a, 2020b). More specifically, the former has concentrated on fostering pilot project using innovative technologies to enhance urban services and improve quality of urban lives in the manner of co-funding with other partners in both the public and private

Table 11.1 Smart city initiatives identified in metropolitan Melbourne

No	Smart city initiatives in greater Melbourne	Description/Challenge	Technology	Objectives/Solution	Types of smart city initiative	Contribution to resilience characteristics
1	Reducing litter Smart Waste Management	400 smart solar bins, 2000 public litter bins and 500 cigarette bins	IoT Network infrastructure	Manage waste and recycling by using smart bins	Infrastructure Waste management	11.3.1.1. Effective management 11.3.4.2. Optimum use of critical infrastructure
2	24-h pedestrian counting system	Sensors and automated counting systems	Sensors Data platform	Understand the pedestrian activities	Infrastructure Mobility Open data	11.3.4.3. Transport network, emergency information, 11.3.4.2. optimum use of critical infrastructure, emergency response plan 11.3.3.2. Policing to promote safety & security
3	Emerging technology test beds	5G and Internet of Things (IoTs)	5G IoT	Within the 5G and IoT test bed, partners will design new partnership and governance models, establish protocols for data sharing, management, privacy and security and explore optimal placement and design of infrastructure	Infrastructure Innovation	11.3.1.2. Leadership and strategy 11.3.3.2. Business development & innovation 11.3.4.3. Optimum use of critical infrastructure
4	CityLab	Utilise the ground floor of the Melbourne Town Hall to work with the community	-	Prototype and examine new ideas and urban services	Community engagement	11.3.1.2. communication between Government and public

(continued)

Table 11.1 (continued)

No	Smart city initiatives in greater Melbourne	Description/Challenge	Technology	Objectives/Solution	Types of smart city initiative	Contribution to resilience characteristics
5	Free Wi-Fi	Free outdoor Wi-Fi for the citizens and visitors	Wi-Fi	Increase the equal accessibility to the Internet for everyone	Community engagement Infrastructure	11.3.4.3. Connectivity 11.3.2.1. Access to basic needs, empowerment
6	Melbourne urban forest	Open data of more than 70,000 trees	Sensors network Online platform	Manage and protect Melbourne's urban forest	Open data Data platform	11.3.4.1. Environmental policy 11.3.4.2. Ecosystem management 11.3.1.3. access to data &monitoring
7	Open Innovation Competitions	Annual competitions for the community to address urban challenges	Open data	Engage with the community (innovators, entrepreneurs, and students)	Open data Community engagement	11.3.1.2. knowledge transfer, empower a range of stakeholders 11.3.3.1. community participation, citizen science
8	Melbourne open data platform	Over 150 publicly available datasets	Data Management (Data platform, open data, data privacy and security, blockchains, standards) Cloud computing Online portal	Improve data transparency, public services, and social engagement	Open data Data platform	11.3.1.3. access to data & monitoring, planning & strategy 11.3.1.1. collaboration & empowerment 11.3.1.2. knowledge transfer
9	BioBlitz	City of Melbourne's first citizen science programme	Data Management Open data	Experts and the community collaborate to formulate documents about different species living in Melbourne	Open data Community engagement	11.3.1.2. Learning Capacity 11.3.4.2. Ecosystem management, Innovation 11.3.3.1. Community participation

(continued)

Table 11.1 (continued)

No	Smart city initiatives in greater Melbourne	Description/Challenge	Technology	Objectives/Solution	Types of smart city initiative	Contribution to resilience characteristics
10	Census of Land Use and Employment (CLUE)	Comprehensive data about land use, demographic structure, economic characteristic	Data Management (Data platform, open data, data privacy and security, blockchains, standards) Cloud computing Online portal	Support in urban studies and planning, decision-making, business activities, and education	Start-up and innovative entrepreneurialism Smart governance	11.3.1.1. Efficiency & Effectiveness of urban service delivery 11.3.1.3. Land Use Planning, access to data & monitoring 11.3.2.2. local business development & innovation
11	Footscray Smart City for Social Cohesion	Build smart infrastructures Construct a platform for the smart city for social cohesion	Internet of Things Wi-Fi Network infrastructure Data Management (Data platform, open data, data privacy and security, blockchains, standards) Wireless sensor networks Environmental sensors IT systems (interoperability) Cloud computing Online portal	Using smart technologies and open data to bring different community values, including city activation, security, economic activity, mobility, and efficiency	Smart lighting Community engagement Public safety Active lifestyle Smart parking Environmental monitoring People movement Visitor experience	11.3.3.1. Social support, community participation, local identity & culture 11.3.2.2. local business development,

(continued)

Table 11.1 (continued)

No	Smart city initiatives in greater Melbourne	Description/Challenge	Technology	Objectives/Solution	Types of smart city initiative	Contribution to resilience characteristics
12	Digital Twin Victoria- pilot: Fisherman Bend	Digital Twin Victoria is an innovative new digital programme led by Land Use Victoria. It brings together rich 3D and 4D spatial data, artificial intelligence and sensor data from across the State to visualise and model places virtually, before investments hit the ground	Internet of Things Wireless sensor networks Applications Online Platform 5G AR Environmental sensors IT systems (interoperability) Cloud computing Online portal	– enable better and faster delivery of construction and infrastructure with 3D building and utilities data – enable intelligent planning and development solutions – unlock greater savings and efficiencies across entire asset life cycles – provide advanced algorithms and artificial intelligence that support faster, more robust regulatory assessments and compliance monitoring, like the eComply project – enrich communication and engagement with our community – fuel the State's start-up ecosystem to create more skills and attract investment	Smart lighting Open data Community engagement People movement Environmental monitoring Public safety	11.3.1.1. Capacity & coordination, Decision-making, collaboration, multi-stakeholder alignment 11.3.1.2. Risk monitoring, communication between government & public, 11.3.1.3. Land use planning, access to data & planning 11.3.2.2. continuity following a shock 11.3.3.1. community participation 11.3.3.2. policing for safety & security, law enforcement 11.3.4.1. Environmental policy, safeguard critical services, 11.3.4.2. ecosystem mng, flood risk mng, opt use of critical infrastructure, emergency response plan, 11.3.4.3. emergency information, transport infrastructure management,

(continued)

Table 11.1 (continued)

No	Smart city initiatives in greater Melbourne	Description/Challenge	Technology	Objectives/Solution	Types of smart city initiative	Contribution to resilience characteristics
13	i-Sense Oakleigh: The Smart Connected Precinct	A smart precinct was constructed by advance sensing technologies and data communication networks	Internet of Things Wireless sensor networks	Monitor, gather and analyse data to enhance the life quality of the community living in the precinct	Smart lightning Smart parking	11.3.4.2. Facility and services management (buildings, energy, waste, utilities) 11.3.2.1. Innovation and economic development 11.3.3.1. Visitor experience
14	Mornington Peninsula Smart Parking and Amenities for high demand areas	The project was launched to decrease the level of traffic congestion and accidents, improve parking activities, and enhance public services in busy urban areas	Internet of Things Network infrastructure Smartphone applications Hardware Wireless sensor networks Environmental sensors IT systems (interoperability) Cloud computing Online portal	Strengthen accessibility and availability of parking using real-time parking data Collecting journey services data to decrease the level of congestion	Smart lighting Smart traffic management Smart parking Smart amenity management Environmental monitoring Visitor experience People movement Other	11.3.3.1. Visitor experience 11.3.2.1. Innovation and economic development 11.3.4.2. Facility and services management (buildings, energy, waste, utilities)
15	My Virtual Moreland	The project uses 3-dimensional mapping, modelling, and other smart technologies of Virtual Reality (VR)	Smartphone applications Hardware 3D modelling Online portal Other	Increase the liveability, affordability, and sustainability by improving urban design and development through smart technologies	Community engagement Smart hub Smart governance Other	11.3.1.3. City and community planning 11.3.2.1. Innovation and economic development 11.3.1.3. Public administration and customer service
16	Smart Technologies – Reinventing Neighbourhoods	To understand traffic flow and people movement across the precinct to make better informed data-driven decisions for improvement and future planning across this neighbourhood activity centre	Internet of Things, Wi-Fi, Wireless sensor networks, Environmental sensors, Smart poles, Cloud computing, Online portal	With IoT-based technology for vehicle parking, pedestrian counting and environmental sensors, and undertaking an Stakeholder Engagement exercise, they will have accurate real-time and time-series data to make informed decisions	Community engagement Smart hub Smart parking Environmental monitoring People movement	11.3.1.1. Decision-Making 11.3.1.2. Stakeholder engagement 11.3.3.1. Community Participation

(continued)

Table 11.1 (continued)

No	Smart city initiatives in greater Melbourne	Description/Challenge	Technology	Objectives/Solution	Types of smart city initiative	Contribution to resilience characteristics
17	Northern Melbourne Smart Cities Network	LoRaWAN IoT-based network that enables the integration of five types of sensors to collect data on a wide variety of aspects of everyday life in the cities. The five sensors are for: counting people, monitoring air quality and environmental factors, monitoring water levels, waste management collection, tracking assets	Internet of Things Network infrastructure Data Management (Data platform, open data, data privacy and security, blockchains, standards) Hardware Wireless sensor networks Environmental sensors Online portal	Sense and enhance the efficacy of public services and provide potential supports for new services Driven by the City of Whittlesea, the project is in partnership with the Banyule City Council, Moreland City Council, Mitchell Shire Council, Nillumbik Shire Council, Royal Melbourne Institute of Technology, La Trobe University, and Minnovation Australia	Smart traffic management Disaster management Visitor experience Public safety Smart waste Smart amenity management Environmental monitoring People movement	11.3.1.1. Effective City Management, Capacity & Coordination, Decision-Making 11.3.1.2. Risk Monitoring, Risk Awareness 11.3.3.2. Promoting safety & Security 11.3.4.1. Environmental Monitoring 11.3.1.3. City and community planning 11.3.1.2. Visitor experience 11.3.3.3. Innovation and economic development 11.3.4.2. Facility and services management (buildings, energy, waste, utilities) 11.3.3.2. Public safety (crime and disaster prevention and management) 11.3.1.3. Education and public health 11.3.4.2. Natural environmental data and measurement (air quality, dust, noise, waterways)

(continued)

Table 11.1 (continued)

No	Smart city initiatives in greater Melbourne	Description/Challenge	Technology	Objectives/Solution	Types of smart city initiative	Contribution to resilience characteristics
18	Smart Redevelopment, Victoria	The project utilises novel online data and spatial analytics planning tools to determine highly potential precinct for redevelopment	Data Management Spatial Analytics	Promote the level of partnership between residents and development agencies to enhance housing supply, diversity, and affordability	Open data Data analytics Planning support tools Community engagement Other	11.3.1.1 Decision-making & leadership, multi-stakeholder alignment 11.3.1.3. Strategy & plans
19	CityLens-Holographic Computing to Decode Our Cities	use of immersive mediums which will allow data such as pedestrian and vehicle traffic, air quality, water usage and other sensor-derived information to be viewed on three-dimensional models	Internet of Things Augmented/Virtual reality Data Management (Data platform, open data, data privacy and security, blockchains, standards) 3D modelling Online portal	The platform was designed for optimising collaboration and engagement, allowing all stakeholders to explore urban smart cities in a novel and beneficial way	City and community planning Innovation and economic development Community engagement	11.3.1.1. collaboration & coordination 11.3.1.2. Communication between government & public 11.3.3.1. Community Participation
20	Smart Sports Field Planning, Monitoring and Management	The project is expected to improve the comprehensibility and management of urban areas for sports	Wi-Fi, CCTV, Network infrastructure, Data Management (Data platform, open data, data privacy and security, blockchains, standards)	Utilise smart technologies to advance sports field planning, availability maximisation, usage patterns, maintenance, and management programme	Active lifestyle Smart amenity management People movement Other	11.3.1.2. Communication 11.3.4.1. provision of alt services, safeguarding critical services, 11.3.4.2. optimum use of critical infrastructure, conservation of assets,

(continued)

Table 11.1 (continued)

No	Smart city initiatives in greater Melbourne	Description/Challenge	Technology	Objectives/Solution	Types of smart city initiative	Contribution to resilience characteristics
21	Interactive City Management in Melbourne	Smart sensor network infrastructure to alleviate reactive or scheduled asset management processes	Internet of Things, Wi-Fi, CCTV, Network infrastructure, Data Management (Data platform, open data, data privacy and security, blockchains, standards), Environmental sensors	Building an interactive platform for data collection, predictive modelling of life cycle performance of assets, and to allow for live community feedback on facilities	Community engagement / visitor experience Smart hub Environmental monitoring Smart waste Smart building Smart amenity management Other	11.3.1.1. Decision, collaboration, coordination 11.3.1.2. knowledge transfer, communication, risk monitoring, 11.3.1.3. Access to data & monitoring 11.3.3.1. community participation 11.3.4.1. alt provision of services, safeguarding critical services,
22	Melbourne City DNA	A large, interactive, 3D-printed city model enriched with data projected onto it, virtual reality immersive experiences, augmented reality demonstrations, a series of interactive data screens with data visualisations and touch-screen maps	AR 5G	The ultimate aim of Melbourne City DNA is to help Melbournians connect on a deeper level to the city: its past, present and possible futures and help shape the decisions that guide its growth	Community engagement/ visitor experience	11.3.1.1 Decision-making, collaboration & empowerment 11.3.1.2 Knowledge transfer, communication between government & public 11.3.1.3. Access to data

(continued)

Table 11.1 (continued)

No	Smart city initiatives in greater Melbourne	Description/Challenge	Technology	Objectives/Solution	Types of smart city initiative	Contribution to resilience characteristics
23	Melbourne Connect	A living lab or a purpose-built innovation district bringing together researchers, government, start-ups designed to help unlock digitally driven, data-enabled and socially responsible solutions to our most pressing challenges		Drawing on the University's expertise across emerging technologies such as AI, robotics, computer-science, cybersecurity and privacy, it is a digital data powerhouse with the people, place and programmes	Collaborative innovation Smart Partnerships Community engagement	11.3.1.1. Collaboration & empowerment 11.3.2.2. Skills & training, local business development and innovation 11.3.2.3. Public health innovation

sectors. There were 81 successful projects in this programme after two selection rounds, including 49 projects in the first round and 32 projects in the second round (Australian Government 2020c). The latter has promised a strengthened partnership between three levels of government to pursue targets around three overarching pillars: "smart investment", "smart policy", and "smart technology". When writing this report, nine deals have been signed under this programme involving seven agreed deals and two announced deals (Australian Government 2020a).

In the study of Dowling et al. (2019), these two mechanisms were also scientifically identified as two preliminary dimensional characteristics of smart city projects across the world. Nevertheless, the authors expressly stated that although the Federal government nationally launched these national strategies, the following schemes do not strictly control how the local authorities take steps to develop smart cities. Furthermore, the local governments have different detailed implementations of smart initiatives outside the national schemes. According to Dowling et al. (2019), over two-thirds of Melbourne's local authorities have actively participated in smart initiatives that have gone beyond the national strategy. Additionally, Melbourne's 18 local governments have utilised numerous smart city initiatives in a "piecemeal" manner (Dowling et al. 2019). In other words, these initiatives were formally adopted by different local governments without having a narrative in common at the national level. Tables 11.1 represents the notable differences between council-scaled (City of Melbourne) and state-scaled (Victoria) smart city initiatives in Melbourne.

11.2.3 Assessing the Potential and Actual Contributions of Smart City Initiatives to Resilience

A comprehensive desktop research was conducted to identify the main smart city projects across metropolitan Melbourne. Table 11.1 in the next section of this chapter shows the name, description, objectives, and the technology used in each project. In order to assess the potential and actual contributions of these initiatives to resilience, the Resilience Melbourne Framework (RMF) was used to analyse how a specific project contributes to each 12 indicators mentioned in this framework. Figure 11.3 shows an example of how and in what ways the Digital Twin Smart City initiative may contribute to resilience of the city.

Considering the city as a complex system interconnected and interdependent components, digital twin helps with project design, planning, implementation, and managing how different factors may affect outcomes from different perspectives for the infrastructure, environment, and the people. Digital twin technologies allow cities to either directly or indirectly contribute and promote resilience attributes by facilitating the optimum decision-making for design, construction, and implementation of the infrastructure assets and brining in lasting benefits for community. Figure 11.3 shows some examples of areas on how digital twin projects may contribute to resiliency through carbon–neutral city design, high efficiency and best use of limited resources, reducing carbon footprint, risk reduction, demographic changes, environmental degradation and climate change. More details for each project will be provided in following section.

11.3 Results and Discussions

A series of smart city initiatives were identified in Metropolitan Melbourne and they were studied against the Resilient Melbourne Framework (RMF) indicators to find if and how they may contribute to the resilience characteristics of the city (Table 11.1). The smart city type is identified based on the seven categories suggested in Fig. 11.2: Energy and Environment, Infrastructure and Buildings, Safety and Security, Governance, Economy, Mobility, Living and Community Engagement. In following subsections, we review which digital solutions used in SC projects contribute to each component of the Resilient Melbourne Framework (Fig. 11.3) and how they may promote resiliency in these areas.

11.3.1 Leadership and Strategy

11.3.1.1 Digital Solutions for Promoting Leadership and Effective Management

Local political leadership is an important catalyst for incorporating resilience and adaptation in urban planning and so promoting leadership and effective management of urban service delivery is one of the main objectives of Resilient Melbourne Strategy (RMS). Other than having the knowledge, solid frameworks, plans, and strategies, it is the leadership prioritisation that eventually determines if the pro-environment or pro-development initiatives will get the major investments or they will only get budget for the incidental and ad-hoc projects.

Fig. 11.2 Categorisation of the studied smart city initiatives

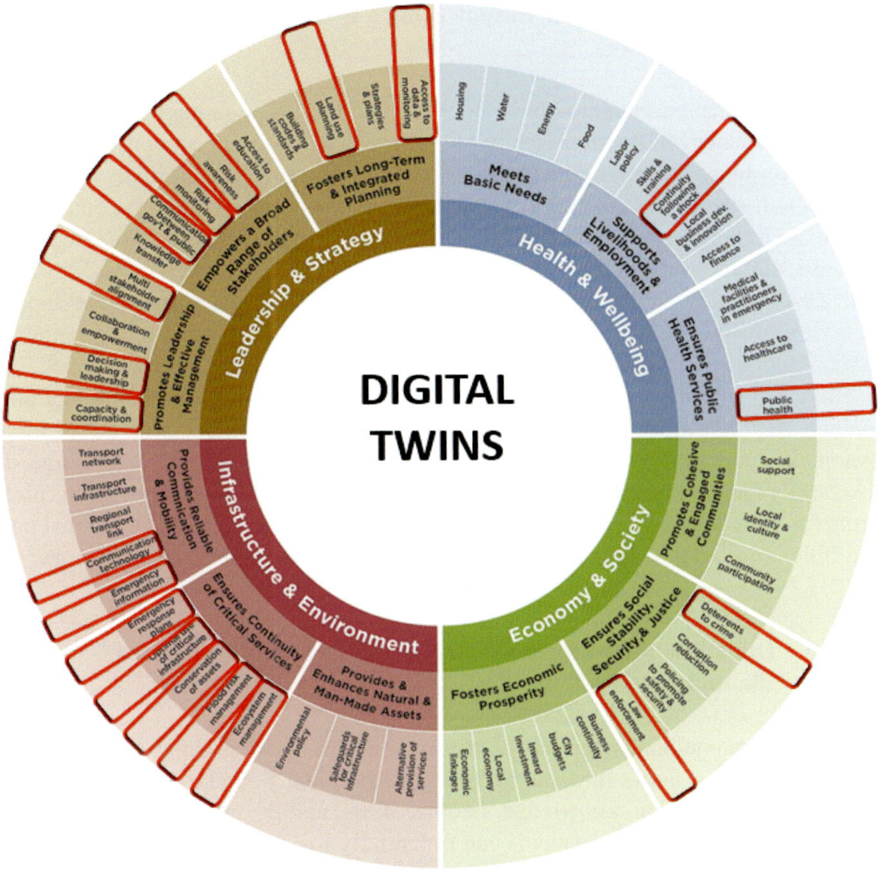

Fig. 11.3 Schematic analysis of contribution of digital twin projects to city resilience using RMF

Many of the smart city initiatives mentioned above utilise open data platforms which contribute to smart and transparent governance by supporting interoperability, open, decentralised, and multi-level governance and providing Decision Support Systems. Interactive City Management is also another key area in these smart city initiatives that can potentially contribute to resilience by supporting connectivity, agility (rapid response), efficiency, foresight capacity, and also adaptive and learning capacity.

11.3.1.2 Digital Solutions for Empowering a Broad Range of Stakeholders

Multi-stakeholder partnerships have been one of the main strategies for building resilience. Greater Melbourne is a city of cities with 32 different councils. The transparent governance and open data platforms provided by smart city initiatives in these councils will enable a broad range of stakeholders to better understand the communities and areas they are working in and so get better integrated plans for preparedness, response, and recovery in case of any disasters.

In a recent study published by Fastenrath et al. (2019), they emphasise the importance of multi-level governance and working collaboratively with a broad range of stakeholders as a key part of RMF implementation. They also suggest reconceptualisation of the RMF strategies as urban transformational innovation strategy and implementation of resilience actions as "governance experiments". Smart city initiatives mentioned in this category in Table 11.1 could empower a broad range of stakeholders and contribute to co-management and partnerships by providing platforms for dialogue and interaction between different stakeholders to build trust and credibility and facilitate a more reciprocal relationship between different actors.

11.3.1.3 Digital Solutions for Fostering Long-Term and Integrated Planning

Access to data, online integrated platforms, and monitoring capabilities provide the means to plan and strategise beyond jurisdictional boundaries, in a "globalized networks of capital, knowledge, people, skills, and resources". Intelligent and integrated decision-making systems can potentially foster integrated and long-term planning, for example, by embedding innovation and smart city principles into long-term tenders and contracts (e.g. waste collection services, asset maintenance, tree watering).

11.3.2 Health and Well-Being

11.3.2.1 Digital Solutions for Providing Basic Needs of Community

One of the main challenges in response to any shock or disasters is continuing providing the basic needs of community members such as access to food, water, housing/temporary shelter, and of course energy. The recent pandemic clearly showed the fragility of our food supply chain and how automation and online digital platform could hugely help in sourcing and delivering product and services to community. Intelligent systems for water and energy usage may enhance the efficiency of using limited resources. Also, all the other e-services that helped community to meet their needs can fall under this category such as e-learning, citizen-services.

11.3.2.2 Digital Solutions for Supporting Livelihoods and Employment

E-commerce and digital tools can offer great economic and productivity advantages to all businesses. The government is investing a lot to support small and medium businesses with access to low cost, high-quality advice on a range of effective digital solutions, and also investing providing access to related digital skills and training. Reliance on these digital technologies would contribute to resiliency to economic, health, and environmental shocks and threats through enabling business continuity, community safety, and swift recovery.

11.3.2.3 Digital Solutions for Ensuring Public Health Services

The purpose-built innovation district called Melbourne Connect, mentioned in Table 11.1, has a centre for digital transformation of health that works on interoperability and effective management of health information systems, sensible and open access to data, and integration of national and local systems, digital health and ICT.

11.3.3 Economy and Society

11.3.3.1 Digital Solutions for Promoting Cohesive and Engaged Communities

One of the main objectives of Melbourne Resilience Strategy is promoting social cohesion and providing adequate and equitable access to infrastructure. And thus many of the smart city activities and initiatives that lead to cohesive, fair, and healthy community will also contribute to better sharing resources and supporting each other in response to shocks, stresses, and disasters (Melbourne 2016).

Historically, Melbourne has been strong in people-cantered and community-based initiatives, and so many of its smart city projects are designed for socially engaged and citizen-driven co-creation and innovation such as Melbourne's innovation district which is a smart living lab jointly created with stakeholders—by and for people—aiming to create healthier, greener, and resilient Melbourne to confront challenges including climate change. The initiatives' vision is completely in line with resilience objectives, better quality of life for all, green infrastructure and biodiversity, improved air and water quality, reduced noise and lower health costs, enhanced mobility conditions, and greater social cohesion. Citizen Science projects like BioBlitz, Social Innovation, and Hackathons are among the other examples of Melbourne's smart city initiatives that aim to co-create smart and resilient city by engaging community.

11.3.3.2 Digital Solutions for Ensuring Social Stability, Security, and Justice

Smart digital technologies such as Facial Recognition, CCTV/ Sensors, Smart Video Analytics, Fraud Detection used in some of the Melbourne's smart city initiatives support resilience through deterring crime, corruption reduction, and even policing and law enforcement in an Integrated Emergency Systems.

11.3.3.3 Digital Solutions for Fostering Economic Prosperity

One of the main focus areas of Melbourne's smart city initiatives is to help start-ups develop smart business models and finance models with smart investment tools. There is various support for entrepreneurship through living labs to encourage continued innovation towards sharing and circular economy.

11.3.4 Infrastructure and Environment

11.3.4.1 Digital Solutions for Reducing Exposure/Fragilities and Enhancing Natural and Man-made Assets

Melbourne Resilience Strategy emphasises on leveraging digital technologies to improve the monitoring, promotion, and evaluation of natural and man-made assets as well as managing them more efficiently and affordably. This has been promoted through initiatives such as Melbourne Urban Forest which is a priority "flagship action" in Resilient Melbourne Strategy and it pursues to develop a metropolitan-wide method to increase the city's tree canopy cover and vegetation through connecting existing urban green infrastructure projects with reforestation and other environmental projects. Other initiatives mentioned above such as smart bins, smart watering, digital twin, Melbourne's innovation district, energy efficiency systems, smart grids, smart lighting, waste management, smart green buildings, innovative materials, and construction systems all have the potential to enhance natural and man-made assets and promote critical infrastructure resiliency.

11.3.4.2 Digital Solutions for Ensuring Continuity of Critical Services

In the wake of a disaster or a shock to city services, it is of most importance to ensure the continuity of critical services such as food supply chain, water, energy. Smart city initiatives such as digital twin and other smart data management platforms can enhance the optimum use of resources, identifying redundant and alternative substitute of services in case of failure. Redundancy is one of the key principles of resilience in complex adaptive systems however, it tends to be ignored in favour of efficiency, optimisation, predictability. Digital solutions mentioned above by providing high modelling and foresight capacity can create a balance between efficiency, optimisation, and redundancy.

11.3.4.3 Digital Solutions for Providing Reliable Communication and Mobility

Melbourne innovation test bed, 5G, IoT projects, digital twin, and other pilot test use of new communication technologies could potentially lay the foundation for a resourceful connected city. This will provide reliable real-time emergency information, transport and mobility networks enhancing the current traffic management system, digital signages, public transport, shared mobility, smart parking and also for future smart initiatives on air mobility, autonomous vehicles, connected vehicles, electric vehicles, logistics and Last Mile Delivery, Micro Mobility.

11.4 Summary and Conclusion

The rapid pace of urbanisation and the increasing challenges that come with it calls for innovative technologies and approaches that can help our cities meet zero carbon emissions targets and accelerate more efficient, liveable, and resilient development. This chapter highlighted the value of data and technology in transforming the ways that our cities are planned, built, and managed by evaluation of resilience measures in Greater Metropolitan Melbourne's smart city initiatives. Twenty-three smart city initiatives were identified and analysed based on their contribution to each resilience indicator in Resilience Melbourne Framework (RMF). The identified smart city solutions have been mostly related to community engagement and technology test beds, but more integrated perspectives are also emerging including Melbourne connect and digital twin initiatives.

This case study showed that Melbourne has a more experimental approach in smart city initiatives with less emphasise on working on a comprehensive smart city plan. While these experiments might be valuable in terms of exploring different possibilities especially that it includes local community in its live smart city experiments. However, a more integrated and comprehensive plan (instead of ad-hoc random projects) can better promote the multiple dividends approach. This means smart investments in digital transformation of cities at the same time can contribute to the other city agendas as much as possible including urban resilience, sustainability, circular, just, and ethical cities. Although this comprehensiveness and interestedness may bring up some top-down vs. bottom-up approaches, but a balance between both seems more beneficial in planning for smart resilient cities. Of course, creating opportunities for genuine community engagement and co-creation are always valuable and of most importance here.

Combination of the concepts of smart city and resilient city also draws increasing attention the challenges that digital technologies may bring into the resilience of cities. Digital divide is one of the common concerns for equitable access to opportunities that smart city projects offer just to those who have the digital literacy and digital devices. E-waste is considered as another challenge of digital world as they are often short-lived and often need upgrading. Some digital components contain toxic materials and their disposal can contaminate environments and threaten human or other species health. Here resilient and sustainable thinking may offer more effective ways to deal with these challenges considering circular design and life cycle assessment for smart city projects. There are several other concerns such as increasing cyber-risks and privatisation of civic spaces as the complex nature of smart city systems and their continuing maintenance may lead to long-term reliance to tech companies, but the major concern here is the level of energy consumption. Cloud computing and other ICT-related tech lead to much higher energy use/demand and consequently more GHG emission. These negative impacts on resilience and sustainability may lead to cancel out all the other positive impacts that we discussed in previous sections.

Considering all the opportunities and challenges that smart city projects contribute to resilience and sustainability, more research is needed to examine how cities can make appropriate trade-off choices considering the DNA of the city, their own vulnerabilities, resources, and priorities. This approach shifts the focus from digital solutions just as a tool to promote economic growth and competitiveness, to a multiple dividend and participatory approach to tackle all the challenges our cities face today including climate change, environmental degradation, and social equity.

And finally, COVID-19 accelerated the take up of digital solutions in many different areas including e-health, e-education, e-governance, e-commerce, remote-working. There is no doubt that digital technology, data, and networks have an extremely critical role in our shared urban future. This future will be determined mainly by our attitudes towards these technologies. We need to ensure that we avoid the short-term flashy projects which will be thrown away when their novelty wears off and focus on designing thoughtful smart city projects that is people and planet-centric.

Acknowledgements This chapter was partially funded by Asia-Pacific Network for Global Change Research (Funder ID: https://doi.org/10.13039/100005536).

References

Allen Consulting Group (2005) Climate change risk and vulnerability: promoting an efficient adaptation response in Australia. Australian Greenhouse Office

Anthopoulos L (2017) Smart utopia VS smart reality: learning by experience from 10 smart city cases. Cities 63:128–148. https://doi.org/10.1016/j.cities.2016.10.005

Australian Government (2020a) City deals. Australian Government Retrieved from https://www.infrastructure.gov.au/cities/city-deals/index.aspx

Australian Government (2020b) Greenhouse gas measurement and reporting. Publications and resources Retrieved 30 May 2020 https://publications.industry.gov.au/publications/climate-change/climate-change/climate-science-data/greenhouse-gas-measurement/publications.html#quarterly

Australian Government (2020c) Smart cities and suburbs. Retrieved from https://www.infrastructure.gov.au/cities/smart-cities/

Bakıcı T, Almirall E, Wareham J (2013) A smart city initiative: the case of Barcelona. J Knowl Econ 4(2):135–148. http://hdl.handle.net/10.1007/s13132-012-0084-9

Burgoyne S, Maalsen S (2017) How smart are Australian cities? Local approaches to adopting smart cities strategies. Retrieved from https://united-states-studies-centre.s3.amazonaws.com/attache/e5/3f/69/53/58/d1/93/83/e5/d8/aa/c9/ab/8a/af/75/2017_How-smart-are-Australian-cities.pdf

City of Melbourne (2020a) Annual plan and budget 2020–2021. Retrieved from https://www.melbourne.vic.gov.au/SiteCollectionDocuments/draft-annual-plan-budget-2020-21.pdf

City of Melbourne (2020b) Melbourne as a smart city. Retrieved from https://www.melbourne.vic.gov.au/about-melbourne/melbourne-profile/smart-city/Pages/smart-city.aspx

Commonwealth of Australia (2007) National climate change adaptation framework. Retrieved from https://www.nccarf.edu.au/sites/default/files/Australian-Government-2007a.pdf

Commonwealth of Australia (2016) Smart cities plan. Retrieved from https://www.infrastructure.gov.au/cities/smart-cities/plan/files/Smart_Cities_Plan.pdf

Commonwealth of Australia (2017) Australia's 7th national Communication on climate change. Retrieved from https://unfccc.int/files/national_reports/national_communications_and_biennial_reports/application/pdf/024851_australia-nc7-br3-1-aus_natcom_7_br_3_final.pdf

Commonwealth of Australia (2019) Australia's fourth biennial report. Retrieved from https://unfccc.int/sites/default/files/resource/Australia%20Fourth%20Biennial%20Report.pdf

Commonwealth Scientific Industry and Research Organisation (2009) CSIRO climate adaptation national research flagship. Retrieved from https://unfccc.int/sites/default/files/nwp_ap_csiro_oct09.pdf

Dowling R, McGuirk P, Gillon C (2019) Strategic or piecemeal? smart city initiatives in Sydney and Melbourne. Urban Policy Res 37(4):429–441. https://doi.org/10.1080/08111146.2019.1674647

Fastenrath S, Coenen L, Davidson K (2019) Urban resilience in action: the resilient Melbourne strategy as transformative urban innovation policy? Sustainability 11(3):693

Green D, Billy J. Tapim A (2010) Indigenous Australians' knowledge of weather and climate. Climatic Change 100(2):337–354

Hollands RG (2008) Will the real smart city please stand up? City 12(3):303–320. https://doi.org/10.1080/13604810802479126

KPMG (2019) Smart cities: A snapshot of Australia in 2019. Retrieved from https://assets.kpmg/content/dam/kpmg/au/pdf/2019/smart-cities-snapshot-of-australia-in-2019.pdf

Kupers R, Foden M (2017) Learning for resilience and complex systems thinking. Agenda Setting Scoping Studies Summary Report. The Resilience Shift

Maalsen S, Burgoyne S, Tomitsch M (2018) Smart-innovative cities and the innovation economy: a qualitative analysis of local approaches to delivering smart urbanism in Australia. J Des, Bus Soc 4(1):63–82. https://doi.org/10.1386/dbs.4.1.63_1

McEvoy D, Fünfgeld H, Bosomworth K (2013) Resilience and climate change adaptation: the importance of framing. Plan Pract Res 28(3):280–293. https://doi.org/10.1080/02697459.2013.787710

Melbourne R (2016) Resilient Melbourne: resilience strategy for greater Melbourne. In: Melbourne: City of Melbourne

Nalau J, Preston BL, Maloney MC (2015) Is adaptation a local responsibility? Environ Sci Policy 48:89–98. https://doi.org/10.1016/j.envsci.2014.12.011

National Climate Change Adaptation Research Facility (2014) NCCARF 2008–2013: The first five years. NCCARF Publication Retrieved from https://www.nccarf.edu.au/sites/default/files/research_content_downloads/NCC030-report%20FA.pdf

Pearce T, Rodríguez E, Fawcett D, Ford J (2018) How is Australia adapting to climate change based on a systematic review? Sustainability 10(9):3280. https://doi.org/10.3390/su10093280

Pettit C, Bakelmun A, Lieske SN, Glackin S, Hargroves KC, Thomson G, Shearer H, Dia H, Newman P (2018) Planning support systems for smart cities. City, Cult Soc 12:13-24. https://doi.org/10.1016/j.ccs.2017.10.002

Prober SM, O'Connor MH, Walsh FJ (2011) Australian aboriginal peoples' seasonal knowledge: a potential basis for shared understanding in environmental management. Ecol Soc 16(2)

Sancino A, Hudson L (2020) Leadership in, of, and for smart cities—case studies from Europe, America, and Australia. Public Manag Rev 22(5):701–725. https://doi.org/10.1080/14719037.2020.1718189

Sinnott RO, the AURIN Technical Team (2016) The Australian data-driven urban research platform: systems paper. Aust Econ Rev 49(2):208–223. https://doi.org/10.1111/1467-8462.12152

State Government of Victoria (2016a) Plan Melbourne 2017–2050: a global city of opportunity and choice. Retrieved from https://www.planmelbourne.vic.gov.au/__data/assets/pdf_file/0009/377127/Plan_Melbourne_2017-2050_Summary.pdf

State Government of Victoria (2016b) Plan Melbourne 2017–2050: metropolitan planning strategy. Retrieved from https://www.planmelbourne.vic.gov.au/__data/assets/pdf_file/0007/377206/Plan_Melbourne_2017-2050_Strategy_.pdf

United Nations (2015a) Australia's intended nationally determined contribution to a new climate change agreement. Retrieved from https://www4.unfccc.int/sites/ndcstaging/PublishedDocuments/Australia%20First/Australias%20Intended%20Nationally%20Determined%20Contribution%20to%20a%20new%20Climate%20Change%20Agreement%20-%20August%202015.pdf

United Nations (2015b) Paris agreement. Retrieved from https://unfccc.int/files/essential_background/convention/application/pdf/english_paris_agreement.pdf

United Nations (2017) New urban agenda. Retrieved from http://habitat3.org/wp-content/uploads/NUA-English-With-Index-1.pdf

United Nations (2020) National inventory submissions 2020. Retrieved 30 May 2020 https://unfccc.int/ghg-inventories-annex-i-parties/2020

Chapter 12
Climate (Un)smart? Case Study of Smart City Projects in Surat, India

Shrutika Parihar

Abstract Indian cities face critical challenges in urban infrastructures with the growing population. It has been reported that urban areas have significantly contributed to increased carbon emissions. India's NDC goals are pursuing efforts to reduce carbon emission aligned to the Paris Agreement that includes the smart city as a mitigation initiative for achieving sustainable development and making climate-resilient cities. Surat city, Gujarat, India, has been selected as a case example to understand the interlinkages between climate resiliency and smart city action. Surat city is exposed to multiple climate risks: flood, heat, sea-level rise, erosion, and biodiversity, and about seventeen diverse organizations are working on adaptation and mitigation measures at the multilateral level. However, there have been limited studies available finding inter-relationship between smartness and resilience at the city level. This chapter explores the framework, methods, approaches, and model for finding solutions for making climate-resilient smart cities. The analysis is broadly undertaken into three steps: developing a conceptual smart city resilience framework, content analysis, and Analytical Hierarchical Process (AHP).

The conceptual smart city resilience framework includes a) four dimensions (criteria) b) fifteen resilience indicators (sub-criteria). Based on content analysis, fifteen identified resilience indicators were mapped with four dimensions to assess city climate adaptation and mitigation policies. The key thrust areas were measured through the triangulation approach using the Analytical Hierarchical Process (AHP) to establish the weights among different indicators for the study. The assessment shows city-focused is more on economic and infrastructure measures. The city climate policy design was set back in terms of targets due to a lack of transparency and a structural review mechanism. The results show key concerns in Surat city are governance, institution, technical learning and information technology, planning system, funding and awareness, and community support system. The outcome suggests that Surat city requires a bottom-up approach in decision planning for addressing key concerns. The proposed research study briefly explained the approach/method/model: Transdisciplinary Approach, Carbon-Centered Comprehensive (3Cs), Knowledge-based solution, Carbon Removal (CR), and Carbon

S. Parihar (✉)
Global Centre for Environment and Energy, Ahmedabad University, Gujarat, India

A. Sharifi and P. Salehi (eds.), *Resilient Smart Cities*, The Urban Book Series,
https://doi.org/10.1007/978-3-030-95037-8_12

Banking System, Community Energy Systems for managing major concerns and critical challenges of the Indian smart city.

Keywords Smart city · Climate-resilient cities · Conceptual smart city resilience framework · Analytical hierarchical process · Approach/method/model

12.1 Introduction

India, a developing economy still has 30% of its current population of 1.35 billion living in urban areas (Census 2011). Cities and urban areas generate about two-thirds of a nation's Gross Domestic Product (GDP) (Statista 2019). Cities offer better infrastructure and economic opportunities, creating challenges of migrating populations from surrounding areas that increase population density, increase the burden on existing resources, and pressure the environmental resources (Sridhar et al. 2013; Thet 2014). These critical challenges stress city managers to manage solid waste, flood, stormwater, waste, and public transportation (Gupta 2020; Rahmasary et al. 2019; Singh 2012). Several of these cities are exposed to climate change, hazards, and risks, specifically floods, heat, cyclones, storm surges, drought, and fires (AghaKouchak et al. 2020; Kumar et al. 2020) which are expected to increase in the near future (IPCC, 2014). By 2050, 53% of the Indian population will be living in the cities (Ritchie and Roser 2018), and hence, climate-resilient planning becomes essential for managing cities. Climate-resilient planning uses different tools such as efficient energy and conservation, retrofitting, Information Communication Technology (ICT), impact assessment of urbanization, and low-carbon green city design (Hussain and Gupta 2020; Kim 2018). These tools are often associated with the smart climate approach to creating opportunities for low-carbon green cities, eco-city, and sustainable cities (Kim 2018). Therefore, building resilient infrastructure plays a vital role in urban planning and management and offers smart investments opportunity for coming decades. This area has been a key focus area of India's Smart City Mission (GOI 2015).

The integrated planning approaches for creating climate-resilient and low-carbon smart cities are current needs for managing urban growth in cities. The urban growth has resulted in reduced agricultural land, forest, and water resources and adding stress on ecosystem services causing harm to the environment by increasing pollution and forming urban heat islands. India's case requires an efficient and sustainable solution for managing activities in urban areas by using smart technology to make a self-sustainable city (Goi 2017; Höjer and Wangel 2014). The inclusion of climate-resilient cities with sustainable smart planning tools will help achieve global net zero-carbon cities (United Nation 2011). Moreover, the emerging technologies powered by big data in smart city trace carbon footprints and increase resource efficiency (Al Nuaimi et al. 2015; Hashem et al. 2016). Thus, climate-resilient smart planning could be introduced in city planning to create more synergies across sectors.

Based on the selected case study, the proposed research highlights the issue pertaining to urban areas and requirements for smart, resilient city planning for climate risk management as an essential component of urban development. Furthermore, to understand how various climate policy interventions reduce climate vulnerability for making climate-resilient cities, the current research study focuses on two aspects (1) How does the smart city initiative support the implementation of climate resilience strategies? (2) How can city-level policies and intervention enhance resilience?

The study looks into current and future smart city, and climate resilience interventions to understand the interlinkages and trade-offs. As a case example, Surat city, located in the Gujarat state in western India, has been selected; the city is the fourth fastest-growing economy globally (City Mayors 2011). Surat city's rate of urbanization is faster than any other Indian city for 2019–2035 (Holt 2018). As a result, the city is already encountering and is also vulnerable to planetary threats of climate change (Blok 2016) like flood, heat, sea-level rise, erosion, and biodiversity in the future (Desai et al. 2015; FSI 2019; Parikh et al. 2018; Rathi et al. 2017; SMC 2018, 2020; SMC and SCCT 2017). For this purpose, the proposed research study develops a *conceptual smart city resilience framework* from an extensive literature review that identifies dimension and resilience indicators. The city urban resilience expert's opinion on indicators helped identify key thrust areas and derive weights for Analytical Hierarchical Process (AHP). The results and findings can help develop a structural review process for addressing the climate concerns for the city with the approach/method/ model. In addition, the outcomes can enable city planners to track the achieved climate risk, identify priorities or thrust areas to transition to low-carbon green cities, and enhance climate-resilient planning in tier 2 and tier 3 Indian cities.

12.2 Background

The challenges considered in the research study are rapid urbanization and how to transform into a smart city by achieving national climate goals to combat climate change and global warming. The city requires comprehensive climate-smart city planning as an integral part of mitigation and adaptation strategies; Mitigation reduces the cause of climate change, and adaptation addresses the local negative impacts of climate change (United Nation 2011). Climate-smart city planning creates new opportunities to improve efficiency, governance, surveillance, social infrastructure, and management, enhancing local ecosystem, supporting urban city development, smart technology for carbon reduction, and climate risk reductions. In addition, climate-resilient planning requires assessing the adoption rate of smart technology and the status of climate action plans in the city. These terms are discussed in the sections below.

12.2.1 *Climate Resilience*

The countries globally anticipated climate agreement on how countries intend to pursue efforts to reduce carbon emission aligned to 1.5°C and 2°C, which drives collective action toward zero-carbon and climate-resilient future (WRI 2020). India's INDC goals are promoting clean energy, renewable energy, energy efficiency, climate-resilient urban centers, sustainable transport, cleaning rivers, and enhance energy and resource efficiency. In addition, India's INDC includes a smart city (hundred smart cities) as one of the mitigation initiatives for achieving sustainable development and building climate-resilient cities (MOEFCC 2015).

Cities are a significant contributor of GHG emissions as producers and consumers of fossil fuel-based energies and other related goods and services (Satterthwaite 2008). Therefore, the city should consider strategies for adaptation and mitigation for socio-economic growth. A city uses energy for heating, cooling, lighting, industry-manufacturing goods and services, building material, and other related operations. Cities emit through wastewater treatment, solid waste decomposition at landfill sites, food production, land use, and energy conversion. The futurist approach of cities aims to achieve climate-neutral cities and climate-proof, or resilient, which is the key strategy to become climate-smart by addressing economic, social, and environmental changes (United Nation 2011). The economic, social, and environmental sectors are constructed at a multi-level by introducing various policies, interventions, and initiatives and play a pivotal role in building and reshaping the cities. The city government has an important role in policymaking and adopting various carbon-resilient initiatives for the city.

12.2.1.1 Government and Resilience Building in the Cities

Several initiatives have been implemented at the national level, such as Smart Cities Mission, Jawaharlal Nehru National Urban Renewal Mission (JNNURM), Atal Mission for Rejuvenation and Urban Transformation (AMRUT), Pradhan Mantri Awas Yojna (PMAY), Swarnim Jayanti Mukhya Mantri Saheri Vikas Yojana, Rajiv Awas Yojna, Swatch Bharat Mission (Clean India Campaign), National Solar Mission, National Wind Mission, National Bio-Energy Mission, Enhanced Energy Efficiency, National Disaster Management Agency (NSDMA), and National River Conservation Plan. As such these initiative helps in achieving Intended Nationally Determined Contributors (INDC) and also helped in addressing climate change issues such as National Action Plan on Climate Change (NAPCC) policies details are given in Table 12.3. These plans cover different timelines ranging from short, medium, and long-term action plans with varied focus areas to achieve comprehensive climate risk reduction and sustainable development in Surat city.

Surat is one of the Indian smart cities where seventeen diverse organizations work together on climate adaptation and mitigation. For more than a decade, the

Surat climate resilience strategy has been built on interventions that entail association within different actors, sectors, institutions, and agencies to address climate change by the city (local), state (subnational), national (central), and international agencies. Most importantly, the city implements and operationalizes climate adaptation and mitigation schemes with various actors, institutions, and government bodies. The list and hierarchy of actors, institutions, and government bodies such as Surat Municipal Corporation, Surat Urban Development Authority (Local); ICLEI-Local Governments for Sustainability (Civil society); Rockefeller Foundation's Asian Cities Climate Change Resilience ACCCRN (Institute); Chamber of commerce, Taru Leading Edge (Private Institute); Asian Development Bank (Multilateral and bilateral agencies) these network supports the city's climate resilience strategies and implemented many pilot projects in Surat city. The city has taken various steps to institutionalize climate adaptation planning by establishing a public trust (Network 2015; Patel 2014; SMC 2011; SMC and SCCT 2017).

The climate change impact assessment has been addressed through national or regional plans. Some of the climate action plans overlap in two domains, city initiative and climate resilience which can be integrated to find synergies and trade-offs (Chu 2016a; City Mayors 2011; Network 2015; Patel 2014; SMC 2011, 2020; SMC and SCCT 2017). The results of this study can propose an integrated approach for city planners to identify maximum synergies in smart city initiatives and climate resilience. However, merging smart planning tools and climate-resilient city concepts incorporates an integrated planning approach to low-carbon smart cities in terms of process and methods (Kim 2018). These processes and methods are addressed by understanding the resilience framework, content analysis, and statistical procedures.

12.2.1.2 Smart City Mission

In 2015, the Government of India launched the "smart city" mission, where 100 cities were shortlisted out of nearly 4000 cities, and towns participated (Smart Net 2015). The Indian government is investing USD 2.7 billion to develop an infrastructure for 100 Indian smart cities by 2024 (Ashwathy et al. 2018; Smart Net 2015). The main objective of the smart city mission is to make a sustainable city, build a clean, eco-friendly environment, provide infrastructure, decent quality of life by smart solution, and promote inclusive city growth. The background of the smart city mission is to draw attention toward making cities resilient. The mission envisages to make a livable, sustainable, and resilient city; there is an urgent need to create affordable housing, slum up-gradation, cleanness program, investment in urban mobility, and creating green infrastructure. National, subnational, international, and non-governmental organizations aim to enhance and promote sustainable and resilient cities (Sharma and Verma 2019). Surat is one of the six cities from Gujarat state which receives funding from the smart city mission of India.

The Surat smart city mission covers six key thrust areas: E-governance, water management, waste management, urban mobility, energy management, and others telemedicine and teleeducation, incubation trade facilitation centers, and skill development centers (Smart Net 2015). The Surat smart city initiatives are people-centric, where projects are evaluated based on people participation. Most projects are devoted to ICT solutions funded by the government (Ashwathy et al. 2018; GOI 2015). The mission encompasses two primary initiatives, "area-based development" and "pan" initiatives. The area-based development covers an 8.77 sq. km area which includes seven town-planning schemes with a budget of 250 million USD focused on improving efficiency through redevelopment, retrofitting, and green field development in Surat city. "Pan" initiatives spread across the entire city boundary with a budget of 110 million USD; it broadly focuses on enhancing the existing infrastructure and services, like smart street lighting, intelligent traffic management, smart waste management systems in a given city (Sheth 2017).

Although smart city missions are encouraging, there are various challenges concerning the deployment of smart city projects. First, there are uncertain returns on investment due to multiple stakeholders' involvement across different sectors. Second, departments and government authorities do not work in an integrated manner. Neither are they proactive in sharing the resources that highlight the need to resolve governance-related issues. Third, social benefits are often difficult to quantify generated by smart cities (United Nation and ESCAP 2019).

This study attempts to understand these challenges in-depth with the case of Surat city and suggests some result-oriented outcomes that would help the policymakers for better decision-making and strategy planning.

12.3 Methodology

The study aims to assess the inter-relationship between smart city initiatives and climate resilience which has not been studied in detail (Sharifi 2021; Kumar et al. 2020). The research methodology (Fig. 12.1) is divided into five stages: data collection, literature review, formulation of conceptual smart city resilience framework, thematic mapping and coding, stakeholder interviews, and establishment of the weightage explained in Sect. 12.3.5.

12.3.1 Data Collection

The study of city climate resilience policies is assessed through content analysis, and a dataset on smart city and climate action plans was compiled and produced. The study compiled the datasets from secondary sources: progress reports of city development plans, research papers, government reports, smart city documents, and articles.

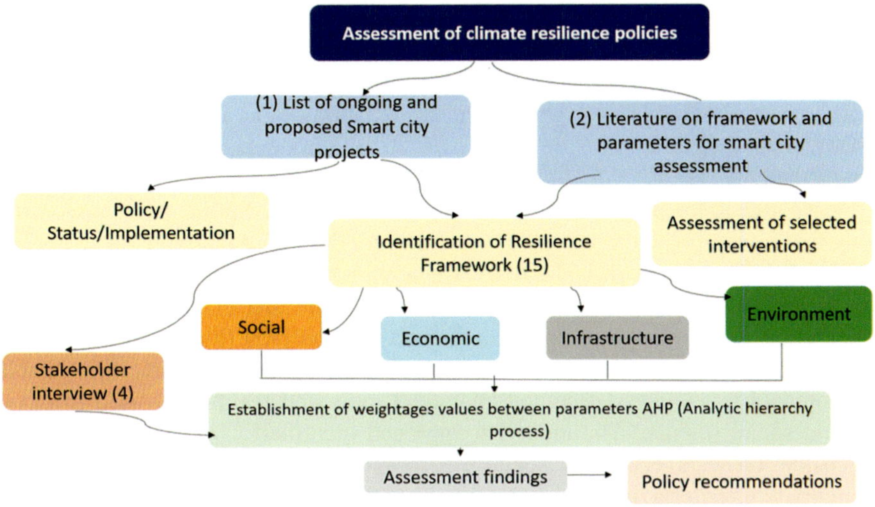

Fig. 12.1 Methodological framework (*Source* Author's analysis)

12.3.2 Literature Review

An extensive literature review was conducted on funded smart city strategies and climate action plans (details are in Sects. 12.4.1). The content analysis of climate policies helped in building various themes, helped in addressing research question one. First, the theoretical literature review on climate ***resilience framework*** helped in developing conceptual smart city resilience framework. Then, fifteen-identified resilience indicators were mapped with four dimensions to assess city climate adaptation and mitigation policies (***resilience indicators***). This review helped in identifying the critical gaps and synergies in the study area.

The assessment of resilience parameters is divided into two subparts—dimension selection (criteria) and selection of resilience indicators (sub-criteria) to evaluate the climate policy implication in Surat city. The identification of parameters was based on an extensive literature review of the various frameworks measuring city and urban resilience (Alshehri et al. 2015; Arup 2014; Cutter et al. 2010; Griffiths et al. 2015, 2017; Gupta et al. 2019; Ohshita 2017; Ostrom 2010; Sharifi 2020; Sustainability Outlook 2014; Tanner et al. 2009; UNDP 2011).

12.3.2.1 Resilience Framework

The various framework on city resilience index like Arup (2014) City resilience Index; Alshehri et al. (2015), the Community Resilience to Disaster (CRD); Cutter et al. (2010) Baseline Resilience Indicators for Communities (BRIC) and Ostrom (2010), Institutional Analysis and Development framework (IAD) were studied and evaluated. But there is no standard procedure and guidelines for developing a composite resilience index due to the diversity of the theoretical foundation (Bosch et al. 2017; S. Zhu et al. 2019; Zhu et al. 2016). Many research studies tend to develop the selection of variables, weighting, and aggregation on the foundation of theoretical understanding of city resilience (Sharifi 2020; Yamagata and Sharifi 2018; Zhu et al. 2016).

Hence, due to the unavailability of a suitable framework for the present study, the study first converts the theoretical framework into the functional practice, based on BRIC and IAD framework.[1] The BRIC framework by Cutter et al. 2010 is the most popular framework used in climate change. It was the first of the kind to develop a theoretical framework into operational practice. The IAD framework by Ostrom (2010) best fits systematic review for policy analysis and outcomes; it is suitable and compatible to assess designed analytical tools and techniques used in physical and social science. The inductive approach identified various themes as four dimensions (criteria) economic, social, environmental, and infrastructure; and fifteen resilience (sub-criteria) indicators classified into three clusters: structural, integrative, and institutional. The themes classified of the four dimensions are as follows:

Economic: It measures the economic strength of communities to provide stability in the face of changes in the external and internal environment which attributes to higher economic returns (Rose 2007), for example, common utility meters and subsidies on solar installation.

Social: This dimension provides the abilities of social entities and mechanisms to effectively predict, mitigate, and respond to disasters and minimize social disruptions (Keck and Sakdapolrak 2013; Saja et al. 2018). This dimension focuses on equity, poverty alleviation, and public health.

Infrastructure: This provides robust solutions that may influence recovery and resistance to adapt to any changes. The city infrastructure includes dams, bridges, embankments, connectivity, and information dissemination capacity through the command-and-control center (Derrible et al. 2020; Griffiths et al. 2015).

Environmental: Environment is the capacity to resist and hold the urban environmental system to respond to disturbance and recover quickly (Meerow et al. 2016). Such available city environmental system includes water reuse and recycle smart waste collection, air and water quality monitoring system, and restoring lakes and structures.

The next step includes identifying the resilience indicator suitable for the study area. In the first stage, twenty-two resilience indicators were shortlisted based on literature. These indicators encompass the environment, governance, finance, education, health, equity, information, and technology (Brown et al. 2018; Engle et al. 2014; European Commission. Directorate-General for the Environment. et al. 2017; Feldmeyer et al. 2019, 2020; ISB and SSEF 2017; Sharifi 2020; Tyler et al. 2016; Zhu et al. 2019). Next, these resilience indicators narrowed to fifteen resilience indicators (details are in Table 12.1) depending on smart city policies, climate-resilient local

---

[1] BRIC and IAD frameworks are referred to select the four-dimension economic, social, environment and infrastructure that are used for assessment for city climate policies and initiatives.

Table 12.1 Conceptual smart city resilience framework

Structural resilience	Integrative resilience	Transformative resilience
Robust	Resourceful	Decentralization
Redundant	Inclusive	Foresight Capacity
Modularity	Connectivity	Creativity
Diversity	Equity	Agility
Adaptative Capacity	Efficiency	Flexible

Source Adapted from (Kupers and Ching 2016; Sharifi 2020)

strategies, and relevance to the study area (Jäger et al. 2015).

12.3.2.2 Resilience Indicators

The economic, social, infrastructure, and environment dimensions become more significant because they encompass a resilient system associated with sustainable development goals, the Paris Agreement, the Sendai Framework, and the new urban agenda (GPDDR 2017). However, there is a gap in the literature studies on conceptual linkage and resilience indicators and a methodology for a comparative assessment of each indicator. Thus, the study attempts to establish these linkages by creating a schema for assessing smart city resilience. The study proposed a schema to review smart city resilience, illustrating a "Conceptual Smart City Resilience framework" based on (Kupers and Ching 2016; Sharifi 2020). The conceptual smart city resilience framework classifies into three clusters: *structural, integrative, and transformative*, based on a typology of each resilience characteristic. The recognized resilience characteristics: robust, redundant, flexible, resourceful, inclusive, diversity, equity, iterative progress, decentralization, adaptive capacity, efficiency, connectivity, agility, modularity, creative details are provided in Chapter 4 and Chapter 6. Each cluster has five different resilience indicators explained in detail below (Table 12.1).

Structural resilience: Structural resilience resists extreme events with minimum threats to people by improving adaptive capacity and creating a system that withholds extreme pressure and demand (Bruneau and Reinhorn 2006). The structural framework involves robust, redundant, modularity, diversity, and adaptive capacity.

Integrative resilience: Integrative resilience creates collaboration within the people and institutions for resource mobilization to meet the demand in shock and stress (Liu et al. 2017). It encourages shared ownership to provide equal access to service and infrastructure. The integrative resilience approach improves the efficiency and performance of the agent that includes inclusiveness, resourceful, connectivity, equity, and efficiency.

Transformative resilience: Transformative resilience is a framework that helps achieve climate-resilient pathways through various institutions and organizations working together to manage the resource allocation uniformly for the different sections of society (Engle et al. 2014). Transformative resilience strengthens the local institution, especially for coping with multiple risks that eventually enhance foresight capacity, creativity, agility, and flexibility in nature.

The structure of the conceptual smart city resilience framework represented as a schema helped compartmentalize the city policies at a detailed sub-level for thematic mapping and coding, which helps understand the drivers and factors affecting city resilience.

12.3.3 Thematic Mapping and Coding

The mapped resilience attributes and dimensions are sub-classified into various themes and coded into city urban development and planning, information dissemination capacity, drought and water security, affordable houses, basic and social infrastructure facilities, stormwater drainage and flood management, restoring lakes and structures. These coded themes and key concepts helped in the formulation of the questionaries for conducting stakeholder interviews. The need for stakeholder interviews is to identify key priorities area for policymakers and city planners.

12.3.4 Stakeholder Interview

Four urban resilience experts were interviewed: academics, the local resilience officer, a city expert, and an international expert. Each stakeholder holds a senior position with work/research experience ranged from 6–30 years on urban resilience for Surat city. The interview duration ranged from 40 to 60 min each by these experts. There were eighteen different questions where the interviewee was allowed to rank different dimensions and resilience characteristics. Each proposed resilience indicator was ranked on a 1 to 5 scale (1 represents the lowest value) and a 1 to 6 scale (1 being the most important element) in the survey questionnaire. The ascending and descending order ranking in the questions removes user biases and helps respondents examine each point more attentively.

The themes generated from the stakeholder interviews, with their experience and advice, highlighted the key thrust areas measured through the triangulation approach using the Analytical Hierarchical Process (AHP). In addition, it helped to establish the weights among different indicators for the study. The detailed process is given in Sect. 12.3.5.

12.3.5 Establishment of the Weightage (AHP)

This study uses the AHP method for deriving the priorities (weights) for the criteria. The AHP is widely used to support the objective of decision-makers; it provides the best optimum solution for resource allocation by prioritizing the alternative options (Bhushan and Rai 2004; Saaty 2008). AHP weighting values derived between four dimensions and 15 resilience indicators grouped into the final three resilience clusters. AHP values were derived from expert ranking and frequency count (refer to Sect. 12.4.2) obtained from thematic coding and mapping exercises. The AHP method is based on Saaty, T. L. (2008) scale of measurement that processes the pairwise comparisons and relies on expert ranking. It helps to indices the judgment of experts to derive a priority scale based on how one element dominates another. The pairwise step was used to rank the four dimensions to determine priorities or weights. For assigning suitable weights, users need to establish a consistency ratio (CR). The consistency ratio helps to cross-validate the assigned weights. It is measured in two steps. The first step includes measurement of inconsistency calculated by the following formula:

$$ConsistencyIndex = \frac{\lambda_{Max} - n}{n - 1}$$

The second step includes the measurement of the consistency ratio.[2]

$$Consistencyratio = \frac{ConsistencyIndex}{RandomIndex}$$

If CR < 0.1, that indicates the determined weights are the best representative numbers.

The validation through the AHP method helped evaluate these policies relative merits, which would help suggest further policy recommendations.

12.4 Case Study

Surat is the second-largest city of the Gujarat state in western India and the 9th largest metropolitan and coastal city. The city's population has been rising in the past two decades, with an annual population growth rate of 5% from 2001–2011 (Census 2011). The city is an important commercial center, has good port accessibility, and proximity to the Indian cities of Mumbai (financial capital of India) and Ahmedabad (financial capital of Gujarat) contributes to high economic growth for small and medium businesses industries. The city is always being a commercial center in national and international territories. As an outcome of urban expansion city is

[2] In the proposed study random index value is considered 0.9 derived from Rao, R. V. (2007).

experienced unprecedented urban and economic development with \$59.8 bn GDP (Haritas 2021). Surat has a tropical savanna climate, and the coast is also an eco-sensitive zone with 11.24% wetland of district geographic area and with mangrove cover of 20.27 Km^2. However, wetland areas are highly affected by coastal inundation and show a negative rate of changes in mangrove cover -0.73 w.r.t 2017 (FSI 2019) due to climate change. Around 5,000 households will be directly at risk along tidal creeks (Bhat et al. 2013). Changes in distribution and survival of aquatic species and algal blooms will occur with expected temperature rise (ACCRN 2010).

Surat being a coastal and riverine city makes it more vulnerable to climate risk. The Tapi River flows from the middle of the city, causing frequent floods and damaging infrastructure (Bahinipati et al. 2015). In addition, the climate models (HADCM3 and CCMA) show higher precipitation and will likely increase in the future (ACCRN 2010). These climate risks are associated with a more considerable degree of loss of life and infrastructure. Therefore, it becomes crucial to understand challenges associated with climate risk for making the city more resilient.

12.4.1 Climate Risk

There is a presence of multiple climate risks on Surat city's resources, primarily concerned with hydro metrological events like flooding, heat stress, storm surges, high erosion, and coastal inundation, water scarcity, and sea-level rise and participation of its external actors in building climate-resilient infrastructure planning. Specifically, flood and heat stress are a major concern for Surat city (SMC 2011).

12.4.1.1 Flood

Almost every year, around 90% of the city area is affected by climate hazards—floods, tides, storms, cyclones, heat calamities, and sea-level rise (Bhat et al. 2013; Parikh et al. 2018). The city witnessed significant floods of varying intensity in 1968, 1994, 1998, 2002, 2006, 2013, and 2018 (Parikh et al. 2018; SMC 2011, 2020; SMC and SCCT 2017). The Regional Climate Model (RCM) and general circulation model (GCM) show a high probability of increased precipitation in the future which ranges from 200 to 450 mm annually by 2070 (Bhat et al. 2013).

12.4.1.2 Heat

Surat has observed a 0.7°C rise in average temperature in the last 30 years (Parikh et al. 2018; SMC 2018). Surat has many mortality cases with a temperature rise (Desai et al. 2015; Rathi et al. 2017). Around 11% of mortality cases are registered when the temperature reaches 40°C. The temperature between 41 and 54°C causes a death rate of 9% (per day), and temperature above 54°C deaths causes an 18% death rate

(Desai et al. 2015; Rathi et al. 2017). The socio-economic scenario suggests future emissions are likely to reach the implied level by RCP 8.5 (OECD 2018). Therefore, climate change will increase heating and cooling requirements in the cities like Surat in the coming years (OECD 2018).

The State Action Plan on Climate Change and the "Smart city mission" have strengthened the city's adaptation and mitigation schemes with various initiatives. As a result of the State Action Plan on Climate Change, Surat became the first city globally by launching an emission trading scheme and reducing pollution levels by 29% (ET energy world 2019). Thus, Surat city sets an ideal example for a case study to analyze convergence and collaboration of policies and institutional frameworks for understanding the climate resilience future of smart cities.

12.4.2 Funding for Climate Adaptation and Mitigation

Surat city is the first city in India to participate in the ACCCRN climate resilience initiative program through a public–private partnership, institutions, integration, and convergence of schemes. However, these schemes exhibit similar concerns in seeking to enhance the adaptative measure for improving city resilience. These climate adaptations and mitigation plans exhibit strong measures on technology to build smart solutions for dealing with urban problems. The policy alignment with urban planning and institutional frameworks is used for practicing urban innovation to deal with the city's rapid urbanization growth. The main expenditure is on improving transportation and infrastructure and smart ICT-based solution. But adaptation and mitigation plans differ in achieving a framework through various policy interventions, implementation, and funding (refer to Table 12.2).

12.4.3 Resilience Strategies Adopted for Mitigation and Adaptation

The mitigation and adaptation initiatives are chiefly downscaled across local, state, national, and international scales in response to associated climate risks to support city development planning. There are eight policies and twenty-two programs, initiatives, schemes, and projects undertaken by the Surat municipal body listed in Annexure Table A1 (ACCRN 2010; Chu 2016a; City Mayors 2011; Patel 2014; SMC 2011, 2019; SMC and SCCT 2017). The Surat resilience synergies developed around twenty goals and sixty-three initiatives/actions for 2015–2025. The mitigation and adaptation plan recognizes eleven initiatives in infrastructure, fourteen initiatives in the environment, thirteen initiatives around social cohesion (includes three on public health), and thirteen initiatives addressing economic welfare (ACCRN 2010; Chu

Table 12.2 Expenditure on Surat climate resilience plans and activities

Sr.No:	Department/Organization	Sub-classification	Objective	Expenditure
1	Surat Municipal Corporation	Municipal government	Basic Services	-
2	Surat Climate Change Trust	Quasi-public agency	Climate resilience	$660,000
3	South Gujarat Chamber of Commerce & Industry	Private	Education and awareness	-
4	Gujarat Department of Climate change	State government	Climate mitigation and adaptation	-
5	Gujarat Disaster Management Authority	State government	Disaster risk reduction	-
6	Gujarat Disaster Management Authority	State government	Heat Action Plan	-
7	Sustainable coastal zone management in Gujarat	State government	The integrated coastal zone management project	-
8	Gujarat Irrigation Department	State government	Flood Control	$5.8 million
9	Swarnim Jayanti	State government	Urban Development	$33 million
10	Jawaharlal Nehru National Urban Renewal Mission	National government	Urban renewal, poverty reduction	$203 million
11	National Sustainable Habitat Mission	National government	Energy efficiency	-
12	National River Conservation Plan	National government	Pollution abatement	$33.5 million
13	Rajiv Awas Yojana	National government	Poverty alleviation	$4.5 million
14	Slum Rehabilitation, Affordable Housing	National government	Poverty alleviation	$62.49 million
15	Centralized Command & Control Centre	National government	Monitoring and Management	$2.71 million
16	Water Supply	National government	Water supply management	$24.18 million
17	Waste Water Management, Drainage System, Storm Water management	National government	Urban Development	$61.81 million
18	Smart Solutions /ICT	National government	Early warning, surveillance, and Redressal system	$91.83 million
19	Transportation and Infrastructure	National government	Smart Urban Mobility	$111.37 million

(continued)

Table 12.2 (continued)

Sr.No:	Department/Organization	Sub-classification	Objective	Expenditure
20	Rockefeller Foundation's ACCCRN	International	Climate resilience	$2.8 million

Source Adapted from Patel (2014), Smart Net (2017) and Smart city (2018)

2016a; City Mayors 2011; Patel 2014; SMC 2011, 2019, 2020; SMC and SCCT 2017).

The Surat's smart city, E-Governance, and citizen services include five major projects on health surveillance, air and water monitoring, intelligent transport connectivity, and smart features. Energy management covers solar, wind, and bioenergy projects and announced special purpose vehicle (SPV) collaborating with local energy agencies. The city has fair coverage on waste management, around 95% underground sewerage system, and 100% door-to-door waste collection. It also includes innovative resilience such as "leakage mapping." In addition, the city has national and state policies on housing and water management and river conservation, etc. These policies are assessed on qualitative and quantitative methods to derive results.

12.5 Results

The study's outcome was derived from content analysis of climate policies, stakeholder interviews, and AHP. The content analysis helped establish synergies and trends-gaps among different indicators. In addition, accuracy assessment based on triangulation by deploying the AHP method and resilience performance was measured.

12.5.1 Mapping and Identification of Resilience Strength

The framework interaction across different dimensions shown in Fig. 12.2 highlights that focus has shifted toward improving adaptive capacity; for example, early warning, surveillance, and redressal system are ICT-based solutions that helped establish air and monitoring systems. The integrative resilience cluster shows high equity, inclusive, and efficiency resilience and holds maximum synergies in social and economic dimensions because these policies focus on affordable housing schemes. On the other hand, city climate policies show low strength on diversity, decentralization, and connectivity because some of the policy's status is unclear. Although many of these policies are overlapping, indicating co-benefits and trend gaps within the sectors explained in Sects. 12.5.2 and 12.5.3.

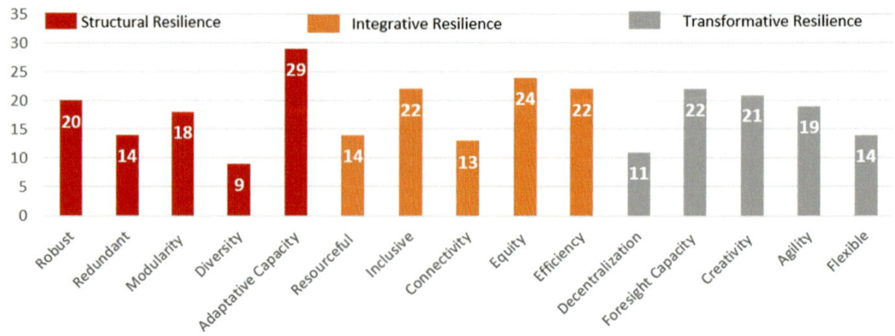

Fig. 12.2 Schematic representation of resilience strength across Surat resilience strategies undertaken (*Source* Author's analysis)

12.5.2 Co-Benefits

The mitigation and adaptation policies helped build themes for addressing the climate change concern of the city for finding synergies and co-benefits. These policies provided climate and hydro metrological information for improving city resilience to support city urban development and planning. Smart technology is used to aid the urban services into an efficient monitoring system and improve the city's information dissemination capacity. The several climate mitigations and adaption programs and initiatives reaped into a couple of co-benefits. Firstly, the initiatives by ACCCRN and experiments helped flood relief programs to deal with public health and climate resilience issues, such as improving vector-borne disease through health surveillance operations (Chu 2016a). Secondly, The Gujarat State Irrigation Department project for building new embankment measures helped reduce drought and water security concerns in Surat and the neighborhood (Patel 2014). Thirdly, JNNURM project funding addressed poverty alleviation and climate resilience. SMC has constructed 60,000 affordable houses, besides 20,000 houses for the urban poor and 2460 units under slum redevelopment scheme i.e., "Basic service of the urban poor (BSUP)." This initiative has covered 370 slums with all basic and social infrastructure facilities (Chu 2016a). This program supported the property registration, and around 18 Lacs properties were registered. It further aided in property tax collection, giving an average rise of 12% annually in the past three years (Smart Net 2015). Fourth, the up-gradation of civic infrastructure by smart features will reduce the Operation and Management cost in the long run. Fifth, improving stormwater drainage and flood management through ICT and remote monitoring would support the city's local ecosystem i.e., restoring lakes and structures (Smart Net 2015).

12.5.3 Trends and Gaps

The Surat resilience efforts of institutionalizing were underpinned mainly by broader planning efforts dependent on the efficient governing systems at the higher levels (Chu 2016b). However, it attained flexibility in pursuing the local objectives, but it lacks long-term sustainability concerns (Patel 2014; SMC 2011). Many city resilience measures are non-structural, focusing on planning, civic engagement, coordination awareness, and designs (ACCCRN 2011, Smart Net 2018, Surat Smart city 2019). These climate-resilient strategies are less on infrastructure and buildings (ACCCRN 2011; Smart Net 2018; SMC 2019; Surat Smart city 2019). Another concern is the lack of flexibility of using funds outside a particular departmental area. In Surat city, the Gujarat State Disaster Management Association (GSDMA) grants (2009) the city-regional emergency response only works on information and capacity building and avoid structural measures. Gujarat State Irrigation Department was unable to integrate with disaster risk reduction for managing city floods (Patel 2014). Thus, the GSDMA failed to address robustness and diversity in city resilience planning in the city. Surat city disseminates the information as part of the heat action plan during high alerts, creating awareness. Still, the heat action plan targets vulnerable sections of society, primarily Low Income Groups (LIG) (SMC 2018). However, these groups had low literacy rates, so there could be a possibility that these groups were not aware of such systems. Another example, ACCCRN organized a design competition on cool roofing and passive ventilation options in Surat city in 2011 (ACCCRN 2011). These competitions outreached a particular group of society. Some of these policies are over-ambitious, Surat city's goal is to attain 100% energy from renewable by 2020, but only 30–40% is commissioned till date (Smart Net 2018) (Table 12.3).

12.5.4 Resilience Performance Analysis

The AHP provides a rational framework for decision-making and policy recommendations. In the proposed study, AHP helped identify key areas or sectors that need more improvement and up-gradation. The AHP weights show each indicator's strength, the identified criteria, and estimating weights (higher the value greater the strength) process is depicted in Table 12.4. Most of the policy criteria focus on achieving economic goals W_e 0.41 in the city. This shows that existing policies are significantly inclined toward aggregate economic welfare and benefits. Despite several environmental policies, W_{en} 0.013. This is because of the city's urban stressors issues i.e., flood, heat, health, and stormwater management. Surat city is witnessing frequent floods, there are twenty-four flood events on record, and it happens after every four years of interval (Stakeholder 3). This flooding problem is interlinked with a lack of robust solutions and the failure of urban planning. In Surat, temperature rising is another growing concern. The heat action plan strengthens transformative resilience, and the city provides ICT-based solutions for managing heat stress. The

12.3 Mapping and interaction of resilience across four different dimensions

Sr. No	Sector	Codes	Mission/Programme/Scheme/Project	Dimensions	Structural Resilience					Integrative Resilience					Transformative Resilience				
					Robust	Redundant	Modularity	Diversity	Adaptative Capacity	Resourceful	Inclusive	Connectivity	Equity	Efficiency	Decentralization	Foresight Capacity	Creative (Innovative)	Agility	Flexible
1	Water	W1	Atal Mission for Rejuvenation and Urban Transformation (AMRUT), Ministry of Urban Development, Government of India	Social	Y	Y			Y		Y		Y	Y		Y			
1.1		W2	Water supply management & quality smart solutions	Social	Y	Y			Y		Y		Y	Y		Y		Y	
1.1.1		W3	24 X 7 water supply & water quality	Social	Y	Y	Y		Y		Y	Y	Y	Y		Y	Y	Y	
1.1.2		W4	Common Utility meter	Economic	Y	Y	Y		Y		Y	Y	Y	Y	Y	Y	Y		
1.1.3		W5	Rain Water Recharging	Economic	Y	Y			Y		Y			Y	Y	Y			
1.1.4		W6	Remodeling & Restructuring of existing creek to create open spaces with smart features	Infrastructure	Y		Y		Y		Y	Y			Y	Y		Y	
2.1	Housing	H8	Pradhan Mantri Awas Yojana (PMAY) - Housing for All Mission, Government of India	Social	Y			Y	Y	Y	Y		Y	Y				Y	
2.2		H9	Affordable Housing (PPP)	Social	Y		Y	Y	Y	Y		Y	Y	Y	Y		Y		Y
2.3		H10	Jawaharlal Nehru National Urban Renewal Mission	Social	Y		Y	Y	Y	Y		Y	Y	Y		Y	Y		Y
2.3		H11	Rajiv Awas Yojana	Social	Y		Y	Y	Y		Y	Y	Y	Y			Y		

(continued)

12.3 (continued)

Sr. No	Sector	Codes	Mission/Programme/Scheme/Project	Dimensions	Structural Resilience					Integrative Resilience					Transformative Resilience				
					Robust	Redundant	Modularity	Diversity	Adaptative Capacity	Resourceful	Inclusive	Connectivity	Equity	Efficiency	Decentralization	Foresight Capacity	Creative (Innovative	Agility	Flexible
4	Waste	WS12	Swatch Bharat Mission (Clean India Campaign), Ministry of Urban Development, Government of India	Social	Y			Y	Y	Y			Y	Y					Y
4.1.1	Waste	WS14	Novation of STPs with SCADA & Energy Generation	Environmental	Y	Y			Y	Y	Y		Y	Y	Y	Y	Y		
4.1.2	Waste	WS15	Recycling/ Reuse of Waste Water	Environmental	Y				Y		Y		Y	Y	Y	Y		Y	
4.1.3	Waste	WS16	Smart Waste Collection System	Environmental	Y				Y	Y				Y		Y			
4.1.4	Waste	WS17	Swarnim Jayanti Mukhya Mantri Saheri Vikas Yojana	Environmental	Y				Y					Y					Y
5	Renewable energy	R18	Renewable energy & energy efficiency initiatives	Environmental															
5.1	Renewable energy	R19	Solar City Programme & Solar Roof Top Scheme, Ministry of New and Renewable Energy, Government of India	Environmental	Y	Y	Y	Y	Y	Y	Y	Y	Y	Y	Y	Y	Y	Y	Y
5.1.1	Renewable energy	R20	Generation Based Incentives (GBI) Scheme, Ministry of New and Renewable Energy, Government of India	Economic	Y	Y		Y	Y	Y	Y	Y	Y	Y	Y	Y	Y	Y	Y
5.1.2	Renewable energy	R21	Solar (1 MW) & Wind Power Generation (2.1 MW)	Environmental	Y	Y		Y	Y	Y	Y	Y	Y	Y	Y	Y	Y	Y	Y
5.1.3	Renewable energy	R22	Biogas plant for organic waste	Environmental	Y	Y		Y	Y	Y	Y	Y	Y	Y	Y	Y	Y	Y	Y
5.1.4	Renewable energy	R23	Smart Street Lighting (LED) & Monitoring System	Environmental	Y	Y		Y		Y	Y						Y		Y

(continued)

12.3 (continued)

Sr. No	Sector	Codes	Mission/Programme/Scheme/Project	Dimensions	Structural Resilience					Integrative Resilience					Transformative Resilience				
					Robust	Redundant	Modularity	Diversity	Adaptative Capacity	Resourceful	Inclusive	Connectivity	Equity	Efficiency	Decentralization	Foresight Capacity	Creative (Innovative)	Agility	Flexible
6	Smart Technology-Early warning, Surveillance, and Redressal system,	S24	Smart city system	Infrastructure			Y	Y		Y	Y						Y	Y	Y
6.1		S25	Advanced Grievance Redressal System	Infrastructure			Y	Y		Y	Y						Y	Y	Y
6.2		S26	Air & Water Quality Monitoring System	Environmental			Y	Y	Y				Y	Y		Y	Y	Y	Y
6.3		S27	Area Surveillance Network	Environmental			Y	Y	Y	Y	Y	Y	Y			Y	Y	Y	Y
6.4		S28	Heat action plan	Environmental			Y	Y	Y	Y	Y	Y	Y		Y	Y	Y	Y	Y
8	Capacity building, information technology, and planning	C31	Surat Climate Change Trust	Environmental			Y	Y	Y		Y	Y	Y	Y		Y	Y	Y	
9		C32	National Sustainable Habitat Mission4	Environmental		Y	Y	Y	Y		Y	Y	Y	Y		Y	Y	Y	
10		C33	National River Conservation Plan	Environmental			Y	Y	Y		Y	Y	Y	Y		Y	Y	Y	
11		C34	Rockefeller Foundation's ACCCRN	Environmental			Y	Y	Y		Y	Y	Y	Y		Y	Y	Y	

(shaded) 'Yes'

Table 12.4 Identified criteria and estimating weights

Sr. No	Dimensions	Weightage priority[a]
1	Social (W_s)	0.090
2	Economic (W_e)	0.416
3	Infrastructure (W_{in})	0.362
4	Environment (W_{en})	0.131

[a]Higher the value greater the strength
Source Author's analysis

early warning system (ICT-based) is used for capacity building in the heat action plan. One initiative was to activate the cooling centers such as temples, public buildings, malls during high (heat waves) alerts. However, this was one of the suggestions from the Ahmedabad heat action plan, but it was not implemented because of "intradomestic comfort" (Stakeholder 3).

The Surat city is fairly working on the social dimension compared to other cities, but W_s (Social weight) ranged 0.09. This is because the city is challenged by migrating population each year. For better management of the social aspect, the city needs a bottom-up approach in the decision-making process. The existing climate resilience policies plus smart city intervention are strengthening economic dimensions such as renewable (30–45%) and water reuse and recycle (80% recharging) (Smart Net 2015). The city is making progress on renewable and innovation "but somehow may be due to political will, or maybe some other external factors, the complete potential has not been tapped yet" (Stakeholder 1).

The four dimensions, social, economic, infrastructure, and environment, are mapped on pairwise comparison with resilience indicator cluster i.e., transformative, structural, and integrated (shown in Table 12.5). The value of integrated (W_i) signifies higher strength as compared to other resilience indicators. This is because most integrated steps come under the smart city projects through SPVs (Smart City 2019); But the smart city projects are in the implementation stage, and their status is unclear. The policy design is set back in terms of targets, lack of transparency, and absence of a structured review mechanism in the city (Stakeholder 2). The scalability of the transformative resilience index can be improved by inclusive growth for all.

Table 12.5 Finalizing criteria and the relative weights (Pairwise)

Sr. No.	Dimension classified resilience cluster	Social	Economic	Infrastructure	Environment	Weightage priority
1	Integrated (W_i)	0.702	0.354	0.632	0.632	0.624
2	Transformative (W_t)	0.206	0.322	0.198	0.198	0.229
3	Structural (W_{st})	0.090	0.322	0.169	0.169	0.145

Source Author's analysis

One of the major challenges in Surat city is equity because of the growing population. Thus, local government adaptation projects are related to water, mobility, environment, and clean air (Stakeholder 4). The city development body should enable equity and inclusive growth together which strengthens the social dimension. The city can adopt role models such as the Delhi government's "Mohalla (Community) clinic," a medical facility accessible to all groups, irrespective of their financial status (Stakeholder 1). This would aid in improving the existing health surveillance system of the city.

Surat city is not doing well on a structured and transformative resilience cluster on the resilience scale with W_s 0.14 and W_t 0.22, respectively. There has been a trade-off among different resilience characteristics, and it shows a low degree of diversity, inter convergence, and decentralization in the system, which slowed the penetration of initiatives taken by the government, as explained in Sect. 12.4.3 and Table 12.3. Recently, the city started enhancing sink with the nature-based solution to strengthen adaptive capacity, diversity, equity, and connectivity. The city is developing a biodiversity park by planting 600 to 1000 native trees on 68 hectares (Stakeholder 4). The nature-based solution will help in mitigating emissions and enhancing the coastal ecosystem.

In the existing city policies, implementing an adaptive strategy through local consensus and technological advancement is lagging. These strategies would help enhance the city planning system. The urban planners need to explore new methods and approaches for improving and enhancing governance, institution, technical learning and information technology, planning system, funding, awareness, and community support system.

12.6 Conclusion

In Surat city, key elements that need improvement for managing climate sustainability are governance, institution, technical learning and information technology, planning system, funding and awareness, and community support system. These elements are facing common challenges in fast-growing other Indian cities too. Each of these identified elements addressed through approach/method/model details are provided in Table 12.6. The research study proposed solutions for identified key elements (a) governance can be addressed by inter- and trans-disciplinary approaches; (b) Strengthen institutions by carbon-centered comprehensive (3Cs); (c) City needs to improve technical learning and information technology by knowledge-based solution; (d) Mobilization of funds by the carbon banking system; (e) Increasing local awareness and developing community support by community energy systems, and (f) Improving urban planning system by implementing mandatory level carbon removal (CR) techniques.

In India, the city government period is five to six years, and about seventeen diverse organizations are currently working in Surat city. There are about ninety-one synergies found among various goals and initiatives led by multiple organizations in Surat

Table 12.6 Policy recommendation through desirable approaches and research model

Sr. No	Key issues		Drivers	Proposed recommendation		Source
	Level	Details		Approach/Method/Model	Details	
1	Governance	The city-level government planned tenure is 5 to 6 years and climate change globally 100 years	Decentralization Flexibility Foresight capacity	Transdisciplinary Approach	The transdisciplinary approach (TDA) integrates different sectors and disciplines for optimizing organizational structure to collaborate and conduct joint research	(ACECC 2016; Tanner et al. 2009)
2	Institution	The existing city system works in silos and lacks a comprehensive centralized system for climate mitigation and adaptation	Foresight capacity Connectivity Creative	Carbon-Centered comprehensive (3Cs)	The 3Cs is a smart planning model that provides digital solution ICT-based networks. This model will trace down the carbon footprint at each stage of the process and offer viability at a holistic level to create awareness, assessment, and action (AAA). Thus, this model in the future will help to achieve connectivity, circularity, sustainability, and efficiency	(Kim 2018)

(continued)

Table 12.6 (continued)

Sr. No	Key issues		Drivers	Proposed recommendation		Source
	Level	Details		Approach/Method/Model	Details	
			Redundant			
3	Technical learning and information technology	The city system has inadequate learning from past climate hazards i.e., city observes frequent floods	Creative	Knowledge-based solution	This approach increases prediction capacity through the development of green and energy big data chain that eventually helps in resource mobilization by reducing demand. It also improves monitoring and surveillance	(Roblek and Meško 2020; Shaqrah 2019)
			Connectivity			
			Iterative progress			
4	Planning System	There is absence of carbon sequestration solutions in the city planning system	Robust	Carbon Removal (CR)	Carbon Removal (CR) is the removal of carbon from the atmosphere through carbon sequestration techniques	(Davies et al. 2011)
			Redundant			
			Inclusive			
			Diversity			
			Adaptive capacity			

(continued)

Table 12.6 (continued)

Sr. No	Key issues		Drivers	Proposed recommendation		Source
	Level	Details		Approach/Method/Model	Details	
5	Funding	The inadequate funding and city resource make difficult to sustain initiative over the long term	Decentralization	Carbon Banking system	The carbon banking concept has been under investigation by turning trees into carbon banks or issuing carbon points with green cards. A green card is issued after more than a 5% average reduction of electricity, gas, and drinking water in the past six months against the average of the past two years. This green card can be used for buying green goods and tickets at national parks. The public also receives points on the purchase of the registered eco-friendly product and opting for public transportation	(Kim 2018)
			Resourcefulness			

(continued)

Table 12.6 (continued)

Sr. No	Key issues		Drivers	Proposed recommendation		Source
	Level	Details		Approach/Method/Model	Details	
			Flexibility			
6	Awareness and community support	The city system lacks a bottom-up approach for resource mobilization and awareness-raising	Decentralization	Community energy systems	The development of a community energy supply (CES) system will optimize city resources, including electricity, water, and gas. Thus, creating opportunities for the new energy market and performing action, assessment, and awareness collectively	(Dincer and Abu-Rayash 2020)
			Modular			
			Flexibility			

Source Author's analysis

city. Thus, inter- and trans-disciplinary collaboration are essential for differentiating between social, environmental, infrastructure, and economic dimensions because no single dimension would cover the complexities of the city's critical urban infrastructural challenges. One of the major concerns is that many of these policies, sectors, and interventions work in silos at each institute in Surat city and lack a comprehensive centralized system for climate mitigation and adaptation. The building 3Cs can be merged into a smart planning model that provides digital solution ICT-based networks that create awareness, assessment, and action (AAA) and help achieve connectivity, circularity, sustainability, and efficiency. In today's world, information and communication centers, like health centers, education centers, water and energy system, fiscal institutions, fire station, and mobility are the backbone of smart cities. Surat smart city is currently investing about $ 204 million in smart solution/ICT and smart urban mobility development. However, these infrastructure centers are at greater risk and highly affected by natural climate calamities due to inadequate learning from past climate hazards i.e., the city observes frequent floods. Surat needs to build a knowledge-based solution for predicting flood and sea-level rise and build technical learning and information technology.

The Surat smart city needs an alternative solution that could be achieved by creating a carbon banking system and community energy system to facilitate funds and optimize resource mobility. Furthermore, the carbon banking system will encourage behavioral changes of the people through the carbon credit system (details in Table 12.6). Thus, the existing Surat smart city system requires more synergies in redundancy, flexibility, efficiency, robustness, decentralization, diversity, modularity, and innovation in practice to sustain in the long run.

In addition, Surat's existing urban planning system lacks robust carbon sequestration solutions observed in many other Indian tier 1 and tier 2 cities despite being an industrial hub. Therefore, the urban planning system should make imperative carbon removal techniques, especially in major industrial cities. Furthermore, in smart cities, resilience planning should improve the feedback system across the temporal scale to help find ways for long-term climate goals as most Indian cities score low on future urban resilience sustainable pathways.

Acknowledgements This chapter was partly funded by Asia-Pacific Network for Global Change Research (Funder ID: https://doi.org/10.13039/100005536). The author would like to thank Dr. Minal Pathak for her technical guidance and Ms. Chaitali Joshi for the operational support. The publication relies on insights from stakeholder discussions. The author bears full responsibility for views expressed and accuracy of statements by the stakeholders.

Appendix

See Tables 12.7 and 12.8.

Table 12.7 Surat resilience policies, programs, schemes, projects, and initiative

Sr. No	Scale	Mission/ Program/ Scheme/ Project	Descriptions
1	Local	Sewerage and waste management projects	• Underground sewerage system 95% coverage • 'Novation of STPs with SCADA & Energy Generation' • One STP and 3 pumping stations is having SCADA • 2 STPs at Anjana and Dindoli will be augmented for quality parameters • automation (SCADA) of STPs and SPS initiated and achieved for about 40% • The city has also initiated the automation system by SCADA for STPs and SPS and achieved 40% efficiency • The scheme targets 50% of water treatment for energy generation from exiting STPs with SCADA • The total waste collection in the city is 1200 TPD to 1600 TPD (2016–2019), and door-to-door waste collection coverage increased from 80 to 100% • Smart Waste Collection System: estimated coverage of 160 smart bins, RFIS tag on 80,000 collection points • The creative (innovative) resilience method is used, such as "leakage mapping," to maintain the current NRW near 20%
2	Local	Smart Street Lighting (LED) & Monitoring System	• SPV would implement projects in coordination with SMC, DGVCL, and Torrent power

(continued)

Table 12.7 (continued)

Sr. No	Scale	Mission/ Program/ Scheme/ Project	Descriptions
3	Local	Smart city system and project	• Strengthen Education and Health infrastructure along with the introduction of Smart features • Air & Water Quality Monitoring System-Proposed for establishing a system for air & wastewater Quality Monitoring System and surveillance • Area Surveillance Network: Improved IT connectivity and smart features • Surat smart city projects align with intelligent transport connectivity, advancement in grievance redressal system, early warning and monitoring through area surveillance network, RFID, VTS technology, and biometric system
4	Sub-national	Solar City Programme & Solar Roof Top Scheme, Ministry of New and Renewable Energy, Government of India (2013–2022)	• Surat city solar program aims at generation of 418 MW solar energy and Net-zero energy emissions by achieving 100% by 2020 • Convergence will be achieved for a rooftop solar scheme by availing permissible subsidy • Surat city solar program has commissioned 18 MW of renewable energy to account for 25% of municipal sector energy consumption • Currently, the city has installed 6 MW (1.39 MW: Rooftop Solar and 0.5 MW recycle) of solar that accounts for 18% of energy generation through renewable energy resources • $92 million project cost and minimum 10% of energy requirement is expected from solar
5	Sub-national	National Wind Mission (2015–2022)	• Wind generation in process of setting up 2.1 MW in wind power in Surat city • Incentives @Rs. 0.50 per unit of wind-electricity fed into the grid will be claimed under the GBI scheme of the MNRE. Energy Cell of SMC will implement in consultation with SPV
6	Sub-national	National Bioenergy Mission (2017–2022)	• New plant of 0.5 MW at Anjana STP
7	Sub-national	Gujarat State Irrigation Department (2013)	• The nodal agency for managing water resources and the department provides aid for flood infrastructure management • Around $5.8 million aid is released to increase the river's carrying capacity and reduce flood risk to neighboring communities

(continued)

Table 12.7 (continued)

Sr. No	Scale	Mission/ Program/ Scheme/ Project	Descriptions
8	Sub-national	The Gujarat State Disaster Management Authority (2001)	• The nodal agency provides aid for reducing and managing the disaster • The primary goal is to make the state "economically vibrant, agriculturally and industrially competitive with improved standards of living and with a capacity to mitigate and manage future disasters." • GSDMA offers grants to the city by making regional emergency response centers • In 2008 GSDMA introduced bilateral and multilateral organizations collaboration, making the system more flexible, transparent, and integrated
9	Sub-national	State Climate Action (2009)	• The department was launched in 2009, "Asia's first dedicated government department for climate change department" (Climate Change Department 2014) • The department coordinate mitigation and adaptation actions include solar, heat action, and biodiversity
10	Local	Surat Heat Action Plan (2018)	• Under the Heat action plan, the city can withdraw 10% of the State government department's funds under the State disaster response protocol • The program aims to alert the population against heat-related illnesses. The plans incorporate awareness to local people through the early warning system • The plan establishes coordination and integration of various agencies to take adaptive measures • Apart from the heat action plan, ACCCRN exhibited awareness camping by organizing competitions for cool roofing and local communities to generate experimentation in planning policy and design
11	National	Jawaharlal Nehru National Urban Renewal Mission (JNNURM)	• The key objective was slum relocation (long-term strategy) and rehabilitation (Mid-term strategy) for people living in low-lying flood-prone areas • JNNURM awarded a grant for the two missions to provide essential urban poor services and improve urban infrastructure. The aid was transferred from the central government on the condition that implementation of governance and fiscal reform by municipal bodies, parastatals, and state government. These reforms scheme enabled the development of "economically productive, efficient, equitable and responsive cities" (MOHUA 2011) • JNNURM has also allowed Surat to strengthen and extend its drainage, sewerage, and solid waste management systems • Surat, JNNURM offered a critical opportunity to address its growing infrastructure deficit (Patel 2014)

(continued)

Table 12.7 (continued)

Sr. No	Scale	Mission/ Program/ Scheme/ Project	Descriptions
12	National lePara>	Swatch Bharat Mission (Clean India Campaign), Ministry of Urban Development, Government of India (Phase I: 2014–2021)	• Open defecation and availability of toilets as per requirement and population are the objectives of the SBM. The same is the essential features to be achieved under the SMART City Mission • Health and Municipal, Solid Waste cell of the SMC will implement this component in consultation with the SMART City SPV
13	National	National Rajiv Awas Yojna (2013–2022)	• This scheme envisages urban poverty alleviation toward "Slum Free India," providing equitable and inclusive city growth by providing affordable housing and subsidies to low-income groups. Most of the slums reside near rivers or coastal areas (low-lying areas) • The city received $4.5 million in funding, and currently, these funds are used for city mapping and surveys. The mapping and surveying of slums might provide prospects for public–private partnerships (Patel 2014). Under the **Zero Slum** program, around 22,000 households relocated and were rehabilitated (Chu 2016a)
14	National	National River Conservation Plan (2009)	• In 2012 Surat city received an amount of $33.5 million under National River Conservation Plan (NRCP) to rehabilitate the Mindhola Creek at the Tapi River bank • The scheme provides a robust solution for the city by constructing gabion pitching to maintain the river's carrying capacity to manage frequent floods • The scheme is also reflected by creating awareness to locals through participation and campaigns to conserve river ecology • Remodeling & Restructuring of the existing creek to create open spaces with smart features • 4.5 km work in progress in Mithi Khadi (Creek) and 5.5 km in Koyali Khadi • Pollution abatement and National River Conservation Plan (NRCP) for the rehabilitation of the Mindhola Creek at the bank of the Tapi River
15	National	Swarnim Jayanti Mukhya Mantri Saheri Vikas Yojana (2011–2021)	• In 2011, state government launched a scheme for urban development that provides a fiscal stimulus for public infrastructure works in the largest municipalities. Surat city received $33 million; however, these funds' allocation varies from year to year • Rising temperature and industrial expansion in Surat city have increased demand by 100 million liters per day with a growing population. Moreover, the future uncertainty of rainfall puts the city system under stress. As a result, the municipal body has initiated water recycling and management, a cost-effective way to meet the current demand

(continued)

Table 12.7 (continued)

Sr. No	Scale	Mission/ Program/ Scheme/ Project	Descriptions
13	National	Pradhan Mantri Awas Yojana (PMAY) - Housing for All Mission, Government of India (2015–2022)	• Development Plan Report (DPR) for 1050 Economically Weaker Section (EWS) and 1950 Low Income Group (LIG) houses have been sent for approval with a request for grant of $ 1.02 billion • Slum up-gradation cell (SUC) of SMC will implement this component in consultation with the SMART City SPV • Affordable Housing with Public–Private Partnership
16	National	AMRUT (2015–2022)	• Atal Mission for Rejuvenation and Urban Transformation (AMRUT) aim is to provide basic services such as water supply, urban transport, sewerage, and parks • Common Utility meter: Proposed a component of 24X7 Quality Water Metered Supply under AMRUT • Surat received 57$ million in funding for providing improve infrastructure for water supply management • Rain Water Recharging at 70 locations in Surat city
17	International	Asian Cities Climate Change Resilience Network (ACCCRN)	• ACCCRN spent $2.8 million granted funds dedicatedly for developing the climate resilience strategy and implementing pilot adaptation projects at the city level (Bhat et al. 2013) • ACCCRN involved with local partners such as TARU leading-edge, Surat Municipal Corporation, the South Gujarat Chamber of Commerce and Industry, Gujarat State Disaster Management Authority and Narmada, Water Resources and Water Supply Department, local academic institutions-SVNIT, Urban Health Climate Resilience Center (UHCRC), Surat Climate Change Trust and the Center for Social Studies (Patel 2014)

Source ACCCRN (2011); Chu (2016a); City Mayors Foundation (2011); Patel (2014); Bhat et al., 2013; MOHUA, 2011; Climate Change Department, 2014; Surat Smartcity (2019)

Table 12.8 List of actors, institutions, and government bodies involved in the Surat resilience strategy

Scale	Decision making	Scope/Program	Multilateral and bilateral agencies	Civil society	Institute	Private institute
Local	City-level administration and governance: Surat Municipal Corporation; Surat Urban Development Authority	Governance and capacity building	-	ICLEI-Local Governments for Sustainability	ACCCRN	Chamber of Commerce
Sub-national	Government of Gujarat: Gujrat state action plan	Vulnerability assessment and adaptation strategy	-	City Managers, Association Gujarat	-	-
National	Government of India: JNNURM, NHMS, NAPCC, and Smart city	National and international collaborations on climate action plans	DFID, GIZ	-	-	Taru Leading Edge
International	United Nations Agencies		Asian Development Bank	World Wildlife Federation	Rockefeller Foundation	Foreign Direct Investment

Source Adapted from Patel (2014)

References

ACCRN (2010) Asian city climate change resilience network: India chapter. 168
ACCCRN (2011) Surat city resilience strategy. Surat Municipal Corporation. http://resiliencestrata.org/uploads/publication/3B07371E-EC6C-47C3-99ED-4A132B9D2772.pdf
ACECC (2016) Transdisciplinary Approach (TDA) for building societal resilience to disasters (TC21Transdisciplinary Approach (TDA) for Building Societal Resilience to Disasters). https://www.acecc-world.org/TC21/publication.files/ACECC_E-Book.pdf
AghaKouchak A, Chiang F, Huning LS, Love CA, Mallakpour I, Mazdiyasni O, Moftakhari H, Papalexiou SM, Ragno E, Sadegh M (2020) Climate extremes and compound hazards in a warming world. Annu Rev Earth Planet Sci 48(1):519–548. https://doi.org/10.1146/annurev-earth-071719-055228
Al Nuaimi E, Al Neyadi H, Mohamed N, Al-Jaroodi J (2015) Applications of big data to smart cities. J Internet Serv Appl 6(1):25. https://doi.org/10.1186/s13174-015-0041-5
Alshehri SA, Rezgui Y, Li H (2015) Delphi-based consensus study into a framework of community resilience to disaster. Nat Hazards 75(3): 2221–2245. https://doi.org/10.1007/s11069-014-1423-x
Arup (2014) City resilience index. https://www.arup.com/perspectives/publications/research/section/city-resilience-index
Ashwathy A, Ajai S, Persis T (2018) An overview of the smart cities mission in India. Centre for Policy Research. https://cprindia.org/system/tdf/policy-briefs/SCM%20POLICY%20BRIEF%2028th%20Aug.pdf?file=1%26type=node%26id=7162
Bahinipati CS, Rajasekar U, Patel M (2015) Flood-induced economic loss and damage to the textile industry in Surat City, India. 43.
Bhat GK, Karanth A, Dashora L, Rajasekar U (2013) Addressing flooding in the city of Surat beyond its boundaries. Environ Urban 25(2):429–441. https://doi.org/10.1177/0956247813495002
Bhushan N, Rai K (2004) Strategic decision making: Applying the analytic hierarchy process. Springer-Verlag. https://doi.org/10.1007/b97668
Blok A (2016) Assembling urban riskscapes. City 20(4):602–618. https://doi.org/10.1080/13604813.2016.1194000
Bosch P, Jongeneel S, Rovers V, Neumann H-M, Huovila A (2017) CITYkeys indicators for smart city projects and smart cities, 305.
Brown C, Shaker RR, Das R (2018) A review of approaches for monitoring and evaluation of urban climate resilience initiatives. Environ, Dev Sustain 20(1):23–40. https://doi.org/10.1007/s10668-016-9891-7
Bruneau M, Reinhorn A (2006) Overview of the resilience concept, 9.
Census (2011) Census of India website: office of the registrar general & census commissioner, India. https://www.censusindia.gov.in/2011census/PCA/pca_highlights/pe_data.html
Chu E (2016a) The political economy of urban climate adaptation and development planning in Surat, India. Environ Plan C: GovMent Policy 34(2):281–298. https://doi.org/10.1177/0263774X15614174
Chu E (2016b) The governance of climate change adaptation through urban policy experiments. Environ Policy GovAnce 26:439–451. https://doi.org/10.1002/eet.1727
City Mayors (2011) City Mayors: world's fastest growing urban areas. http://www.citymayors.com/statistics/urban_growth1.html
Climate Change Department (2014) Gujarat state action plan on climate change. Climate Change Department, Government of Gujarat, India. https://ccd.gujarat.gov.in/Images/Gujarat-State-Action-Plan-on-Climate-Change.pdf
Cutter SL, Burton CG, Emrich CT (2010) Disaster resilience indicators for benchmarking baseline conditions. J Homel Secur Emerg Manag 7(1). https://doi.org/10.2202/1547-7355.1732
Davies ZG, Edmondson JL, Heinemeyer A, Leake JR, Gaston KJ (2011) Mapping an urban ecosystem service: quantifying above-ground carbon storage at a city-wide scale. J Appl Ecol 48(5): 1125–1134. https://doi.org/10.1111/j.1365-2664.2011.02021.x

Derrible S, Chester M, Guikema S (2020) Infrastructure resilience to climate change. J Infrastruct Syst 26(2):02020001. https://doi.org/10.1061/(ASCE)IS.1943-555X.0000532
Desai VK, Wagle S, Rathi SK, Patel U, Desai HS, Khatri K (2015) Effect of ambient heat on all-cause mortality in the coastal city of Surat India. Curr Sci 109(9): 1680–1686.
Dincer I, Abu-Rayash A (2020) Chapter 5—community energy systems. In: Dincer I, Abu-Rayash A (eds) Energy sustainability. Academic Press, pp 101–118. https://doi.org/10.1016/B978-0-12-819556-7.00005-X
Engle NL, de Bremond A, Malone EL, Moss RH (2014) Towards a resilience indicator framework for making climate-change adaptation decisions. Mitig Adapt Strat Glob Chang 19(8):1295–1312. https://doi.org/10.1007/s11027-013-9475-x
ET energy world (2019) World's first emissions trading scheme for particulate pollution starts in Surat. EPIC. https://epic.uchicago.edu/news/worlds-first-emissions-trading-scheme-for-particulate-pollution-starts-in-surat/
European Commission. Directorate General for the Environment, Intrasoft International and University of the West of England (UWE). Science Communication Unit (2017) Indicators for sustainable cities. Publications Office. https://data.europa.eu/doi/10.2779/121865
Feldmeyer D, Wilden D, Jamshed A, Birkmann J (2020) Regional climate resilience index: a novel multimethod comparative approach for indicator development, empirical validation and implementation. Ecol Indic 119:106861. https://doi.org/10.1016/j.ecolind.2020.106861
Feldmeyer D, Wilden D, Kind C, Kaiser T, Goldschmidt R, Diller C, Birkmann J (2019) Indicators for monitoring urban climate change resilience and adaptation. Sustain 11(10):2931. https://doi.org/10.3390/su11102931
FSI (2019) India state of forest report 2019. Forest Survey of India. https://fsi.nic.in/isfr19/vol1/cover-page.pdf
GOI (2015) Smart cities: mission statement & guidelines. Ministry of Urban Development, Government of India. https://smartcities.gov.in/themes/habikon/files/SmartCityGuidelines.pdf
Goi C-L (2017) The impact of technological innovation on building a sustainable city. Int J Qual Innov 3(1):6. https://doi.org/10.1186/s40887-017-0014-9
GPDDR (2017) Plenary coherence between the sendai framework, the 2030 agenda for sustainable development and climate change. Global Platform for Disaster Disk Reduction. https://www.unisdr.org/files/globalplatform/592361be6e1b3Issue_Brief_-_Global_Platform_Plenary_on_Coherence_30.pdf
Griffiths K, Boyle C, Henning T (2015, June 10) Infrastructure sustainability rating tools—How they have developed and what we might expect to see in the future
Griffiths K, Boyle C, Henning T (2017) Comparative assessment of infrastructure sustainability rating tools
Gupta AK, Mani N, Sarkar BB, Singh S, Katyal S (2019) Training module for urban local bodies, including contexts of climate risk and children's resilience (No. 978-81-933285-3-8). National Institute of Disaster Management New Delhi, India and United Nations Children's Fund (UNICEF), India. https://reliefweb.int/sites/reliefweb.int/files/resources/capres_DDRRC.pdf
Gupta K (2020) Challenges in developing urban flood resilience in India. Philos Trans R Soc A: Math, Phys Eng Sci 378(2168):20190211. https://doi.org/10.1098/rsta.2019.0211
Haritas B (2021) Richest cities of India. BW Businessworld. http://businessworld.in/article/Richest-Cities-Of-India/28-06-2017-121011
Hashem IAT, Chang V, Anuar NB, Adewole K, Yaqoob I, Gani A, Ahmed E and Chiroma H (2016) The role of big data in smart city. Int J Inf Manag 36(5):748–758. https://doi.org/10.1016/j.ijinfomgt.2016.05.002
Höjer M, Wangel J (2014) Smart sustainable cities: Definition and challenges. In Advances in intelligent systems and computing, vol 310, pp 333–349. https://doi.org/10.1007/978-3-319-09228-7_20
Holt R (2018) Which cities will be leading the global economy in 2035? 15.

Hussain A., Gupta A (2020) Planning for low-carbon smart cities in India. In: Ahmed S, Abbas SM, Zia H (eds) Smart cities—opportunities and challenges. Springer, pp 189–206. https://doi.org/10.1007/978-981-15-2545-2_18

IPCC (2014) Mitigation of climate change summary for policymakers (Contribution of Working Group III to the Fifth Assessment Report). Intergovernmental Panel on Climate Change. https://www.ipcc.ch/site/assets/uploads/2018/03/ipcc_wg3_ar5_summary-for-policymakers-1.pdf

ISB and SSEF (2017) Smart cities index: a tool for evaluating cities. Indian School of Business and Shakti Sustainable Energy Foundation. https://shaktifoundation.in/wp-content/uploads/2017/11/Smart-Cities-Index.pdf

Jäger J, Rounsevell MDA, Harrison PA, Omann I, Dunford R, Kammerlander M, Pataki G (2015) Assessing policy robustness of climate change adaptation measures across sectors and scenarios. Climatic Change 128(3):395–407. https://doi.org/10.1007/s10584-014-1240-y

Keck M, Sakdapolrak P (2013) what is social resilience? lessons learned and ways forward. Erdkunde 67(1):5–19

Kim K-G (2018) Low-carbon smart cities: tools for climate resilience planning. Springer International Publishing. https://doi.org/10.1007/978-3-319-59618-1

Kumar A, Diksha, Pandey AC, Khan ML (2020) Urban risk and resilience to climate change and natural hazards. In Techniques for disaster risk management and mitigation. John Wiley & Sons, Ltd., pp 33–46. https://doi.org/10.1002/9781119359203.ch3

Kupers DR, Ching SH (2016) A resilience framework for smart cities, 14

Liu JJW, Reed M, Girard TA (2017) Advancing resilience: An integrative, multi-system model of resilience. Pers Individ Differ 111:111–118. https://doi.org/10.1016/j.paid.2017.02.007

Meerow S, Newell JP, Stults M (2016) Defining urban resilience: a review. Landsc Urban Plan 147:38–49. https://doi.org/10.1016/j.landurbplan.2015.11.011

MOEFCC (2015) India's intended nationally determined contributions—towards climate justice. Ministry of Environment, Forest and Climate Change, Government of India. http://moef.gov.in/wp-content/uploads/2018/04/revised-PPT-Press-Conference-INDC-v5.pdf

MOHUA (2011) Jawaharlal Nehru national urban renewal mission: toolkit. Ministry of Housing and Urban Affairs, Government of India. https://mohua.gov.in/upload/uploadfiles/files/7Preface_English(1).pdf

Network A (2015, January 6) India [Text]. ACCCRN Network. https://www.acccrn.net/country/india

OECD (2018) Climate-resilient infrastructure: policy perspectives (ISSN 2309-7841). Organisation for Economic Co-operation and Development. https://www.oecd.org/environment/cc/policy-perspectives-climate-resilient-infrastructure.pdf

Ohshita S (2017) Resilient urban energy: making city systems energy efficient, low carbon, and resilient in a changing climate. https://www.eceee.org/library/conference_proceedings/eceee_Summer_Studies/2017/3-local-action/resilient-urban-energy-making-city-systems-energy-efficient-low-carbon-and-resilient-in-a-changing-climate/

Ostrom E (2010) Response: the institutional analysis and development framework and the commons. SSRN Electron J. https://doi.org/10.2139/ssrn.1615582

Parikh K, Parikh J, Kumar M (2018) Vulnerability of Surat, Gujarat to flooding from Tapi river: a climate change impact assessment. /paper/Vulnerability-of-Surat%2C-Gujarat-to-Flooding-from-A-Parikh-Parikh/d093b55b327fd1d69a1b919a5ef28f720588af5f

Patel T (2014) Funding for adaptation to climate change: the case of Surat. Department of Urban Studies and Planning, Massachusetts Institute of Technology. https://dspace.mit.edu/bitstream/handle/1721.1/90100/890144160-MIT.pdf?sequence=2&isAllowed=y

Rahmasary AN, Robert S, Chang I-S, Jing W, Park J, Bluemling B, Koop S, van Leeuwen K (2019) Overcoming the challenges of water, waste and climate change in Asian cities. Environmental Management 63(4):520–535. https://doi.org/10.1007/s00267-019-01137-y

Rathi S, Desai V, Jariwala P, Desai H, Naik A, Joseph A (2017) Summer temperature and spatial variability of all-cause mortality in Surat City, India. Indian J Community Med: Off Publ Indian Assoc Prev & Soc Med 42(2), 111–115. https://doi.org/10.4103/0970-0218.205216

Ritchie H, Roser M (2018) Urbanization. Our World in Data, University of Oxford. https://ourworldindata.org/urbanization
Roblek V, Meško M (2020, June 15). Smart city knowledge management Holistic review and the analysis of the urban knowledge management. https://doi.org/10.1145/3396956.3398263
Rose A (2007) Economic resilience to natural and man-made disasters: multidisciplinary origins and contextual dimensions. Environ Hazards 7(4), 383–398. https://doi.org/10.1016/j.envhaz.2007.10.001
Saaty TL (2008) Decision making with the analytic hierarchy process. Int J Serv Sci 1(1):83. https://doi.org/10.1504/IJSSCI.2008.017590
Saja AMA., Teo M, Goonetilleke A, Ziyath AM (2018) An inclusive and adaptive framework for measuring social resilience to disasters. Int J Disaster Risk Reduct 28:862–873. https://doi.org/10.1016/j.ijdrr.2018.02.004
Satterthwaite D (2008) Cities' contribution to global warming: notes on the allocation of greenhouse gas emissions. Environ Urban 20(2):539–549. https://doi.org/10.1177/0956247808096127
Sharma VR, Verma C (eds) (2019) Making cities resilient. Springer International Publishing. https://doi.org/10.1007/978-3-319-94932-1
Shaqrah AA (2019) Future of smart cities in the knowledge-based urban development and the role of award competitions [Chapter]. Smart Cities and Smart Spaces: Concepts, Methodologies, Tools, and Applications; IGI Global. https://doi.org/10.4018/978-1-5225-7030-1.ch024
Sharifi A (2020) A typology of smart city assessment tools and indicator sets. Sustain Cities Soc 53:101936. https://doi.org/10.1016/j.scs.2019.101936
Sharifi A (2021) Co-benefits and synergies between urban climate change mitigation and adaptation measures: a literature review. Sci Total Environ 750:141642. https://doi.org/10.1016/j.scitotenv.2020.141642
Sheth KN (2017, April 6) Surat smart city. Surat Smart City, Surat Municipal Corporation, Gujarat, India. https://doi.org/10.13140/RG.2.2.33336.34560
Singh SK (2012) Urban transport in India: issues, challenges, and the way forward (SSRN Scholarly Paper ID 2465552). Social Science Research Network. https://papers.ssrn.com/abstract=2465552
Smart City (2018) Fund released status till 2031. Ministry of Housing and Urban Affairs, Government of India. https://smartcities.gov.in/upload/uploadfiles/files/Fund%20released%20status%20till%2031%20March%202018.pdf
Smart City (2019) Guideline for framing Human Resource (HR) Policy for the Special Purpose Vehicles (SPVs) under Smart Cities Mission. 124.
Smart Net (2015) India smart city mission. Government of India. https://smartnet.niua.org/sites/default/files/resources/SCP_%20SURAT.pdf
Smart Net (2017) PPP projects in smart cities. National Institute of Urban Affairs, Ministry of Housing and Urban Affairs, Government of India. https://smartnet.niua.org/sites/default/files/resources/PPP-Projects.pdf
Smart Net (2018) Surat IT-MAC: integrated transport-mobility administration centre. Ministry of Housing and Urban Affairs, Government of India. https://smartnet.niua.org/sites/default/files/resources/1_1.pdf
SMC (2011) Surat city resilience strategy. Surat Municipal Corporation. http://resiliencestrata.org/uploads/publication/3B07371E-EC6C-47C3-99ED-4A132B9D2772.pdf
SMC (2018) Surat heat wave action plan 2018. Surat Municipal Corporation and Gujarat State Disaster Management Authority. http://resiliencestrata.org/uploads/publication/1D97E6B4-57F2-4371-9DB7-7DF79D93BA21.pdf
SMC (2019) Surat smart city development limited (Annual Report for Financial Year 2018–2019). https://www.suratmunicipal.gov.in/Home/SmartCitySurat
SMC (2020) Surat smart city development Ltd. https://www.suratmunicipal.gov.in/Home/SmartCitySurat
SMC and SCCT (2017) Surat resilience strategy. Surat Municipal Corporation and Surat Climate Change Trust. https://resilientcitiesnetwork.org/downloadable_resources/Network/Surat-Resilience-Strategy-English.pdf

Sridhar KS, Reddy AV, Srinath P (2013) Is it push or pull? Recent evidence from migration into Bangalore, India. J Int Migr Integr 14(2):287–306. https://doi.org/10.1007/s12134-012-0241-9
Statista (2019) India—urbanization 2019 [Economy & Politics and International]. India: Degree of Urbanization from 2010 to 2020. https://www.statista.com/statistics/271312/urbanization-in-india/
Sustainability Outlook (2014) Shaping new age urban systems energy, connectivity & climate resilience. Summit of the sustainable business leadership forum. https://india.smartcitiescouncil.com/system/tdf/india/public_resources/Shaping-New-Age-Urban-Systems-Energy_0.pdf?file=1&type=node&id=2242&force
Tanner T, Mitchell T, Polack E, Guenther B (2009) Urban governance for adaptation: assessing climate change resilience in ten Asian cities. IDS Working Papers 2009(315):01–47. https://doi.org/10.1111/j.2040-0209.2009.00315_2.x
Thet KK (2014) Pull and push factors of migration: a case study in the urban area of monywa township, Myanmar. 14
Tyler S, Nugraha E, Nguyen HK, Nguyen NV, Sari AD, Thinpanga P, Tran TT, Verma SS (2016) Indicators of urban climate resilience: A contextual approach. Environ Sci & Policy 66:420–426. https://doi.org/10.1016/j.envsci.2016.08.004
UNDP (2011) Paving the way for climate-resilient infrastructure: guidance for practitioners and planners. https://www.uncclearn.org/wp-content/uploads/library/undp_paving_the_way.pdf
United Nation (2011) Climate neutral cities: how to make cities less energy and carbon intensive and more resilient to climatic challenges. United Nations, Geneva, Switzerland. https://sustainabledevelopment.un.org/content/documents/808climate.neutral.cities_e.pdf
United Nation and ESCAP (2019) The future of Asian and pacific cities 2019: transformative pathways towards sustainable urban development. Economic and Social Commission for Asia and the Pacific. https://www.unescap.org/publications/future-asian-and-pacific-cities-2019-transformative-pathways-towards-sustainable-urban
WRI (2020) Enhancing NDCS: a guide to strengthening national climate plans by 2020. World Resource Institute and United Nation Development Program. https://files.wri.org/d8/s3fs-public/enhancing-ndcs_0.pdf
Yamagata Y, Sharifi A (eds) (2018) *Resilience-oriented urban planning: Theoretical and empirical insights*. Springer International Publishing. https://doi.org/10.1007/978-3-319-75798-8
Zhu S, Li D, Feng H (2019) Is smart city resilient? evidence from China. Sustain Cities Soc 50:101636. https://doi.org/10.1016/j.scs.2019.101636
Zhu Z-T, Yu M-H, Riezebos P (2016) A research framework of smart education. Smart Learn Environ 3(1):4. https://doi.org/10.1186/s40561-016-0026-2

Chapter 13
The Contributions of Smart City Initiatives to Urban Resilience: The Case of San Francisco, California, United States

Alison Grovert, Carmela Sambo, Briana Meier, and Yekang Ko

Abstract This chapter evaluates five smart technology projects that have been implemented in the City of San Francisco and the surrounding San Francisco Bay Area in California to assess the extent to which the projects support urban resilience. Increasingly, cities worldwide are transforming their systems through smart technologies. Emerging smart technologies are supporting efforts to reduce emissions, address social inequities, and build economic security. San Francisco leads US cities in its efforts adopt smart technologies for improving urban resilience. Major risks that immediately threaten San Francisco include earthquakes, fire, tsunamis, flooding, extreme heat, droughts, terrorism, cyber terrorism, and communicable diseases. As a result, San Francisco's private and public sectors are funding smart technology in transportation, waste management, social, government, and economic realms to improve long-term resilience and sustainability. This study selected five key smart technology projects that seek to improve urban resilience including micro-grids, connected and automated vehicles, earthquake alerts, digital platforms, and air quality monitoring. We evaluated these projects against fourteen principles for resilience developed by Sharifi and Yamagata (Renewable and Sustainable Energy Reviews 60:1654–1677, 2016). This chapter also examines San Francisco's contribution to climate resilience planning and highlights the importance of innovation through collaboration with startups. Results will contribute to a smart city resilience assessment toolkit used to investigate actual and potential contributions of smart city initiatives to resilience in the United States.

Keywords Smart technology · Urban resilience · Climate action · Energy · Transportation · Air quality

A. Grovert · C. Sambo · Y. Ko (✉)
Department of Landscape Architecture, School of Architecture and Environment, 5249 University of Oregon, Eugene, OR 97403-5249, USA
e-mail: yekangko@uoregon.edu

B. Meier
History, Anthropology, and Science and Technology Studies, Massachusetts Institute of Technology, 77 Massachusetts Avenue Room E51-163, Cambridge, MA 02139, USA
e-mail: bkmeier@mit.edu

A. Sharifi and P. Salehi (eds.), *Resilient Smart Cities*, The Urban Book Series,
https://doi.org/10.1007/978-3-030-95037-8_13

13.1 Introduction

This chapter analyzes five projects that utilize smart technologies in the San Francisco Bay region in California, a region well-known for its leadership in adopting smart technologies for improving urban resilience. The projects we analyze range in focus from energy to transportation, disaster, innovation, and health. Please confirm the section headings are correctly identified throughout the chapters Stakeholders involved across these projects include emergency departments, first responders, commuters, city-level lawmakers, technology startup firms, and the broader public. As the regional population continues to grow, it becomes increasingly vital to adopt measures in support of urban resilience such as the earthquake planning and air quality management responses included in our analysis.

Cities are the locations and populations that are simultaneously large contributors to global climate change and that bear much of the impacts of increasing climate instability. The intricate causal web of relationships between cities and issues of climate change are made even more complex with increasing urbanization and climate instability (Sharifi 2016). Many cities are now experiencing an array of climate-related disasters. Municipal level efforts to support urban resilience are emerging rapidly in response to increasing instability.

Urban resilience refers to the ability of individual humans, as well as social, ecological, and economic systems to adapt in response to shocks, adversity, or stressors. In these contexts, resilience can be defined as "the ability to absorb, adapt, and respond to changes in an urban system" (Desouza & Flannery 2013, p. 89) or "the capacity to buffer perturbations, self-organize, learn, and adapt" (Folke et al. 2002, p. 51). Resilience theory is a widespread method for urban planning and disaster planning that works in a systems approach across multiple temporal and geographical scales and includes individuals and collective local capacity to cope with and adapt to stress and change. Krasny and Tidball (2009) note that resilience can apply across both ecosystems and communities, and, using examples of urban community gardens, suggest that resilience at either the social or ecological level can help buttress resilience within the other. As defined by Sharifi and Yamagata, resilience is that which enables the system to bounce back and also bounce forward (Sharifi and Yamagata 2016). Bouncing forward is to learn from the past and build back better by using the rebuild to resolve existing infrastructural, social, and environmental issues in the city, as well as anticipating novel future scenarios.

Smart technologies are an increasingly salient tool in the growing efforts to cultivate urban resilience in the face of accelerating climate change.

Smart technologies include a variety of sensors, controls, and automated mechanisms, as well as networked data systems, all of which are intended to increase the adaptability of the variety of systems to which they are applied (Holnicki-Szulc 2008). Smart technologies are intended to improve efficiency by allowing devices to communicate with one another, providing decision-makers with real-time metrics, and monitoring environmental conditions such as air quality with high accuracy. Smart technologies are quickly transforming cities by increasing the speed, accuracy, variety, quantity, and accessibility of information available to municipal governments and residents.

A "smart city" employs a variety of digital technologies to build data and other connections across multiple urban domains. Definitions of smart cities are multiple and nebulous, as noted by Albino et al, in general a smart city "connects physical infrastructure, IT infrastructure, social infrastructure, and business infrastructure to leverage the collective intelligence of the city" (Albino et al. 2015, p. 4). The transformation of pre-digital cities to so-called smart cities is a long-term and dynamic process, requiring analysis, vision, and flexibility. A smart, sustainable city is one that harnesses both smart technologies and sustainable practices across three crucial levels: economic, environmental, and social (Ibrahim et al. 2018). In assessing a city's potential for smart, sustainable transformations, it is important to gauge the city's overall readiness for change (Ibrahim et al. 2018) in terms of technological capacity and the willingness of firms, municipal governments, and urban denizens to participate.

13.1.1 Local Climate Actions in the U.S. and California

The United States' withdrawal from the 2015 Paris Agreement prompted greater commitments to address climate change at local and regional levels. Even prior to the US withdrawal—and since the renewed the US commitments—many cities have pledged to achieve the local emissions goals outlined in the Paris Agreement (Hultman 2019). The U.S. Conference of Mayors' Climate Protection Agreement, consisting of 407 US Mayors representing 70 million Americans, commits to adopting, honoring, and upholding Paris Climate Agreement goals by reducing their cities' carbon emissions below 1990 levels. These commitments adhere to standards set decades earlier by the Kyoto Protocol (Hultman 2019).

In the Southwest region of the United States, climate change continues to impact cities, agricultural land, and the region's ecosystems (Carter et al. 2014). According to the third U.S. National Climate Assessment, the decline of snowpack and streamflow due to global warming is reducing available surface water resources for urban areas. This decrease is exacerbated by agricultural and environmental water use. This water shortage affects the entire country since over half of the high-value specialty crops in the US are produced in the southwestern region. Many specialty crops are vulnerable to new climatic extremes. California has recorded drier than usual conditions since 2000.

Although California is home to the highest population of any state in the United States, it maintains one of the lowest per capita energy consumption rates. California also has the world's fifth largest economy, making it a major world player in issues of greenhouse gas emissions. In response to climate change impacts, the State of California has implemented adaptation and mitigation strategies and actions to protect water infrastructure, maintain air quality, and protect stream flows and wetlands (Albino et al. 2015). Wildfire is also a pressing climate issue in California, demonstrated by its inclusion in the state's 2009 Climate Adaptation Strategy. The document aims to strategically decrease wildfire risk and vulnerability for current and future generations of Californians (Albino et al. 2015).

13.2 Study Area: San Francisco Bay Area, California

13.2.1 Location and Demographics

The nine-counties that border the San Francisco Bay and compose the San Francisco Bay Area include Alameda, Contra Costa, Marin, Napa, San Francisco, San Mateo, Santa Clara, Solano, and Sonoma counties (Fig. 13.1). In this diverse region, the cities of Oakland, San Jose, and San Francisco rank higher for cultural diversity than other major American cities like Boston, Orlando, San Diego, and Los Angeles

Fig. 13.1 Study Area: San Francisco Bay Area. The 9 California counties that comprise our study area and make up the "San Francisco Bay Area"

(McCann 2021).The City of San Francisco is 46.89 square miles with a population of 805,235 within city limits (2010 US Census). The broader Bay Area is known for tech innovations in Silicon Valley and a strong culture of startups and risk-taking (Funk and Deininger 2018).

Geographically, seven fault lines, including the San Andreas, fault crisscross their way through the region. This makes the threat of earthquakes high after decades of rapid urban growth. In addition, the region's dry climate lends itself to wildfires. In 2020, the region experienced a record-breaking heat wave and record-size wildfires, further complicated by the novel coronavirus. A booming economy and jobs market contributes to rapid population growth of the Bay Area, leading to increasingly challenging traffic congestion in recent years. The congestion is in part due to the ongoing housing affordability crisis, which leads to large numbers of commuters. These high volumes of traffic negatively impact from the air quality in San Francisco, an issue that was made apparent in 2020 when clear skies emerged following the regional COVID-19 shelter-in-place orders.

13.2.2 Climate Resilience Plans and Smart Technologies in San Francisco

To combat the risks of climate-related disasters, the City of San Francisco has implemented a Hazard and Climate Resilience Plan. The plan's vision is to "make San Francisco resilient to immediate and long-term threats of climate change and natural hazards through actions to mitigate risks, adapt built and natural assets, and build a more equitable and sustainable city" (The City and County of San Francisco 2020). The plan also coordinates with the 2020 Climate Action Strategy, aiming to achieve zero carbon emission by 2050.

In addition to the Climate Resilience Plan and the Climate Action Plan, more than 25 other recent plans, assessments, and visions aim to boost climate resilience in the Bay Area, including the 2016 strategic vision entitled "Resilient SF: Stronger Today, Stronger Tomorrow," which established the Office of Resilience and Capital Planning, the 2020 "Sea Level Rise Vulnerability and Consequences Assessment," and the 2019 "Hazards and Climate Resilience Plan."

Furthermore, San Francisco has begun to implement smart technologies to prevent, adapt to, and counter its social and environmental threats. Private and public sectors are funding smart technology for transportation, waste management, social, government, and economic realms, with a goal to improve long-term resilience and sustainability.

For example, the city sets a national example in terms of waste management, aggressively pursuing goals of zero-waste, achieving an 80% diversion of waste from landfills in 2012 (Airhart 2018). The city has implemented eco-districts aimed at reducing emissions, waste, and consumption of water, as well as projects that facilitate smart transportation. Proposals for four varieties of "eco-districts" were proposed in 2012 by the San Francisco Planning department. Goals of these proposals included reducing carbon emissions and decreasing waste volumes, water usage, and energy use. In the transportation realm, the city received $11 million from the Department of Transportation (DOT) in 2016 to initiate projects combatting traffic, vehicle collisions, transit travel times, and emergency vehicle response times (Funk & Deininger 2018). Using DOT funds, a smart parking management program called SF Park implemented wireless sensors to support demand-responsive pricing (Funk and Deininger 2018). Also, in accordance with demands for clarification of data privacy regulations, San Francisco hired its first Chief Data Officer (Funk and Deininger 2018).

To achieve a waste-free city, San Francisco implemented an online tool to provide residents with resources on recycling, reuse, and disposal options through "RecycleWhere." San Francisco also managed to reduce its average water consumption to 49 gallons per day, compared to the national average of 80–100 gallons per day. The city provides free and accessible water-saving devices including high-efficiency shower heads. These devices can reduce water use in both residential and commercial settings.

13.2.3 Smart City Movements Within the US and San Francisco

The U.S. Department of Transportation launched a Smart City Challenge in 2015, which challenged mid-sized American cities to propose plans for "an integrated, first-of-its kind smart transportation system that would use data, applications, and technology to help people and goods move more quickly, cheaply, and efficiently" (Smart Columbus, n.d.). Of the 78 applicants, San Francisco placed as one of seven finalists. San Francisco envisioned increased access to shared vehicles for people of all backgrounds to travel easily without owning a car (PolicyLink and PERE 2017). The City proposed a mobile device application through which users can route, schedule, and pay for selected modes of transportation. San Francisco also sought to provide low-income residents with access to smartphones and free, public wireless internet connection. The City also proposed a partnership with UC Berkeley and industry leaders to test, analyze, and disseminate best practices (PolicyLink and PERE 2017). Currently, the State of California continues to lead the country in terms of stricter emissions standards across many sectors (Climate Action Tracker, n.d.).

13.3 Methods

13.3.1 Selecting a Smart City in the US

To select which US city's smart technology efforts to analyze, we conducted a literature review using a keyword search on both academic and non-academic search engines using terms related to smart cities. We addressed the question: "Which city in the United States is leading in smart city technologies?" We consulted the following databases to identify possible cities to analyze: Google, Google Scholar, University of Oregon LibrarySearch, and Web of Science. The search strings included the following terms: smart technology; Internet of Things (IoT); smart city. The searches, conducted in July 2020, yielded 17 documents (8 academic articles and 9 non-academic articles). Of the 17 documents, San Francisco was mentioned the most.

13.3.2 Smart Technology Initiatives in the San Francisco Bay Area

After identifying San Francisco as a leading city in the adoption of smart city technologies, we then conducted a review of key literature and recent media to identify emerging smart city projects in the region. From this review, we selected five projects for analysis in this chapter. We analyzed these smart city projects against a set of principles for resilient urban energy systems developed by Sharifi and Yamagata (2016) and our own criterion that considers the extent to which the project contributes to reducing greenhouse gas emissions. We selected fourteen relevant principles from the complete set of seventeen principles developed by Sharifi and Yamagata. While the principles presented by Sharifi and Yamagata were developed for resilient urban energy systems, the authors signal their broader applications, noting that they can serve as general criteria for assessing any resilient system, including the broader urban resilience goals of the projects we analyze here. The principles against which we assess each project are listed below. Our own terminology for the principles are listed in plain text, while the terms drawn directly from resilience principles developed by Sharifi and Yamagata (2016) are indicated in italics:

Robustness (page 1663),
Diversity (page 1665),
Redundancy (page 1665),
Connectivity (*Coordination Capacity*, page 1665),
Flexibility (page 1665),
Resourcefulness (page 1665),
Rapid response (*Agility*, page 1666),
Efficiency (page 1665).
Learning capacity (*Adaptability*, page 1666),

Modularity and self-organization (*Independence*, page 1665),
Innovation (*Creativity*, page 1666),
Equity (page 1666),
Inclusiveness (participatory) (*Collaboration*, pages 1665–1666), and
Foresight capacity (page 1665).

In addition to these principles drawn from Sharifi and Yamagata, as mentioned, we assess the extent to which the project is "low-carbon" or contributes to reducing GHG emissions.

13.4 Results

The following section introduces five smart technology projects that are being implemented in the San Francisco Bay region and evaluates the extent to which each project contributes to improving urban resilience against each of the principles listed above.

13.4.1 Project 1: Gridscape Microgrids

Bringing Microgrids to California Fire Stations

In Fremont, California, a collaboration between startup Gridscape Solutions, the City of Fremont, and the Fremont Chamber of Commerce has led to increased energy resilience by way of solar microgrids (Thurston 2019). The project, which went online in 2019, brings clean energy to three of the city's fire stations. The project is subsidized by the California Energy Commission (CEC). The project is the first-of-its kind, providing power to fire stations via solar microgrids and battery storage, paving the way for fire stations across the country to increase their resilience. Other California cities that are considering adding solar microgrids to their fire stations include Redwood City and Montecito. In addition to acting as an off-grid power source when wildfires encroach on the area, a common occurrence in recent years across California, solar power replaces fossil-fueled power, which reduces the carbon footprint of the fire stations (Gore 2019).

Low carbon

The City of Fremont has a climate action plan which calls for reduction of greenhouse gases and zero net energy use in government buildings (California Energy Commission 2019).

In 2018, the State of California's top five in-state generated energy sources were natural gas (47%), solar photovoltaic (13%), large hydro (11%), nuclear (10%), and wind (6%). Notably, quantities of in-state oil, coal, and petroleum coke generation were well under 1% combined. California imports energy from out of state as well, which increases its fossil fuel intake.

For Fremont's fire stations, using solar microgrids daily and as their main backup source rather than the prevalent diesel generators helps Fremont achieve its climate goals. This clean energy option reduces the amount of greenhouse gases emitted into the atmosphere by 142,000 pounds carbon dioxide annually. The microgrid is owned and operated by Gridscape Solutions, which utilizes a "power purchasing agreement," allowing the City of Fremont to reap the cost savings and clean energy benefits (Gore 2019).

Robustness
Fire stations need to operate 24/7, especially during an emergency event such as a wildfire. As the electricity grid poses threats to the spread of the wildfire, strategies to control the spread involve shutting off power to affected areas. Sparks from electricity lines can and have ignited flames and caused harm to California communities. However, power is necessary for firefighters to track the burn and deploy critical resources to places in need. Gridscape offers a solution to this issue by providing fire stations across California with solar powered microgrids and battery storage, affording them an islandable backup power supply that allows the system to respond quickly and bounce back.

Diversity
By reducing reliance on the energy grid and backup diesel generators, the microgrid system increases the variety of energy sources used to power fire station operations. The fire stations can still use power from the energy grid. However, by increasing its reliance on solar power, the project will save the fire station an estimated 350,000 USD on utility fees over 10 years. This project has allowed the fire stations to achieve an 80% cost savings over traditional diesel generators (Gore 2019).

Redundancy
In emergency operations, solar batteries allow the fire station's facilities to keep working during a power outage. Fire stations in the area are required to have a backup source of energy in order to operate without disruption during normal events. By using the energy grid, the solar power, and the backup solar storage, Fremont fire stations have at least three methods for generating power. If one were to fail, the other would kick in.

Flexibility
The project utilizes technology based on the concept of the Internet of Things (IoT) as well as "smart inverters" and various digital controls (Gore 2019) for managing energy flows. Gridscape microgrids are capable of providing clean power on or off the grid, depending on the situation. To make this possible, the company employs an "EnergyScope Microgrid Controller" which enables the solar microgrid to optimize timing and direction of flow among batteries and the main energy grid (Gridscape Solutions, n.d.) This smart technology is called an automated energy control system. Microgrids help increase grid reliability and reduce grid congestion, minimizing

utility load of electrical substations at peak hours (California Energy Commission 2019). During a power outage, the fire station can still operate at least 3 h per day.

Agility

Wildfire, power outages, and storms are the fire station's main concerns and reasons for installing the battery backup. The more resources that fire stations have access to during disaster, the more speed and dexterity they can have in deploying their forces. Having a reliable power source during a disaster enables the fire stations and first responders to provide quick, informed, and coordinated services (Stark 2019).

Efficiency

Gridscape's microgrid project makes better use of locally available energy, therefore, reducing the need for energy import from distant locations. This is especially important in the case that wildfire or earthquakes sever the connection between the fire stations and outside energy sources. By utilizing an existing, readily available resource, the project reduces Fremont's reliance on outside energy sources and therefore makes the fire stations more resilient.

Modularity

These solar microgrids are islandable, offering a decentralized energy option during a disaster. Independence from the central energy grid means a more resilient energy system. In addition, using a local energy source rather than relying on electric transmission lines means that the energy system is more self-sufficient than before.

13.4.2 Project 2: Iteris and CAV

Connected and Automated Vehicles (CAV) in the Bay Area

The San Francisco Metropolitan Transportation Commission (MTC) is preparing San Francisco for a connected and automated vehicle (CAV) future with the help of Iteris, a Los Angeles-based transportation tech company with projects across the country and abroad. Iteris has partnered with MTC to expand their 511 real-time travel alert system and implement and advance the Bay Area's intelligent transportation systems, fulfilling the goals of the MTC's Innovative Deployments to Enhance Arterials (IDEA) program. Readying the roads for CAV futures means installing sensors, cameras, and artificial intelligence in street infrastructure so that vehicles can communicate with a network of responsive technologies. Seventy-nine intersections across the Bay Area cities of Los Gatos, Emeryville, Walnut Creek, Concord, and Dublin will receive upgrades including applications for connected vehicles, automated vehicles, and bicycle detection signals. It's up to local governments to decide how to implement CAV technologies. San Francisco is pushing to be a leader in CAV technology, which means its population is projected to be safer, smarter, and more efficient.

Within the San Francisco Bay Area, Silicon Valley has become known as a hub for development and testing of CAV products and policies. By 2017, over 30 automakers, suppliers, tech companies, and electric vehicle companies were testing their autonomous vehicles on public streets in the Bay Area. The collaboration between the MTC and Iteris that began in 2019 builds on the existing strength of transportation tech in the area including the presence of ride hailing (Uber, Lyft), microtransit, and ridesharing companies (Botello et al. 2019). A two-hour drive from Silicon Valley sits GoMomentum Station, the country's largest facility for test-driving connected and automated vehicles (CAV). The Bay Area is pushing its innovators and collaborators to make San Francisco a leader in CAV futures.

Low Carbon

Although connected and automated vehicles (CAV) will not necessarily be electric, there are benefits to the technology that may reduce transportation-related emissions in the Bay Area. These include powertrain electrification, the light-weighting of vehicles, platooning, and eco-driving (U.S. Energy Information Administration 2017). However, the ease, efficiency, and safety of driving in a CAV atmosphere may increase the number of cars and drivers on the road, which would increase vehicle emissions in California.

Connectivity

Enhancing efficiency and effectiveness of transportation operations is made possible by the communication between vehicles utilizing CAV technologies. When cars and other entities are able to communicate with one another, there is potential for increases in safety, faster travel times, and reduced delays due to vehicle collisions. During a disaster, this system connectivity can increase response time and may reduce congestion for mandatory evacuations.

Automated vehicles use in-vehicle sensors, cameras, LiDAR data, network analysis, artificial intelligence, and machine learning[1] to understand the environment in and around the vehicle, but they do not require a response from other vehicles or infrastructure. On the other hand, connected vehicles use a two-way exchange of information via a wireless connection between vehicles and objects including other vehicles (V2V) like cars, trucks, and buses; infrastructure (V2I) such as traffic signals and lane markings; and anything else (V2X). Increasing road safety is the main goal of connected and automated vehicle applications. These applications can take many forms such as alerting drivers to the presence of heavy braking in the traffic ahead (Hartman, n.d.).

Flexibility

Changing conditions are frequent on major travel corridors in the Bay Area. With so many ways to monitor travel, this project is capable of sensing threats and alerting travelers in real-time. Adopting the IoT concept, connected and automated vehicles (CAV) will have the ability to "speak" with one another, making communication and rerouting of traffic smoother. This project has the ability to do all of this quickly and smoothly once brought online. Shifting between different traffic configurations is essential to any flexible system, and Iteris is bringing this functionality onboard.

Resourcefulness
Sensing vehicles, walkers, cyclists, and other entities requires a massive amount of data collected by various technologies involved in CAV road design. This data could be used by planners, officials, and individuals to make decisions about the built environment. By collecting information about different types of travelers, objects, and entities in the five cities involved, this project provides planners with a large volume of information on which they may base decisions.

Agility
This effort to upgrade Bay Area road systems is centered around safety for drivers, pedestrians, and cyclists. As CAV technologies become more widespread, they have the potential to drastically decrease response time during an emergency to prevent congestion, crashes, and wait times that could create cascading failures of the whole MTC transportation system.

13.4.3 Project 3: ShakeAlert

ShakeAlert, an earthquake early warning system, detects significant earthquakes along with alerts residents before shaking occurs. Because the San Francisco Bay Area is vulnerable to natural disasters, such as earthquakes, this technology provides potentially life-saving services for residents. ShakeAlert's system addresses the following resilience characteristics: adaptive/learning capacity, connectivity, efficiency, and equity.

Adaptive/Learning Capacity
Earthquakes pose a national challenge because more than 143 million Americans live in areas of significant seismic risk across 39 states. In California, significant earthquakes of magnitude 6.7 or higher are expected within the next few decades (USGS Earthquake Early Warning 2016). The ShakeAlert system was developed to provide earlier detection for earthquakes and to notify people in affected areas more quickly (USGS Earthquake Early Warning 2016). ShakeAlert has also helped improve the city's capacity to learn from a disaster, which could reduce its vulnerabilities and increase its adaptive capacity to changing conditions.

Connectivity
UC Berkeley's ShakeAlert started in collaboration with Caltech, University of Washington, and USGS. Project partners have expanded to include FEMA, cellular network carriers, and other private firms specializing in public notifications (USGS Earthquake Early Warning 2016).

The program monitors quake activity in the state of California using a network of sensors that relays data to a computing network accessible by scientists, transportation agencies such Bay Area Rapid Transportation (BART) and emergency responders. The system is supported by a network of approximately 800 sensors positioned across the state (Calma 2019). The early warning system reaches the public through an application called MyShake and through existing wireless emergency alerts that sound an alarm on cellular devices. The message delivery is entirely cloud-based to keep information consistent in the event of an earthquake (Calma 2019). Since January 2012, the system has sent live alerts to users in California. It has also sent live alerts to users in the Pacific Northwest since February 2015.

ShakeAlert technologies are being integrated with other regional seismic networks in California and along the west coast, including the Advanced National Seismic System (ANSS), the Pacific Northwest Seismic Network (PNSN), and the California Integrated Seismic Network (CISN) (USGS Earthquake Early Warning 2016). CISN provides monitoring and notification for earthquake prone communities in southern and northern California. This integration across various detection and alert networks allows a variety of agencies to benefit from public investments in monitoring and notification systems and technologies. Beyond the west coast, through ANSS, the ShakeAlert systems are connected to a nationwide network of earthquake detection and warning systems including the National Earthquake Information Center (NEIC), the National Strong Motion Project, and other regional seismic networks (Shakealert: earthquake early warning, n.d.). In addition to alerting residents, this integrated network supports the efforts of emergency response (Shakealert: earthquake early warning, n.d.).

Efficiency

ShakeAlert integrates data from on-site stations with regional networks in order to maximize accuracy and warning times. ShakeAlert also leverages data and infrastructure from existing investment in earthquake monitoring on the west coast. By using sensors and other infrastructure that already exists, the firm was able to reduce its own startup costs while providing early access to system users. Integration between ShakeAlert systems and existing regional sensor networks allows for efficiency gains for multiple (Shakealert: earthquake early warning, n.d.).

Equity

To support equitable access to the warning system, the ShakeAlert application is available in both Spanish and English. Efforts are in development to provide warnings people without cellphones, to improve ADA accessibility, and to expand the number of languages for which the application is available. The USGS and partners are also working on a comprehensive education and training program to increase public awareness of how to respond to alerts (Shakealert: earthquake early warning, n.d.).

Foresight Capacity

Foresight capacity consists of the ability to anticipate uncertain future (Sharifi 2016). While it would not be possible to provide early warning at an earthquake epicenter, increasing warning times of even a few seconds can allow for important adjustments such as slowing (Shakealert: earthquake early warning, n.d.). There is also a possibility that the system might send false warnings or detection warnings for miniscule earthquakes (Shakealert: earthquake early warning, n.d.). On the other hand, the system could also fail to send a warning to its users or even send alerts too late for residents to act upon. These "failure modes" are possible and may be highly likely, which makes it important for ShakeAlert to provide in-depth planning and testing to help minimize them.

13.4.4 Project 4: StiR

Startup in Residence Program (STiR) is a digital platform designed to connect private technology firms with public institutions in service to a variety of urban. STiR was launced in 2013 as a special project within the San Francisco Mayor's Office. In 2015, STiR received a federal grant from the Department of Commerce to grow regionally to Oakland, San Leandro, and West Sacramento, and eventually expanded in 2017. By 2018, STiR became part of City Innovate which expanded the program beyond California. City Innovate is a registered non-profit on a mission to work with the public and private sector to address urban problems. STiR's innovative program addresses the following resilience characteristics: creativity, equity, and inclusiveness.

Creativity
STiR's vision is that civic government technology should be cutting edge, which can be achieved by partnerships between startup firms and public organizations. Over 30 cities use STiR digital tools and over 100 startups have applied to the program. Partnerships that have arisen include Spiral Scout and San Francisco Public Works, who partnered to develop a mobile platform for conducting post-disaster evaluation and reporting on city buildings. In 2019, Actionable Science joined STiR to help Bay Area Rapid Transit (BART) improve its Information Technology. CityDash also joined to help the City of San Jose develop digital platforms available on city street light poles. GovIQ is another startup working to develop a new emergency call center tool for the City and County of San Francisco that will include voice recognition for non-English language calls. Another organization, Gruntify, is working with the City of San Jose to improve disaster response. In 2017, a firm called URBANSIM worked with the San Francisco Mayor's Office of Housing and Community Development to develop software that can evaluate the suitability of either land parcels or buildings for affordable housing. LOTADATA, a startup that joined City Innovate in 2017, provides software allows San Leandro's Recreation and Human Services Department to track the demographics of city facilities and parks users as well revenue at specific facilities. Lastly, a firm named SPIRALSCOUT assisted the San Francisco Public Works in developing software to assist the City in post-disaster evaluation for public buildings and structures (City Innovate, n.d. : Quaintance 2019).

Equity
STiR has adopted corporate responsibility commitments partnering with startups that address social equity and justice issues related to disaster response, developing quicker payment and refund systems for low-income renters, and supporting improved public transportation on urban streets (Quaintance 2019).

Inclusiveness
The purpose of CityInnovate is to bridge the gap between community stakeholders and create a more inclusive govtech system. CityInnovate exists to help emerging technology companies solve challenges in government. This allows government officials to not settle for solutions that do not entirely meet their needs (City Innovate, n.d.). Startups can help build technology for the good of the public.

13.4.5 Project 5: PurpleAir

PurpleAir, an online mapping platform made by a grassroots group in Salt Lake City, provides real-time air quality data from sensors around the region. This small sensor is essential in helping California residents deal with toxic wildfire smoke. PurpleAir's smart technology addresses the following resilience characteristics: low carbon, connectivity, creativity, and equity.

Low Carbon
PurpleAir's quality data collector promotes a low carbon environment through real-time measurements made available to the public. By obtaining real-time information on air quality in specific neighborhoods, the public and air quality enthusiasts can protect their well-being and raise awareness to others.

Connectivity
Currently, the Bay Area Air Quality Management District (BAAQMD) has 30 stations spread around the region to monitor a variety of types of pollution and are located at fixed positions miles apart (Stark, Arcuni, and Brooks 2019). PurpleAir sensors, on the other hand, are placed by anyone at any location around the region. This provides more air quality information than BAAQMD. The numerous air sensors have allowed residents to feel more connected and have proven to be beneficial in a natural disaster. Recent damages and disturbances in California that were caused by wildfires have forced residents to become more aware of their neighborhood's air quality. The wildfires, which burned more than 1.2 million acres of land, led San Francisco residents to prioritize PurpleAir data to check air quality within their neighborhoods.

Creativity
PurpleAir's data differs from the official air quality readings in that its readings are consistent, reliable, and are higher than monitors used by the government during

a wildfire (Peters 2020). The Environmental Protection Agency (EPA) metrics for air quality are based on the weight of rather than number of particles. In contrast, PurpleAir has created an air quality measurement tool based on lasers that measure airborne particles (Peters 2020).

To enhance public connectivity and awareness, PurpleAir monitors are also combined with Wi-Fi technology to help report data to the PurpleAir map. By using data from low-cost sensor networks, it could give citizens more accurate data during wildfires in real-time.

Equity

PurpleAir promotes citizen science by providing resources for ordinary individuals. The company attempts to improve equitable access by developing devices that are affordable to area residents and that can be adopted by public agencies (Stewart 2019). The company has developed a publicly available map that provides information to citizen scientists as well as air quality professionals, as it can be viewed by anyone.

Wildfire smoke poses numerous threats to life and property and sometimes requires to flee their homes with short notice (Peters 2020). Wildfire smoke also causes multiple sorts of health problems (Stewart 2019. Knowing when it's relatively safe to go outside is crucially important, and, to this end, PurpleAir provides a vital resource for residents. Even if a resident does not own a PurpleAir device, data can still be viewed using the freely available PurpleAir mapping system.

Table 13.1 summarizes the five smart technology projects that were evaluated above and indicates how each project contributes to each resilience principle.

13.5 Discussion and Conclusion

As the city of San Francisco, California strives to build a resilient future, and smart technologies are helping the city achieve its goals. The City's Climate Action Plan clarifies its pursuit of net-zero emissions by 2050 through four sub-goals: "zero waste, 80% sustainable trips, 100% renewable energy, and urban greening for carbon sequestration" (Climate plans and reports, n.d.). Programs such as Startup in Residence (STiR), alongside technologies like Gridscape microgrids, Iteris transportation upgrades, PurpleAir, and ShakeAlert provide important data on designing streets for disaster resilience and efficiency. STiR provides opportunities for public/private collaboration while offering startup companies the opportunity to creatively reduce greenhouse gas emissions, prioritize low-income neighborhoods, and strengthen community ties. With the goal of 100% renewable energy by 2030, San Francisco's startups and creative collaborations push the city to the forefront of resilience in the US. Aggressive climate goals set by San Francisco set an example for the rest of the country in building resilience. As a leader in resilience planning, San Francisco needs to continue innovating, experimenting with safe-to-fail projects, and piloting futuristic technology to bring the region and the country closer to a carbon–neutral future.

Table 13.1 Summary evaluation of smart technology projects in San Francisco Bay Area

	Gridscape Microgrids energy	**Iteris and CAV** Transportation	**ShakeAlert** disaster	**STiR** innovation	**PurpleAir** health
Low carbon (contribution to reducing GHG emissions)	✔	✔			✔
Robustness	✔				
Diversity	✔				
Redundancy	✔				
Connectivity		✔	✔		✔
Flexibility	✔	✔			
Resourcefulness		✔			
Agility (rapid response),	✔	✔			
Efficiency	✔		✔		
Adaptive/ Learning capacity			✔		
Modularity (independence and self-organization)	✔				
Creativity (innovation)		✔		✔	✔
Equity			✔	✔	✔
Inclusiveness (participatory)				✔	
Foresight Capacity			✔		

Smart technologies can help San Francisco achieve its climate goals, even when projects address only a handful of resilience characteristics. Looking at the city as a whole, projects like the five highlighted in this analysis show how each supports a slightly different aspect of resilience in the San Francisco region. This diversity of projects is important for resiliency, as it means the city does not rely too heavily on any single project, partner, or collaboration. The quantity and richness of each project expresses the resilience characteristics of redundancy, diversity, flexibility, and modularity. The city's resources in Silicon Valley allow it access to world-class thinkers, innovators, and risk takers.

As a smart city, San Francisco could lean more heavily on resources from Silicon Valley to create stronger and smarter climate and disaster solutions. Collaborations with startups have been proven successful through programs such as STiR and through Gridscape's microgrid collaboration with the City of Fremont. The city can also benefit from collaborations with private companies like PurpleAir, for instance, to provide residents with real-time data on current air quality measurements. Data acquired through these sorts of partnerships can immediately inform and protect residents' health, especially during this time of COVID-19. This low-cost system could be implemented in school institutions and public buildings to increase data collection especially in low-income neighborhoods. This private company has proven to be effective during times of disaster, with metrics weighing toward low carbon, connectivity, and equity.

With hotter summers and larger impacts from wildfires, the city is in dire need of more creative thinking from startup companies to introduce smart technologies into the state's governmental systems. Our evaluation shows that smart technologies are supporting the San Francisco region's efforts toward resilience and that the most resilient projects are neighborhood-based, as shown in San Francisco microgrids.

Based on our results, smart technologies in San Francisco can support achieve resilience, but there is still room for innovation and improvement. Transportation innovations such as the effort to upgrade street intersections with connected and automated vehicle technologies should be wary of possible decreases to resilience that may surface under this new technology. Keeping the most vulnerable and least carbon intensive users in mind, such as pedestrians and cyclists, will determine the success of such projects.

In addition, San Francisco's utilization of foresight capacity across its smart city programs is debatable. Our results show that ShakeAlert emphasizes foresight capacity, due to its disaster-focused research in urban areas. However, San Francisco has an opportunity to improve its ability to predict and prepare for novel futures and disasters. Such preparation would take into consideration funding from various stakeholders, the urgent realization that San Francisco Bay Area could benefit from a framework that consists of the proposed resilience characteristics.

Acknowledgements This chapter was partially funded by Asia-Pacific Network for Global Change Research (Funder ID: https://doi.org/10.13039/100005536).

References

Airhart E, (2018) San Francisco's dream of 'Zero Waste' lands in the Dumpster. Wired, Conde Nast. Retrieved August 2020, from www.wired.com/story/san-franciscos-dream-of-zero-waste-lands-in-the-dumpster/

Albino V, Berardi U, & Dangelico RM, (2015) Smart cities: Definitions, dimensions, performance, and initiatives. J Urban Technol 22(1):3–21. https://doi.org/10.1080/10630732.2014.942092

Botello B, Buehler R, Hankey S, Mondschein A, & Jiang Z, (2019) Planning for walking and cycling in an autonomous-vehicle future. Transportation Research Interdisciplinary Perspectives, 1. 100012. https://doi.org/10.1016/j.trip.2019.100012

California Energy Commission. (2019). City of Fremont Fire Stations Microgrid Project. Retrieved August 2020, from http://innovation.energy.ca.gov/SearchResultProject.aspx?p=30084

Calma J, (2019, October 17). California's earthquake warning app: how it works and what comes next. Retrieved August 28, 2020, from https://www.theverge.com/2019/10/17/20919639/california-earthquake-early-warning-system-app

Carter LM, Jones JW, Berry L, Burkett V, Murley JF, Obeysekera J, Schramm PJ, & Wear D, (2014) Ch. 17: Southeast and the Caribbean. Climate Change Impacts in the United States: The Third National Climate Assessment. In JM, Melillo Terese TC, Richmond & GW Yohe. (Eds.), U.S. Global Change Research Program 396–417. https://doi.org/10.7930/J0NP22CB.

City Innovate (n.d.). Retrieved August 27, 2020, from https://www.cityinnovate.com

Climate Action Tracker (n.d.). Climate Action Tracker: USA. Retrieved August 2020, from https://climateactiontracker.org/countries/usa/

Climate Plans and Reports (n.d.). Retrieved August 2020, from https://sfenvironment.org/climate-plans-reports

Desouza K, & Flanery T, (2013) Designing, planning, and managing resilient cities: A conceptual framework. Cities 35:89–99. https://doi.org/10.1016/j.cities.2013.06.003

Folke C, Carpenter S, Elmqvist T, Gunderson L, Holling CS, Walker B, Bengtsson J, Berkes F, Colding J, Danell K, Falkenmark M, Gordon L, Kasperson R, Kautsky N, Kinzig A, Levin S, Mäler KG, Moberg F, Ohlsson L, Olsson P, Ostrom E, Reid W, Rockström J, Savenije H, & Svedin U, (2002) Resilience and sustainable development: Building adaptive capacity in a world of transformations. World Summit on Sustainable Development, April 16.

Funk K, & Deininger N, (2018) Five innovative examples of Smart cities in the U.S. Retrieved August, 2020, from https://bipartisanpolicy.org/blog/five-innovative-examples-of-smart-cities-in-the-u-s/

Gore V, (2019) Solar Emergency Microgrids for Fremont Fire Stations. California Energy Commission (CEC-500–2019–054). California Energy Commission. Retrieved from https://ww2.energy.ca.gov/2019publications/CEC-500-2019-054/CEC-500-2019-054.pdf

Gridscape Solutions (n.d.). Renewable Microgrid. Retrieved August 2020, from https://grid-scape.com/renewable-microgrid

Hartman K, (n.d) Intelligent Transportation Systems - CV Pilot Deployment Program. Retrieved May 28, 2021, from https://www.its.dot.gov/pilots/cv_pilot_apps.htm

Holnicki-Szulc J, Motylewski J, & Koakowski P, (2008) Introduction to smart technologies. Smart Technologies for Safety Engineering, 1–9. https://doi.org/10.1002/9780470758595.ch1

Hultman N, et al. (2019) Accelerating America's Pledge: Going all in to build a prosperous, low-carbon economy for the United States. Retrieved August, 2020, https://www.bbhub.io/dotorg/sites/28/2019/12/Accelerating-Americas-Pledge.pdf

Ibrahim M, El-Zaart A, & Adams C, (2018) Smart sustainable cities roadmap: Readiness for transformation towards urban sustainability. Sustainable Cities and Society 37:530–540. https://doi.org/10.1016/j.scs.2017.10.008

Krasny ME, & Tidball K, (2009) Applying a resilience systems framework to urban environmental education. Environmental Education Research, *15*(4), 465–482.

McCann A, (2021) 2021's Most & Least Ethnically Diverse Cities in the U.S. WalletHub. https://wallethub.com/edu/cities-with-the-most-and-least-ethno-racial-and-linguistic-diversity/10264.

Peters A, (2020, August 27) How this small sensor startup became essential to helping California deal with toxic wildfire smoke. Retrieved August 28, 2020, from https://www.fastcompany.com/90543956/how-this-small-sensor-startup-became-essential-to-helping-california-deal-with-toxic-wildfire-smoke

PolicyLink and PERE (2017) An Equity Profile of the Nine-County San Francisco Bay Area Region. The San Francisco Foundation. https://nationalequityatlas.org/sites/default/files/Final_9_County_BayAreaProfile.pdf

Quaintance Z, (2019, January 11) STiR Program Announces 40 Participating Startups for 2019. Retrieved August 26, 2020, from https://www.govtech.com/civic/stir-program-announces-40-participating-startups-for-2019.html

San Francisco Department of Energy Management (n.d.) Our Hazards: Not-So-Everyday Emergencies. Retrieved August 2020, from https://sfdem.org/our-hazards

Sharifi A, (2016) A critical review of selected tools for assessing community resilience. Ecological Indicators, 69:629–647. https://doi.org/10.1016/j.ecolind.2016.05.023

Sharifi A, & Yamagata Y, (2016) Principles and criteria for assessing urban energy resilience: A literature review. Renewable and Sustainable Energy Reviews, 60:1654–1677. https://doi.org/10.1016/j.rser.2016.03.028

Smart City SF (2016) Smart City Challenge: How San Francisco is changing the way we move. Retrieved August 2020, from https://smartcitysf.com

Smart Columbus (n.d.) Retrieved August 2020, from https://smart.columbus.gov/

Stark (2019) This East Bay Energy Startup Is Building Microgrids for California's Fire Stations. Retrieved August 2020, from https://www.greentechmedia.com/articles/read/startup-microgrids-fire-stations

Stark K, Arcuni P, & Brooks J, (2019) Californians Turn to Low-Cost Sensors for Highly Local Air Quality Data. Retrieved August 28, 2020, from https://www.kqed.org/science/1950648/californians-turn-to-low-cost-sensors-for-highly-local-air-quality-info

Stewart L, (2019, February 1) All Particulate Is Local: New Tech Helps Map Community Air Quality. Retrieved August 27, 2020, from https://bayareamonitor.org/article/all-particulate-is-local-new-tech-helps-map-community-air-quality/

The City and County of San Francisco (2020) Hazards and Climate Resilience Plan. https://onesanfrancisco.org/hazard/overview

The United States Conference of Mayors (n.d.) US Mayors climate Protection Center. Retrieved August 2020, from http://www.usmayors.org/programs/mayors-climate-protection-center/

Thurston Charles W, (2019) Fremont, California, Fire Station Is First In US With Solar Microgrid. CleanTechnica. Retrieved August 2020, from https://cleantechnica.com/2019/04/05/fremont-ca-fire-station-is-first-in-us-with-solar-microgrid/.

United States Energy Information Administration (U.S. EIA) (2017) Study of the Potential Energy Consumption Impacts of Connected and Automated Vehicles. US Department of Energy. https://www.eia.gov/analysis/studies/transportation/automated/pdf/automated_vehicles.pdf

United States Environmental Protection Agency (2016) Adapting to Climate Change: Southwest. Retrieved August, 2020, from https://www.epa.gov/sites/production/files/2016-10/documents/southwest-fact-sheet-arcx_update.pdf

United States Geological Survey (USGS) Earthquake Early Warning (2016) ShakeAlert® | Earthquake Early Warning. (n.d.). Retrieved August 28, 2020, from https://www.shakealert.org/

Chapter 14
Data-Sharing Approaches for Achieving Resilient Smart Cities: A Case of Smart City R&D Project in Daegu, South Korea

Yesuel Kim, Sunghee Lee, Ayyoob Sharifi, and Youngchul Kim

Abstract This study explores recent nationwide projects, including those related to smart cities, climate change, urban regeneration, and the K-New Deal, and in particular analyzes how the national smart city R&D project instills resilience in a smart city. This study analyzes a government-funded smart city R&D project in Daegu, South Korea with a focus on three main topics: the effects of the system, the main items that should be considered by planners and decision makers, and ways to ensure participation from diverse groups of citizens. Advanced smart city technologies and services are being adopted as part of the smart city R&D project, such as deep learning-based civil motion recognition, advanced technology for intelligent disaster prediction, and warning technologies for heatwaves, heavy rain, slope collapses, etc. Our analysis of the smart city R&D project according to the analytics framework shows that the Daegu smart city R&D project has sought to consider 15 indexes of resilience and include the three main topics mentioned above. The list of resilience indicators presented in this study can be used as an assessment toolkit that comprehensively considers various parts of the city, such as technology/services, planners/decision makers, and citizens, all of which make up a smart city. This checklist provides a means of evaluating various stages of smart city projects that aim to increase resilience.

This work was supported by Asia-Pacific Network for Global Change Research (CRRP2019-03SY-Sharifi) and the Korea Agency for Infrastructure Technology Advancement (KAIA) grant funded by the Ministry of Land, Infrastructure, and Transport of Korea (21NSPS-B149840-04 and Innovative Talent Education Program for Smart City).

Y. Kim · S. Lee · Y. Kim (✉)
KAIST Urban Design Lab, KAIST Smart City Research Center, Department of Civil and Environmental Engineering, KAIST, Daejeon, South Korea
e-mail: youngchulkim@kaist.ac.kr

Y. Kim
e-mail: yesuel21@kaist.ac.kr

A. Sharifi
Graduate School of Humanities and Social Sciences, Hiroshima University, Higashihiroshima, Japan
e-mail: sharifi@hiroshima-u.ac.jp

A. Sharifi and P. Salehi (eds.), *Resilient Smart Cities*, The Urban Book Series,
https://doi.org/10.1007/978-3-030-95037-8_14

14.1 Introduction

The Korean central government actively promotes nationwide leading initiatives and projects, including those related to smart cities, climate change, urban regeneration, and the K-New Deal. This study explores recent initiatives and projects, and among various efforts, seeks to analyze how the national smart city research & development (R&D) project instills resilience in a smart city. This study consists of three parts: an overview of nationwide smart city initiatives and projects, a discussion of the characteristics of smart city R&D projects, and an analysis of those projects with an eye toward resilience.

This section gives an overview of nationwide smart city initiatives and projects in Korea. Four key initiative-related keywords have become dominant in recent decades: smart city, climate change, urban regeneration, and K-New Deal. Section 14.1.1 briefly demonstrates relevant initiatives, policies, and projects corresponding to the four keywords. Following that, Sect. 14.1.2 will provide a brief overview of the current status of smart city projects in Korea.

14.1.1 Korea's Climate Change-Related Initiatives and Trends in Development

In Korea, Ubiquitous City (U-City) projects have been carried out since the 2000s based on various urban informatization projects. A U-city project is begun on the basis of public initiatives focusing on new cities; the "Ubiquitous City Construction Act" was enacted in 2008 in order to efficiently construct and manage ubiquitous cities. Some U-City projects are led by the Ministry of Land, Infrastructure, and Transport, while local governments and the Ministry of Science, ICT, and Future Planning have expanded upon related projects. In U-City projects, the physical methods of solving urban problems through large-scale infrastructure and resource input have reached their limit. To address this problem, initiatives have changed their focus from building U-Cities by expanding intelligent infrastructure to building smart cities that solve urban problems through efficient use of urban infrastructure. A smart city is generally defined as a city that uses information and communications technology (ICT) such as the Internet of Things (IoT) to solve urban problems, improve the city's competitiveness, and improve the quality of life of its citizens, while also pursuing urban sustainability. The transition to the smart city trend is a revolutionary worldwide phenomenon that has been promoted in various fields such as energy, environment, transportation, and crime prevention in Korea. Today, the concept of a smart city in Korea is no longer a public-led new city project. It has become a key tool for efficient management and improvement of urban problems in existing cities as well as new cities and has become a common goal of Korean cities. In addition, relevant projects have been promoted to further the national interest and address the phenomenon of climate change, which is a global issue. In particular, projects focusing on low-carbon green growth are being implemented. A recent example in Korea is an R&D project called "Development of National Land and Urban Space Planning and Management Technology for Greenhouse Gas Reduction." This R&D project aims to develop a nationwide greenhouse gas planning and management of land use, land-use change, and forestry (LULUCF) system for statistical calculations and reporting and aims to develop land and urban space planning and management technology for greenhouse gas (GHG) reduction.

Urban regeneration is also an important focus of Korea's smart city initiatives. In June 2013, Korea enacted the Special Act on Revitalization and Support of Urban Regeneration (hereinafter referred to as the Special Act on Urban Regeneration) and announced the "Basic Policy for National Urban Regeneration" to promote urban regeneration in a comprehensive, planned and efficient manner. These initiatives promote government-funded regeneration projects. An "urban regeneration project" refers to a creative project that revitalizes a city by introducing or creating new functions in an existing city that is declining due to various causes, such as changes in industrial structure and urban expansion centered on new cities. Urban regeneration projects in Korea have been established with a focus on regeneration plans and solutions to fundamental problems, such as industrial, economic, cultural, and civil society-related aspects of urban decline, in areas that have been overlooked. Given the lack of fundamental consideration for areas in urban decline, it is difficult to effectively cope with the problems that arise in such areas. Therefore, Urban Regeneration New Deal projects have been promoted to support existing urban regeneration projects, improve the living conditions of urban residents, and resolve urban decline, which has become a national problem.

Urban Regeneration New Deal projects in Korea aim to comprehensively regenerate cities by improving the physical environment and enhancing residential capacity in response to urban decline. Such projects focus on improving the environment in older residential areas and restoring urban competitiveness. Some important goals of urban regeneration projects include realizing residential welfare and restoring urban competitiveness, social integration, and job creation. Urban regeneration-related R&D projects are currently being carried out in Korea. One example is "Technology development to analyze urban space risks and improve urban resilience in declining areas," which is related to the Urban Regeneration New Deal. This aims to create a safe environment and promote resilience by developing disaster risk analysis methods in urban spaces in declining areas, along with resilience improvement techniques. Accordingly, this project focuses on risk analysis, disaster prevention, and urban resilience, and encompasses the concepts of "climate change" and "resilient city."

In July 2020, the Korean government held the Korean version of the New Deal National Report Conference at the 7th Emergency Economic Conference, presided over by the President, and announced the "Korean Version of the New Deal Comprehensive Plan." Such a plan was necessary because low growth, intensification of polarization, and the unexpected impact of COVID-19 have led to an unprecedented recession that is the worst economic environment since the Great Depression. The goals of the Korean version of the New Deal (hereinafter referred to as K-New Deal) promoted by the Korean government can be divided into three main categories. First, the K-New Deal intends to reliably overcome economic and social crises and expand the employment and social safety nets to maximize stability in people's lives. In particular, it aims to reinforce efforts to create high-quality jobs that support structural transformation, such as jobs in the digital and green economy, rather than basic low-skilled public jobs. Second, the K-New Deal seeks to make Korea the world's best advanced economic powerhouse through the Digital New Deal and Green New Deal, both of which play to worldwide demands and the strengths of Korea's current workforce. Third, the K-New Deal intends to make Korea a leading country in the effort to counteract the economic and social effects of COVID-19. In particular, it actively promotes the enactment and amendment of laws and regulations to vitalize the digital and green economies, provides employment insurance for the entire country, and establishes a net-zero carbon basis.

According to the main components of the K-New Deal Comprehensive Plan announced by the Korean government, the K-New Deal Comprehensive Plan has three objectives. These are the Digital New Deal, the Green New Deal, and the reinforcement of the safety net. The Digital New Deal aims to advance the strength of information and communication technology (ICT) in Korea, such as e-government infrastructure and services. The Green New Deal aims to accelerate the transition to a green economy with eco-friendly initiatives and low-carbon development. The safety net reinforcement section aims to alleviate unemployment insecurity and income gaps in an era of uncertainty due to economic restructuring. Among 20 Digital and Green New Deal projects, the K-New Deal Comprehensive Plan includes a strategy to select 10 representative projects that create a significant number of jobs and new industries and to use them as the initial center of influence.

With the four initiatives of smart city, climate change, urban regeneration, and K-New Deal (Fig. 14.1), the Korean government has established a foundation to promote new urban developments and improvements. In particular, Korea's government has

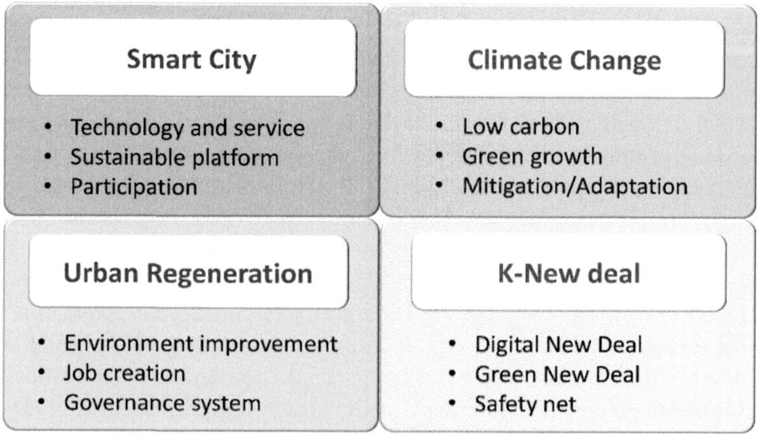

Fig. 14.1 Four nationwide leading initiatives in Korea

continuously recognized the significance of environmentally friendly and low-carbon strategies by implementing nationwide Green New Deal initiatives. In recent years, since the beginning of the Fourth Industrial Revolution, digital and smart technologies have become effective solutions for urban problems. The adoption of advanced digital and smart technologies in Korea has become a dominant strategy in the most recent smart city planning and development, and R&D projects.

14.1.2 *Current Status of Smart City Initiatives in Korea*

The definition of a smart city varies depending on the conditions of each country, but in general, a smart city aims to improve the quality of life of its citizens, improve the sustainability of the city, and foster new industries by utilizing the innovative technologies that are available in this era of the Fourth Industrial Revolution (Sharifi, 2019). It also should have a concrete platform and clear goals. In Korea, strategic projects have continuously been carried out to achieve sustainable smart cities by building urban infrastructure that relies on both construction and information and communication technologies. Our analysis of the latest trends in Korea's Smart City initiatives is as follows.

National Smart City Pilot Projects

In Korea, national smart city pilot projects have been launched to demonstrate and incorporate smart city technologies. These projects are pursued with the goal of presenting a leading model of future smart cities by creating an innovative industrial ecosystem that can realize creative initiatives. Two representative sites are Sejong City's 5–1 district and Busan City's Eco-Delta City (Fig. 14.2).

(1) Sejong 5–1 District

(a) (b)

Fig. 14.2 National smart city pilot projects in Korea. (a) Sejong 5–1 District Smart City; (b) Busan Eco-Delta Smart City (*Source* Smart City Korea)

Fig.14.3 Sejong 5-1 District pilot smart city (*Source* Smart City Korea).

Sejong 5-1 District is creating an artificial intelligence (AI)-based smart city that changes citizens' daily lives through innovations in seven crucial areas, including mobility, health care, education, and energy (location of the district is shown in Fig. 14.3). For example, smart transportation reduces travel time and costs by optimizing traffic flow through AI-based analysis of shared modes of transportation and traffic flow data and introduces various future modes of transportation such as autonomous vehicles. The future city designed by Sejong City pursues post-materialism, which is both human-centered and environmentally friendly at the same time, and decentralization, which respects diversity and promotes citizen participation. Furthermore, it aims to utilize smart technologies that prompt innovation based on data and AI.

(2) Busan Eco-Delta City

Busan Eco-Delta City is creating a state-of-the-art waterfront smart city through 10 innovative elements, such as robots, water, and energy. It is known as a city specialized in water (Fig. 14.4); it applies smart water management technology to the entire process of urban water circulation. The city planners in Busan are building a zero-energy city that achieves 100% energy independence using renewable energy from nature, such as water and solar power. With this plan, Busan City aims to ensure sustainable life for future generations and maximize social public value. In particular, by realizing the "Smart Life, Smart Link, Smart Place" model of future cities, it aims to create high-quality jobs that secure economic sustainability for the country

Fig. 14.4 Busan pilot city—10 areas that are the targets of innovation (*Source* Smart City Korea)

by creating a cluster of five innovative industries (public autonomous innovation, healthcare/robot, hydrothermal energy, water energy science, and VR/AR).

Smart City Innovation Growth R&D

In Korea, government-funded smart city R&D projects are being implemented to realize a data-based smart city innovation model that allows sustainable growth and improves residents' quality of life. The smart city innovation growth R&D projects have three objectives: to develop a smart city model and base technology, to foster solutions to urban issues through the development of smart city services, and to develop a new living lab-based model of sustainable urban development. An R&D project in Korea usually consists of three stages: basic technology development (1st stage), verification of the developed technology (2nd stage), and technology use and stabilization (3rd stage). The smart city R&D project is currently in the second stage. It is largely composed of three core tasks. A brief overview of the three core tasks is as follows (Table 14.1).

Smart Challenge Project

The Smart Challenge is a competition-based public offering project that was introduced in light of the Challenge Project conducted in the United States in 2016. The Korean Smart Challenge Project is a project that reflects the characteristics of the US Smart City Challenge Project and the European Horizon 2020 project. It is divided into Smart City Challenge, Smart Town Challenge, and Smart Solution Challenge according to the size, support method, and type of the project. The Smart City Challenge Project effectively addresses urban problems with creative ideas developed by the private sector and effectively supports the demonstration and diffusion of corporate solutions. The Smart Town Challenge Project is a group of initiatives that discover and integrate solutions for local problems through regional governance operations such as the Living Lab. The Smart Solution Challenge Project develops region-specific solutions, such as approaches that improve citizenship and ameliorate discomfort in small areas in existing cities.

Table 14.1 Overview of key tasks for smart city R&D in Korea.

Task	Project title	Overview
Task 1	Development of smart city model and base technology	– Development of a "data hub model" for the integrated operation and management of information generated through data hubs, massive IoT, digital twins, etc. and lay the foundation for leading the 4th Industrial Revolution – Main Technology: Data Hub, Massive IoT, Digital Twin, Semantic Data, PPP Management
Task 2	Citizen service enhancement (Use Case model)	– Site: Daegu city – Demonstration of the use case type by focusing on the development and verification of services and technologies to solve urban problems (transportation, stability, administration, etc.) that citizens are experiencing discomfort – Main services and technologies: traffic, safety, city administration
Task 3	Technology innovation & Initiatives intelligence (Living Lab model)	– Site: Siheung city – Demonstration of city lab (including many living labs) by combining economic factors for urban growth and regeneration, environmental/energy factors for sustainability – Major services and technologies: environment, energy, living/welfare

(1) Smart City Challenge Project

The Smart City Challenge Project is a competition-based public offering project based on the Challenge Project conducted in the United States in 2016. While the U.S. Challenge Project was limited to the transportation field, the Korean Challenge Project was introduced to solve urban problems in various fields, such as transportation, energy, the environment, and safety. It is expected to present ideas and to demonstrate and spread new smart solutions. Representative projects include the Alive and Vivid Smart City Biz model city, Daejeon City, and the "Solving urban and social problems through social economic models and sharing platforms" project in Bucheon, Gyeonggi-do.

(2) Smart Town Challenge Project

The Smart Town Challenge Project promotes the integration of smart technologies into existing urban areas with an eye toward increasing regional competitiveness. This project aims to discover and introduce smart services optimized for small-scale regional demands, centering on local governments and local living labs. The 2020 Town Challenge was applied by a total of 20 municipalities. Four representative projects are as follows: "Jinhae Marine Park Smart Tourism Town" in Changwon, Gyeongsangnam-do, "Industrial complex safety, transportation and environmental services" in Seosan, Chungnam, "Transportation and safety services connecting universities and communities" in Gangwon, Wonju, and "Circular Eco-Point Platform" in Jeonnam, Gwangyang.

(3) Smart Solution Challenge Project

The Smart Solution Challenge Project supports the introduction and spread of a single smart solution that will directly impact citizens' lives, such as smart solutions for school headquarters, energy production, the environment, and safety. The effort is centered on small living areas, and it aims to solve urban problems and improve citizens' well-being using smart solutions owned by the private sector. The 10 main projects being implemented in 2020 include a "Virus, fine dust blocking solution" in Gangdong, Seoul and a "Smart bus stop" in Wanju, Jeollaabuk-do.

Smart City Urban Regeneration

> "Smart urban regeneration" refers to an effort by the government to integrate smart technology with urban regeneration initiatives. For example, drones are used to monitor roads at night and back roads, and smart parking lots are constructed to provide convenient transportation for residents. In these ways, smart technology is being introduced to areas undergoing urban regeneration. Some representative projects include Sejong City's "Original City Restoration Project" and Incheon Bupyeong's "Building Economic Ecosystem through Urban Regeneration New Deal."

Overall, this section showed that many initiatives are being undertaken in Korea to promote smart city development. Figure 14.5 shows the status of smart city projects that are led by different local governments.

14.2 Materials and Methods

In this study, we applied a smart solution to a specific case (city/project unit) in Korea and assessed whether the smart solution improved resilience of the area according to a variety of metrics. The analysis measured the degree to which the technology contributed to increased resilience. This section introduces the materials used in the case analysis process and discusses the major analytical methods that were applied, then goes on to explain the results of the analysis for a specific case. For the analysis, indexes of resilience characteristics were collected and utilized based on major climate change research in Korea, smart city-related reports, various references, and official portal sites.

When selecting the indicators of resilience, given our emphasis on developing smart solutions that could reduce disasters related to climate change, the following three main topics were judged, and related indicator items were collected: "System effects," "Key considerations for planners and decision makers," and "Ensuring various citizens' participation." We determined that the effects of using smart technology in general and the conditions that planners and decision makers should

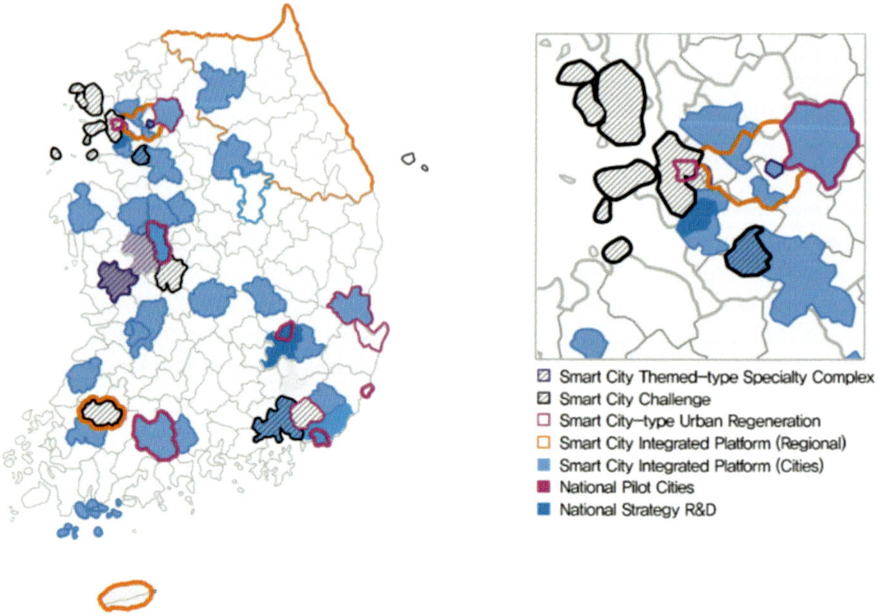

Fig. 14.5 Status of smart city projects by local governments (*Source* Smart City Korea)

consider in the process of applying the smart technology are factors that directly contribute to resilience; thus, these were selected as the main focus. In addition, since Korea's smart cities are subject to a system of cooperative governance in which related ministries, local governments, companies, and citizens all participate, the degree of participation of various citizens was also taken as a major focus. The list of resilience indicators according to the three main focuses is shown in Table 14.2. These indicators are adapted from Sharifi and Yamagata (2016) and detailed explanations of them are given in the Appendix.

As shown in Table 14.2, nine indicators were used to measure the effects of the system, which are low-carbon, robustness, agility, flexibility, efficiency, adaptive/learning capacity, modularity, creativity, and foresight capacity. Four indicators, diversity, redundancy, connectivity, and resourcefulness were identified as major items that planners and decision makers should consider. In terms of the degree of participation of various citizens, a resilience characteristic index list consisting of a total of 15 indicators, including indicators of equity and inclusivity, was created. In Sect. 14.3, a selected case is analyzed in terms of the list of resilience indicators, and the results are described in Sect. 14.4.

Table 14.2 List of resilience indicators

Main focus of index selection	Resilience characteristic indicators
System effect: effects of applying technology	1. Low-carbon (contribution to reducing GHG emissions) 2. Robustness 3. Agility (rapid response) 4. Flexibility 5. Efficiency 6. Adaptive/learning capacity 7. Modularity (independence and self-organization) 8. Creativity (innovation) 9. Foresight Capacity
Planner decision maker's key considerations	10. Diversity 11. Redundancy 12. Connectivity 13. Resourcefulness
Diverse citizen participation	14. Equity 15. Inclusivity (participation)

14.3 Description of the Selected Case

In this section, a target region and case to be analyzed are selected, and a brief introduction to the region and the initiatives is given.

14.3.1 Region Selection: Daegu

Daegu Metropolitan City was selected as the target site for the analysis of the resilience indicators in Sect. 14.2 because of its environmental characteristics and plans for incorporating smart technologies. Daegu Metropolitan City is a medium-sized city with an infrastructure base of 885.56 km^2 and a population of 2.47 million people. It is a good city to demonstrate the use of R&D initiatives. The environmental characteristics of Daegu City were examined. As the outskirts are surrounded by mountains, it has an inland basin-type climate that is cold in winter and hot in summer. There is a large difference in temperature by season, and the average monthly rainfall is much higher in summer than in the other seasons. Due to its basin-type topography, air circulation is poor, and an industrial complex that generates a lot of pollutants is located in the northwest region, so there is a high risk of air pollution in the city due to the main wind direction in winter. Given these environmental characteristics, the region is particularly vulnerable to disasters caused by climate change relative to other cities in Korea, and plans are being made to respond to this issue. In recognition of the importance of local governments in responding and adapting to climate change, Daegu has established 16 core policies in five areas, including setting GHG reduction targets (30% in 2020). In addition, to reduce GHG emissions, the city is promoting increased energy efficiency, eco-friendly means of transportation, use of landfill gas resources, and citizen participation. Daegu City is attempting to incorporate various smart technologies at the city level (Fig. 14.6). In particular, various smart projects are being promoted in consideration of the specific characteristics of Daegu City by region (district). Regarding established smart city services, a smart park has been created, and the safety, convenience, and eco-friendliness of the park have been established. Three main types of implementation strategies are used in Daegu: (1) Advancement of smart city services and creation of businesses, (2) Plans for linking and spreading smart city services, and (3) Enactment of ordinances for the creation and operation of smart cities and the institutionalization of cooperative systems. Daegu plans to advance existing smart city services and infrastructure such as transportation/safety/administration, etc., to develop use cases, and to enhance the status of cooperative systems by organizing the Smart City Initiatives Council.

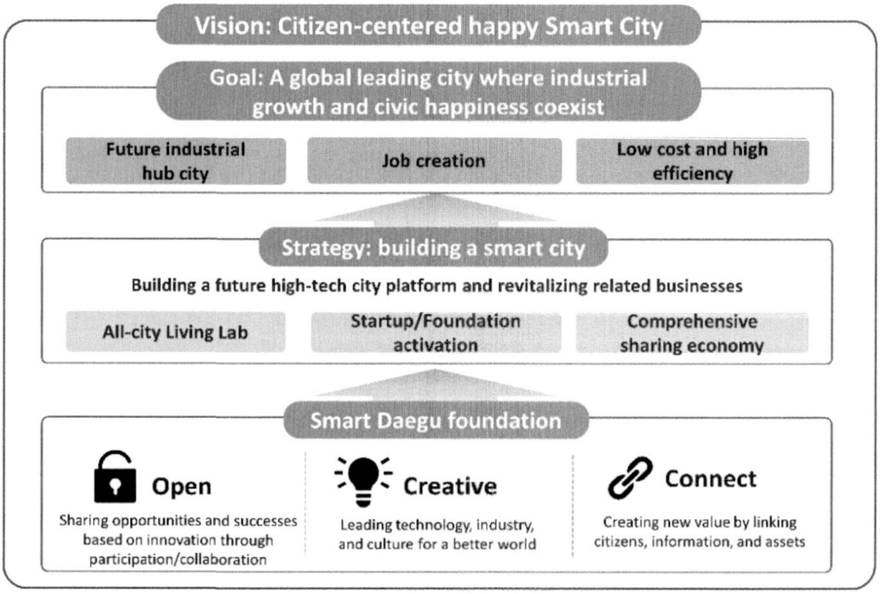

Fig. 14.6 Establishing a smart city in Daegu (*Source* 2030 Daegu Urban Master Plan)

14.3.2 Case Selection: Development of Safety Technology that Protects Urban Residents from Natural Disasters and Urgent Rescue Technology to Ensure Social Safety through the Sharing of Data

One of the smart city R&D projects in Korea which is being demonstrated in Daegu was selected for analysis. As explained earlier, there are three core projects in Korea's smart city R&D projects. We selected the 2nd core project with Daegu City as the region to be demonstrated as an analysis case. The title of the selected project is the "Citizen Service Enhancement (Use Case model)." Here, "Use Case" refers to a prototypical city that focuses on the development and verification of services and technologies to solve problems relevant to citizens in a medium-sized or larger city with an infrastructure base. This core project deals with the content of major services and technologies related to transportation, safety, urban administration, and data hub centers (Fig. 14.7) and is classified into four detailed tasks (2–1 to 2–4). Task 2–1 aims to develop smart mobility and parking space sharing support technology. Task 2–2 aims to develop urban disaster safety and social safety emergency rescue technology through data sharing. Task 2–3 aims to develop advanced technologies

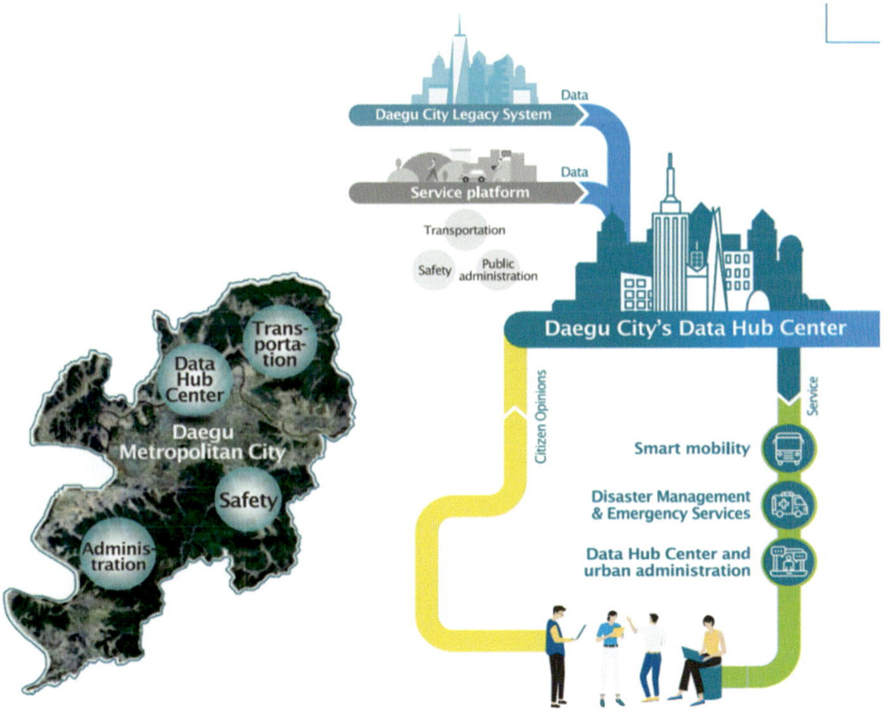

Fig.14.7 Daegu smart city initiatives concept diagram (*Source* Smart City Korea)

for data hub centers and urban administration services. Finally, Task 2–4 aims to develop data-based smart city use cases for civic participation-type urban problem solving.

We selected "Development of safety technology from natural disaster in urban area and urgent rescue technology for social safety through sharing data" which is a Task 2–2 subproject focusing on "disaster" as a detailed analysis case. This project is a demonstration of the use of smart tools to respond to natural disasters (slope collapse, flood, etc.) and social disasters (accidents and fires). It is divided into four stages: development, demonstration, verification, and stabilization of use-case-based technology. Since it deals with major issues related to climate change, it was judged suitable for analysis according to the resilience characteristics in Sect. 14.2.

The case we selected consists of two main objectives: (1) Development of real-time data collection systems for disaster response, real-time prediction/analysis/sharing systems and rescue services, (2) Improvement of rescue response through civic data sharing. The contents of research and development for each are shown in the Table 14.3.

As shown in Table 14.3, Daegu is using citywide data to establish a system to reduce and manage urban problems including various disasters caused by climate

Table 14.3 Research aims and contents of the selected case

Research aims	Research contents
(1) Development of disaster real-time data collection, real-time prediction·analysis·sharing system and rescue services	– Construction of heatwave reduction service – Data filtering and management system for risk factors for slope collapse – Pre-detection of slope collapse and development of disaster signal linkage technology – Detailed location-based (LBS) real-time disaster/accident information provision technology development – Development of urban flood situation awareness system to respond to urban flood disasters and production of flood maps based on converged data – Rainwater pumping station pump/sluice automatic operation system construction – Development of technology to reduce disasters (heat waves, etc.)
(2) Advancement of urgent rescue response through civic data sharing	– Development of safety system for citizen data generation and analysis management based on sensor value – Development of crisis warning level management system and electronic manual to support decision-making – Development of intelligent dispatch order technology using spatial big data – Big data-based data sharing-type safety platform development

change, such as heatwaves, slope collapses, and water-related disasters. In addition, the information in Table 14.3 confirms that promotion of citizen participation and support for decision-making are important; this includes efforts to use citizen data when developing safety systems and to develop a safety platform that relies on data sharing.

The quantitative research objectives of this case are as follows. With the provision of urban flood response services, this project aims to contribute to a reduction of up to 20% of human and property damage caused by flood disasters. In addition, the project intends to contribute to reducing the number of casualties caused by heat waves by 20% by reviewing the expansion of application and commercialization through operation/verification of heat wave reduction services. It aims to contribute to an 80% responder arrival rate at every Daegu accident site (within the first hour after an accident, referred to as the "golden hour") through the creation and analysis of a management safety system based on accident data (falls, traffic accidents, fires, etc.). The project aims to reduce the time it takes for emergency medical services to arrive in vulnerable areas by 10% to achieve that 80% arrival rate through the development of a service module for emergency dispatch instructions. It aims to contribute to that 80% on-site arrival rate within the golden hour by reducing the time it takes for local government officials to make decisions by 40%. As a result, the project aims to contribute up to a 20% reduction in property damage and personal damage caused by the collapse of slopes, water disasters, heat waves, etc., which are disasters caused by climate change in the disaster and disaster safety field, through use case services for each field. In addition, the project aims to contribute to achieving an arrival rate of 80% within the golden hour in the field of social safety emergency rescue.

In this case, the expected benefits of technology utilization were found in all four aspects that were considered: technological, economic, social and environmental, and policy. We analyzed the performance of each technology and service that will be developed through the project according to the two project aims described above. The expected effects of each technology and service according to the research aim (1) in Table 14.3 are explained in Table 14.4.

To summarize the contents of Table 14.4, in general, technology is expected to allow for real-time service on the basis of various technologies including smart technologies, and to increase the speed and efficiency of work. In terms of the economy, we can expect job creation and GDP growth due to a reduction in the number of casualties and the extent of property damage following a disaster because of an increase in the rapidity of response to that disaster, and because of the application of the developed technologies to smart cities. In addition, the expected effects in the social and environmental aspects were minimization of the vulnerable groups and vulnerable areas in the disaster field, and an improvement in citizens' trust through the establishment of a participatory disaster management system. Lastly, in terms of policy, inducing citizens' direct participation through the establishment of a community page related to the development system and the establishment of a natural disaster reduction plan using converged data are expected to have an effect.

Table 14.4 Expected benefits of each technology and service to be developed

Technology and service	Benefits
Heat wave reduction service	- Technical aspect: Provides a base environment for sharing and using information by collecting and storing real-time local weather information and heat waves - Economic aspect: Expectation to create high-quality jobs in various fields through information convergence and utilization through urban data-based big data construction - Social/environmental aspects: Measures vulnerable to heat waves and minimizes vulnerable areas by expanding test-bed construction and applicability review - Policy aspect: Expected to be used to support decision-making on the preferential application of heat-reducing facilities to vulnerable spots
Disaster (heatwave, etc.) reduction device operation management system	- Technical aspect: temperature reduction through installation of mist spraying streetlights combined with smart technology/development of real-time heatwave alarm service through data collection - Economic aspect: budget reduction effect and maximization of budget utilization through job creation and cost-effective identification of adaptation countermeasure technology – Social and environmental aspects: efficient energy use through sensor and data sharing technology – Policy aspect: The disaster management agency expects efficient urban and disaster management in terms of local government administration through prompt information acquisition
Slope collapse prediction and warning service	– Technical aspect: Increased the speed and efficiency of disaster prevention work by establishing a participatory disaster management system, improving public confidence – Economic aspect: Reduce casualties and property damage through linkage with location-based services (LBS) – Social/Environmental Aspects: Increased the speed and efficiency of disaster prevention work by establishing a participatory disaster management system, improving public confidence – Policy aspect: Securing safety time according to improvement of national and local government slope management laws considering the introduction of real-time data collection system

(continued)

Table 14.4 (continued)

Technology and service	Benefits
Detailed location-based (LBS) real-time disaster/accident information service	– Technical aspect: real-time user location-based disaster/accident response message delivery service construction – Economics: Minimizes damage to life and property through prompt delivery of customized disaster accident messages – Social and environmental aspects: Improve the reliability of disaster messages by delivering customized disaster messages in real time based on user location – Policy aspect: Prepare a system for delivering alerts from various media by linking the alert systems used by local governments (civil defense, disaster alerts, etc.)
IoT-based rainwater pumping station automatic operation system	– Technical aspect: Demonstration of water level prediction technology and automatic operation technology – Economic aspect: Business and job creation in related fields by applying smart city development technology – Social and environmental aspects: Developing techniques and systems that can integrate and utilize individual element technologies accumulated in existing domestic research – Policy aspect: Induce citizens' direct participation and improve the environment by configuring a system-related community page
Urban flood response service	– Technical aspect: Proactive flood response by developing a system capable of prompt flood forecasting and warning issuance – Economic aspect: GDP growth by implementing a smart city that is organically linked with urban flood management infrastructure – Social and environmental aspects: improving citizens' ability to respond to floods, securing social safety nets and realizing digital social welfare – Policy aspect: Establishment of natural disaster reduction plan using fusion data-based flood map

Next, the details of the expected effects for each technology and service according to research aim (2) in Table 14.3 are explained in Table 14.5.

To summarize the contents of Table 14.5, in terms of technology, establishing a safety platform and applying it to the Living Lab project can be expected to occur via improving citizens' situational awareness using citizen-sourced data. In terms of economics, the effort is expected to revitalize the use case-type smart service development industry and reduce the social costs associated with damage recovery in the

Table 14.5 Expected effect by technology and service

Technology and service	Benefits
Sensor value-based citizen data generation and safety analysis management system	– Technical aspect: Establish a safety platform by increasing the situation awareness rate using accumulated citizen sensor data – Economic aspect: revitalization of the use case-type smart service development industry using new devices – Social/Environmental Aspect: Increased individual's satisfaction with city life by ensuring safety in everyday life space – Policy aspect: Securing the willingness to spread local governments by presenting safety service scenarios in the city
Development of intelligent dispatch instruction technology using spatial big data	– Technical aspect: Applicable to Living Lab project for realization of new use case in the future by reducing emergency rescue response time through derivation of optimal route – Economical: Improved reliability and accuracy through continuous real-time traffic information data collection by building a platform – Social and environmental aspects: improving the quality of life and establishing a safe society by reducing response time for emergency rescue services, mitigating regional imbalances – Policy aspect: Systematic supplementation of operation management aspect when continuing management of traffic information data through smart city safety platform
Crisis warning level management system	– Technical aspect: The development system automatically determines the level of crisis warning and disseminates it to the relevant person in charge, contributing to quick decision-making and more efficient disaster response – Economic aspect: Expanded application to other local governments to create a market for automated systems in the field of disaster response – Social and environmental aspects: To improve the quality of life of the people, such as increasing public confidence in the disaster response process and reducing damage – Policy aspect: Expect efficient city management and disaster management in terms of local government administration

(continued)

Table 14.5 (continued)

Technology and service	Benefits
Safety platform and mobile app SW	– Technical aspect: Real-time operation management system and safety platform linkage technology development, living lab-oriented technology development – Economic aspect: Contributing to reduction of rescue time, reducing social costs for recovery from damage in the safety field – Social and environmental aspects: Equal service provision to all citizens, resolving inequality among service beneficiaries – Policy aspect: Enhancing civic services in the safety field by establishing a system that allows citizens to voluntarily participate

safety field. In addition, the expected effects in terms of social and environmental aspects include improving the quality of life of the residents, easing regional imbalances, and resolving beneficiary inequality. Finally, in terms of policy, the system is expected to improve civic services in the field of safety and the efficiency of disaster management by establishing a system that allows citizens to voluntarily participate, such as an SNS (social network service).

In conclusion, as summarized in Tables 14.4 and 14.5, this analysis case is expected to have positive effects in terms of technical, economic, social, environmental, and policy aspects in the areas of disaster and social safety through the development of various technologies and services. In particular, it is expected to contribute to the formation of a sustainable city by improving the resilience of Daegu City, a prototype smart city, through integration of various smart technologies.

14.4 Results

This section describes the evaluation of the cases in Sect. 14.3 based on the list of indicators of resilience in Sect. 14.2 (see Fig. 14.8). To help understand the characteristics of resilience, the 15 indicators are defined in terms of smart cities. In addition, in order to evaluate how much a particular smart city solution contributed to an increase in resilience, a detailed description of the results of the analysis for each item is given.

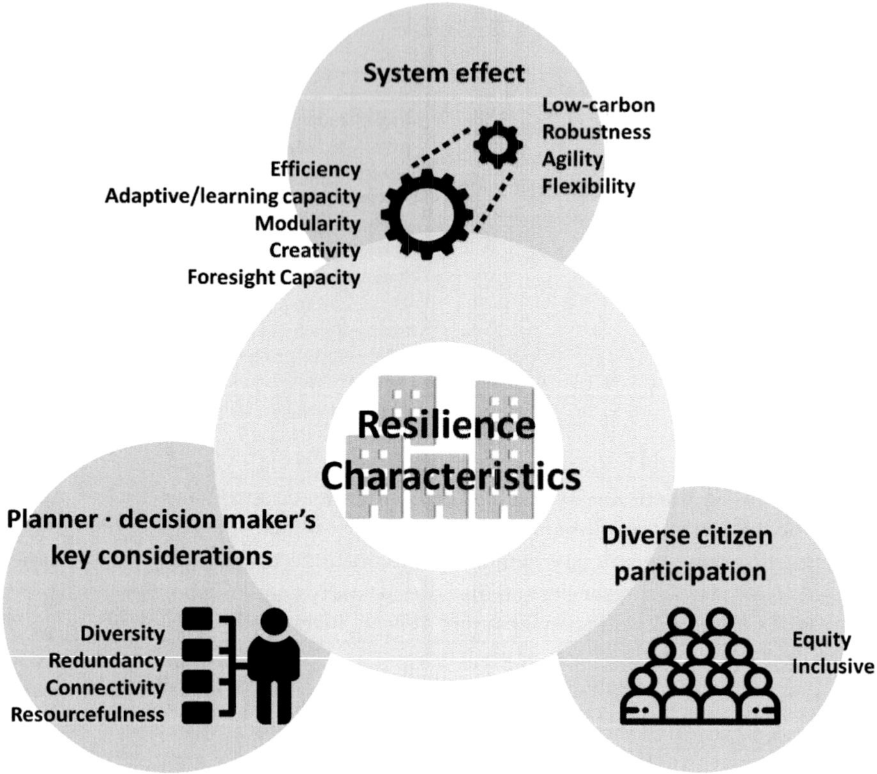

Fig. 14.8 Resilience-related characteristics of a smart city

14.4.1 Low-carbon

Low-carbon emissions can generally be realized through the development of carbon reduction devices and energy-efficient systems. The case in Sect. 14.3 aims to develop various technologies and services based on the use of city data and sharing of citizen data. Grafting of the developed technology contributes to a reduction in disasters and improves citizens' reliability in the disaster process. It is possible to provide a reliable analysis and to efficiently use information by applying the developed technologies and services to test areas and establishing a linkage system between various projects. In particular, systems developed by grafting smart technology improve the speed and accuracy of work, minimize resource waste, and increase energy efficiency. Due to these effects, a carbon reduction effect can be expected, which contributes to resilience.

14.4.2 Robustness

Robustness refers to the strength of a system that can withstand sudden external shocks without serious damage or deterioration of major systems. From the perspective of a smart city responding to climate change, robustness is obtained by establishing a smart city platform and a systematic management system. In the case in Sect. 14.3, the city is aiming to improve global technology systems such as real-time IoT and massive IoT networks by developing technology for collecting city-related and citizen data using smart city solutions. Therefore, it is possible to increase the speed and accuracy of disaster recognition when a dangerous situation occurs and to create a more robust safety platform by using data on various causes of disasters that is accumulated through the standardized linkage system of the smart city. Therefore, an improvement in robustness can be expected when smart city technologies are implemented; given such positive effects, the use of smart solutions contributes to resilience in the analysis case.

14.4.3 Diversity

Diversity, from a smart city perspective, refers to the inclusion of various options (resources, tools) to preserve one unique function in a particular system. The case analysis in Sect. 14.3 includes the establishment of a data sharing-type safety platform through efficient linkage of various smart city technologies. This makes it possible to minimize the effect of disasters in disaster-prone areas in cities, such as areas at risk of collapse and areas that experience frequent flooding. In particular, various data, technologies, and services aimed at predicting and responding to disasters that constitute the safety platform improve the efficiency of business-related activities. We can conclude that the analysis case in Sect. 14.3 contributes to diversity among the indexes of resilience.

14.4.4 Redundancy

Redundancy means that, in the systems that make up a smart city, there are multiple elements with similar functions and elements that can be replaced if they malfunction. In most smart cities, a service management system with excellent durability is possible through convergence and integration of various smart solutions. In particular, the analysis in Sect. 14.3 connects the operations of different disaster management systems by combining various smart solutions. Therefore, it is possible to expand applications between systems. Thus, we can conclude that the pattern of use of smart technologies in Daegu contributes to redundancy, which is expected to improve the resilience of the city in terms of climate change.

14.4.5 Connectivity

Connectivity enables connection and coordination of activities between systems to improve the efficiency of system operation from a smart city perspective. Smart cities with good connectivity are capable of more accurate and efficient resource utilization in terms of disaster preparedness. In the analysis case in this study, the linkage between technologies and services developed by combining various smart solutions is considered. The system is also designed to enable organic sharing between systems using city and citizen data. As a detailed development example, there is "advanced technology for emergency rescue response through sharing citizen data," an initiative that aims to provide accident-related information in connection with a central control system such as a firefighting headquarters in a safety platform or data hub. The accident-related information may be linked with dispatch support technology and accident site information sharing technology within the safety platform. These characteristics contribute to rapid decision-making and efficient response by providing and utilizing effective (functional) information in the event of a disaster scenario. Therefore, we can conclude that this case reflects the connectivity characteristic due to its technical benefits.

14.4.6 Flexibility

Flexibility refers to the ability of a system to adapt to changing conditions. A flexible system can immediately detect risks in the event of a disaster and make immediate changes in small subsystems to maintain overall system performance without disruption. The analysis case is capable of linking independent safety field element technologies based on the platform. In the event of a disaster, some of the various field element technologies, including smart technologies that are configured in the platform, immediately detect the risk. In addition, system expansion or service combination is expected through cross-connectivity between technologies. This leads to efficient system linkage so that it is possible to adapt to changing conditions. In particular, the developed technologies and services can be expected to be applied to the Living Lab project to realize new use cases in the future. Furthermore, it contributes to technological competitiveness for the expansion of smart cities in Korea as well as other countries. Customized application of smart technologies is possible according to the needs of various projects and environments. Therefore, we can conclude that the smart solution contributed to an improvement of resilience of the case in Sect. 14.3

14.4.7 Resourcefulness

Resourcefulness is a term relating to the adequacy of resources that urban planners and decision makers can use to identify, respond to, and recover from potential disasters. Resources, from a smart city perspective, include the ability to understand the status quo and patterns in the data, as well as the ability to identify potentially risky situations. These characteristics are representative items that planners and decision makers should consider in the process of applying smart technology. In the case analysis in Sect. 14.3, a new service is provided through the establishment of a database for various disasters. This contributes to reducing communication costs and establishing a country-centered data collection system by establishing a self-network via linkage of massive IoT network technologies when developing heatwave reduction technology. By combining these advanced technologies, the collection database can be expected to provide basic data that will be useful in the decision-making process. Therefore, we can conclude that the research contents of the analysis case contributed to the improvement of resilience.

14.4.8 Agility

Agility refers to the rate of response and recovery following a disaster in a smart city. Agility refers to the ability of a system to mobilize necessary resources and return to normal functions within an acceptable time frame in the event of a disaster. In the analysis of the case in Sect. 14.3, it is possible to switch to an online-clockwise system by developing a smart solution-based disaster risk factor measurement technology and establishing a data collection system. Disaster management according to the online-constant clock system of such risk factors can minimize the damage caused by the disaster through rapid situational awareness and good judgment. In addition, the system is expected to reduce rescue time and damage recovery costs in the safety field by establishing a base technology that improves the accuracy of judgment of accidents. Therefore, the analysis case contributed to the improvement in resilience.

14.4.9 Efficiency

Efficiency prevents resource waste in order to improve the productivity of resource use and the ratio of energy input to resources used. The case in Sect. 14.3 contains information on the development of forecasting and warning technologies and services for various disasters (heatwaves, heavy rainfall, slope collapses, etc.) caused by climate change. In the process of developing these technologies, various smart solutions, such as machine learning and deep learning, are applied. In particular, use of self-learning technologies reduces rescue time by establishing a base technology that improves the accuracy of smartphone pattern analysis and accident judgment; as a result, it can be expected to reduce social costs for damage recovery in the safety field. By synthesizing these contents, we can conclude that the analysis case contributed to the improvement in resilience.

14.4.10 Adaptive/learning capacity

Adaptive/learning capacity can be interpreted as the ability to adapt to disaster vulnerability and the ability of urban systems to learn from previous disasters. The analysis case in Sect. 14.3 includes the implementation of pattern recognition-based technology through self-learning of various sensor data to build a safety platform. To this end, various types of wireless communication technologies and data communication linkage technologies are used. Specific examples based on pattern recognition include "Deep Learning-based Citizen's Motion Recognition Advanced Technology" and the "Accident Signal Pattern Analysis Algorithm." The safety platform built with these smart and progressive solutions supports the provision of reliable services to citizens. Considering that the platform was built on the basis of this learning ability, we can conclude that this analysis case reflects the adaptive/learning capacity characteristic on the list of resilience characteristics.

14.4.11 Modularity

Modularity is generally interpreted to mean independence and self-organization. From a smart city point of view, modularity refers to the system's ability to be self-reliant; in order be modular, a system should include reinforced regional community-based management platforms that can respond independently to disasters without relying on a centralized system. In the case in Sect. 14.3, the stage of a situation is established according to disaster type and organizational system management system. In detail, the plan includes contextual information management and radio wave management module development. As a result, system expansion or integration of services becomes possible due to the establishment of an organizational system management system based on such module development. This is expected to increase work efficiency and reduce unnecessary social costs. As another example, the contents of building independent safety field element technologies include a platform-based linkage system. This supports flexible combination of services as needed in various scenarios when a disaster occurs. Based on this analysis, the case in Sect. 14.3 reflects the characteristics of modularity.

14.4.12 Creativity

From a smart city perspective, creativity entails innovation in the management, planning, and design of urban systems in order to reach a more advanced state. In smart cities, innovation that improves system resilience is essential so the system is not overwhelmed by diverse and constantly changing risks. In the disaster field in smart cities, various environmental variables occur and non-linear patterns are involved in the occurrence and response to disasters, so it is important to continuously update the latest information in the system. Innovative technologies are used in this process. In the case in Sect. 14.3, data was extracted using GIS-based information collection software. In addition, the urban system is innovatively planned and managed by equipping the smart city urban disaster situation awareness system with technology capable of automatic extraction of urban information and disaster prediction techniques. Therefore, this case reflects the creativity characteristic.

14.4.13 Equity

Equity plays an essential role in achieving resilience from a smart city perspective. In particular, it is essential to ensure that all citizens of a city can use each service according to their needs and to ensure the participation of the poor and marginalized classes in society. According to the case in Sect. 14.3, smart technologies mitigate inequality for service beneficiaries by providing the same services to all citizens without being limited to specific affiliations/organizations; technologies that are important for equity include mobile app SW in safety fields such as safety platforms. Therefore, we can conclude that this case reflects the equity characteristic due to the effects of this social aspect.

14.4.14 Inclusivity

From a smart city perspective, inclusivity means a participatory structure system in which various decision makers participate in the planning and decision-making process. Inclusivity has a positive effect on improving the adaptability of smart cities. The case in Sect. 14.3 contributes to the enhancement of participatory civic services by establishing a system that allows citizens to voluntarily participate through the development of smart services such as safety platform and mobile app software. In addition, participation in civic administration was encouraged by introducing IoT Living Lab and citizen data. Furthermore, by including a community page in the developed system homepage, direct participation by citizens and environmental improvements can be expected. Accordingly, this case reflects the characteristics of inclusivity, a social factor.

14.4.15 Foresight Capacity

From a smart city perspective, foresight capacity refers to the ability to predict future risks. This characteristic enables prediction of uncertainty and non-linear impact behavior through the application of smart city solutions. In the case analyzed, smart technologies developed in the 4th industrial revolution such as machine learning methods and hybrid numerical models were introduced to predict the risk of natural disasters (heatwaves due to climate change, heavy rain, slope collapse, etc.), and these technologies are used to provide a highly accurate disaster warning service. In addition, foresight capacity improves the trust among citizens in disaster scenarios by delivering customized disaster-related messages based on user location through a detailed location-based (LBS) real-time disaster/accident information service. As a result, we can conclude that foresight capacity contributes to the resilience of the system in terms of climate change response in the case in Sect. 14.3.

14.5 Discussion and Conclusions

In Korea, various social and environmental problems such as fine dust, traffic congestion, water shortages, and natural disasters have emerged due to climate change and rapid urbanization. Since these problems pose a serious threat to urban sustainability, government-level efforts have been made to address them. Recently, the Korean government has been attempting to create and spread smart cities with visions of improving citizens' quality of life, improving urban sustainability, and cultivating new industries by utilizing innovative technologies in the era of the 4th Industrial Revolution. The government has continuously implemented active policy supports, such as expanding financial investment and drastic regulatory improvements. On the basis of these efforts, many local governments are promoting the construction of smart cities. Each local government that promotes smart city projects in Korea is spurred to create a smart city that fits the unique characteristics of the region. In addition, the Korean government is promoting the establishment of various cooperative governance systems. In order to foster new industries of the 4th Industrial Revolution, various cooperative governance systems have been implemented, in which ministries, local governments, companies, and citizens participate based on the smart city platform.

A trend analysis and case study of developed businesses was conducted given the continuous national interest in smart cities. Analyzing the development trend of smart cities in Korea in Sect. 14.1 showed that Korea has continuously made efforts to solve urban problems related to climate change disasters by developing advanced technologies and services and implementing various projects. In particular, concepts such as the Green New Deal, which focus on eco-friendliness and low-carbon emissions, have been continuously developed in major national projects. Thus, Korea has an advantage in that it is rich in the infrastructure necessary for disaster adaptation and response. As a result, implementing smart city technologies that are organically linked up with the nation's robust infrastructure was expected to produce positive effects in terms of technological, economic, social, and environmental factors as well as policy.

Section 14.2 shows a list of indicators of positive resilience that can be improved upon with smart solutions, according to three main focuses. The derived list is composed of various characteristics related to smart cities, which range from characteristics related to system performance, such as low-carbon output and robustness, to indicators related to citizen participation, such as equity and inclusivity (participation). A case study was used to analyze a prototypical Korean smart city project using the list of resilience indicators.

Section 14.3 analyzes the contents of "Development of safety technology from natural disaster in urban area and urgent rescue technology for social safety through sharing data," which is a climate change disaster-related Korean smart city R&D project. Section 14.4 describes the results of that analysis. According to the results, all 15 characteristics were improved due to the incorporation of smart and progressive solutions.

The list of resilience indicators could be understood as a checklist for evaluating the overall benefits of the technologies that compose a smart city, the aspects that are important to planners and decision makers, and the aspects related to citizen participation. Since the 15 characteristics are not limited to specific stages of a smart city project, but instead relate to various stages from planning to management, it is expected that the list can be used to evaluate different stages of various smart city projects in the future. Korea will continue to develop smart solutions. Looking beyond the limited development in Korea, the Korean government is building a global network of smart cities in collaboration with countries around the world to usher in the future of sustainable smart cities. Such efforts involve sharing of the policies and experiences developed in each country and include a variety of projects, ranging from person-to-person exchanges to joint project discovery, and practical cooperation, such as demonstration of solutions. Therefore, it is expected that global knowledge and technology/service sharing will be facilitated, as will efficient cooperation between countries, which will positively contribute to resilience on a global scale.

Acknowledgements This chapter was partially funded by Asia-Pacific Network for Global Change Research (Funder ID: https://doi.org/10.13039/100005536) and supported by the Korea Agency for Infrastructure Technology Advancement (KAIA) grant funded by the Ministry of Land, Infrastructure and Transport of Korea (22NSPS-C149830-05).

Appendix

Definitions of resilience indicators (Adapted from Sharifi and Yamagata [2016])

Robustness	Robustness refers to a system's strength to withstand short-term(sudden), acute internal and external shocks without suffering from major degradation of the main functions. To achieve this and enhance system security, the system needs to have the ability to counteract and/or absorb the disturbance
Diversity	Refers to the degree to which multiple distinct functions, that can be used simultaneously, are included in the system. The aim of this principle is to hedge against supply disruptions and ensure that a variety of options (resources, instruments, etc.) for dealing with disturbances and ensuring functionality exist in an urban system
Redundancy	Redundant capacity refers to the availability of (substitutable) components with similar (even overlapping) functions in the urban system to enhance its adaptive capacity and ability to absorb shocks, give it reserve capacity for problem solving, and ensure that uncertain events causing the failure or displacement of one component would not result in the failure of the whole system. In a system featuring redundant capacity, exclusion of an element should not result in significant loss of functioning
Connectivity	An urban system includes multiple subsystems. A resilient system should be capable of establishing connections between those subsystems and coordinate their activities in order to enhance effectiveness and efficiency of operations. Without this capacity the existing resources would not be effectively utilized to prepare for the disaster, the system will not be able to achieve its full absorptive capacity, and consequently there would be procrastination in the recovery efforts
Flexibility	Flexibility means that a system should have the ability to "adapt to changing conditions" and undergo a safe failure by changing its configuration. A flexible system is capable of sensing threats, immediately detecting the failure and making prompt changes at smaller scales of its subsystems and thereby maintain overall performance during disaster. In the context of energy systems, this could (for example) refer to the ability to shift between different energy configurations or adjust regulations or prices according to changing conditions
Resourcefulness	Relates to the adequacy of resources at the disposal of urban planners and decision makers to appropriately identify, prepare for, respond, and recover from potential disruptions. This includes having appropriate capacity to understand status quo and identify patterns, potential threats, and contingencies

(continued)

(continued)

Agility (rapid response),	Represents the system's capacity to mobilize the resources necessary for recovery and return to normal functioning within an acceptable time frame. Agility is essential for avoiding cascading failures that can result in the disruption of other functions in the system
Efficiency	Means that the proportion of energy and resources provided by an urban system to the resources given to it as input should be positive to improve resource use productivity and avoid waste
Adaptive/learning capacity	Refers to an urban system's capacity to learn from the disaster to reduce its pre-disturbance vulnerabilities and enhance its capacity to adapt to the changing conditions. Adaptability implies recognition of the inherent vulnerability of the system components, availability of appropriate knowledge and assignment of authority to prioritize tasks at the time of crisis, and ability to respond with rapidity in order to facilitate a "safe-to-fail" (or at least "soft-fail") urban system. A resilient urban system should entail "adaptive" cycles that "alternate between long periods of aggregation and transformation of resources and shorter periods that create opportunities for innovation," thereby ensuring survivability of the system
Modularity (independence and self-organization),	A resilient system should possess a "certain degree of self-reliance that gives it the ability to maintain a minimum acceptable level of functioning (without external support) when influenced by disturbance." A self-organized system discourages centralization of resources and authorities and should involve community-based management characterized by strengthened local communities capable of independently responding to disaster, cross-scale partnerships, and "horizontal" and "vertical" institutional connections that provide direct feedback to the system and enable better informed decision-making. Furthermore, it should entail the ability to build upon and strengthen networks established to respond to an earlier disturbance
Creativity (innovation)	This principle represents the "urban system's ability to use the disruption as an opportunity to attain a more advanced state." This requires utilizing innovation (both technological and non-technological) in management, planning, and design of urban system. Innovation is essential to enhance various resilience abilities and avoid being overwhelmed by the constantly changing nature of risks

(continued)

(continued)

Equity	Equity plays an essential role for achievement of resilience. This is to ensure that all urban citizens have the ability to utilize services to prepare/plan for, cope with and recover from disruptions. Also, justice is needed in terms of exposure to adverse impacts. This is to ensure that marginalized and poor people do not bear the brunt of those impacts
Inclusive (participatory)	Engagement of various stakeholders in planning and decision-making processes enhances social capital and improves planning, absorption, recovery, and adaptation capacities
Foresight Capacity	Any resilient system must be able to face the uncertainty and relativity of the future conditions. The concept of disaster is entangled with uncertainty and nonlinearity of the impacts and behaviors of a portfolio of endogenous and exogenous forces that can potentially become sources of disturbance in the system. This principle is essential for disaster preparation and also absorption of initial shocks. It implies that only preparation based on shortcomings exposed by past events is not enough and forecasting methods should also be applied in preparation to respond to newer risks that may unfold in the future

References

2014 Urban Regeneration Information System. Urban Regeneration New Deal. (2018, september 20). https://www.city.go.kr/portal/policyInfo/newDeal/contents03/link.do

Daegu Metropolitan City (2018) 2030 Urban Master Plan, Daegu. http://www.daegu.go.kr/build/index.do?menu_id=00933155

Korea Agency for Infrastructure Technology Advancements (2020~2024) Development of green-house gas (GHG) estimation and space planning & management technology to reduce GHG within the settlements in LULUCF sector

Korea Agency for Infrastructure Technology Advancements (2019~2022) Technology development to analyze urban space risks and improve urban resilience in declining areas.

Korea Agency for Infrastructure Technology Advancements (2018~2020) Development of safety technology from natural disaster in urban area and urgent rescue technology for social safety sharing data.

Ministry of Land, Infrastructure and Transport (2020) Smart City Innovation Growth Engine R & D. Retrieved From the Smart City Korea. https://smartcity.go.kr/

Ministry of the Interior and Safety (2020) Korean New Deal Comprehensive Plan. https://www.gov.kr/portal/ntnadmNews/2207711

Presidential Commission on Architecture Policy (2016) Research on policy measures to strengthen Smart City competitiveness. http://www.ndsl.kr/ndsl/commons/util/ndslOriginalView.do?dbt=TRKO&cn=TRKO201900001113&rn=&url=&pageCode=PG18

Sharifi A, & Yamagata Y, (2016) Principles and criteria for assessing urban energy resilience: A literature review. Renewable and Sustainable Energy Reviews, 60:654–1677. https://doi.org/10.1016/j.rser.2016.03.028

Sharifi A, (2019) A critical review of selected smart city assessment tools and indicator sets. J Cleaner Prod, 233:1269–1283. https://doi.org/10.1016/j.jclepro.2019.06.172

Shin W, et al. (2015) Comparative analysis research on the difference between U-City and smart city for the establishment of international competitiveness of U-City - By a comparative analysis of smart cities index indicators and U-City plans established by the local governments in Korea. J Urban Des Inst of Korea, 16(5):5–16.

Chapter 15
Urban Resilience in the Fourth Industrial Revolution: Transformative Digitalization in European Smart Cities to Address Climate Change

Abdul-Lateef Balogun, Himanshu Shekhar, Paulina Budryte, Olasunkanmi Habeeb Okunola, Teslim Abdul-Kareem, Ismaila Rimi Abubakar, Yusuf A. Aina, Abdulwaheed Tella, and Shamsudeen T. Yekeen

Abstract Climate change has been and is still affecting every region in Europe, with varying impacts across the continent. While some cities are generally resilient to CC impacts, other cities are not necessarily as fortunate. Promoting policies that build resilience enhances cities' capabilities to cope with acute shocks and chronic stresses, adapt well to changing climate conditions, and ultimately transform to resilient cities. Conversely, the weak or absence of urban resilience increases the vulnerability of the urban poor to risks. It is therefore imperative to rapidly enhance urban resilience practices. To date, only 26% of 885 cities in Europe have viable adaptation plans, highlighting an imbalance in the adaptation and resilience progression across countries. Although digital transformation through increased data availability and the use of digitalization instruments has the potential to improve the rate of achieving the adaptation strategies, not much has been documented in this regard. Thus, this chapter examines the potentials of digitalization in accelerating adaptation and boosting resilience in selected European cities. Case studies are analysed through a systematic literature search, and evidence of fruitful cases are presented. We conclude by discussing some challenges of digitalization and make recommendations for future works.

A.-L. Balogun (✉) · A. Tella · S. T. Yekeen
Geospatial Analysis and Modelling (GAM) Research Laboratory, Department of Civil & Environmental Engineering, Universiti Teknologi PETRONAS (UTP), 32610 Seri Iskandar, Perak, Malaysia

H. Shekhar
Institute for Environment and Human Security (UNU-EHS), United Nations University, UN Campus, Platz der Vereinten Nationen 1, 53113 Bonn, Germany
e-mail: shekhar@ehs.unu.edu

P. Budryte
Faculty of Social Sciences, Arts and Humanities, Kaunas University of Technology, A. Mickevicius st. 37, LT-44244 Kaunas, Lithuania
e-mail: paulina.budryte@ktu.lt

A. Sharifi and P. Salehi (eds.), *Resilient Smart Cities*, The Urban Book Series,
https://doi.org/10.1007/978-3-030-95037-8_15

Keywords Climate Change · Digitalization · Europe · Transformative adaptation · Urban resilience

15.1 Introduction

Globally, many cities face myriad challenges, such as rapid urbanization, political crises, financial challenges, climate change, energy crises, food insecurity, terrorism, and rising inequality that potentially serves as major threats to the resilience and cohesiveness of urban areas and the people who work and live there (Bush and Doyon 2019; McPhearson et al. 2015). In an increasingly interconnected and globalized world, the advent of the deadly COVID-19 pandemic has also compounded the problems facing urban areas all over the world (Sharifi and Khavarian-Garmsir 2020; Sheller 2020). Consequently, lives and livelihoods in urban communities, especially informal settlements, are at risk as a result of limited coping, adaptive, and transformative capacities (Okunola 2019; Satterthwaite et al. 2020). This necessitates the need to respond more quickly and more effectively for long-term urban resilience and management interventions.

Over the years, several studies have posited that the most complimentary and viable solution to preventing and managing the plethora of risks facing cities of the world lies in understanding, supporting, and developing urban resilience (Folke 2006; Meerow and Newell 2015). Urban resilience in this context refers to the capacity of urban structure and its socio-technical and socio-ecological networks across spatiotemporal scales to swiftly revert to or maintain its desired status despite

O. H. Okunola
Federal University of Technology, Institute for Land and Community Resilience, Minna, Nigeria

Global Change Institute, University of Witwatersrand, Johannesburg, South Africa

T. Abdul-Kareem
Georgia Department of Natural Resources, Environmental Protection Division, Atlanta, GA, USA
e-mail: teslim.abdul-kareem@dnr.ga.gov

I. R. Abubakar
College of Architecture and Planning, Imam Abdulrahman Bin Faisal University, P.O. Box 1982, Dammam 31441, Saudi Arabia
e-mail: irabubakar@iau.edu.sa

Y. A. Aina
Department of Geomatics Engineering Technology, Yanbu Industrial College, Yanbu, Saudi Arabia
e-mail: ainay@rcyci.edu.sa

A. Tella
Earth, Environment and Space Division, Foresight Institute of Research and Translation, Ibadan, Nigeria

S. T. Yekeen
Department of Geography, Environment, and Geomatics, University of Guelph, Guelph, ON, Canada

disorder, to adapt and rapidly transform urban systems that inhibit short or long-term adaptive capacity (Simone et al. 2021; Woodruff et al. 2018). Implementation of urban resilience is seen by stakeholders across the world as a viable approach to manage multiple threats such as ageing infrastructure, equity concerns, economic instability, extreme weather, and shifting climate (Woodruff et al. 2018). Urban resilience, therefore, offers an optimal approach to overcome different contextual and developmental challenges by using available internal and external resources. It involves an effective and coherent level of coordination across and between multiple levels of government, diverse humanitarian organizations and industries, as well as various perspectives, disciplines and mechanisms, which combine and transform different methods for viable transition pathways of urban centres (Bush and Doyon 2019; Meerow et al. 2016).

The importance of building urban resilience cannot be overemphasized. It provides an opportunity to address understudied components of cities, understand the historical and socio-political processes that create and maintain social vulnerabilities, and to sustain the livability and economic competitiveness of cities (Weichselgartner and Kelman 2014). In line with this, investment and implementation of urban resilience policies better equip cities to cope with chronic stress and shocks, adapt to changing environments, and transform cities into resilience hubs. On the other hand, Rentschler (2013) argues that the weak or absence of urban resilience exposes urban poor living in unsafe areas to various types of risks and suffers excessively when the risk materializes and perpetuates poverty traps to increase urban vulnerability to hazards of different magnitudes. Hence, efficient and rapid urban resilience practice is urgently required for effective adaptation and transformation of cities.

Much scholarly attention has focused on various practices of enhancing urban resilience in the cities of global south and north (Galderisi et al. 2020; Staddon et al. 2018; Zhou et al. 2017). For instance, constructed wetlands are used as nature-based wastewater treatment in Kruger National Park, South Africa, where 365 mega-litres of water is treated and returned annually. The wetlands comply with the regulatory authority's guidelines and are ecologically suitable for releasing treated water into aquatic ecosystems (Staddon et al. 2018). In furtherance of the use of green infrastructure for urban resilience (Morin et al. 2021) noted that the delivery of environmental, commercial, and social infrastructure across a development of 6000 units of carbon zero houses in Bicester, England, was greatly assisted by green infrastructure. To ensure total compliance, 40% of the total site area is required to be reserved as green space based on the government policies such as Delivering Sustainable Development and Planning Policy Statement 1. Consequently, the value of the project lifecycle demonstrates the resilience of the development and the effectiveness of using green infrastructure. Another typical example of common resilience practice in cities is the creation of devoted Resilience Offices in the 100 Resilient Cities Initiative and engagement of stakeholders which represent an avenue to encourage a more efficient and promising collaboration among different segments of local administration and multi-sectoral cooperation in Athens and Rome (Galderisi et al. 2020).

Even though urban resilience has contributed immensely to the development of cities worldwide (especially European cities), scholars have identified various limitations and challenges of implementing and mainstreaming resilience in European cities. For instance, Kourti et al. (2019) suggested that to foster urban resilience in European cities, it is imperative to resolve three fundamental challenges. These include a lack of simple yet effective models that appeal to policymakers, easy and quick access to simplified datasets, and a hindrance to sharing information, design strategies, and analysis of outputs between the government and private sectors. This argument was also shared by Komendantova et al. (2016) who stated that resilience strategies in European cities could be hampered or facilitated by an inclusive and holistic stakeholders' approach for risk prevention and mitigation at all levels of governance, and existing institutional frameworks which involve issues such as the centralization of decision-making process. To this end, this chapter offers some critical reflections on urban resilience in the fourth industrial revolution with a specific focus on the impacts of climate change in Europe, adaptation and transformative initiatives, as well as digitalization as a transformative adaptation tool.

15.2 Climate Change (CC) in Europe

15.2.1 Brief Intro on CC

From about two centuries ago when Joseph Fourier published his article on the earth's natural greenhouse effect, much has been discovered about how the climate is changing. As a result, numerous measures have been put in place globally, regionally, and locally to mitigate it—the world population has multiplied, the United Nations Framework Convention on Climate Change (UNFCCC) entered into force, the Intergovernmental Panel on Climate Change (IPCC) has produced five Assessment Reports, and the effects of CC have become more apparent, to mention a few.

Various definitions of CC have been documented in the literature. The common component to these definitions is a change to the climate over a long duration. According to the IPCC usage, CC is 'a change in the state of the climate that can be identified (e.g. using statistical tests) by changes in the mean and/or the variability of its properties, and that persists for an extended period, typically decades or longer' (Balogun et al. 2020). Even though the definition of the UNFCCC in 1992 included both direct and indirect human activities, in addition to natural climate variability, the increasing human influence on the climate system was confirmed in IPCC's Fifth Assessment Report with a 95 per cent certainty that the main cause of current global warming is humans (Chelleri and Baravikova 2021).

15.2.2 Impact of CC on European Cities

CC has been and is still affecting every region in Europe, however, the impact across the continent that varies. While some cities are generally regarded to be resilient to CC impacts, other cities are not necessarily as fortunate. Studies have also shown that what may be regarded as the level of impacts and vulnerability nationally could be different at certain regional and local levels (Chelleri and Baravikova 2021; Coaffee 2008).

CC impact across European cities, such as droughts, forest fires, sea level rise, increases in atmospheric temperatures, heavy rain, and change in storm patterns (Chelleri and Baravikova 2021), have been multisectoral.

15.2.3 Nexus Between CC and Urban Resilience

Globally, levels of urbanization have been increasing for decades, and the trend is projected to continue through 2050 Chelleri and Baravikova (2021). As of 2020, urban areas accounted for approximately 75% Europeans and the growth is expected to reach as high as approximately 84% by 2050.

Many urban areas have been stretched over time due to increase in influx; increasing use of energy, traffic congestion, elevated levels of emissions, etc. are just a few of the impacts felt in these areas. As a result of factors including high population and infrastructure density, cities are usually vulnerable to CC impacts (Galderisi et al. 2020). Therefore, not having adequate receptors and infrastructure in place to handle the impending shocks from CC will not only cause huge impacts, but it will also lead to huge costs (Bodansky 1993). A more proactive approach, such as designing cities for resilience, is not merely an option but it is turning out to be the way forward (Pachauri et al. 2014).

For resilient cities to meet the challenges of CC, the approach will need to be adaptive and sustainable. This will include adaptive strategies for water management (Ionita et al. 2020) by reducing impervious surfaces and channelling floods and stormwater to areas they are the most required (Weilnhammer et al. 2021), digitalization (Desa 2014) to include bridging the digital divide between governments, real-time data usage for traffic monitoring and heating/cooling systems, (Lindberg et al. 2013) effective policies, and public participation (Hunt and Watkiss 2011).

Some region-wide efforts to address CC issues in urban areas of Europe include the 'Pact of Amsterdam' (Coaffee 2008) and '10 years of the Covenant of Mayors'. Meerow et al. (2016) in 2016, an urban agenda for the European Union (EU) was established—the 'Pact of Amsterdam' (Pact)—to address urban challenges. Intended to enhance the quality of living of the community, the Pact notably listed climate adaptation (including green infrastructure initiatives) as part of its major themes, and the objectives are to take appropriate action to prevent the damage of CC by anticipating its adverse effects. These initiatives foster the kind of collaboration that is useful for cities to become resilient in tackling CC.

15.3 Climate Change Adaptation Initiatives in Europe

The European countries had not considered CC adaptation strategies until the mid of 2000s (Biesbroek et al. 2010). Prior to this period, efforts were mainly focused on CC mitigation. However, due to the realization that climate change impacts cannot be rapidly reversed (IPCC-WG 2001) and the UNFCCC, different countries started exploring adaptation options alongside mitigation efforts (Biesbroek et al. 2010). The adaptation strategies started with the adoption of National Adaptation Strategies (NASs). Biesbroek et al. (2010) showed that between 2005 and 2008, about nine European countries including the United Kingdom, Netherlands, France, and Germany established NASs. The EU-wide strategies commenced with the adoption of a white paper that presented the framework for climate change adaptation in 2009 EC (European Commission) (EC 2009). In 2013, an EU adaptation strategy was adopted (EC 2013) which has culminated in the inauguration of a new EU adaptation strategy in February 2021 (EC 2021). The EU realized that despite the strategies about EUR 450 billion was lost to climate-related events in the EU indicating that the strategies need to be improved (EC 2018). The new strategy which builds on the 2103 strategy aims to 'realise the 2050 vision of a climate-resilient Union by making adaptation smarter, more systemic, swifter, and stepping up international action' (EC 2021).

15.3.1 An Overview of the Adaptation Strategies in Europe

The white paper on EU adaptation strategy highlighted a framework of 'incremental adaptation' in two phases (EC 2009). The first phase was a foundation for the second phase which would commence in 2013. The focus of the first phase was on ensuring coherence in the EU's response to climate change across multiple sectors and levels of governance, making climate adaptation part of EU policies, establishing a knowledge base and increasing the resilience of the agriculture, health, and water and coastal and marine sectors (Biesbroek et al. 2010). In the 2013 document, the strategies were to encourage the adoption of comprehensive adaptation strategies by EU countries, build resilient infrastructures, implement adaptation in vulnerable sectors, provide funding, and close the knowledge gap (EC 2013). The latest strategy involves improving the knowledge base, promoting digital transformation, making Climate_ADAPT the platform for adaptation knowledge and information, supporting adaptation strategies at all levels, ensuring swift actions in reducing risks and vulnerability, and promoting international action on climate resilience (EC 2021). These strategies indicate that the general framework is incremental (building on the previous strategies) promotes multilevel governance and digital transformation. Ishtiaque (2021) asserted that due to the multidimensional nature of climate impacts and the involvement of multiple actors, adaptation strategies should adopt the multilevel governance concept.

15.3.2 Limitations of Existing Adaptation Initiatives

Rutherford et al. (2020) posited that few studies have critically reviewed or comparatively evaluated the adaptation strategies. One of the early studies by Biesbroek et al. (2010) concluded that most NASs focused on their local problems without regard to the EU. Furthermore, they posited a lack of acceptable mechanisms, including funding for implementing the strategies. Some of the limitations of the previous strategies have been addressed by the later ones especially the new 2021 adaptation strategies. However, the new document highlights that the knowledge gap in adaptation strategy remains (EC 2021). It also emphasized the need to improve local resilience and integrate EU strategies with individual strategies. Grafakos et al. (2020) identified capacity gaps in the implementation of adaptation strategies by the local governments in Europe. They also found that mitigation strategies are more common in climate change plans than adaptation strategies. Rutherford et al. (2020) showed that there is still an imbalance in the adaptation progression across countries. Also, Aguiar et al. (2018) concluded that though local adoption of adaptation is increasing, there are differences in the adaptation plans and adaptive capacity at the local level. In the study by Reckien et al. (2019), only 26% of 885 cities in Europe have adaptation plans. Klostermann et al. (2018) highlight the need for a common framework for assessing the achievement of adaptation plans to improve adaptation monitoring. Digital transformation through the increase in the availability of data and the use of instruments such as Digital Twins (EC 2021) might help improve the rate of achieving the adaptation strategies.

15.4 Transformative Adaptation

Transformative adaptation can be defined as 'a strategy that aims to reduce the root causes of vulnerability to climate change in the long-term by shifting systems away from unsustainable or undesirable trajectories' (Fedele et al. 2019). According to the World Resource Institute (WRI), it is a systemic approach for protecting development gains, maximizing resilience investments, and reducing climate change risks (https://www.wri.org/our-work/project/transformative-adaptation). The approach emerges based on the premise that incremental adaptation is insufficient in addressing climate change because it is short-term and small-scale in nature (Wilson et al. 2020). Transformative adaptation, on the other hand, involves fundamental transformation within and across systems by altering their key attributes such as values, institutions, regulations, policies, practices, financing, bureaucracy, technologies, biological process, lifestyles, and thinking (Lonsdale et al. 2015). It portends a long-term strategy requiring complex learning to achieve non-linear and radical changes (Pelling et al. 2015). Such changes can be at the macro-level (international/national), meso-level (region/community), and micro-level (household/individual). Internal or external shocks such as climate change and disasters are the key circumstances that prompt the need for transformative transformation,

Colloff et al. (2017) classified transformative adaptation into three broad types. First is the ecosystem transformation, which deals with strategies to achieve long-lasting change leading to a resilient and stable condition, valued and governed by society, including ecosystem's use and management. The second type is the transformation of decision context involving a significant change in the social arrangements, the knowledge, values and belief systems, formal and informal laws that dictate how power is codified, allotted, and utilized. Lastly, the governance transformation deals with the capability to develop 'adaptive, transformative governance for whom, to enable what kinds of changes in governance systems, for what purpose' (p. 88).

Efforts to employ transformative adaptation are undermined by the lack of awareness, community resistance, limited funding, and political uncertainty about their costs, benefits, and risks (Clarke et al. 2018; Colloff et al. 2017). Thus, disseminating evidences of fruitful cases of transformative approaches are vital to surmounting adaptation obstacles (Thaler et al. 2019). Global agreement and action are also vital for enabling and fostering transformative adaptation, especially in production and consumption processes, which are the major contributor to environmental degradation and CC (Revi et al. 2014). Also, there is the need for collaboration among domestic organizations and developing novel approaches to evaluating and managing trade-offs between the existing socio-ecological systems and accepting fundamental modifications to the systems (Colloff et al. 2017).

Transformative adaptation is not only essential, but it is also feasible. It allows implementing anticipatory strategies in which fresh alternatives are co-created and tried, leading to adaptive capacity, resilience, and sustainability through action–learning adaptation cycles (Abubakar and Aina 2019; Wilson et al. 2020). This kind of adaptation is effective in managing the growing CC impacts (Chowdhury et al. 2020; Fedele et al. 2019) and confronting urbanization challenges (Dano et al. 2020) by ensuring that communities most affected by climate change have an active voice in preparing and implementing adaptation plans and policies. It allows decision-makers and residents to develop an effective and lasting approach to achieve transformative outcomes, such as saving limited land and water resources and lessening the risks of climate-induced conflicts (Abubakar 2021). The approach elevates the visibility of climate resilience, adaptation, and commitments and sharing solutions at different scales and across borders (Wilson et al. 2020). It helps tackle short- and long-term climate change threats, uncertain and non-linear changes that transcend sectors and scales (Colloff et al. 2017).

In Europe, transformative adaptation is largely through bottom-up initiatives occurring at local levels. In Norway, for example, the country's multilevel and bottom-up governance system facilitates its implementation of transformational changes. The local public agency addresses climate change adaptation by focusing on the production and use of knowledge and collaboration. The approach involved developing and implementing urban policies and plans to reduce greenhouse gas emissions and preserve open spaces, blue-green structures, and green roofs for handling surface water and floods. The collaboration is between local public agencies and external research institutes, and consultancy companies (Orderud et al. 2020). Similarly, Thaler et al. (2019) investigated transformation approaches in natural hazard management in Austria. The study found that local initiatives by local governments, community organizations, and citizens have complemented the conventional hazard management policies on mitigating flood and avalanche hazards.

15.5 Digitalization as a Transformative Adaptation Tool

Explanation of Digitalization

Digitalization as a concept is broadly understood as 'the development and application of digital and digitalized technologies that augment and dovetail with all other technologies and methods' WBGU—German Advisory Council on Global Change (2019). It can have profound impacts on the economic and social systems and restructure the domains of social life (Brennen and Kreiss 2016). This potential makes digitalization one of the most transformative forces of the modern world.

How Digitalization Supports Transformative Adaptation to Climate Change in European Cities

As Kates, Travis, and Wilbanks (2012) pointed out, the transformations could be an extrapolation from smaller to larger scale or intensity, true innovation, and transform/shift regions/locations. The first transformation often happens when some locally tested innovative new ideas provide significantly positive results and present scalability. In this regard, cities often act as the testing hub, and if they prove to be beneficial, they could be spread through the entire region or country. Such projects are often termed lighthouse projects. True innovations address an existing or emerging issue in unprecedented ways thus setting the stage for replication. Though such transformation seldom happens, cities as hotbeds of innovation and technology are fertile incubation hubs for such path-making approaches. The transformative approach can be a very useful strategy for the cities with limited resources as they can learn from others and select the best ideas and implement them as per their local context.

European cities across the continent are increasingly exploring digitalization as a tool to reduce their carbon footprint, respond to the emerging risks from climate change, and become more sustainable. The European Commission repository on smart cities reports as many as 82 projects incorporating various facets of digitalization that span across 122 cities.

Some European countries, like Estonia, do not stop their ambitions at the city level but aim to transform the entire country by leveraging upon digitalization. Capitalizing on their relatively smaller population (1.32 million, 2019), it aims to have an ambitious roll-out of transformative projects to make the entire nation more resilient. This was epitomized in their response to the COVID-19 pandemic, where unlimited and free educational tools (apps, platforms, etc.) were provided across the country (News report on PRNewswire). Such approaches demonstrate the transformative potential of digitalization across scales. Additionally, digitalization enables cities to establish integrated platforms, thus promoting economies of scale and strengthening their purchasing power. For example, in Lithuania, all calls higher than 10,000 EUR, must be organized via the national platform. From 2014, in this platform, an option to select criteria for sustainability has been established and promoted, which has allowed cities to work together as a cohesive group and efficiently use available resources.

A sustainable approach to urbanization can also help settlements in addressing climate change. Furthermore, sustainability encompasses economic and societal considerations in addition to environmental and other perspectives. The often used top-down approach to resilience and sustainability often suffers from poor implementation and acceptance by the end-user, in part due to their centralized and non-participatory design. Many countries are using digital tools to not only improve the design of such ‘top-down’ approaches but also to make the entire process, across its life cycle, more participatory. In certain cases, such as in Poland and Lithuania, cities have established open digital platforms where anyone can find relevant information about spatial development projects (promoted visions, conceptual ideas, or practical solutions) as well as share their suggestions (e.g. the citizen could point out tree-planting sites or where infrastructure damages exist) (see Table 15.1).

Digital tools are increasingly being used to improve various community services and amenities (like drinking water supply, waste management, street management, etc.). Apps are increasingly used to inform about shortcomings, damages, or order services or pay for these services. This can help in transforming the transparency and trust level in such provisions and optimize resource utilization.

There are multiple ways to use digitalization as a transformative tool towards resilience and sustainability, which often depends on not just the technical aspects of the project but also on the socio-economic, environmental, efficiency, and political aspects. In this chapter, we have picked two clusters of approaches to digitalization in cities. Firstly, a centralized, top-down approach where digitalization is used as an equalizer to promote resilience and enables everyone to participate efficiently in it, as shown by the examples from Helsinki and Rotterdam. Secondly, a bottom-up participatory approach that builds upon the usage of digital means to foster collective actions and involvement in the design and implementation of such projects, as shown by examples from Vilnius, Gdansk, and Kaunas (see Table 15.1). Here we present selected digitalization measures undertaken in these five cities, which have the potential to bring about transformative changes in their urban resilience quest.

From the examples listed in the table, it is easy to point out that their local and modular design can facilitate scalability and contextualization across Europe and

Table 15.1 Examples of digital tools in the cities

S/N	City	Challenge addressed	Digitalization means adopted	Transformative potential
1	Helsinki, Finland	High carbon footprint and energy consumption	The creation of the 'energy and heat atlas' which includes building specific information such as heating systems, conducted renovation and energy certification. In addition, the Atlas includes information on the consumption of electricity, district heating and water in rental buildings for certain city neighbourhoods. The Atlas also allows users to examine the solar energy potential of buildings	The capital city of Finland is continuously ranked among the best cities to live in the world. The Atlas project is a part of the larger digitalization and smart city projects undertaken to make Helsinki not only a very liveable place but also to make it more resilient and sustainable. It is based on 3D technology, which can enable accurate, updated model for existing and future building and development plans to make the process comprehensive across the city and anticipate challenges (such as traffic flow) before they arise and allow each resident to reduce their carbon footprint through the utilization of this 'top-down' information portal and tailor it as per individual yet community synchronized needs and designs. Such a model can bring transformation across scales and regions as they reduce piecemeal approaches and enable comprehensive collective efforts that are resource-efficient and quick

(continued)

Table 15.1 (continued)

S/N	City	Challenge addressed	Digitalization means adopted	Transformative potential
2	Vilnius, Lithuania	Centralized planning and involvement of local communities	To facilitate easy reporting and sharing of citizen concerns, the municipality of Vilnius switched to a digital medium which was made accessible via an internet platform. Through an interactive map, citizens have detailed information about spatial planning projects at their disposal and they can also provide feedback using the website	Vilnius is the Lithuanian capital and home to approx. 0.5 million people, is the main destination for in-country migration. Such a rapidly growing city required an easy, dynamic and transparent medium to engage with its ever-growing population and involve them in the planning process, particularly those related to the social and environmental aspects. The facilitation of this interaction through the website, helped in exponentially increasing outreach and communication. It additionally enables interactive communication and updated information sharing with society The interactive internet platform contributed to establishing transparency in public administration. Following the dissolution of the Soviet Union, the corruption was high (the first time Transparency International evaluated the level of corruption in Lithuania was in 1999 and placed it at 50th place. A digital and far accessible tool helped in establishing trust and transparency

(continued)

Table 15.1 (continued)

S/N	City	Challenge addressed	Digitalization means adopted	Transformative potential
3	Kaunas, Lithuania	Reduction of individual carbon footprint through smart transportation and commute	*Žiogas* (eng. Grasshopper) - Public transportation mobility app. It seeks to provide current information on public transportation and suggests best suitable route for the individual commuter	Kaunas city is the second biggest city in Lithuania (population approx. 0.3 mln. people). It is the hub of local industries, several universities. Located on the European crossroads, it holds massive potential for commerce The number of private cars per 1000 population rose from 372 in 2015 to 432 in 2019 in Kaunas city. Meanwhile, the popularity of public transport has substantially waned. To reverse this trend and promote sustainable transportation, the *Žiogas* app was launched. By providing real-time data on buses and suggesting the most suitable commute, it aims to make individual transportation smart and green This app is part of the comprehensive urban mobility strategy of the city (Sustainable Mobility Plan in Kaunas City). *Žiogas* app changes the way people experience public transportation, enhances access and makes people feel connected and active participants in the sustainability process

(continued)

Table 15.1 (continued)

S/N	City	Challenge addressed	Digitalization means adopted	Transformative potential
4	Rotterdam, Netherlands	Socio-economic resilience to adapt to a digital world economy and reduce its carbon footprint	Digital datahub for the city	Rotterdam, the largest maritime port in Europe is one of the global leaders in the urban climate change and sustainability quest. It is one of the first cities to align its vision to the 2015 Paris Climate Agreement. It has also embraced the smart city vision by using digitalization across a spectrum of functions and services such as the Rotterdam port, the Innovation District, smart industry and smart governance. Rotterdam, Glasgow and Umeå are three lighthouse cities of the EU supported RUGGEDISED project that aims to create scalable smart and resilient cities across Europe by combining ICT, e-mobility and energy solutions. Within this Rotterdam is particularly focusing on its socio-economic sustainability and resilience by transitioning from an Industrial economy to a green economy. Through a set of 13 smart solutions that integrate digital tools with its context, Rotterdam aims to create a blueprint for cities across Europe to comprehensively synchronize their vision, planning and management with digitalization. Centred around a 'digital datahub' for the city, these solutions range from creating the open-source 3D operational model of cities that by analysing the real-time data flow from grids to end-users, would optimize energy consumption to smart solid waste management that would provide real-time waste collection information through sensors resulting in optimized route planning for waste collection

beyond. Transferability of digital tools usually is very high since the contextual description could be changed inside the programme/platform/app/etc. The concern of these tools lies more in the local culture to use them for daily needs and purposes. This contextualized replication potential of these digital innovations sets them apart from traditional incremental measures and makes them truly transformative.

15.6 Challenges to Digitalization for Transformative Adaptation in Europe

Despite the merits of digitalization in building cities' resilience, such rapid transformation can disrupt existing regulations and policies. For instance, the significant shifts in the functions, primary attributes, and socio-ecological interactions, that is, land use and land cover change (Fazey et al. 2018; Fedele et al. 2019) can affect agroecological processes. If proper care is not taken, it can alter the social and ecological factors thereby redefining and disrupting the whole system (Colloff et al. 2017), which then necessitates a holistic approach to fix by combining several disciplinary perspectives (Fedele et al. 2019).

Although smart grids (SG) are digital transformative tools that can support CC adaptation, however, the deployment of SG can further lead to the absorption and emission of radiant energy (greenhouse gas) (Giesbrecht 2016). Another major challenge with the application of smart grid is the scale of data and processing extent (Balogun et al. 2020). Low-quality data do affect the accuracy of the network forecast, which means, there is need for large volume of data for accurate prediction and precision of different nodes in the grid. The monetization and paucity of free big data (Loebbecke and Picot 2015) also hinder access to data for effective adaptation planning.

Although the Helsinki Energy and Climate Atlas (Table 15.1) offers timely information to minimize carbon emissions (Table 15.1), it has limited capability to provide updated information about carbon footprint. Further, the atlas caters to only electricity and heat emission without consideration of other crucial sources of emission such as deforestation, land-use changes, vehicular and industrial emission (Cui et al. 2018; Ma et al. 2019). Another challenge of the energy and climate atlas is that it cannot be relied upon to exploit the geothermal heat due to its limited coverage (Helsinki 2020).

The reduction of carbon emission from the transportation sector leading to the invention of *Žiogas* (eng. Grasshopper) (Table 15.1) may help reduce congestions but it does not provide a complete solution. For example, users' responses show that the application still has many limitations which may cause a loss of interest. Therefore, its functionality to reduce carbon emissions may not be realized or delayed. Similarly, the sustainability of the Netherlands' green infrastructure, smart design promotion, streamlining industries, and data leveraging for solving climate change and urban resilience problems is threatened by excessive network traffic on the sole source of the internet (Amsterdam Internet Exchange [AMS-IX]) (Smartdc 2021).

15.7 Conclusion

Although the devastating impact of climate change in Europe has been well documented, there is limited information on how digitalization tools are transforming the resilience of European cities to the constant pressure of the rapidly changing climate. The consideration of several resilience concepts has become an important scientific research area to improve sustainability and minimize vulnerability to CC impacts in European cities. Over the past few years, there has been a considerable increase in the number of digital technologies that have been developed to foster resilience across different scales. In this chapter, four of such digitalization technologies are analysed. The energy and heat atlas of Helsinki Finland, centralized planning and engagement of local communities in Vilnius Lithuania through interactive maps, the public transport mobility app at Kaunas, Lithuania and the city digital datahub at Rotterdam, Netherlands are enabling climate change adaptation and boosting resilience. Despite the early gains, the complexities of smart cities encompassing the interaction of the environmental, social-economic, physical, and political dimensions of a city pose challenges to successful digitalization. Disruption of existing urban plans that could cause the alteration of social and ecological spaces, emission of GHG, processing of big digital data and technical limitations have all been documented. Further research is required to identify sustainable solutions to overcome challenges to the widespread adoption of digitalization in order to address pertinent climate change problems and boost urban resilience.

References

Abubakar IR (2021) Predictors of inequalities in land ownership among Nigerian households: implications for sustainable development. Land Use Policy 101:105194. https://doi.org/10.1016/j.landusepol.2020.105194

Abubakar IR, Aina YA (2019) The prospects and challenges of developing more inclusive, safe, resilient and sustainable cities in Nigeria. Land Use Policy 87:104105. https://doi.org/10.1016/j.landusepol.2019.104105

Aguiar F, Bentz J, Silva J, Fonseca A, Swart R, Santos F, Penha-Lopes G (2018) Adaptation to climate change at local level in Europe: an overview. Environ Sci Policy 86:38–63

Balogun A, Marks D, Sharma R, Shekhar H, Balmes C, Maheng D, Arshad A, Salehi P (2020) Assessing the potentials of digitalization as a tool for climate change adaptation and sustainable development in urban centres. Sustain Cities Soc 53:101888. https://doi.org/10.1016/j.scs.2019.101888

Biesbroek G, Swart R, Carter T, Cowan C, Henrichs T, Mela H, Morecroft M, Rey D (2010) Europe adapts to climate change: comparing national adaptation strategies. Glob Environ Chang 20(3):440–450

Bodansky D (1993) The United Nations framework convention on climate change: a commentary. Yale J Int Law 18:451
Brennen J, Kreiss D (2016) Digitalization. In: The international encyclopedia of communication theory and philosophy, pp 1–11
Bush J, Doyon A (2019) Building urban resilience with nature-based solutions: how can urban planning contribute? Cities 95:102483
Chelleri L, Baravikova A (2021) Understandings of urban resilience meanings and principles across Europe. Cities 108:102985
Chowdhury M, Rahman S, Abubakar I, Aina Y, Hasan M, Khondaker A (2020) A review of policies and initiatives for climate change mitigation and environmental sustainability in Bangladesh. Environ Dev Sustain 23:1133–1161
Clarke D, Murphy C, Lorenzoni I (2018) Place attachment, disruption and transformative adaptation. J Environ Psychol 55:81–89
Coaffee J (2008) Risk, resilience, and environmentally sustainable cities. Energy Policy 36(12):4633–4638
Colloff M, Martín-López B, Lavorel S, Locatelli B, Gorddard R, Longaretti P, Walters G, van Kerkhoff L, Wyborn C, Coreau A (2017) An integrative research framework for enabling transformative adaptation. Environ Sci Policy 68:87–96
Cui Y, Li L, Chen L, Zhang Y, Cheng L, Zhou X, Yang X (2018) Land-use carbon emissions estimation for the Yangtze River delta urban agglomeration using 1994–2016 landsat image data. Remote Sensing 10(9). https://doi.org/10.3390/rs10091334
Dano UL, Balogun AL, Abubakar IR, Aina YA (2020) Transformative urban governance: confronting urbanization challenges with geospatial technologies in Lagos, Nigeria. GeoJournal 85(4):1039–1056. https://doi.org/10.1007/s10708-019-10009-1
Desa U (2014) World urbanization prospects, the 2011 revision. Department of Economic and Social Affairs of the United Nations Secretariat
EC (2009). White paper-adapting to climate change: towards a European framework for action. J European Commission. http://eur-lex.europa.eu/LexUriServ/LexUriServ.do
EC (2013) Strategy on adaptation to climate change. J Communication from the Commission to the European Parliament, the Council, the Europcan Economic Social Committee the Committee of the Regions, COM, 216
EC (2018). On the implementation of the EU strategy on adaptation to climate change. EU Commission, Brussels, Belgium
EC (2021) Forging a climate-resilient Europe—the new EU strategy on adaptation to climate change. European Commission. J Communication from the Commission to the European Parliament, the Council, the European Economic Social Committee the Committee of the Regions, COM, 216
Fazey I, Moug P, Allen S, Beckmann K, Blackwood D, Bonaventura M, Burnett K, Danson M, Falconer R, Gagnon A (2018) Transformation in a changing climate: a research agenda. Climate Dev 10(3):197–217
Fedele G, Donatti C, Harvey C, Hannah L, Hole D (2019) Transformative adaptation to climate change for sustainable social-ecological systems. Environ Sci Policy 101:116–125. https://doi.org/10.1016/j.envsci.2019.07.001
Folke C (2006) Resilience: the emergence of a perspective for social—ecological systems analyses. Glob Environ Chang 16(3):253–267
Galderisi A, Limongi G, Salata K (2020) Strengths and weaknesses of the 100 resilient cities initiative in southern Europe: Rome and Athens' experiences. City, Territory and Architecture 7(1):1–22
Giesbrecht J (2016) Smart grid deployment and climate change response: evaluating climate change integration in Ontario's smart grid deployment regime. University of Waterloo
Grafakos S, Viero G, Reckien D, Trigg K, Viguie V, Sudmant A, Graves C, Foley A, Heidrich O, Mirailles J (2020) Integration of mitigation and adaptation in urban climate change action plans in Europe: a systematic assessment. Renew Sustain Energy Rev 121:109623
Helsinki (2020) Helsinki energy and climate Atlas. https://kartta.hel.fi/3d/atlas/

Hunt A, Watkiss P (2011) Climate change impacts and adaptation in cities: a review of the literature. Clim Change 104(1):13–49

Ionita M, Nagavciuc V, Kumar R, Rakovec O (2020) On the curious case of the recent decade, mid-spring precipitation deficit in central Europe. npj Clim Atmos 3(1):1–10

IPCC-WG I (2001) Climate change 2000, third assessment report. Cambridge University Press, Cambridge, UK

Ishtiaque A (2021) Multilevel governance in climate change adaptation: conceptual clarification and future outlook. In: Climate change and extreme events. Elsevier, pp 171–185.

Kates RW, Travis WR, Wilbanks TJ (2012) Transformational adaptation when incremental adaptations to climate change are insufficient. Proc Natl Acad Sci 109(19):7156–7161. https://doi.org/10.1073/pnas.1115521109

Klostermann J, van de Sandt K, Harley M, Hildén M, Leiter T, van Minnen J, Pieterse N, van Bree L (2018) Towards a framework to assess, compare and develop monitoring and evaluation of climate change adaptation in Europe. Mitig Adapt Strat Glob Change 23(2):187–209

Komendantova N, Scolobig A, Garcia-Aristizabal A, Monfort D, Fleming K (2016) Multi-risk approach and urban resilience. International Journal of Disaster Resilience in the Built Environment 7(2):114–132

Kourti N, Kempner T, Marin Ferrer M, Luoni S, Antofie T, Georgios T, Negro P, Giannopoulos G, Galbusera L, Krausmann E (2019) Strategies for improving urban resilience in Europe

Lindberg F, Grimmond C, Yogeswaran N, Kotthaus S, Allen L (2013) Impact of city changes and weather on anthropogenic heat flux in Europe 1995–2015. Urban Climate 4:1–15

Loebbecke C, Picot A (2015) Reflections on societal and business model transformation arising from digitization and big data analytics: a research agenda. J Strateg Inf Syst 24(3):149–157

Lonsdale K, Pringle P, Turner B (2015) Transformative adaptation: what it is, why it matters and what is needed.

Ma X, Wang C, Dong B, Gu G, Chen R, Li Y, Zou H, Zhang W, Li Q (2019) Carbon emissions from energy consumption in China: its measurement and driving factors. Sci Total Environ 648:1411–1420. https://doi.org/10.1016/j.scitotenv.2018.08.183

McPhearson T, Andersson E, Elmqvist T, Frantzeskaki N (2015) Resilience of and through urban ecosystem services. Ecosyst Serv 12:152–156

Meerow S, Newell J, Stults M (2016) Defining urban resilience: a review. Landsc Urban Plan 147:38–49

Meerow S, Newell J (2015) Resilience and complexity: a bibliometric review and prospects for industrial ecology. J Ind Econ 19(2):236–251

Morin S, Samacoïts R, François H, Carmagnola CM, Abegg B, Demiroglu OC, Pons M, Soubeyroux J-M, Lafaysse M, Franklin S, Griffiths G, Kite D, Hoppler AA, George E, Buontempo C, Almond S, Dubois G, Cauchy A (2021) Pan-European meteorological and snow indicators of climate change impact on ski tourism. Clim Serv 22:100215. https://doi.org/10.1016/j.cliser.2021.100215

Okunola O (2019) Residents vulnerability analysis and explanation of development-induced disasters in a Sub-Saharan African city. International Journal of Disaster Resilience in the Built Environment 11(1):71–84. https://doi.org/10.1108/IJDRBE-05-2019-0024

Orderud G, Naustdalslid J (2020) Climate change adaptation in Norway: learning—knowledge processes and the demand for transformative adaptation. Int J Sust Dev World 27(1):15–27

Pachauri R, Allen M, Barros V, Broome J, Cramer W, Christ R, Church J, Clarke L, Dahe Q, Dasgupta P (2014) Climate change 2014: synthesis report. Contribution of Working Groups I, II and III to the fifth assessment report of the intergovernmental panel on climate change. IPCC.

Pelling M, O'Brien K (2015) Adaptation and transformation. Clim Change 133(1):113–127

Reckien D, Salvia M, Pietrapertosa F, Simoes S, Olazabal M, Hurtado S, Geneletti D, Lorencová E, D'alonzo V, Krook-Riekkola A (2019) Dedicated versus mainstreaming approaches in local climate plans in Europe. Renew Sustain Energy Rev 112:948–959

Rentschler J (2013) Why resilience matters-the poverty impacts of disasters (6699)

Revi A, Satterthwaite D, Aragón-Durand F, Corfee-Morlot J, Kiunsi R, Pelling M, Roberts D, Solecki W, Gajjar S, Sverdlik A (2014) Towards transformative adaptation in cities: the IPCC's fifth assessment. Environ Urban 26(1):11–28

Rutherford V, Hills J, Le Tissier M (2020) Comparative analysis of adaptation strategies for coastal climate change in North West Europe. Mar Policy 111:102478

Satterthwaite D, Archer D, Colenbrander S, Dodman D, Hardoy J, Mitlin D, Patel S (2020) Building resilience to climate change in informal settlements. One Earth 2(2):143–156

Sharifi A, Khavarian-Garmsir A (2020) The COVID-19 pandemic: impacts on cities and major lessons for urban planning, design, and management. Sci Total Environ 749:142391

Sheller M (2020) Reconstructing tourism in the Caribbean: connecting pandemic recovery, climate resilience and sustainable tourism through mobility justice. J Sustain Tour 29(9):1436–1449

Simone C, Iandolo F, Fulco I, Loia F (2021) Rome was not built in a day. Resilience and the eternal city: insights for urban management. Cities 110:103070

Smartdc (2021) Rotterdam as digital data hub. Retrieved April 26, 2021, from https://www.smartdc.net/news/rotterdam-as-a-digital-data-hub/

Staddon C, Ward S, De Vito L, Zuniga-Teran A, Gerlak A, Schoeman Y, Hart A, Booth G (2018) Contributions of green infrastructure to enhancing urban resilience. Environ Syst Decis 38(3):330–338

Thaler T, Attems M, Bonnefond M, Clarke D, Gatien-Tournat A, Gralepois M, Fournier M, Murphy C, Rauter M, Papathoma-Köhle M (2019) Drivers and barriers of adaptation initiatives—how societal transformation affects natural hazard management and risk mitigation in Europe. Sci Total Environ 650:1073–1082

WBGU—German Advisory Council on Global Change (2019) Towards our common digital future [summary]. WBGU. https://www.wbgu.de/fileadmin/user_upload/wbgu/publikationen/hauptgutachten/hg2019/pdf/WBGU_HGD2019_S.pdf

Weichselgartner J, Kelman I (2014) Challenges and opportunities for building urban resilience. A|Z ITU J Fac Archit 11(1): 20–35

Weilnhammer V, Schmid J, Mittermeier I, Schreiber F, Jiang L, Pastuhovic V, Herr C, Heinze S (2021) Extreme weather events in europe and their health consequences—a systematic review. Int J Hyg Environ Health 233:113688

Wilson R, Herziger A, Hamilton M, Brooks J (2020) From incremental to transformative adaptation in individual responses to climate-exacerbated hazards. Nat Clim Chang 10(3):200–208

Woodruff S, Meerow S, Stults M, Wilkins C (2018) Adaptation to resilience planning: alternative pathways to prepare for climate change. J Plan Educ Res. https://doi.org/10.1177/0739456X18801057

Zhou X, Li Z, Staddon C, Wu X, Song H (2017) Issues and challenges of reclaimed water usage: a case study of the dragon-shaped river in the Beijing Olympic Park. Water International 42(4):486–494

Chapter 16
Wielding a Concept with Two Edges: How to Make Use of the Smart Cities Concept and Understanding Its Risks from the Resilient Cities Perspective

Roman Serdar Mendle and Anina Hartung

Abstract Smart Cities and Resilient Cities are two normative theoretical concepts within urban development research and practice. Using a dialectic approach, we take a closer look at both concepts, identify their core ideas and examine two opposing narratives about how they interact: the thesis of Smart Cities supporting Resilient Cities, and the antithesis that they clash and contradict each other. We examine practical examples for Smart City applications to improve resilience, including sensor networks in early warning systems in Japan, Blockchain applications for disaster relief fund distributions in Vanuatu and using Artificial Intelligence to better understand social resilience in Haiti. We also look at examples of new hazards and risks for cities such as cyber security, privacy issues and threats to economic and material resilience of a city coming from digitalization. As a synthesis, we propose that both concepts are used deliberately and consciously as conceptual *lenses* to guide urban development practice. We propose that Smart Cities are best to be used as a tool for urban development, not as a normative goal or vision. We suggest that the Resilient City concept provides a lens that can help set the goals and targets for Smart Cities, but also needs to consider new types of risks in what we call *digital resilience.*

16.1 Introduction

When a large fire destroyed one of four OVHcloud server centers and damaged another in Strasbourg on March 10, 2021, "Millions of websites" went offline, and an estimated 2% of all sites with an.fr domain were affected (Satter 2021). Not all data and cloud services affected had disaster recovery plans or backups (Sverdlik 2021), and this was not the first calamity this campus experienced either. In 2017, a power outage had brought down the entire campus (Sverdlik 2017). If nothing

R. S. Mendle (✉)
ICLEI—Local Governments for Sustainability (European Secretariat), Freiburg, Germany
e-mail: roman.mendle@iclei.org

A. Hartung
ICLEI—Local Governments for Sustainability (World Secretariat), Bonn, Germany

A. Sharifi and P. Salehi (eds.), *Resilient Smart Cities*, The Urban Book Series,
https://doi.org/10.1007/978-3-030-95037-8_16

else, this case serves as a reminder that cloud services, web services, and digitalization, as one of the most revolutionary developments of our century, are tied to physical infrastructure, require resources, and come with their own set of risks for external shocks. Resilience, therefore, is a highly relevant concept in the space, and it becomes increasingly important as global digitalization and our dependence on services provided by new technologies increases. This is not only true for commercial web services, but even more within the space of urban development and urban systems. On the other hand, Smart City research and practice have brought forward a large range of ways that can aid cities in becoming more resilient, prepare them better for external stress, or improve their ability to adapt and grow while decreasing systemic vulnerabilities. Some recent research suggests that Smart Cities may overall have a positive effect on resilience, even though "different types of urban resilience are affected differently" (Zhou et al. 2021). We argue that there is a positive narrative of how smart technologies can be used to improve the resilience of a city, either deliberately or as a side effect, and a critical narrative of contradictions and friction between Smart Cities and Resilient Cities as two opposing concepts of urban development. Taking both perspectives into account, we re-examine Smart Cities from the perspective of the Resilient Cities concept in order to get a better and more nuanced understanding of how to use (or not to use) Smart Cities as a tool for urban development.

16.2 The Concepts of Smart Cities, Resilient Cities, and Critical Views on Both

There is a vast body of academic literature out there on Smart Cities and Resilient Cities, as well as other related concepts such as sustainable cities, green cities, digital cities, and intelligent cities, which has been thoroughly reviewed and analyzed as to whether both concepts implicitly share a common paradigm of urban development (de Jong et al. 2015). One of the most important results of this analysis is, that at least in its origin, Resilient Cities "established a theoretical branch of its own which has steadily been gaining in academic popularity since 2006 [and] is only connected with the 'sustainable city' and no other categories," while the Smart City can "be seen as a more advanced successor to, the older 'information city,' 'digital city' and the 'intelligent city' categories." Smart Cities and Resilient Cities stem from different conceptual discourses in academic literature and are defined differently by scholars from various disciplines. A similar phenomenon can be observed at urban development and sustainability forums and events in the practitioner's space. For example, events like the Barcelona Smart City Expo and World Congress (since 2011), ICLEI's Resilient Cities Congress series (2009–2019), urban climate adaptation focused events within the framework of the annual UNFCCC's Conference of Parties (COPs), the Adaptation Futures, or European Climate Change Adaptation Conference tend to attract different ecosystems of urban experts and stakeholders, and as such, despite some overlaps on the topics and objectives, the discourses elaborated in these events differ in norms, values, and vocabulary. For this reason, it is worth unpacking the Smart City and Resilient City concepts to some extent, before diving deeper into how they relate to one another and inform the urban development as a whole.

16.2.1 The Smart Cities Concept

Smart Cities have been defined and redefined numerous times, so that now there is a large and diverse set of definitions (or at least descriptions of) Smart Cities available (Antwi-Afari et al. 2021; Bagloee et al. 2021; Colding et al. 2020; de Jong et al. 2015; Yigitcanlar et al. 2019). It is hard to adequately cover and represent this body of available definitions and even harder to do proper justice to the complexity and nuances of thought they represent. Even more difficult is representing this daily practice beyond the academic space, where Smart Cities have long become a mainstream urban development concept, widely used by practitioners of all sectors. By and large, the main focus of Smart Cities is the deployment and integration of technologies, particularly information and communication technologies (ICT) in cities and their effect on urban systems (Araral 2020; Bagloee et al. 2021; Yigitcanlar et al. 2019). In addition, Smart Cities can be seen as an inherently normative concept, pursuing varying sets of goals of urban development attributed to the use of technologies (Bjørner 2021; Cardullo and Kitchin 2019; de Jong et al. 2015), such as better urban services, increased efficiency of financial and material resource use, the optimization of urban flows and processes, urban sustainability, the betterment of the local economy and community, and/or leveraging some form of collective intelligence. Smart Cities are, at the heart, a positively framed agenda for urban development through technological progress that pursue some form of vision for a "better city" by optimally and efficiently using all the technologies, tools, ideas, and forms of capital available to them. As such, Smart Cities are also an attractive concept for private business, private investment, technology providers, and entrepreneurs. They conceptually welcome the expected higher efficiency and innovation capacity of markets and private business, value technological solutions, and often support or encourage private investment. This is especially true for entrepreneurship, particularly start-ups and small and medium sized enterprises (SMEs) in support of diverse local economies and smart city ecosystems (Ooms et al. 2020), along with business models as a supplement or alternative to the "traditional" urban service provision led by the public sector. When it comes to the role of citizens, Smart Cities often emphasize that the residents are at the center of their efforts (Ji et al. 2021; Trencher 2019) for improved urban services or improved urban economies and quality of life (as "end users" or collectively as participants in "living labs").

16.2.2 The Resilient Cities Concept

Summarizing the essence of Resilient Cities is arguably not any easier, as the resilience concept has undergone an even longer process of evolution, reaching back to at least the 1970s (Manca et al. 2017). Originally concerned with the ability of—in particular, ecological (Biggs et al. 2015)—systems to "bounce back" after experiencing external shocks, resilience is a concept that is traditionally more oriented toward decreasing vulnerability and strengthening the abilities of systems to prepare for, survive, withstand, recover, and even emerge stronger from disasters or stress exposure. Applied to cities, resilience has a strong connection to adaptation for climate change, as cities cope with disasters such as extreme weather events, flooding, rising sea levels, heat waves, or the degradation of their local environment and life support systems. The breadth of issues, potential threats, and risks to urban systems are subject to a wider discourse on resilient cities; however, they are by no means limited to the climate or even environmental space. The global COVID-19 pandemic is a key example of a health-related external shock concerning a global dimension that affects nearly all types of urban systems from transportation to education, from electricity and data use, to urban retail and the demand for green infrastructure in cities. Likewise, the scope of resilience has evolved over time from physio-ecological systems to include aspects like social resilience or economic resilience, urban finance, and even incorporates elements of digital resilience (Bizzotto et al. 2019). "Building back better" as a mantra catch-phrase cited in the Sendai Framework for Disaster Risk Reduction 2015–2030 (UNDRR 2015) illustrates another aspect of the evolution of the resilience concept from the original idea of "bouncing back" and restoration of systems to an earlier and potentially idealized "snapshot" state. A more contemporary understanding of increasing urban resilience is therefore to improve cities' inherent adaptation capacity itself, or "bouncing forward" in order to stay resilient in the face of constant global change and evolving unknowns. As Colding et al. puts it, "Urban resilience refers to the ability of an urban system - and all its constituent socio-ecological and socio-technical networks across temporal and spatial scales-to maintain or rapidly return to desired functions in the face of a disturbance, to adapt to change, and to quickly transform systems that limit current or future adaptive capacity" (Colding et al. 2020).

16.2.3 Critical Views on Smart Cities and Resilient Cities

While there are vast bodies of literature directly or indirectly supporting Smart Cities and Resilient Cities, neither concept is uncontested. In fact, highlighting critical arguments against Smart Cities from a sustainability and accountability perspective has been one of three major sub-themes of Smart City literature (Araral 2020). Despite their frequently proclaimed people-centric approach (Ji et al. 2021), Smart Cities are criticized for following an open or hidden neoliberal agenda of urban development that not only runs the risk of disenfranchising people in cities from their rights and roles as citizens and drivers of governance, but also perpetuates chronic local austerity politics in a push to increase privatization of services under the cover of "solutionism" (Cardullo and Kitchin 2019; Willis 2019). Furthermore, Smart Cities are dependent on critical raw materials, the access to which is governed by extensive cooperation and global geopolitics, rather than urban development principles (David and Koch 2019), and which often cause severe sustainability challenges of their own (Mendle 2011). On the other hand, Resilient Cities are also being criticized for instilling normative imperatives on urban dwellers in practice and confining their agency to redetermined systemic goals dictated to cities (Cardullo and Kitchin 2019), despite the emphasis on governance and power relations as an important aspect of social resilience concepts (Manca et al. 2017). The ever-present mixing, muddling, clutching together of both Smart City ideas with Resilient City ones, and many other normative urban development concepts (safe, green, inclusive, intelligent, circular, etc.) can undermine their very own goals and purposes. Kaika (2017) argues that "Smart cities and ICTs cannot be the solution because, in fact, they are part of the [socio-ecological] problem," and that there is "ample evidence that 'green' development agendas have been driving new forms of displacement and 'environmental/ecological gentrification' in the global South."

In this chapter, we deliberately refrain from endorsing or rejecting Smart Cities or Resilient Cities as a whole. Instead, we acknowledge and build upon both the supportive and critical work of improving the utility of Smart City and Resilient City concepts in supporting and guiding urban development practice, while "disenchanting" the discourse around them. In particular, we want to highlight the distinct relationship of Resilient Cities and Smart Cities with one another, since the literature review here has shown that both concepts have common ground in their systemic approach to urban development and are also united by critical arguments against them. More specifically, we will analyze the question of how key aspects of Smart Cities, such as the role of new technologies, synergize, or clash with Resilient Cities, and how in turn risk management and other key elements of Resilient Cities contest or support Smart Cities. We do so by looking at both the evidence in support of the optimistic *thesis* that Smart City approaches can synergize with Resilient Cities and vice versa, as well as the evidence in support of the critical *anthesis* that there are indeed conceptual and practical clashes between both concepts that inevitably lead to trade-offs in urban development.

16.3 The Thesis: Smart Cities Support Resilient Cities

16.3.1 The Positive Narrative of Smart Technology Applications for Resilient Cities

According to the positively framed narrative of Smart Cities, digitalization and the rapid technological progress of recent decades has led to an opportunity of using technology as a way of "mak[ing] cities more intelligent in a variety of aspects including efficiency, effectiveness, productivity, transparency, and sustainability," frequently through the use of various industrial and ICT technologies (Gil-Garcia et al. 2015). In practice, Smart Cities are often comprised of initiatives undertaken by local governments and other public and private stakeholders to solve pressing city-level problems arising from urbanization in addition to the development, maintenance, and improvement of critical infrastructure (e.g., communications, water, energy, mobility, buildings), while improving public services to the urban population and increasing living standards (Silva et al. 2018). They also improve the information about urban systems through novel forms of data generation and analysis to better inform policy and decision-makers and to ultimately automate some urban processes to increase performance and efficiency. This, relating to the argument, supports urban resilience in that it can better inform policymaking, help identify risks, break administrative silos, integrate infrastructures for better efficiency, and provide an avenue for cities to "build back better" in the aftermath of a disaster.

Major categories of technology featured in Smart Cities involve various forms of digital connectivity including the Internet of Things (IoT), cloud computing, data analytics, sensors and—increasingly—automated actuators, Blockchain, deep learning, and artificial intelligence, with these different types of technology being applied to a vast spectrum of cases (Toh et al. 2020). Typical urban challenges such as security and crime, pollution, climate change impacts, economic fluctuations, social conflict, and terrorism have the opportunity to be tackled from a different angle by deploying scalable technological solutions. For this reason, it can be argued that Smart Cities contribute to cities' mitigation and adaptation of many forms of disaster risk and stress by protecting critical infrastructure and monitoring the probability of hazards, which in turn, informs decision-makers and other relevant stakeholders (Zhu et al. 2019).

The Internet of Things (IoT) is a key technological cornerstone of Smart Cities (Colding and Barthel 2017). It can be understood as "a radical evolution of the current Internet into a ubiquitous network of interconnected objects that not only harvests information from the environments (sensing) and interacts with the physical world (actuation), but also uses existing Internet standards to provide services for information transfer, analytics, and applications" (Jin et al. 2014). As projected by the UN, it is estimated that by 2050, 68% of the global population will be living in cities (UN. Population Division 2019). The consequence of an expanding urban population into becoming an increasingly digitalized one under global trends includes a surge of smart devices, which will consecutively aid in the development of the Internet of Things (Zhang et al. 2021). In addition, IoT sensors can be used in Smart Cities to gather high volumes of data to monitor the structural integrity of buildings and other infrastructure and thus, increase the overall resilience of a city in the event of a hazard (Zhu et al. 2019). For example, they are effective for monitoring temperature, air pollution and noise levels, humidity, smoke, monitoring traffic, or even detecting the movement and behavior of crowds to identify dangerous situations in public spaces. Such data can be analyzed to recognize, predict, and even prevent hazardous events from happening.

Data analysis itself has also radically evolved and now holds enormous potential in being used to support the resilience of cities. *Big data* approaches for combining different types of information have the ability to reveal patterns across different datasets, cross-referencing with social media traffic and content, along with other ways to better understand the multi-dimensional and complex phenomena of cities. While this increases the amount of data to be handled beyond the capacity of conventional data interpretation, *artificial intelligence (AI)*—and its technological subsets like machine learning and deep learning—have become a promising technology for transforming the way cities can analyze data generated by IoT (Bhattacharya et al. 2020). AI deep learning enables machines to mimic human intelligence and decision-making capabilities, and these skills can rapidly increase systemic response time to identified hazards or stresses. Deep learning can be used to increase cybersecurity, and therefore improve digital resilience of cities (Chen et al. 2021) as a new form of risk that, as we will argue, increases proportionately with the promotion of Smart Cities.

Blockchain, often used synonymously with distributed ledger technology (Treiblmaier et al. 2020), is another key technology for Smart Cities with multiple applications and use cases (Bagloee et al. 2021; Bhushan et al. 2020; Hughes et al. 2019). Arguably, Blockchain was invented to foster resilience of the global financial system, as its origins lie in Bitcoin, a virtual currency proposed by an anonymous individual or group under the pseudonym Satoshi Nakamoto as a response to the 2008 financial crisis (Nakamoto 2008). Its essential value lies in creating trust for transactionary peer-to-peer systems without the need for a centralized regulatory agent, by replacing it with a transparent record of transactions replicated throughout the system. In other words, it allows decentralized systems that can be used to create transparency and democratic ownership, while establishing effective public service delivery and trust in governance bodies (Bagloee et al. 2021). It has the potential to provide entirely new forms of e-governance, public services, and share information in a way that is credible and reliable for citizens (Zhu et al. 2019) while protecting their privacy. Blockchain solutions can allow citizens to provide sensitive personal data in confidence, rest assured that it will not be at risk from imminent security threats (Bagloee et al. 2021), and it has the ability to ensure that data records are secure and not apt to being tampered with. For example, data on water and energy use, along with medical data relevant for pandemic response measures, can be shared securely and privately while ensuring the integrity of data used for decision-making (Lin et al. 2018).

Following this logic and narrative of innovative solutions available under the conceptual umbrella of Smart Cities, many local governments around the globe are already implementing and experimenting with such initiatives. As further illustration, we provide an ensemble of case stories from Japan, Vanuatu, Haiti, and India.

16.3.2 Case Examples to Support the Thesis

One older example of an IoT-powered approach to improve resilience toward natural disasters can be found in Japan, where in February 2007, the Japanese government introduced an emergency warning system for natural hazards, along with the early warning system involving seismometers to monitor seismic activity and detect shockwaves. Primary shockwaves (P-Waves) are analyzed by computers in order to forecast how powerful the secondary waves (S-Waves) will be. If the system predicts the S-Waves will be above a magnitude of 5.0, a warning alert is released and response mechanisms are triggered. Four years after this implementation, the country suffered a 9.0 magnitude earthquake which was followed immediately by a tsunami, threatening the Tokyo metropolitan region among others. Springing into action, the system triggered a range of response mechanisms and government reactions which limited the fatalities and economic damage. The warning system, for example, caused the Shinkansen bullet trains to automatically come to a halt when the first tremors were felt and, for this reason, no trains derailed, thus protecting all passengers on board. Such passengers in Tokyo's subways were also evacuated, surgical operations were paused, gas was disconnected, and nuclear reactors were shutdown. Furthermore, all runways at Haneda and Narita Airport were closed, and 86 flights were safely diverted. On the ground-level, the Tokyo Electric Power Company (TEPCO) prepared for a power outage by prioritizing hospitals, power and nuclear plants, traffic control agencies, and the government bodies tasked with collecting and analyzing data. In terms of communicating this emergency with the population, the major mobile networks delivered an SMS warning which was sent out via the Japanese Broadcasting Corporation (NHK). This example demonstrates the importance of a city's capability to collect, process, analyze, and distribute information compiled from sensors as this can dramatically increase the resilience of a city during a time-sensitive event (GSMA 2013). This case also illustrates that Smart City technologies have been successfully used to increase urban resilience in the metro-area and even regional and country level, and are, in some areas, well-established and far beyond experimental.

Blockchain technology, on the other hand, is arguably still in its infancy as far as real-world examples are concerned, and even more so when it comes to illustrating its use to improve the resilience of cities. A range of examples from cities, such as Dubai and Singapore, has not yet moved far beyond the strategy level as of now and are largely following procedural efficiency improvement agendas, such as reducing paper documents in administrative processes or creating new payment systems (Manushaqa et al. 2019). However, there are some interesting examples emerging from cooperatives between non-governmental organizations that are more in line with resilience challenges. Oxfam, for example, has piloted a Blockchain-powered system for disaster relief spending, cash, and voucher assistance in Port Vila, Vanuatu (Rust 2019). The Unblocked Cash pilot investigated whether Blockchain can reduce the cost, transaction time, improve transparency, security, and overall user experience for relief hand-outs. The project distributed "966,443 Vanuatu Vatu to 187 households and 29 vendors via the Ethereum token DAI, 'wrapped' in a Crypto Collateralized Voucher and issued near-field communication cards designed for low-connectivity environments as a means of payment for goods through a local vendor network." This resulted in modest cost and time savings, and reduced recipient enrollment time in the support scheme to a few minutes compared to over an hour during previous cash assistance programs. More importantly, the Blockchain system also eliminated the need for identity verifications and reduced dependency on post offices or banks to deposit checks and/or withdraw cash. This example demonstrates how well Blockchain can create faster or at least redundant transactional infrastructures that may provide a necessary alternative to analogue infrastructures, especially in the face of systemic stress situations. Other promising examples for Blockchain applications include applications for aggregating and integrating medical data in a centrally accessible place to improve cities' responses to health-related emergencies, including pandemics. Sharing highly sensitive data in an accessible way without violating privacy of individuals is a major issue that can be resolved by distributed ledger technology which creates anonymity and trust. The MIT Media Lab has been developing a Blockchain solution that allows medical data to be shared by practitioners and healthcare institutions efficiently (Treiblmaier et al. 2020). Such systems could in future also inform policy-makers, for example, in deciding on vaccination priority schemes or for a more nuanced lock-down and social distancing policies which address outbreak hotspots rather than subjecting entire cities or countries to rigid lock-down measures. These examples illustrate that there is emerging evidence for how Blockchain technologies serve to make cities more resilient.

In a recent example, *artificial intelligence* and *machine learning* have been used in Port-au-Prince and other Haitian cities and communities to analyze mobile phone generated and other remote sensing data in order to map the estimated income poverty, inequality, and other social vulnerability factors (Pokhriyal et al. 2020). Unlike other Smart City initiatives that are intended to replace analogue systems in pursuit of systemic or process efficiency gains, this initiative explicitly aims to create a redundant form of information gathering and is complementary to regular household surveys and census. In the report of this case, the Interamerican Development Bank explains that this approach addresses a major information gap about the social resilience in many developing country cities, who may not be able to afford the 186 USD survey cost per household and, as a consequence, often do not have sufficient and up-to-date information on their level of poverty relating to social resilience. The tested machine learning solution showed that some "communes located in the Sud departments became increasingly more deprived than the rest between 2014 and 2019, likely due to Hurricane Matthew in late 2016," and provided important insight into the effects of environmental disasters in cities like Haiti. This goes to show that there are clear uses for novel data analysis mechanisms in support of resilience, even in vulnerable and developing communities, which go beyond more well-known Smart City examples such as early warning systems for fire hazards (Zhang et al. 2021) and improved traffic management or smart grids, which are suitable for more affluent cities (Ullah et al. 2020).

Lastly, the Smart Cities concept has been used in India in order to drive and accelerate urban development in general, rather than focusing on technological aspects. Responsible for one third of the global South's population (Prasad and Alizadeh 2020), the Indian government in 2015 announced an ambitious Smart Cities Mission to invest and develop 100 Smart Cities until 2020 (Khare 2019). The City of Bhubaneswar was among the first beneficiary communities of the program and subsequently ranked as the country's smartest city (Pradhan 2016). An impact assessment on the 2019 cyclonic storm "Fani" in Bhubaneswar and its vulnerability across a range of social, financial, and physical infrastructure indicators found that India's "smartest" city was indeed "more capable than vulnerable" (Kawyitri and Shekhar 2021) in light of extreme events and external shocks. It is important to highlight, however, that the measures that led to this result were not attributed to the use of technologies like IoT, Blockchain, or artificial intelligence. Cited examples of the same study with a beneficial effect for resilience included a "'Smart Park' project, [in which] three existing city parks were redeveloped to allow more social gatherings for public walking, open gym practices, yoga practice, etc.," and a project that could have increased the resilience if it was not for a lack of progress at the time was the "'Integrated Improvement Infrastructure', [which included works] like installing efficient utility services for water supply, sewage management, underground electric power lines, and solid waste management" (Kawyitri and Shekhar 2021). The City of Bhubaneswar potentially illustrates a trend among India's Smart Cities that includes technological infrastructure at times, but does not prioritize it. In fact, one study shows that only 8 out of 20 examined Indian Smart Cities actually feature technological infrastructure, but all 20 of them feature elements of improving basic services and transportation infrastructure (Prasad and Alizadeh 2020). This example illustrates how the hype about Smart Cities as a normative concept has been expropriated in practice to fund and advance traditional urban development.

16.4 The Antithesis: Smart Cities Harm Resilient Cities

16.4.1 The Critical Narrative of New Urban Hazards as a Result of Digitalization

Following a critical narrative, digitalization and technological development as a driving force for Smart City opportunities also give rise to a new set of serious risks and challenges to cities that are perpetuated by Smart Cities or at least come as "part of the package." Smart City technologies from IoT to cloud computing and artificial intelligence to e-government and digital services, create dependencies on new types of critical infrastructure, perpetuate social vulnerabilities, and increase the spending of public funds on expensive technological gadgets rather than focusing on solving more pressing urban development matters. New risks introduced range from physical damage to critical infrastructure like server centers as described in the introduction to this chapter, to cybersecurity issues and cyberattacks (Chen et al. 2021; Vitunskaite et al. 2019), or privacy issues of protecting citizen data in light of ever increasing amounts of urban data mining and difficult legal and regulatory issues, such as the question of regulating urban air space for the emerging use of drones or "air taxis" in cities (Straubinger et al. 2020). Creating new services that require access to certain devices like smartphones and a certain amount of technological literacy creates and increases a socio-economic and age-related digital divide, along with governance changes in unintended ways based on the shifting use from media and interaction channels between government and citizens (Johnson et al. 2020; Lam and Ma 2019; van Dijk 2006). Since all of these risks and challenges are directly or indirectly related to global digitalization, we proceed to call the need for this entirely new type of resilience in cities—and elsewhere—*digital resilience.*

Cybersecurity issues or cyber risk are perhaps the first group of challenges that come to mind in regards to digital resilience. Cyberattacks have been a concern and reality in cities for quite some time (AlDairi and Tawalbeh 2017; Chen et al. 2021; Kitchin and Dodge 2019), and there are a variety of attacks that can be made on Smart Cities. Ransomware attacks have successfully led to blackmailing public authorities into paying large sums of money in exchange for regaining access to stolen or blocked off data ("Florida Cities Pay up After Crippling Ransomware Attacks," 2019). Other forms of attacks cripple IoT structures and stop them from functioning and providing services (Altaf et al. 2021). As such, "IoT devices can be compromised in a number of ways e.g. connected into a botnet, made inoperable by a worm or used to penetrate the inner networks and systems" (Vitunskaite et al. 2019). With the global Smart City market being estimated to reach a volume of over 150 billion USD as early as 2022, an even higher amount of financial resources is needed to ensure cybersecurity of Smart City systems to make sure they are secure, reliable, and scalable (Chen et al. 2021). Given the global uptake of implementing smart city technology, evidence suggests that cyberattacks to Smart Cities are increasingly common and can result in considerable damage (Vitunskaite et al. 2019).

Social risks and challenges tied to digital resilience include threats to surveillance and violation of *privacy*, issues of socio-economic inequality creating—and created by—*digital divides, governance* issues, and *loss of citizen agency* (Bettencourt 2019) as traditional formal and informal governance institutions, democratic decision-making processes, and public negotiation are changed or replaced by digital supported ones, often with an underlying neoliberal agenda driven by corporate technology providers (Cardullo and Kitchin 2019). Interlinked with this is a certain bias toward socio-economic formality, in the sense that those often excluded from Smart City initiatives are the urban poor, actors within the informal economy, or inhabitants of informal settlements (Willis 2019).

Finally, there is a largely unexplored dimension of *resource dependency and supply chain vulnerability* of digital technologies which are increasingly becoming a hazard to cities the more they follow the Smart City approach. Within this space, a more well-known and documented element is the rapid increase of power demand from global digital infrastructure. While there are some arguments for global energy savings through the deployment of technological solutions to increase energy efficiency, evidence suggests that global power demand has been growing at 3% per year and "that the production and operation of ICT will rise to 21% of global electricity consumption by 2030" (Morley et al. 2018). What is more, there is an emerging body of research looking into the role of Smart Cities in increasing the demand for critical raw materials (David and Koch 2019; King 2021; Wäger et al. 2015), which are irreplaceable and simultaneously highly dependent on a well-functioning and uninterrupted global supply chain. Smart Cities therefore tend to erode the *economic and material resilience*, reducing economic and resource autonomy of urban systems and making them vulnerable to geopolitics and global material shortage crises.

Following this logic and narrative of new and existing hazards and risks introduced or perpetuated by Smart Cities, local governments around the globe need to carefully consider the adverse effects of Smart City solutions on the resilience of their urban systems. As further illustration, we provide examples from the United States, Canada, Japan, and a current global resource shortage phenomenon.

16.4.2 Case Examples Supporting the Anthesis

In 2018, Atlanta, Georgia was targeted in the worst cyber assault of any U.S. city up to that date (Kearney 2018). The incident that held more than one third of the city's 424 software programs hostage by encrypting decades worth of city council member's correspondence, preventing residents from using online billing services, crippling access to legal documents, and delaying court case and warrants from being processed, sent city officials into a frenzy, thereby "essentially forc[ing] an aspiring Smart City to revert to pen and paper" (Poon 2018). The infiltration by the cybercriminals, Samsam, directed their assault on compromising vulnerable, outdated systems with weak passwords before a breach was made (Freed 2018). In return for the files, $51,000 in Bitcoin cryptocurrency was demanded; an anonymous and secure way for the culprits to cover their tracks. For weeks, the city government was woefully disorganized and struggled to back up the missing files, while the newly elected mayor admitted to severely overlooking cybersecurity and addressing outdated software. Instead of paying the meager ransom, despite their $2.1 billion annual budget, the city was frantically investing in IT personnel that amounted to $2.7 million in emergency contracts. By doing this, Atlanta's city government and FBI sent a clear message that such an easy and tempting way for cybercriminals to make financial gains will not be tolerated, while encouraging other vulnerable cities to not give in to malware ransomware. However, fears surrounding the use of digitalization and momentum of Smart Cities have brought unique concerns of similar incidents potentially involving the hostage of smart grids or other smart devices in return for cryptocurrency. As a result, the need for up-to-date security measures and internal education with city employees will be necessary to avoid such cyberattacks.

Protecting *privacy*, citizen interests, and the integrity of democratic governance in local communities while ensuring citizen inclusion is another challenge in Smart Cities. This became evident in a large-scale Smart City experiment by Sidewalk Labs in 2019 (Berger 2020). As a sister company of Google, Alphabet's Sidewalk Labs outlined an ambitious plan in Toronto, Canada to construct a neighborhood comprised of commercial, residential units, and office buildings, with the infrastructure company hailing it as a "neighborhood built from the internet up" and "the most innovative district in the world" (Hawkins 2019). Such innovations in this 1.3 billion USD investment involved the use of mass timber to construct eco-friendly and affordable housing in the mixed-use community, which could have had the potential to spearhead similar initiatives in the construction industry for green housing (Berger 2020). In addition, the initiative planned for the installation of public Wi-Fi, a pneumatic garbage collection system, favoring walkability and bicycling over car usage by implementing street systems to limit traffic with a promise to slash greenhouse gas emissions by 89%, and a list of other technology use cases. Sidewalk Labs also estimated that the project would lead to 44,000 job opportunities, 4.3 billion USD in annual tax revenue, and 38 billion USD of private sector investment over the course of twenty years. In May 2020, Sidewalk Labs suddenly announced that the project had been discontinued, officially due to the economic impacts of COVID-19. However, by then, there had been a large number of criticisms and warnings voiced about privacy concerns and corporate interest taking over the urban space and governance in Toronto ("Surveillance Capitalism," 2019). Via public Wi-Fi and other sensors, the neighborhood Sidewalk Labs had been collecting "urban data" from, insisting that they were needed to provide informative housing, traffic, and other urban system assessments. Critics were worried that Google would ultimately use these datasets to collect further personal information and opposed the notion of big technology companies profiteering, gentrifying, and experimenting on the city. Soon after, plans of Sidewalk Labs were published that expressed the company's interest in expanding the project from a 12-acre lot to 190-acres, public backlash intensified and the project was stopped ("Google Affiliate Sidewalk Labs Abruptly Abandons Toronto Smart City Project," 2020). Arguably, such a case would not have been framed and publicly discussed in this fashion within most other countries around the world, as there is little global consensus on privacy and citizen rights to the same. Nevertheless, this case vividly illustrates the democratic deficit that comes not from the application of smart technology itself, but from the neoliberal and corporatist agenda often embraced by the Smart Cities concept (Cardullo and Kitchin 2019; Kaika 2017; Willis 2019).

Finally, in regards to *economic and material resilience*, or rather the dependency of Smart Cities on supply chains, documented case examples to support the antithesis do not yet serve justice to the full extent of the risk provided. Some evidence from the Fukushima disaster in Japan shows that the power outage caused by the Tsunami in 2011 had a cascading effect on some of the smart disaster response systems (Sakurai and Kokuryo 2015). Backup generators were either not in place or not functional at the time, causing them to fail and in turn increase the extent of the disaster. Overall, the potential risks of resource shortages that could affect Smart Cities are, however, much more severe and perhaps best illustrated by the 2021 semiconductor shortage crisis (the last but less severe shortage of this sort also happened ten years earlier as another result of the disaster in Fukushima) (SCMP Reporters 2021). Semiconductors are the backbone of any microchip and needed to power anything that is "smart" about ICT and, therefore, Smart Cities. The shortage affected most areas of electronics from IoT devices, smart phones, computers, and car manufacturing industries and caused significant production interruptions and delays (Sweney 2021). Arguably, a delay of the newest generation smart phone to enter the market may hardly represent a major threat to cities, and more research on the effects of the semiconductor shortage on Smart Cities is needed before detailed insights and evidence can be provided. However, the shortage experience already adds to resource security issues debated by national governments, as semiconductor production is largely dominated by China, Taiwan, and South Korea (CSIA 2020), and to a global struggle among major corporations for priority access to critical supply chains (Sweney 2021). Material and supply chain resilience is a key factor that is largely ignored by both research and urban development practice on Smart Cities.

16.5 Toward a Synthesis: Discussion and Conclusions

There is a lot of common ground between Smart Cities and Resilient Cities. Both concepts treat the city as a multi-dimensional "system of systems" or draw parallels to ecosystems in the way that cities integrate social, environmental, institutional, economic, physical and digital processes, and infrastructures into a holistic and highly complex whole (Colding et al. 2020). Both concepts seek to create positive transformative effects throughout all these individual domains by improving, strengthening, and changing individual systemic aspects through a perpetual process of innovation and adaptation. In the examples from Tokyo and Port-au-Prince, this is achieved through increasing the amount of available data, aggregating, and processing information via machine learning and automation. The Port Vila example demonstrates how technologies like Blockchain can be used to overcome trust issues to allow for more effective allocation of relief fund distribution. Arguably, all the case examples simultaneously represent good examples of both Smart City and Resilient City initiatives. The apparent difference here between both concepts is that while the Smart City component in these examples lies in the application of a technological use case, the Resilient City aspect lies in the goal setting; the resilience lens allows for the identification of a threat (e.g., an earthquake in a tectonically active region of Japan) or an opportunity to improve certain systemic coping abilities (e.g., improved health information systems). A first observation therefore is that while both Smart Cities and Resilient Cities have normative components, and resilience provides a concept more focused on a beneficial goal or outcome, while Smart Cities are more focused on tools, instruments, and applications. In other words, if Smart Cities are about providing solutions, Resilient Cities are about identifying problems to solve.

On the other hand, the pursuit of Smart Cities as a goal in itself holds much potential clash with the goals of Resilient Cities. The cases from Atlanta and Toronto clearly show that cybersecurity risks are a new and serious type of hazard to cities, with the neoliberal nature of normative aspects for the Smart City concept eroding and contorting social resilience and local governance. Therefore, it is important to point out that the concept of Smart Cities, despite its powerful and necessary normative narratives that speak to its benefits of technological progress, becomes problematic if it is used for goal setting and decision-making on urban development itself. There is a conceptual difference by design between Smart Cities and Resilient Cities regarding their inherent relationship to risk. The Resilient City concept sees the identification, management, and even utilization of risk and awareness of vulnerability to systemic shock and stress factors, as a main motivation for decision-making and local action. For Resilient Cities, potential risks and vulnerabilities or experienced stresses are a guideline for decision-making and a driver for change. Increasing cybersecurity, overcoming a digital divide, or adapting to climate change induced heat waves becomes a call to action and, in addressing them, an opportunity to simultaneously generate additional positive effects ("build back better"). Smart Cities may share this notion, as long as it provides a task or a challenge that can be solved with the technological approaches and the fruits of digitalization. However, a potential conflict between both concepts arises when digitalization and technological applications themselves become the cause of new risks or increased vulnerabilities, as is the case with cybersecurity of IoT applications, the digital divide that prevents universal access to services, and weakened social resilience and integrity, and the largely underestimated resource dependencies introduced by Smart City solutions, which we are just beginning to experience in practice. In these cases, risks become an obstacle for the advancement of the Smart City ideal for a technologically advanced and innovative city. Furthermore, accepting risk is part of the pioneering and experimentation idea inherent to the Smart City concept, and the Sidewalk Labs case from Toronto clearly follows this logic of deliberately trading the privacy of inhabitants for the sake of advancing technological maturity. For this reason, a shortcoming of the Smart City concept from a resilience perspective is a possible "blind spot" for detrimental side effects of technological solutions.

Therefore, we argue that Smart Cities should be seen as a "toolbox" concept that is fit to help achieve targets set via a different normative paradigm. They should not be used as an urban development paradigm in and of their own. This does not, however, render the values and ideals of the Smart City concept useless or to be disregarded. Private sector and private capital involvement in urban development, the importance to test and experiment with technologies in real urban contexts, and a positive fascination for technological progress itself are a necessity in order to improve Smart Cities, and they help inform urban development policies on conditions and trade-offs necessary when applying ICT technologies in cities. But they should be taken with a grain of salt and not be misused to derive some deductive ideological imperative from the Smart Cities concept that propagates the use of technology for technology's sake. Smart City practitioners should be fully aware of the conceptual limitations of Smart Cities. In turn, critical views on Smart Cities should not reject the concept outright, but rather help to objectively weigh the cost of Smart City propositions against the benefits they provide. The Resilient City concept should not reject or overlook technological solutions valuable for increasing the adaptation capacity and sustainability of cities in the long run. Rather it may weigh support in evaluating them in comparison with other non-technological approaches, such as nature-based solutions or social innovation.

Adapting established resilience principles (e.g., Biggs et al. 2015) to digital and technological systems, and incorporating resilience principles into the Smart Cities concept promises a way to improve the practical implementation of the Smart City concept and increase its maturity. To this end, Colding et al. (2020) propose the adoption of seven resilience principles for what they call Digital Cities, i.e., maintain diversity and redundancy, manage connectivity, manage slow variables and feedbacks, foster complex adaptive systems thinking, encourage continuous learning, broaden participation, and promote polycentric governance. This proposal is a major conceptual advancement toward improving Smart Cities. However, the subject of *digital resilience* we propose here in this chapter needs much more attention, in research and even more so in urban development practice and policy. Raising awareness and building the capacity of local governments, decision-makers, the private sector, and citizens through case studies, aids in making abstract resilience issues of Smart Cities tangible and could improve the application of these principles outside the academic discourse.

Digital resilience in Smart Cities should not be limited to the technical aspects of technologies and new infrastructure themselves. Rather, it is important for cities to take into account wider social, environmental, and economic risks stemming from digitalization as a global trend and which are perpetuated by Smart Cities in the current form of the concept. The unhampered increase in demand of electricity and critical raw materials, the digitalization of all aspects of urban life and the erosion of established, formal and informal governance, and civic institutions are a major issue that cities need to understand and address. The Smart City concept is not fit for purpose to address these issues, and even resilience is only able to highlight and explain parts of these phenomena. Circular City concepts, equitable and social development, and a renaissance of the much watered down and eroded discourse of the still relevant-as-ever urban sustainability itself are perhaps the best conceptual lenses to address the complexity of these issues.

In conclusion, we argue in this chapter that both Smart Cities and Resilient Cities should be seen as two well-established concepts of urban development theory, which can provide valuable insights for research and practice. Rather than heralding one concept as a panacea or superior ideology for how best to change and shape our cities, they should be used as lenses or filters to help make differentiated decisions and improve decision-making and governance in cities. This also means that it is not advisable to muddle, relativize, or attempt to "fix" Smart Cities or Resilient Cities as a concept in an attempt to make one or the other a catch-all term for better urban development. Rather, the conceptual clarity and the focus of Smart Cities and

Resilient Cities as clearly distinguishable "black-and-white" concepts improves their utility as tools for urban development practice, planning, and decision-making.

References

AlDairi A, Tawalbeh L (2017) Cyber security attacks on smart cities and associated mobile technologies. Procedia Comput Sci 109:1086–1091. https://doi.org/10.1016/j.procs.2017.05.391

Altaf A, Abbas H, Iqbal F, Khan MMZM, Rauf A, Kanwal T (2021) Mitigating service-oriented attacks using context-based trust for smart cities in IoT networks. J Syst Arch 115:102028. https://doi.org/10.1016/j.sysarc.2021.102028

Antwi-Afari P, Owusu-Manu D-G, Simons B, Debrah C, Ghansah FA (2021) Sustainability guidelines to attaining smart sustainable cities in developing countries: a Ghanaian context. Sustain Futur 3:100044. https://doi.org/10.1016/j.sftr.2021.100044

Araral E (2020) Why do cities adopt smart technologies? Contingency theory and evidence from the United States. Cities 106:102873. https://doi.org/10.1016/j.cities.2020.102873

Bagloee SA, Heshmati M, Dia H, Ghaderi H, Pettit C, Asadi M (2021) Blockchain: the operating system of smart cities. Cities 112:103104. https://doi.org/10.1016/j.cities.2021.103104

Berger B (2020) Sidewalk Labs' failure and the future of smart cities (Global media platform). TriplePundit, June 16. https://www.triplepundit.com/story/2020/sidewalk-labs-failure-smart-cities/120616

Bettencourt LMA (2019) The digital transformation of metropolises. Metropolis. https://www.metropolis.org/sites/default/files/resources/Observatory_Issue-Paper-08_Digital-transformation-metropolises.pdf

Bhattacharya S, Somayaji SRK, Gadekallu TR, Alazab M, Maddikunta PKR (2020) A review on deep learning for future smart cities. Internet Technol Lett:e187. https://doi.org/10.1002/itl2.187

Bhushan B, Khamparia A, Sagayam KM, Sharma SK, Ahad MA, Debnath NC (2020) Blockchain for smart cities: a review of architectures, integration trends and future research directions. Sustain Cities Soc 61:102360. https://doi.org/10.1016/j.scs.2020.102360

Biggs R, Schlüter M, Schoon ML (2015) Principles for building resilience: sustaining ecosystem services in social-ecological systems. Cambridge University Press, Cambridge

Bizzotto M, Huseynova A, Estrada VV (2019) Resilient cities, thriving cities: the evolution of urban resilience. ICLEI—Local Governments for Sustainability. https://e-lib.iclei.org/publications/Resilient-Cities-Thriving-Cities_The-Evolution-of-Urban-Resilience.pdf

Bjørner T (2021) The advantages of and barriers to being smart in a smart city: the perceptions of project managers within a smart city cluster project in Greater Copenhagen. Cities 114:103187. https://doi.org/10.1016/j.cities.2021.103187

Cardullo P, Kitchin R (2019) Smart urbanism and smart citizenship: the neoliberal logic of 'citizen-focused' smart cities in Europe. Environ Plan C: Polit Space 37(5):813–830. https://doi.org/10.1177/0263774X18806508

Chen D, Wawrzynski P, Lv Z (2021) Cyber security in smart cities: a review of deep learning-based applications and case studies. Sustain Cities Soc 66:102655. https://doi.org/10.1016/j.scs.2020.102655

Colding J, Barthel S (2017) An urban ecology critique on the "Smart City" model. J Clean Prod 164:95–101. https://doi.org/10.1016/j.jclepro.2017.06.191

Colding J, Colding M, Barthel S (2020) Applying seven resilience principles on the Vision of the Digital City. Cities 103:102761. https://doi.org/10.1016/j.cities.2020.102761

CSIA (2020) Building a more resilient ICT supply chain: lessons learned during the COVID-19 pandemic. Cybersecurity and Infrastructure Security Agency. https://www.cisa.gov/publication/ict-supply-chain-lessons-learned-covid-19

David M, Koch F (2019) "Smart is not smart enough!" Anticipating critical raw material use in smart city concepts: the example of smart grids. Sustainability 11(16):4422. https://doi.org/10.3390/su11164422

de Jong M, Joss S, Schraven D, Zhan C, Weijnen M (2015) Sustainable–smart–resilient–low carbon–eco–knowledge cities; making sense of a multitude of concepts promoting sustainable urbanization. J Clean Prod 109:25–38. https://doi.org/10.1016/j.jclepro.2015.02.004

Florida cities pay up after crippling ransomware attacks (2019) Comput Fraud Secur 2019(7):1–3. https://doi.org/10.1016/S1361-3723(19)30067-3

Freed B (2018) Atlanta was not prepared to respond to a ransomware attack (News organization). Statescoop, April 24. https://statescoop.com/atlanta-was-not-prepared-to-respond-to-a-ransomware-attack/

Gil-Garcia JR, Pardo TA, Nam T (2015) What makes a city smart? Identifying core components and proposing an integrative and comprehensive conceptualization. Inf Polity 20(1):61–87. https://doi.org/10.3233/IP-150354

Google affiliate Sidewalk Labs abruptly abandons Toronto smart city project (2020) The Guardian, May 7. http://www.theguardian.com/technology/2020/may/07/google-sidewalk-labs-toronto-smart-city-abandoned

GSMA (2013) Smart city resilience: learning from emergency response and coordination in Japan. Groupe Speciale Mobile Association, pp 1–5. https://www.gsma.com/iot//wp-content/uploads/2013/02/cl_SmartCities_emer_01_131.pdf

Hawkins AJ (2019) Alphabet's Sidewalk Labs unveils its high-tech 'city-within-a-city' plan for Toronto (News). The Verge, June 24. https://www.theverge.com/2019/6/24/18715486/alphabet-sidewalk-labs-toronto-high-tech-city-within-a-city-plan

Hughes L, Dwivedi YK, Misra SK, Rana NP, Raghavan V, Akella V (2019) Blockchain research, practice and policy: applications, benefits, limitations, emerging research themes and research agenda. Int J Inf Manage 49:114–129. https://doi.org/10.1016/j.ijinfomgt.2019.02.005

Ji T, Chen J-H, Wei H-H, Su Y-C (2021) Towards people-centric smart city development: investigating the citizens' preferences and perceptions about smart-city services in Taiwan. Sustain Cities Soc 67:102691. https://doi.org/10.1016/j.scs.2020.102691

Jin J, Gubbi J, Marusic S, Palaniswami M (2014) An information framework for creating a smart city through Internet of Things. IEEE Internet Things J 1(2):112–121. https://doi.org/10.1109/JIOT.2013.2296516

Johnson PA, Robinson PJ, Philpot S (2020) Type, tweet, tap, and pass: how smart city technology is creating a transactional citizen. Gov Inf Q 37(1):101414. https://doi.org/10.1016/j.giq.2019.101414

Kaika M (2017) 'Don't call me resilient again!': the New Urban Agenda as immunology … or … what happens when communities refuse to be vaccinated with 'smart cities' and indicators. Environ Urban 29(1):89–102. https://doi.org/10.1177/0956247816684763

Kawyitri N, Shekhar A (2021) Assessing vulnerability and capacity of Bhubaneswar as a progressive smart-city: an empirical case study of Fani cyclone impact on the city. Int J Disaster Risk Reduct 56:101986. https://doi.org/10.1016/j.ijdrr.2020.101986

Kearney L (2018) Atlanta officials reveal worsening effects of cyber attack (News organization). Reuters, June 7. https://www.reuters.com/article/us-usa-cyber-atlanta-budget/atlanta-officials-reveal-worsening-effects-of-cyber-attack-idUSKCN1J231M?feedType=RSS&feedName=technologyNews

Khare V (2019) India election 2019: have 100 'smart cities' been built? (News organization). BBC News, March 25. https://www.bbc.com/news/world-asia-india-47025472

King A (2021) This is not new. A short history of materials criticality and supply-chain challenges. In: King A (ed) Critical materials. Elsevier, Cambridge, MA, pp 19–51, Chapter 2. https://doi.org/10.1016/B978-0-12-818789-0.00002-5

Kitchin R, Dodge M (2019) The (in)security of smart cities: vulnerabilities, risks, mitigation, and prevention. J Urban Technol 26(2):47–65. https://doi.org/10.1080/10630732.2017.1408002

Lam PTI, Ma R (2019) Potential pitfalls in the development of smart cities and mitigation measures: an exploratory study. Cities 91:146–156. https://doi.org/10.1016/j.cities.2018.11.014
Lin Y-P, Petway JR, Lien W-Y, Settele J (2018) Blockchain with artificial intelligence to efficiently manage water use under climate change. Environments 5(3):34. https://doi.org/10.3390/environments5030034
Manca AR, Benczur P, Giovannini E (2017) Building a scientific narrative towards a more resilient EU society. Part 1: a conceptual framework (No. JRC106265; JRC Working Papers). Joint Research Centre (Seville site). https://ideas.repec.org/p/ipt/iptwpa/jrc106265.html
Manushaqa L, Grant M, Baliakas P, Holotescu T, Amellal J (2019) Blockchain implementation in smart cities
Mendle RS (2011) Selten Unnachhaltig – Seltene Erden und Umweltverschmutzung in China. Stimmen aus China. https://www.stimmen-aus-china.de/2011/04/26/selten-unnachhaltig-seltene-erden-und-umweltverschmutzung-in-china/
Morley J, Widdicks K, Hazas M (2018) Digitalisation, energy and data demand: the impact of Internet traffic on overall and peak electricity consumption. Energy Res Soc Sci 38:128–137. https://doi.org/10.1016/j.erss.2018.01.018
Nakamoto S (2008) Bitcoin: a peer-to-peer electronic cash system. Satoshi Nakamoto Institute, October 31. https://nakamotoinstitute.org/bitcoin/
Ooms W, Caniëls MCJ, Roijakkers N, Cobben D (2020) Ecosystems for smart cities: tracing the evolution of governance structures in a Dutch smart city initiative. Int Entrep Manag J 16(4):1225–1258. https://doi.org/10.1007/s11365-020-00640-7
Pokhriyal N, Zambrano O, Linares J, Hernández H (2020) Estimating and forecasting income poverty and inequality in Haiti using satellite imagery and mobile phone data. Interamerican Development Bank. https://publications.iadb.org/publications/english/document/Estimating-and-Forecasting-Income-Poverty-and-Inequality-in-Haiti-Using-Satellite-Imagery-and-Mobile-Phone-Data.pdf
Poon L (2018) Why are cities so vulnerable to cyber attack? (News organization). Bloomberg CityLab, March 30. https://www.bloomberg.com/news/articles/2018-03-30/atlanta-s-ransomware-attack-could-have-been-much-worse
Pradhan A (2016) Bhubaneswar ranked country's smartest city (Newspaper). The Times of India, January 29. http://timesofindia.indiatimes.com/articleshow/50766312.cms?utm_source=contentofinterest&utm_medium=text&utm_campaign=cppst
Prasad D, Alizadeh T (2020) What makes Indian cities smart? A policy analysis of smart cities mission. Telematics Inform 55:101466. https://doi.org/10.1016/j.tele.2020.101466
Rust B (2019) Unblocked cash: piloting accelerated cash transfer delivery in Vanuatu. Oxfam Australia. https://policy-practice.oxfam.org/resources/unblocked-cash-piloting-accelerated-cash-transfer-delivery-in-vanuatu-620926/
Sakurai M, Kokuryo J (2015) Municipal government ICT in 3.11 crisis: lessons from the Great East Japan Earthquake and Tsunami crisis. Berkman Klein Center. https://cyber.harvard.edu/publications/2012/Municipal_Government_ICT_in_3_11_Crisis
Satter MR (2021) Millions of websites offline after fire at French cloud services firm. Reuters, March 11. https://www.reuters.com/article/us-france-ovh-fire-idUSKBN2B20NU
SCMP Reporters (2021) Everything you need to know about the global semiconductor shortage. South China Morning Post, May 12. https://www.scmp.com/tech/tech-war/article/3133061/why-there-global-semiconductor-shortage-how-it-started-who-it-hurting
Silva BN, Khan M, Han K (2018) Towards sustainable smart cities: a review of trends, architectures, components, and open challenges in smart cities. Sustain Cities Soc 38:697–713. https://doi.org/10.1016/j.scs.2018.01.053
Straubinger A, Rothfeld R, Shamiyeh M, Büchter K-D, Kaiser J, Plötner KO (2020) An overview of current research and developments in urban air mobility—setting the scene for UAM introduction. J Air Transp Manag 87:101852. https://doi.org/10.1016/j.jairtraman.2020.101852

'Surveillance capitalism': critic urges Toronto to abandon smart city project (2019) The Guardian, June 6. http://www.theguardian.com/cities/2019/jun/06/toronto-smart-city-google-project-privacy-concerns
Sverdlik Y (2017) OVH to disassemble container data centers after epic outage in Europe. Data Center Knowledge, November 11. https://www.datacenterknowledge.com/uptime/ovh-disassemble-container-data-centers-after-epic-outage-europe
Sverdlik Y (2021) Fire has destroyed OVH's Strasbourg Data Center (SBG2). Data Center Knowledge, March 10. https://www.datacenterknowledge.com/uptime/fire-has-destroyed-ovh-s-strasbourg-data-center-sbg2
Sweney M (2021) Global shortage in computer chips 'reaches crisis point'. The Guardian, March 21. http://www.theguardian.com/business/2021/mar/21/global-shortage-in-computer-chips-reaches-crisis-point
Toh CK, Sanguesa JA, Cano JC, Martinez FJ (2020) Advances in smart roads for future smart cities. Proc R Soc A: Math, Phys Eng Sci 476(2233):20190439. https://doi.org/10.1098/rspa.2019.0439
Treiblmaier H, Rejeb A, Strebinger A (2020) Blockchain as a driver for smart city development: application fields and a comprehensive research agenda. Smart Cities 3(3):853–872. https://doi.org/10.3390/smartcities3030044
Trencher G (2019) Towards the smart city 2.0: empirical evidence of using smartness as a tool for tackling social challenges. Technol Forecast Soc Chang 142:117–128. https://doi.org/10.1016/j.techfore.2018.07.033
Ullah Z, Al-Turjman F, Mostarda L, Gagliardi R (2020) Applications of artificial intelligence and machine learning in smart cities. Comput Commun 154:313–323. https://doi.org/10.1016/j.comcom.2020.02.069
UNDRR (2015) Sendai framework for disaster risk reduction 2015–2030. United Nations Office for Disaster Risk Reduction. https://www.undrr.org/publication/sendai-framework-disaster-risk-reduction-2015-2030
UN. Population Division (2019) World urbanization prospects: the 2018 revision. Population Studies. Department of Economic and Social Affairs, Population Division, United Nations, pp 1–99. https://population.un.org/wup/Publications/Files/WUP2018-Report.pdf
van Dijk JAGM (2006) Digital divide research, achievements and shortcomings. Poetics 34(4):221–235. https://doi.org/10.1016/j.poetic.2006.05.004
Vitunskaite M, He Y, Brandstetter T, Janicke H (2019) Smart cities and cyber security: are we there yet? A comparative study on the role of standards, third party risk management and security ownership. Comput Secur 83:313–331. https://doi.org/10.1016/j.cose.2019.02.009
Wäger PA, Hischier R, Widmer R (2015) The material basis of ICT. In: Hilty LM, Aebischer, B (eds) ICT innovations for sustainability. Springer, Cham, pp 209–221. https://doi.org/10.1007/978-3-319-09228-7_12
Willis KS (2019) Whose right to the smart city? In: Cardullo P, Di Feliciantonio C, Kitchin R (eds) The right to the smart city. Emerald Publishing Limited, Bingley, pp 27–41. https://doi.org/10.1108/978-1-78769-139-120191002
Yigitcanlar T, Han H, Kamruzzaman Md, Ioppolo G, Sabatini-Marques J (2019) The making of smart cities: are Songdo, Masdar, Amsterdam, San Francisco and Brisbane the best we could build? Land Use Policy 88:104187. https://doi.org/10.1016/j.landusepol.2019.104187
Zhang Y, Geng P, Sivaparthipan CB, Muthu BA (2021) Big data and artificial intelligence based early risk warning system of fire hazard for smart cities. Sustainable Energy Technol Assess 45:100986. https://doi.org/10.1016/j.seta.2020.100986
Zhou Q, Zhu M, Qiao Y, Zhang X, Chen J (2021) Achieving resilience through smart cities? Evidence from China. Habitat Int 111:102348. https://doi.org/10.1016/j.habitatint.2021.102348
Zhu S, Li D, Feng H (2019) Is smart city resilient? Evidence from China. Sustain Cities Soc 50:101636. https://doi.org/10.1016/j.scs.2019.101636

Chapter 17
Resilient Smart Cities in South America: City Diplomacy for Sustainable Urban Development

Rodrigo Perpétuo, Mariana Nicolletti, Pedro Jacobi, and Armelle Cibaka

Abstract The rapid process of urbanization on a global scale and the centrality of the sustainable development agenda in the international context constitute two important inflections in the contemporary world. As a result, the prominence of cities in current international debates has grown so that cities exercise their diplomacy and implement internationally aligned policies through global agreements. In this context, this chapter analyzes the phenomena of city diplomacy and resilience in the sustainable development agenda of South American municipalities. Resulting from the discussion on the smart cities approach in the context of the region, it recognizes the uncertainty, complexity, interdependence, and multidimensionality of contemporary phenomena from a resilience perspective. The exercise of international relations at the municipal level in South America also has the support of local government networks and associations, such as ICLEI—Local Governments for Sustainability, which with the Smart Cities for Climate Analytical Report, published in 2017, has enabled the materialization of the perspective of convergence between the climatic action of cities and the strengthening of their governance and institutional capacity, as well as their capacity for innovation. This convergence is illustrated by the case of the Metropolitan Region of Campinas, which combines a regional program to enhance biodiversity and its socioeconomic benefits, a long-term vision of resilience and sustainable development, and a municipal Smart City Strategic Plan.

R. Perpétuo (✉)
Regional Director, ICLEI South America, Sao Paulo, Brazil
e-mail: rodrigo.perpetuo@iclei.org

M. Nicolletti
Low Carbon and Resilience Regional Manager, ICLEI South America, Sao Paulo, Brazil
e-mail: mariana.nicolletti@iclei.org

P. Jacobi
President of the Regional Directive Council, ICLEI South America, Sao Paulo, Brazil

A. Cibaka
Knowledge Management and Youth Regional Coordinator, ICLEI South America, Sao Paulo, Brazil
e-mail: armelle.cibaka@iclei.org

A. Sharifi and P. Salehi (eds.), *Resilient Smart Cities*, The Urban Book Series,
https://doi.org/10.1007/978-3-030-95037-8_17

Keywords City diplomacy · Resilient development · Adaptation · Smart cities · South America

17.1 Introduction to City Diplomacy for Resilient Development in South America: International Society and City Diplomacy for Resilient Development

The contemporary world is experiencing two complementary inflections in its international ecosystem. The first concerns the rapid urbanization process, which, although consolidated in some parts of the world, such as South America, is already resonating as an inevitable phenomenon globally. The second refers to the centrality that the sustainable development agenda also occupies in the international context. Global agreements such as the Aichi Biodiversity Targets (2010), the Paris Agreement (2015), the pacts around the United Nations Sustainable Development Goals (2015), and the guidelines for a New Urban Agenda (2016) illustrate this movement.[1]

One of the main consequences of these inflections is that cities occupy an important centrality in current international debates, although not always aware of this phenomenon. Local authorities are called, therefore, to the constant exercise of diplomacy, on the one hand, and to contribute to the implementation of policies aligned with such global agreements for sustainability, on the other hand.[2]

However, the classical theoretical reading of international relations presents concepts and approaches that limit not only the perspective of cities' participation in the international scene but also, as a result, the use of diplomacy as an instrument in their favor.

A clear example is Martin Wight (2002), who presents the traditional functions of diplomacy (representation, communication, and negotiation) and the diplomatic career categories in his classic book, The Politics of Power. He refers to the Congress of Vienna (1814/1815) as a milestone for this instrument of international relations, although marking the limitations of the use of diplomacy by the nation-state and safeguarding the diplomatic profession monopoly to collaborators of corporations linked to the state apparatus.

In the same publication, Wight (2002) presents his best conceptual reading for International Society, based on four characteristics that distinguish it from other societies and domestic societies: (i) single society, composed by States; (ii) the number of members is small and relatively stable; (iii) although there is a standard format, the members of the international society (the States) are heterogeneous; (iv) the members of the International Society, are, as a whole, immortal.

---

[1] https://www.cbd.int/sp/targets/, https://unfccc.int/process-and-meetings/the-paris-agreement/the-paris-agreement, https://sdgs.un.org/goals, https://habitat3.org/the-new-urban-agenda/.

[2] Escobar, F. S. (2017). *Un lugar en la mesa global: Los gobiernos locales como tomadores de decisiones en la agenda mundial. Proyecto Allas.* https://proyectoallas.net/wp-content/uploads/2018/01/Un-lugar-en-la-mesa-global-1.pdf.

It turns out that the twenty-first century makes room for less simplifying approaches that recognize uncertainty, complexity, interdependence, and multidimensionality of facts as relevant. Thus, it requires an approach to analysis distinct from classical and traditional approaches.

In this context, the post-normal science approach seems to impose itself as the most adequate approach for analyzing contemporary phenomena, such as the climate issue, and its interrelation with the urban issue, from the resilience perspective.

It is worth highlighting two key elements referenced by this approach: (i) the management of uncertainty, recognizing that it is more than a technical issue or a methodological question; (ii) the management of the plurality of perspectives in the scientific context and beyond (Petersen et al. 2010).

The link between the possibility of recognizing cities as legitimate actors on the international arena, capable of using the instrument of diplomacy in the face of the agenda to tackle the climate crisis, especially from the perspective of urban resilience, will be the subject of this chapter.

Thus, a conceptual reading about the understanding that should be built on the contemporary International Society is necessary. The approach proposed by Olsson (2006), at the same time, denies the simplifying concept:

> The nation-state model, self-determined as territorially well-delimited, fully in control of the destinies of its citizens, and equally sovereign over other counterparts in an anarchic universe, is no more than a Weberian ideal model challenged by contemporary reality. The belief that today's world can be boiled down to a balance of power or a security agenda is a rather unsatisfactory explanation for today's complex relationships, played out by diverse actors with very unique roles. (Olsson 2006, p. 127).

Moreover, Olsson (2006), quoting Dietrich Jung, proposes an approach that prioritizes the approach that recognizes complexity, uncertainty, and plurality:

> In this track, Dietrich Jung complements that international society cannot be limited to a '[...] territorially or functionally constricted entity whose integration mechanism is based on consensus of norms and interests.' The author proposes, then, a conception of society that is at the same time much broader and more complex that rescues the idea of social reproduction. This scholar argues that, from a 'holistic' perspective, '[...] social reproduction theoretically fulfills three elementary functions that all empirical societies must satisfy - the control of physical force, the assurance of material means, and the production and preservation of symbolic means of orientation.' Within the framework of a 'totality of social reproduction', this theorist summarizes that as a systematic concept, '[...] world society implies the ideal construction of a global integration of these three elementary functions according to modern patterns of consociation, while the evolution of a global society represents a contingent historical process'. (Jung, 2001, p. 452, as cited in Olsson 2006, p. 132)

Furthermore, and complementing this understanding of international society proposed by Olsson (2006), one should consider conceptual approaches from authors such as Hocking (1993), who advocate multi-level governance. Also, Murray (2008), who, resorting to Sharp, alerts to the need to recognize the variety of definitions for diplomacy and the diversity of actors that can exercise it. According to the author, diplomacy is:

> The way in which relations between groups that regard themselves as separate ought to be conducted if the principle of living in groups is to be retained as good, and if unnecessary and unwanted conflict is to have a chance of being avoided. (Sharp, as cited in Murray 2008, p. 24)

Thus, one can arrive at the definition of city diplomacy, a relatively recent concept, proposed by the Netherlands Institute of International Relations, which can be understood as:

> The institutions and processes by which cities or local governments in general engage in relations with other actors at the international political level, in order to represent their own interests vis-à-vis each other. City diplomacy is a professional, pragmatic and growing activity that is transforming and will continue to transform current diplomatic processes. (Melissen and Pluijm, 2007, p. 11, as cited in Perpétuo 2010 p. 33)

In South America, several cities are active in the international context. Metropolises such as Bogotá,[3] Buenos Aires,[4] São Paulo,[5] and Mexico City[6] are examples of municipalities that are very active on the global scene. When it comes to the climate issue, all of them are signatories of the Global Covenant of Mayors for Climate and Energy,[7] and have not only participated in the UNDRR's "Making Cities Resilient" campaign in its first phase but have also indicated their willingness to commit to and take the lead in the new phase of the campaign, starting in 2021.[8]

Other effective instruments for the exercise of international relations at the municipal level are the networks and associations of local governments. Characterized as spaces for exchanging experiences, building knowledge and capacities, and articulating projects and resources, the networks materialize the perspective of multilateral action by cities.

Three of them stand out in the context of climate action in South America: ICLEI[9]—Local Governments for Sustainability; C40[10]; and the Global Resilient Cities Network.[11] In South America, ICLEI connects over 80 associated governments in eight countries, C40 is present in 10 cities, and the Global Resilient Cities Network is also present in 10 cities. Therefore, it can be seen that the phenomenon of city diplomacy is becoming increasingly vigorous in South America. Besides, cities use this instrument to strengthen their capacity to take action against the climate crisis

---

[3] https://bogota.gov.co/internacional.

[4] https://www.buenosaires.gob.ar/jefedegobierno/secretariageneral.

[5] https://www.prefeitura.sp.gov.br/cidade/secretarias/relacoes_internacionais/.

[6] https://www.cgaai.cdmx.gob.mx/.

[7] http://pactodealcaldes-la.eu/pt-br/pacto-convoca-signatarios-para-atualizar-informacoes-de-contato-2/.

[8] https://program.unisdr.org/campaign/resilientcities/home/article/making-cities-resilient-2030-mcr2030-initial-proposal.

[9] ICLEI is a global network of over 2,500 local and regional governments in more than 125 countries committed to sustainable urban development. https://iclei.org/.

[10] C40 is a global network of megacities committed to tackling climate change by reducing GHG emissions and climate risks. https://www.c40.org/.

[11] https://www.rockpa.org/project/global-resilient-cities-network/.

to search for better possibilities for resilience to safeguard their routine functions and the sustainable development of their territory.

This movement opens space for cities to become important innovation drivers. Innovation can arise in the form of more efficient public policies, in the form of the application and use of technology to improve the efficiency of public policies, in the form of improving instruments of the relationship between public authorities and local society, in the perspective of refining data collection and analysis, and, without exhausting the possibilities of creativity and innovation, in the form of the implementation of new methodologies and social technologies.

In the next section, we will see how South American cities have sought to take advantage of the opportunities related to smart cities, from a conceptual reading of this phenomenon, concomitant to the others already mentioned at the beginning of this chapter.

17.2 Smart Cities and Resilience in South America

Due to its magnitude and the uncertainties inherent in risk projections, climate change is considered one of the most significant challenges of our time. Phenomena such as heatwaves, water shortages, floods, sea level rise, desertification, and landslides threaten cities around the world on a daily basis. The identification of such phenomena can be made from the elaboration of climate risk and vulnerability assessments such as those carried out by Urban-LEDS II,[12] a project developed by ICLEI in partnership with UN-Habitat and funded by the European Commission currently operating in 8 countries. In the scope of this project, reports were elaborated for the cities of Sorocaba (São Paulo, Brazil) and Betim (Minas Gerais, Brazil), where flooding, landslides, heat waves, and proliferation of vector-borne diseases were detected as the main consequences of the population's exposure to climate change effects.[13]

To face these challenges, local authorities must understand how the climate crisis, from extreme natural events or changes in weather patterns, strongly affects the dynamics of their territories, usually with a high compromise of built infrastructure, causing substantial financial losses, and, dramatically, displacing people and taking lives.

South America, for example, has presented multiple environmental stress factors, mainly derived from significant changes in land use in the region and exacerbated by variations in the weather pattern. As competing activities such as livestock, food production, and bioenergy increase, climate variability increasingly affects social and natural systems, and in cases of Nature's extreme events, has affected large

[12] The Urban-LEDS initiative "Accelerating Climate Action through the Promotion of Urban Low Emission Development Strategies" aims to make low emission development strategies a crucial part of urban policy and planning in cities.

[13] ICLEI, Prefeitura de Sorocaba, Urban Leeds (2020). Análise de risco climático: Sorocaba. ICLEI, Prefeitura de Betim, Urban Leeds (2020). Análise de risco climático: Betim.

regions. According to data from the IPCC (2014, p. 1504), 613 extreme climatological and hydro-meteorological events occurred from 2000 to 2013 in the regions corresponding to Central and South America, resulting in 13,883 deaths, 53.8 million people affected, and economic losses of USD 52.3 billion.

In the face of this, turning cities into 'smart cities' has been taking as all-in-one solution by experts, in both academia and industry, to deal with the complex challenges that occur from the combination of population growth, environmental pollution, climate events in face of an increasingly defied public management (Shah et al. 2019). The belief is that smarter the city, better its livability, workability, and sustainability (MACKE et al. 2018), and higher its resilience (Velásquez et al. 2018; Shah et al. 2019). Nevertheless, from the Information and Communication Technology solutions and products to the real improvement in the quality of life of the citizens (taken as 'users' in this 'new' paradigm) there is a long way to go. Right from the beginning, the latter scenario requires a holistic perspective over smart cities, which would allow that the planning and implementation of those solutions are based on the understanding of the relations between humans, citizenship, and the technological devices and networks (Macke et al. 2018).

As such, the panacea of the Smart City shows to be a more potential instrument of urban sustainable and resilient development than a silver bullet. In this sense, there is a convergence trend between the growing approach called smart cities, often dressed up as market offers of little or no tested technologies that do not add much for local governments, and the demand for greater knowledge and better tools for the integrated management of resilience in cities. Being a means, the smart infrastructure has to be part of a robust strategy.

It is not negligible what connected devices and networks can offer regarding early warnings, monitoring and predicting disasters to minimize fatalities and other losses by generating information and insights for the concerned authorities. As Shah et al. (2019, p. 2) point out: "the availability and integration of information from heterogeneous data sources and its coordination and understanding with decision makers, emergency responders, governments and also citizens will be the core ideology of this new and highly needed disaster management." Nevertheless, for that to become a pillar for resilience, some conditions have to be fulfilled.

Especially in developing countries, such as Latin American ones, the 'smart city' label relates to the increased capacity to deal with the most prominent problems in place while improving the populations' quality of life and increasing the opportunity for citizen participation in public policies processes through the implementation of adapted ICT technologies (Calderon et al. 2018; Albino et al. 2015). Among the streams of action to build the capacity for technology to be effectively used in face to structural and complex urban issues, Fernandes et al. (2019) highlight: (i) improving urban management; (ii) building a participatory governance; and (ii) developing human capital, ICT infrastructures, and active citizenship.

In this way, becoming a smart city refers to a process of institutional change based on organizational, political, and technological innovations toward a smart government and governance[14] (Ojo et al. 2016). This process of change lies upon innovative strategies to build more effective and resilient government structures, the review and enhancement of collective arrangements of governance and intersectoral capacity building (Fernandes et al. 2019).

That is why it is important to qualify the smart city understanding from the perspective of technology as an instrument and not as an end in itself. With the lens of the Paris Agreement, the smart city concept gains a clear purpose and is transformed, as proposed by ICLEI South America, in the Smart Cities for Climate Analytical Report (2017).[15]

Perpétuo and Ades (2020) qualify the guiding principles of a city that intends to use this instrument. They are the following: (i) the principle of inter-sectoriality, which, if practiced continuously, allows the integration of the various policies in the territory; (ii) the principle of information, which advocates the development and use of indicators that allow the constant monitoring and evaluation of public policies, giving efficiency to the planning process; (iii) the principle of transparency, so that information is shared with society as much as possible; (iv) the principle of encouraging participation, since local governments must maintain permanent and institutional channels for social dialogue; and last but not least, (v) the principle of identity, since it is of utmost importance to take into account the history and cultural DNA of each city.

In a year when cities are called upon by parties to raise their climate ambition either through the race to zero (toward carbon neutrality) or the race to resilience,[16] led by the champions of COPs 25 and 26, convergence between the use of technologies and commitment to the climate agenda is key.

It is also important to elucidate ways to materialize this convergence in the territory. ICLEI South America developed and presented an initiative that intends to achieve precisely this objective: ICLEI Innovation.[17]

As a pilot initiative, the first call for startup acceleration launched by ICLEI Innovation, run in South America by ICLEI and Publicae, focused on the theme of green urban areas and tree-planting policies. The theme was defined after consultations with the Brazilian Capitals Environment Secretaries Forum, the CB27.[18] Fifty-seven companies, NGOs, and universities registered, demonstrating an intense activity in developing new technologies in the scope of resilience and adaptation to climate change. Eleven of the applications were selected, and eight completed the acceleration process. The technologies selected illustrate the potential in this universe that contemplates the pressing needs for a more sustainable urban development and the supply of good technologies, for instance: the mapping of soil quality (*Verde Drone*);

---

[14] Smart governance can be defined as the interaction of technologies, people, policies, practices, resources, social norms, and information supporting the city's governance activities (Fernandes et al. 2019).

[15] https://issuu.com/icleisams/docs/relatorio_analitico_paginas_simples.

[16] https://racetozero.unfccc.int/, https://racetozero.unfccc.int/race-to-resilience.

[17] https://iclei-innovation.com.br/.

[18] http://www.forumcb27.com.br/.

the planting and restoration of native vegetation (*Verde Novo*); and the management of arboreal individuals in the urban environment (Anubdz, Arbolink, and Exati[19]).

It is interesting to see how the diversity and multiplicity of actors supports the concept of a contemporary diplomacy, where non-state actors are protagonists, together with their partners, with a well-set common target. The challenge to test and scale up those technologies is attached to the one that aims to foster climate and resilience compliance policies at the local level.

The next session will discuss the resilient development of cities in South America, followed by a concrete case that illustrates how the integration between governance, capacity building, and technology, combined with the commitment of local authorities to global sustainability agendas and international cooperation, can generate efficient and transformative results in the territories.

17.3 Climate Change Adaptation and Resilient Development in South America

The integration of climate change adaptation and resilience approaches is especially critical in the context of Latin American cities. If cities are at the center of the discussion on how global society and each country and territory will deal with climate change in the coming years, effective policies[20] for adaptation and resilience in the region's urban centers, marked by abysmal socioeconomic inequalities,[21] cannot be exempt themselves from addressing the vulnerabilities derived from situations of poverty, informality, lack of access to public services and facilities, and disorderly use of land and natural resources (Hardoy and Lankao 2011; UN HABITAT 2012).

Exposure and vulnerability to extreme weather events have intensified in recent decades in Latin American and Caribbean cities due to the increased frequency of these events and the accentuation of the impacts of human activity on natural dynamics (Carrizosa et al. 2019). The social construction of vulnerability is still ongoing with the perpetuation of development and urbanization models that have become hegemonic since the nineteenth century, marked by segregation between people and nature, soil sealing, closure of water bodies, and concentration of populations in unsafe areas (Carbone et al. 2020). Adaptation and resilience, in this context,

---

[19] https://anu.bz/, https://arbolink.com.br/, https://exati.com.br/.

[20] Effective public policies "can be called effective if it achieves a desired change in a situation considered to be a problem" (Mayntz 1983, pp. 126–127). Public policies of adaptation to climate change are considered effective when they reduce impacts and risks by acting on vulnerability factors of populations, organizations, and territories, either by acting on exposure and sensitivity to climate events or by strengthening adaptive capacities.

[21] Given the socioeconomic inequalities that shape life and social reproduction in cities and are reflected in urban landscapes, social groups in poverty are more affected by climate change. If effective public policies for adaptation and resilience do not reach urban territories, a distorted cycle of worsening inequalities and accentuation of vulnerability factors is formed.

fundamentally represent a new perspective on this model of development and urbanization and translate into a necessary and urgent agenda for the realization of Human Rights (UN, 1948) for all and the 17 Sustainable Development Goals (UN Global Compact) in the coming decades.

From the perspective of evolutionary resilience, the inevitable changes in urban structures in the face of climate threats and impacts open opportunities for previously unmet social needs to be met by the social and environmental system in an inclusive and sustained manner over time. It is crucial, therefore, not only that adaptation and resilience are combined in Latin American cities but also that they are embedded in a broader public discussion about the directions of urban changes, transformations,[22] and transitions[23] that we want to promote in the short, medium, and long term in relation to latent social interests and needs. In this sense, Porter and Davoudi (2012) raise the questions '*resilience for what? directed by what decisions and by whom?*' and tie the notion of resilience in the context of public policy planning to the debate on social justice and opportunities for participation in the formulation and prioritization of problems and actions.

The investments in adaptation and resilience needed in 'developing countries' are estimated between US$ 280 and 500 billion per year until 2050 (UNEP 2016). Whether originating from public, private, or international agencies sources, the results achieved by these investments will directly depend on the regulatory framework, policies, instruments, and institutional arrangements established at different government levels. Such political-regulatory and institutional bases for adaptation and resilience are not limited to the climate change agenda. However, they need to be considered transversally to the various impacted and relevant areas, among which we highlight housing, sanitation, drainage, public health, transportation, and mobility. Therefore, urban planning and land use planning instruments are pivotal to the holistic and comprehensive approaches that should guide sustainable urban development policies in the coming decades.

While the urgency of these new instruments and policies is already known, there are no ready-made solutions. They emerge, are built and tested, in experiments considered niches of innovation and social learning[24] anchored in active and diverse social participation (Ansell and Geyer 2017). In this sense, the smart technologies

---

[22] Transformations in urban systems in the context of the climate agenda are processes of change both in the technological sphere—infrastructure, equipment—and in the social sphere and its relationships. It aims to cut GHG emissions, remove carbon from the atmosphere, and increase the capacity of systems to cope with climate events, reducing vulnerabilities and strengthening resilience (Bulkeley et al. 2011).

[23] Geels and Kemp (2007) proposed three categories of change processes: reproduction, transformation, and transition. The latter refers to significant long-term changes through which a different systemic configuration is achieved, composed of new social functions or new dynamics that feed social functions (Geels 2002).

[24] Reed et al. (2010, p. 6) define social learning as a "change in understanding that goes beyond the individual to become situated within wider social units or communities of practice through social interactions between actors within social networks." Social learning consists of learning by doing, from which people learn from each other in ways that benefit broader social-ecological systems and promote social change (Van Epp and Garside 2014). Central to these processes is social

can be applied to promote social participation and governance improvement toward resilience; however, the new call for "safe, resilient, sustainable and inclusive cities" has been anchored, so far, in techno-managerial solutions detached from the territorial dynamics (Kaika 2017).

Brazilian regions, such as the Metropolitan Region of Campinas, coexist with adaptation and resilience approaches directed to specific risks, focused on engineering solutions, and those aimed at strengthening adaptive capacities and reducing systemic vulnerabilities through learning processes and social innovation. Thus, urban territories in the region serve as a stage for experiences to be expanded and replicated considering the results already achieved and the potential to inaugurate, through bottom-up processes, new logics of territorial planning, relationships between individuals and social groups, and between them and Nature, using technology as a means.

17.4 The Case of the Metropolitan Region of Campinas (São Paulo, Brazil)

In the spatial context, the Campinas Metropolitan Region is the second-largest metropolitan region in the State of São Paulo in terms of population, with 3,792 km^2, home to more than 3.1 million inhabitants. An important economic center, the Region, accounts for 3% of the national Gross Domestic Product (GDP). It has a diversified industrial park, significant agricultural and agro-industrial activity, specialized tertiary activities, research and innovation centers, in addition to Viracopos Airport, the second largest in the country in cargo transportation (Emplasa 2018).

In the normative scope, the city of Campinas is influenced by international guidelines for sustainable development. Among them, the mechanisms established by the United Nations Conference on Environment and Development of 1992 (Eco-92), where a series of agreements were signed, such as the United Nations Framework Convention on Climate Change, the Convention on Biological Diversity (CBD), and Agenda 21. Other global milestones of influence on sustainability policies are the Strategic Plan for Biodiversity and its 20 proposals to manage development policies in biodiversity, called Aichi Targets, the Paris Agreement, the Sustainable Development Goals (SDGs), and the New Urban Agenda (NAU).

These global milestones have also contributed to the adoption of sustainable guidelines at the Brazilian national level that impacts local governments' decisions. At the national level, the city of Campinas follows the guidelines of the City Statute (2001). Despite being dated before the NAU propositions, the Statute is an urban development policy law that ensures the protection of the environment in a balanced and sustainable way and the preservation of urban land and the natural environment. Following established international standards, the National Biodiversity Strategy and Action

participation, the foundation for building shared knowledge and understandings (Lave and Wenger 1991).

Plan (Epanb) was created in 2017 by the Brazilian Ministry of Environment, aiming to foster the fulfillment of Aichi Target 17,[25] commit to the CBD agenda, and set new biodiversity targets for 2020 in the country. At the state level, the state of São Paulo stands out as a pioneer in implementing the Targets by mapping existing initiatives in the state and the actions needed to meet them (ICLEI 2021b).

At the local level, the city of Campinas has a Municipal Secretariat of International Cooperation, responsible for attracting investments for its projects and identifying institutions for potential cooperation in the private, educational, and research sectors. Campinas also acts through decentralized diplomatic actions, such as establishing cooperation liaisons with its 21 sister-cities present in the American, African, Asian, and European continents, as well as by the performance of its five international offices, four of which in China and one in Italy. As part of the efforts to meet the international guidelines for resilient development, the city of Campinas is part of International Networks of cities such as ICLEI, where it is associated since 2015, UCLG—United Cities and Local Governments[26] and *Mercociudades*.[27]

In sintony with the "paradigm shift" hailed in the Habitat III Conference's New Urban Agenda toward the Sustainable Development Goals, the city also counts with the Campinas Smart City Strategic Plan (PECCI, 2019–2029). The document adopts innovation, resilience, and sustainability as guiding concepts for a smart city, and places people at the center of development through collaborative planning and citizen participation (Campinas City Hall 2019).

The PECCI, 2019–2029 guidelines are being adopted to guide the digital transformation of the municipality into a smart, human and sustainable city for the next ten years. It is worth pointing out that the plan highlights that technology is not an end in itself, nor suffices to make the city 'smart,' but as a means, which integrates a broader strategy for implementing the 2030 Agenda for Sustainable Development. In this way, key elements of the plan are governance and participation, which were already reflected in the process of the PECCI elaboration through a working group that organized the discussions on major themes, defining macro guidelines that were successively detailed with the participation of several actors from the Science, Technology, and Innovation ecosystem of Campinas.

Currently, Campinas is already considered a reference in Technology and Innovation in Brazil due to its technological parks and incubators, but also a bunch of solutions, mediated by innovative technologies, in place to improve citizens quality of life and enhance the quality and coverage of public services provided to the population. Among them, there is the On-Line Zoning Portal, which allows citizens to know and monitor, in a clear, accessible, and interactive way, the law of use and occupation of the city's land. The integrated and strategic use of information and

[25] Target 17 states that by 2015, each Party should develop, adopt as a policy instrument, and begin implementing a participatory and updated national biodiversity strategy and action plan.

[26] https://www.cglu.org/.

[27] https://mercociudades.org/.

communication technology also makes possible the understanding of the city development process within the Metropolitan Region of Campinas (RMC—in its original acronym).

The RMC, formed by 20 municipalities, makes up the Region's Development Council, which, supported by the Metropolis Statute, is responsible for preparing proposals of collective interest for planning and land use, housing, basic sanitation, and environment. Also, in 2017, with the support of ICLEI, the city of Campinas created the Program RECONECTA[28] RMC, which aims to subsidize a line of regional action under the environmental bias, focusing on the preservation and recovery of protected areas in the region (ICLEI 2021a).

The combination of economic growth, urbanization, environmental management, and resilience is a challenge faced in the Region, which is reflected, for instance, in the peripheral demographic expansion, reduction of rural areas, pressure on water resources, loss of biodiversity, soil sealing in urban areas and heat islands (Carmo and Hogan 2006). This challenge is accentuated by the local and regional effects of climate change, such as changes in rainfall patterns and increased risk of water shortage, and is enhanced by the loss of biodiversity and degradation of ecosystem services.

Although the RMC has a rich biodiversity, supported by 17 Conservation Units (CU) of Atlantic Forest and remaining areas of Cerrado, concentrated mainly in the city of Campinas, studies on the region point out that the existing green areas are not enough to conserve biodiversity and springs. This is primarily due to the fragmentation between these areas and the lack of integrated management concerned about the fauna, flora, and their relations with human populations (ICLEI 2021b).

In this context, the RMC and two other metropolitan regions in Brazil, Londrina and Belo Horizonte, were supported by the INTERACT-Bio Project,[29] led by ICLEI South America and designed to recognize and strengthen ecosystem services and Nature-based Solutions to the complex social and environmental issues faced in metropolitan regions. With the support of INTERACT-Bio, there is a significant increase in the contribution of subnational governments in the implementation of their countries' National Biodiversity Strategies and Action Plans. Not coincidentally, Brazil, India, and Tanzania, countries in which INTERACT-Bio has been implemented, are considered pioneers in implementing the Aichi Biodiversity Targets (ICLEI 2021a).

Therefore, joining forces with the RECONECTA RMC Program, the INTERACT-Bio Project unfolded, in 2018, in the proposal for the creation of the Connectivity Area (CA) intending to promote the (re)connection between forest remnants, protected

---

[28] An initiative of the Campinas City Hall that integrates the Campinas Municipal Green Plan and articulates the 20 cities of the Metropolitan Region of Campinas to establish mutual cooperation for actions of recovery and conservation of fauna and flora focusing on animal protection, the recovery of permanent preservation areas, and the strengthening of existing protected areas and creation of new protected areas (Prefeitura De Campinas, n.d.).

[29] In Brazil, the project is led by ICLEI South America and funded by the International Climate Initiative (IKI) of the German Federal Ministry for the Environment, Nature Conservation, and Nuclear Safety.

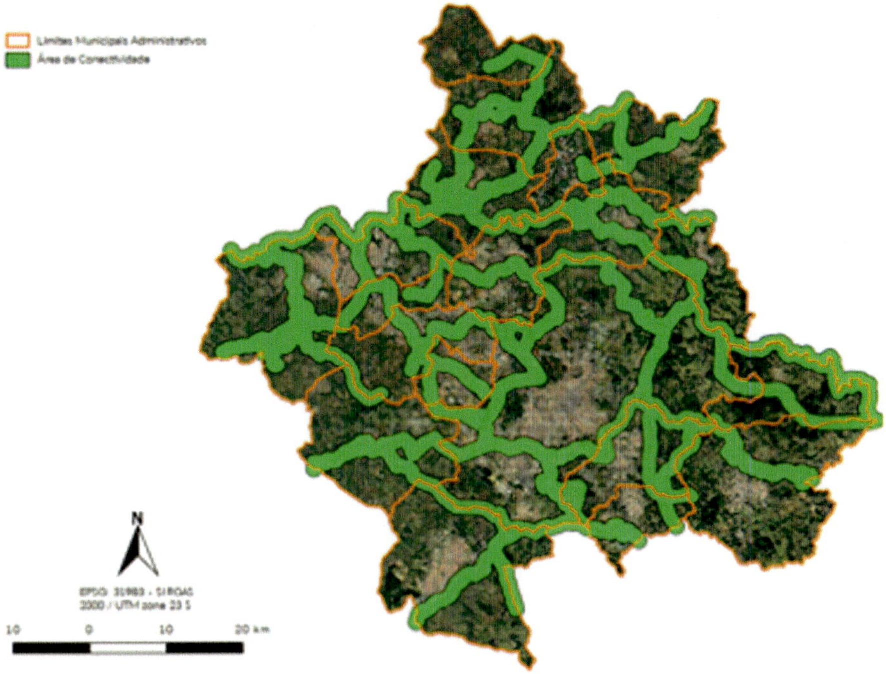

Fig. 17.1 Final layout of the connectivity area drawn in the metropolitan region of Campinas (*Source* RECONECTA RMC Program)

areas, and urban green areas, restoring the landscape connection and facilitating the gene flow between populations of fauna and flora, promoting sustainable productive practices and transforming the territorial management. This is based on six pillars of action: urban forestation; linear parks; ecological corridors; wildlife; regulation, inspection, and compensation; and articulation and communication. Figure 17.1 shows the final layout of the Connectivity Area in the RMC (ICLEI 2021b).

The CA results from an unprecedented consultation between the 20 municipalities of the RMC and was registered in the Action Plan for Implementation of the Connectivity Area in the Metropolitan Region of Campinas. It brought a strategic vision on biodiversity and ecosystem services as a basis for sustainable regional development and was included in the preliminary notebook of proposals of the Integrated Urban Development Plan (PDUI—*Plano de Desenvolvimento Urbano Integrado*). The CA contribution to the region's resilience comes not only from the reduction of the biophysical vulnerabilities of the territories to climate change but also through the process of formulation and implementation of the CA by strengthening the interaction between local governments, civil society, and non-governmental organizations, setting up an experiment on new forms of territorial planning and management that constitute process innovations for this region.

To design the CA proposal, coordinated decision-making and capacity-building processes between local governments and engaged organizations were shaped in the framework of INTERACT-Bio. To this end, a participatory diagnosis of the pressures operating on ecosystem services and the production of reliable and accessible information about priority areas for conservation and necessary restoration efforts were fundamental. The result of the diagnosis was represented in the Ecosystem Services Map, allowing the dynamic visualization of the data plotted on the territory and strengthened the understanding of local actors about the dynamics, flows, and policies that influence and regulate Nature in the cities of the region. (ICLEI; CBC 2020; ICLEI 2021b).

By facilitating a series of workshops and virtual meetings, the collaborative construction of the action plan for implementing the CA, throughout 2020, relied on the equal participation of different sectors and actors, reflecting the multiple interests, needs, and realities present in the territories of the RMC. As a result, there are 19 strategic objectives prioritized in the plan, which unfold in targets, actions, and indicators, in addition to the definition of those responsible for each action. The objectives are directly related to the Sustainable Development Goals (SDGs) of the United Nations (UN), among which stand out SDG 6—Clean Water and Sanitation, 11—Sustainable Cities and Communities, 15—Life on Land, and 17—Partnerships and Means of Implementation.

The plan is a future vision grounded on medium and long-term policies and reestablishing the social pact between the different segments of society and public power. The bet is that the CA experience will usher in a paradigm shift in planning and regional development, adopting innovative public policy processes and instruments based on the integration between restoration and biodiversity conservation, integration between natural and built environments, and improving the quality of life and well-being of the entire population.

Among the lessons learned in the experience of planning and implementing the CA in the RMC are:

Explicitly integrating the biodiversity and climate change (mitigation and adaptation) agendas strengthens the initiatives involved and the power to mobilize people and resources for the project as a whole. Moreover, the integration of analyses and potential benefits related to the two agendas fosters systemic approaches that consider the long-term relationships and dynamics underway in the territories, critical for policies and actions to strengthen resilience.

The engagement and participation of several sectors and actors, with effective opportunity to influence the process and the decisions, are indispensable for the resulting instruments for development and territorial planning to be in line with local realities and support their implementation. In the composition of the collective instances, there must be representation from different levels of government, with a clear definition of roles, processes, and mandates. In the case of the CA in the RMC, the involvement of regional bodies and state institutions was essential for articulating policies and instruments and making project implementation feasible.

Articulation and collaboration among projects underway in the region increase the potential for impact and optimize resources and efforts. In the case of the CA, the alignment between INTERACT-Bio and RECONECTA RMC made it possible to leverage synergies to strengthen technical knowledge and financial support and broaden the range of actors involved.

Collective knowledge production, based on the integration between scientific and empirical knowledge and translation into accessible and suitable materials for non-specialist actors, is vital to ensure ownership of the initiative and its results, broadening awareness about complex problems, vulnerability factors, and potential solutions. The production of illustrated maps, for example, facilitated the process of knowledge production and dissemination throughout the CA planning.

The institutionalization of collective planning processes and products is an effort that should permeate all stages of the project to ensure its perenniality and the achievement of long-term results given the changes in the political scenario. To this end, in the case of CA, the establishment of long-term cooperation agreements, the connection of the initiative with regional policies, such as through the Integrated Urban Development Plan (PDUI), and the involvement of São Paulo State authorities, promoting multi-level cooperation, contributed to this.

The Connectivity Area in the Metropolitan Region of Campinas illustrates how sustainable urban development can gain materiality and be aligned with the process of strengthening long-term resilience through regional planning, once it is anchored in the integration of public policies, in multi-level and intersectoral governance arrangements and the participation of the various actors and social groups. Moreover, the experience is an example of how Nature-based Solutions can foster such elements given their potential to generate multiple benefits and to promote understanding of the dynamics and patterns of biophysical, socioeconomic, and political issues at play in the territories. By becoming the first metropolitan region in Brazil to launch a regional strategy for biodiversity, designed collaboratively by the municipalities, the RCM serves as a reference and should inspire other municipalities and regions in the country and in Latin America.

In this sense, the results of this project are considered innovative not only for the collaborative mechanisms used to foster qualified exchanges and search for efficient and shared solutions but also for the promotion of recognized cutting-edge techniques to, for example, intelligently incorporate Nature-based Solutions in the RMC. As such, the project results were widely disseminated, allowing opportunities to foster knowledge exchanges, best practices, and technical cooperation between the cities participating in the project and other Brazilian and international cities. Among other events where the processes and results were presented and discussed, we can mention BIO 2020: Brazilian perspectives on the post-2020 Global Biodiversity Framework, III International Seminar on Nature-based Solutions, Greening Cities-Shaping Cities Symposium of Milan, Urban Nature Forum—ICLEI World Congress, and the nrg4SD Conference of Cuenca.

17.5 Discussion

In the context of South America, and perhaps in all the developing Global South, innovation and smart solutions are viewed differently than in already developed countries. Innovation is of course recognized as important and somehow salutary for the much sought progress and development. However, in the challenged socioeconomic context that characterizes the region, a rational use of the limited resources will not choose to acquire new technologies over addressing much more urgent needs of the populations.

Apart from the reduced financial resources, other challenges, such as limited knowledge and poor constructive exchanges between universities, private and public sector, make it difficult to co-create and integrate local smart solutions in the construction and administration of territories. It is mostly to respond to this problem that this paper presents City Diplomacy as an important first step toward smart solutions to promote resilience in South America. City Diplomacy allows local authorities to see a broader range of solutions and success cases through experience sharing opportunities with other cities. Thus, through their contacts with the international society, local governments are also exposed to the experiences of more developed territories and get inspired to adopt new policies. Cities international interactions as already mentioned are possible from city to city or within organizations and networks that may have a focus on specific topics such as the resilient agenda or broadly urban sustainable development.

More access to information, more knowledge, and more capacity-building opportunities are imperative actions to consolidate innovation and to implement smart solutions to solve local issues. In the context of South America, these actions should come before the race to acquire the latest technologies and still constitute innovation themselves. This paper understands smart cities as the ones committed to progress and equitative development by finding and implementing solutions that carry innovation, responsible use of resources, and efficiency in their core.

This understanding makes innovation possible not only by the use of technology but also through innovative processes and methods leaving no one behind as advocated by the Brazilian Letter for Smart Cities (2020) that claims that connections and innovation can be done in different manners than through digital technologies. In this sense, smart cities are human constructions and will present different structures influenced by the diversity of the local contexts and inspired by international standards.

The case of Campinas is an inspiration in all those senses. The city is very active internationally especially through its participation in ICLEI, therefore exercising diplomacy. The recent Mayor's participation at the launch of MCR2030 and the leadership implementing the InteractBio Project mentioned in the case study above are proof of good use of City Diplomacy for Resilience. The unprecedented processes and methods used during the elaboration of the Action Plan for the Implementation of the Connectivity Area in the Metropolitan Region of Campinas constitute innovations that permitted the adoption of this instrument that is also pioneer and innovative by itself. The processes and methods were paired with strong capacity-building sessions, leaving Campinas and the other participant cities prepared to now integrate adequate new technologies to enhance the effectiveness of the plan. All this was only possible because of Campinas active city diplomacy through ICLEI.

17.6 Conclusions and Recommendations

The international ecosystem is changing. Security and sovereignty will find new shapes and this article suggests that urbanization and the centrality of sustainability issues are driving this transition.

The first highlight that deserves attention is what the local authorities, as mentioned in the introduction, are calling "a seat at the global table." Here they are to stay. Players of diplomacy, members of the international society. But now what should they bring to the table? What should they ask? How should they behave? Without neglecting the uncertainties which are a strong part of this process, the answer resides in the capabilities of the international city and local governments networks to strengthen and qualify this representation. Also, it is worthwhile mentioning that the answer will be as good as the synergies and complementarities that those associations can generate among themselves.

Secondly, it is clear that climate change is already considered a crisis, an emergency, and therefore, an international security issue. The nature of this thread, though, differs a lot from the traditional ones. It has no solution without a common action from all. And that includes all levels of governments. Better resilience policies are key to address this challenge successfully. And since it is a much more urbanized world, the consequences of climate change are being felt in the cities and, therefore, should have the cities and local authorities as the first layer of the structural modifications that society urgently needs.

Thirdly, but not less important, technology is there to support and accelerate the required action! But to become a smart city is much more than to use technology properly. To become a smart city is to know, understand, and manage data; it is to foster civic participation and appropriation at the deepest level possible; it is to value and preserve local history and culture, therefore identity; it is to participate internationally, with clear interests; and it is also to foster diversity, creativity, and innovation, allowing better use of IT.

To achieve solid smart transformations in sustainable development, local governments all over the world but especially in the Global South should understand and look forward to this contemporary triad: diplomacy, resilience, and innovation. These elements, combined in the right way, may constitute the basis for a successful urban sustainable development.

References

Albino V, Berardi U, Dangelico RM (2015). Smart cities: Definitions, dimensions, performance, and Initiatives. J Urban Technol 22(1):3–21. http://dx.doi.org/10.1080/10630732.2014.942092. Available from: https://www.tandfonline.com/doi/full/10.1080/10630732.2014.942092

Ansell C, Geyer R (2017) 'Pragmatic complexity' a new foundation for moving beyond 'evidence-based policy making'? Policy Stud 38(2):49–167. https://doi.org/10.1080/01442872.2016.1219033

Bulkeley H, Castán Broto V, Hodson, M (2011) Cities and low carbon transitions. Routledge, Abingdon

Calderon M, Lopez G, Marin G (2018) Smartness and technical readiness of Latin American cities: A critical assessment. IEEE Access 6:56839–56850. http://dx.doi.org/10.1109/ACCESS.2018.2864218

Campinas City Hall (2019) PECCI 2019–2029. Strategic Planning Campinas Smart City 2019–2029. Municipal Secretariat for Economic Development, Social and Tourism and Municipal Council for Science, Technology and Innovation

Carbone A, Sanches D, Maglio I, Uehara V, Locosselli G (2020) O futuro das cidades: Um olhar emergente para os rios urbanos e a infraestrutura verde. Jornal da USP, 25/11/2020

Carmo RL, Hogan DJ (2006) Questões ambientais e riscos na Região Metropolitana de Campinas. In: Cunha JM (ed) Novas metrópoles paulistas: população, vulnerabilidade e segregação. Núcleo de Estudos de População/Universidade Estadual de de Campinas, pp 581–604

Carrizosa M, Cohen M, Gutman M, Leite F, López García D, Nesprias J, Orr B, Simet L, Versace I (2019) Facing risk. New urban resilience practices in Latin America. CAF, Caracas

Emplasa (2018) Plano de Desenvolvimento Integrado. Região Metropolitana de Campinas. Relatório 2: Diagnóstico Preliminar dos Problemas Metropolitanos

Fernandes RAS, Queiroz AO, Wilmers JTAVS, Hoffman WAM (2019) Urban governance in Latin America: Bibliometrics applied to the context of smart cities. Transinformação, 31:e190014, November 11. https://dx.doi.org/10.1590/2318-0889201931e190014

Geels F (2002) Technological transitions as evolutionary reconfiguration processes: A multi-level perspective and a case-study. Res Policy 31:1257–1274

Geels F, Kemp R (2007) Dynamics in socio-technical systems: Typology of change processes and contrasting case studies. Technol Soc 29(4):441–455

Governo do Brasil (2020) Carta Brasileira Cidades Inteligentes

Hardoy J, Lankao PR (2011) Latin American cities and climate change: challenges and options to mitigation and adaptation responses. Environ Sustain 3:158–163

Hocking B (1993) Localizing foreign policy: Non-central governments and multilayered diplomacy. Sr. Martin's Press

ICLEI (2021a) Interact-Bio. ICLEI - Governos Locais pela Sustentabilidade

ICLEI (2021b) Plano de Ação para Implementação da Área de Conectividade da Região Metropolitana de Campinas - Resumo, São Paulo, Brasil

ICLEI; CBC (2020) Mapping the Campinas Connectivity Area. A case of policy and planning integration for biodiversity conservation in Brazil. ICLEI South America, INTERACT-Bio

ICLEI, Embaixada Britânica (2017) Relatório analítico: Cidades inteligentes pelo clima

ICLEI, Prefeitura de Betim, Urban Leeds (2020) Análise de risco climático: Betim

ICLEI, Prefeitura de Sorocaba, Urban Leeds (2020) Análise de risco climático: Sorocaba

IPCC (2014) Climate Change 2014 impacts, adaptation, and vulnerability. Part B: regional aspects. Cambridge

Kaika M (2017) Don't call me resilient again!': the New Urban Agenda as immunology … or … what happens when communities refuse to be vaccinated with 'smart cities' and indicators. Environ Urban 29(1):89–102

Lave J, Wenger E (1991) Situated learning. Legitimate peripheral participation. Cambridge University Press, Cambridge

Losada AF, Zapata-Garesché E, Llamas F, Perpétuo RO (2020) How mayors are ignoring dysfunction and handling COVID-19 among themselves. Americas Quarterly, Novembro 16. https://americasquarterly.org/article/how-mayors-are-ignoring-dysfunction-and-handling-covid-19-among-themselves/
Macke J, Casagrande RM, Sarate JAR, Silva, KA (2018) Smart city and quality of life: Citizens' perception in a Brazilian case study. J Clean Prod 182:717–726, May 1
Mayntz R (1983) The conditions of effective public policy: A new challenge for policy analysis. Policy & Politics 11(2):123–143
Murray S (2008) Consolidating the gains made in diplomacy studies: A taxonomy. Int Stud Perspect
Ojo A, Dzhusupova Z, Curry E (2016) Exploring the nature of the smart cities research landscape. In: Gil-Garcia et al. JR (eds) Smarter as the new urban agenda. A comprehensive view of the 21st century city, vol 11. Springer, Cham, Switzerland. Technology, Public Administration and Information, pp 23–47. http://dx.doi.org/10.1007/978-3-319-17620-8_2
Olsson G (2006) Poder Político e Sociedade Internacional Contemporânea: Governança Global com e sem Governo e seus desafios e possibilidades [Tese de Doutorado, UFSC]
ONU (1948) Declaração Universal dos Direitos Humanos. https://declaracao1948.com.br/declaracao-universal/declaracao-direitos-humanos/?gclid=/CjwKC/Ai/A_eb-BRB2EiwAGBnXXpXwWvtn4RAqsFToee7ygfb5nUVVaml7Ft2N449Ie8zaKU6wZzNH8RoCv-wQAvDBwE
Perpétuo RO (2010) A cidade além da nação: a institucionalização do processo de internacionalização de Belo Horizonte [Dissertação de Mestrado, Pontifícia Universidade Católica de Minas Gerais]
Perpétuo RO, Ades DI (2020) Cidades Inteligentes pelo clima: inovação e sustentabilidade a serviço da cidadania. In Diego de Melo Conti, Vinnicius Lopes Ramos Vieira (Orgs.), O futuro das cidades: sustentabilidade, *inteligência urbana e modelos de viabilidade utilizando PPPS e concessões*. CD.G Casa de Soluções e Editora, pp 117–126
Petersen A, Cath AHM, Kunseler E, Van der Sluijs J (2010) Post-normal science in practice at the Netherlands Environmental Assessment Agency. Sci Technol Hum Values
Porter L, Davoudi S (2012) The politics of resilience for planning: A cautionary note. Plan Theory Pract 13(2):329–333
Prefeitura De Campinas (s.d.) Reconecta RMC. http://www.campinas.sp.gov.br/governo/meio-ambiente/reconectaRMC.php
Reed M, Evely A, Cundill G, Fazely IR, Glass J, Laing A, Newig J, Parrish B, Prell C, Raymond C, Stringer LC (2010). What is social learning? Ecol Soc 15(4)
Shah SA, Seker DZ, Rathore MM, Hameed S, Ben Yahia S, Draheim D (2019) Towards disaster resilient smart cities: Can Internet of Things and big data analytics be the game changers? IEEE Access 7:91885–91903. https://doi.org/10.1109/ACCESS.2019.2928233
UN HABITAT (2012) And Caribbean cities 2012 State of Latin American: Towards a new urban transition
UNEP (May de 2016) Adaptation Gap Report. https://climateanalytics.org/media/agr2016.pdf
Van Epp M, Garside B (2014) Monitoring and evaluating social learning: A framework for cross-initiative application. CGIAr Research Program on CCAFS
Velásquez W, Munoz-Arcentales A, Yanez W, Salvachúa J (2018) Resilient smart cities: An approach of damaged cities by natural risks. In: IEEE 8th Annual Computing and Communication Workshop and Conference (CCWC), pp 591–597. https://doi.org/10.1109/CCWC.2018.8301649
Wight M (2002) A Política do Poder/ Martin Wight (1913–72) - Prefácio de Henrique Altemani de Oliveira; Trad. C. Sérgio Duarte (2a. edição) Editora Universidade de Brasília, Instituto de Pesquisa de Relações Internacionais, Imprensa Oficial do Estado de São Paulo

Chapter 18
The Role of Smart Cities in Building the Resilience of Vulnerable Communities: Three Case Studies from Europe, Asia, and Africa

Prakash Kamtam, Pourya Salehi, Amy Jones, and Asad Asadzadeh

Abstract The smart city is an emerging dominant urban development paradigm worldwide. Many local governments are using technology solutions to address their challenges. The growing and disproportionate threat of climate-induced disasters is posing multiple risks to cities. Resilience building is a complex phenomenon and it has been increasingly featured in policy agenda, specifically for urban risk management. This paper explores the resilience strategy within the smart city paradigm. The focus is more on how cities can build the resilience of vulnerable communities using smart governance and digital innovations. The three different case studies from different geographies discussed in this paper highlight possible solutions with justification for linkage and integration of smart city approaches and urban resilience. However, each city is unique, requiring local innovation and appropriate technology solutions.

Keywords Urban resilience · Smart city · Vulnerable communities · Data

18.1 Acronyms

AI:	Artificial Intelligence
GLA:	Greater London Authority

P. Kamtam
University of Hertfordshire, Hatfield, UK
e-mail: p.kamtam@herts.ac.uk

P. Salehi (✉) · A. Jones
ICLEI—Local Governments for Sustainability (World Secretariat), Bonn, Nordrhein-Westfalen, Germany
e-mail: pourya.salehi@iclei.org

A. Jones
e-mail: amy.jones@iclei.org

A. Asadzadeh
University of Bonn, Bonn, Germany
e-mail: asad.asadzadeh@uni-bonn.de

A. Sharifi and P. Salehi (eds.), *Resilient Smart Cities*, The Urban Book Series,
https://doi.org/10.1007/978-3-030-95037-8_18

ICLEI: Local Governments for Sustainability
ICT: Information and Communication Technologies
IoT: Internet of Things
SMS: Short Message Service
TfL: Transport for London
UN DESA: United Nations Department of Economic and Social Affairs
UNCDF: United Nations Capital Development Fund
UNCTAD: United Nations Conference on Trade and Development
UNDRR: UN Office for Disaster Risk Reduction
UN-Habitat: United Nations Human Settlements Programme
UN-SDG: United Nations-Sustainable Development Goals

18.2 Introduction

The twenty-first century poses two significant challenges to humankind: rapid urbanization and frequent disasters causing increased risk to millions of people (Correa et al. 2011; World Bank 2016). At present, "more people live in urban areas than in rural areas, with 55 percent of the world's population residing in urban areas" (UN DESA 2018, p. xix), and this trend will continue in the coming decades, as "by 2050, the population of the world is projected to be 68 percent urban, with urban dwellers numbering 6.7 billion" (UN DESA 2018, p. 10). In addition, according to a more recent report by the UN DESA, nearly 1.3 million fatalities and 4.4 billion injuries were due to natural disasters (Gu 2019). Based on this report, "some 60 million people were affected by extreme weather in 2018" (UN 2019, para. 1), leaving millions homeless, displaced, or under emergency assistance, while economically speaking, UNISDR released a report "highlighting that 91 percent of disaster events are climate related. These account for 77 percent of the US$3 trillion in recorded economic losses from climate-related and geophysical disasters over the last 20 years" (UNISDR 2018, p. 82).

Furthermore, disaster risk is large and growing in urban areas (Lall and Deichmann 2010) due to the high population density and location of many cities around low-lying coastal or riversides. UN DESA (2018) also states that globally, three in five cities with a population of 500,000 inhabitants are prone to natural disasters and vulnerable to either flood, cyclones, earthquakes, droughts, landslides, or volcanic eruptions—or a combination of them. When combined, these cities host approximately one-third of the world's urban population (1.4 billion people).

Due to the increase in urban disasters, the Global Average Annual Losses (AAL) in the urban built environment is estimated to increase to $415 billion by 2030 (World Bank 2016). Therefore, developing resilient cities is now more critical and imminent than ever as climate change is further increasing the disasters in cities, and incidentally, the cities have a more vulnerable population likely to be impacted. Otherwise, if neglected, the disasters may adversely affect the vulnerable population and could even reverse many development gains achieved previously, thus forcing tens of millions of populations back again into poverty and despair. For example, the disaster losses in the Global South are approximately $862 billion, which is equivalent to one-third of international development aid (HABITAT III 2015).

In an urban context, there is a desperate need and urgency for both a pragmatic and scientific approach to building a resilient city (Moraci et al. 2018), and more so due to its disproportionate impact on low income and maintaining the most vulnerable population. Resilience is now increasingly applied in urban contexts to build scientific and political strategies for cities toward proactively dealing with natural and climate-related disasters and their associated challenges (Béné et al. 2017; Fekete et al. 2020). While finding a consensus on the concept of the term resilience is challenging, many recent attempts have made the case that new resilience initiatives should concentrate on transformative capacities instead of coping or incremental adaptive capacities (Hölscher et al. 2019; Romero-Lankao et al. 2018; Ziervogel et al. 2016; Pelling et al. 2015). The implications of this paradigm shift in resilience are twofold. First, the conventional disaster risk management processes and frameworks are insufficient to deliver long-term resilience pathways to effectively address the increasing complexities and uncertainties of rapid urbanization and climate change dynamics (Torabi et al. 2018; Birkmann et al. 2014). Second, new methods and tools are needed to capacitate urban governance and planning to guide the transition toward transformative capacities for resilience (Borie et al. 2019; Coaffee et al. 2018). However, it is not possible for resilience to be enhanced by just a single program or sector, it relies upon the inclusion of relevant actors (rich, poor, men, women, old, young, among others), sectors, and programs to increase the capacity of urban systems in times of adverse shocks or stresses (Borie et al. 2019; Hölscher et al. 2019; Ulrichs and Slater 2016).

Further resilience should not be looked at from a narrow perspective of climate change adaption; instead, it should include a comprehensive and holistic fabric of social, economic, political, technological, and environmental aspects, including equity, social justice, and multicultural dynamics (ICLEI 2019). Therefore, the key message of urban resilience thinking is to avoid possible disruptions, cope with disasters and their associated consequences, respond to impacts, restore the city's previous function, and adapt to new changes (Traballesi et al. 2019).

ICT can be fundamental to the prevention, mitigation, and management of disasters (Berawi 2018). Emerging technologies such as AI, data analytics, cloud computing, IoT, mobile communications, and social applications seem to have a critical role in supporting urban resilience. However, according to Jeffrey Chan and Ye Zhang (2019), the link between the technological capacity of cities and their ability to recover from shocks is less studied, researched, and needs more focus, particularly due to the current trend of increasing adaption of smart governance and technology solutions in smart cities. In this context, the study of the relation and linkage between urban resilience and the smart city's current dominant urban development paradigm seems to be very appropriate and relevant.

In line with this backdrop, this paper explores the potential linkage between the urban resilience concept and smart cities model with a key focus on vulnerable communities living in urban areas by narrating three case studies of London,

Bengaluru, and Kampala. The case studies identify best practices and highlight the contribution of technology, in particular the adoption of the smart city planning framework in addressing the resilience of vulnerable communities living in cities.

This research employs a theoretical research method, specifically a literature-based research approach, to identify a knowledge gap in the extant literature on the resilience of vulnerable communities and its relation and linkage to emerging smart city discourse. Since resilience, smart city, and the associated concepts and tools are context-dependent phenomena, the case study method can present learnings from the local context on the use and application of smart city strategy, tools, and techniques in building resilient communities, along with the required capacities, systems, and institutions. Furthermore, the insights may reveal those positions within the present systems and processes of the smart city concept that open up or constraint opportunities to deliver more transformative pathways to resilience. The paper, therefore, identifies the potential areas of linkage between the concepts of smart cities and resilience and presents relevant recommendations for designing smart and resilient cities of the future.

18.3 Positioning of Vulnerable Communities in Urban Resilience

Several categories of people in cities experience exclusion from mainstream development and are often identified based on their gender, age, race, ethnicity, religion, class, and persons with disabilities, including migrants, refugees, and others (UN-SDG 2015). The excluded groups are most vulnerable with minimal capacity to cope with disasters, shocks, and stresses because they are poor, living in high-risk areas with limited access to resources and institutions, including social capital, and often hidden from view (Pinchoff 2018).

Older people living in cities, who are often excluded and marginalized, are currently estimated at 500 million (OECD 2018) and this number is expected to double to over one billion by 2050 (HelpAge International 2021). According to the World Report on Disability (2011), approximately 15 percent of the global population live with some form of disability or impairment, and by 2050 nearly one billion people in this category will be living in cities (World Health Organization 2011). Another vulnerable group is refugees; Edwards (2016) states that 65.3 million people are forcibly displaced worldwide, out of which 60 percent live in urban areas, and more than half of all refugees are school-aged children. The UN-Habitat (2004), in addition to these estimates, suggests that over one billion people reside in urban slums (informal settlements), further expected to rise to 2 billion by 2030. Today a quarter of children globally live in these informal urban settlements, where they are subject to social exclusion and deprived of access to essential services and opportunities for a better future (UNICEF 2012).

Globally, there are 2.5 billion adults who do not possess formal bank accounts and experience financial exclusion (Demirgüç-Kunt and Klapper 2013). The recent estimates of UN-Habitat (2020) show that "while more than 50 percent of the world's population is online, there are still 3.6 billion people without affordable access to the internet" (UN-Habitat 2020, p. 1). The global digital divide across countries and regions also contributes to the inequality of access to goods and services available through technology. Likewise, the International Monetary Fund (2020) argues that the internet poses a danger, as it could exacerbate the existing socioeconomic imbalance between those with and without access. Therefore, there is a critical need to act at the subnational, national, and international levels to erase the worldwide digital divide.

Against this backdrop, urban shocks and stresses, including natural and artificial disasters, disproportionately impact these vulnerable communities (Pinchoff 2018; HABITAT III 2015). The Population Reference Bureau (PRB-USA) states that viewing disasters from the lens of the vulnerability perspective implies that the adverse impacts of disasters are huge when it strikes a deprived population or community. For this reason, understanding various degrees of vulnerability is crucial in planning a resilient city (Landry et al. 2016), yet the vulnerability in many cities is both large and diverse, and the needs among these populations are intricate and multisectoral (Patel et al. 2016).

Building resilience within vulnerable communities, such as the elderly, people with disabilities, women, children, poor, migrants, refugees, and others is a challenging task and needs a targeted and specific approach. The key challenges and obstacles in building the resilience of vulnerable communities from across cities and countries are identified from the literature survey and shown in Table 18.1 below.

18.4 Smart City as Part of the Solution for Boosting the Resilience of Vulnerable Communities!

Over the last decade, smart city concepts have been increasingly addressed to shape urban resilience planning and initiatives (Vermiglio et al. 2020; Van Winden and Carvalho 2017). Smart cities mainly focus on developing and integrating new technological solutions in urban and regional systems to pave the way to more liveable communities (Melkonyan et al. 2020). The notion of resilience, on the other hand, denotes that it is still possible to influence the unpredictable future of a city (Moglia et al. 2018). Despite the differences in concepts and notions, they have some common characteristics. For example, both terms focus on increasing the capacities of urban systems and the associated elements to either sudden shocks or chronic stresses (Jovanovic 2018). Furthermore, the concept of resilience is characterized by the notions of complexity, learning, sustainability, and decentralization, which is in line with the evolutionary smart city approach; smartness, forward oriented, and inclusiveness (Tzioutziou and Xenidis 2021). Building resilience is a multi-level governance exercise (Kubicki 2017), and the management of urban resilience

Table 18.1 The key challenges and obstacles in building the resilience of vulnerable communities

Key challenges and obstacles in building the resilience of vulnerable communities	*Literature evidence*
The narrow approach to disaster vulnerability: *Lack of holistic view, integrated and systems approach, and working in silos have a severe impact on urban resilience, particularly on the vulnerable groups who are already deprived and underprivileged of fundamental rights*	Marana et al. 2019; Patel et al 2016; Pasteur and McQuistan 2016; HABITAT III 2015
Difficulty in the identification of vulnerable populations and lack of data: *The vulnerable population is loosely spread and much diversified. There is very little subnational and local data available to measure the exposure and risk of vulnerable and marginalized populations, and where data exists, it is not disaggregated by critical measures of vulnerability like sex, age, poverty status,* etc	Pinchoff 2018; Patel et al. 2016
Complexity: *Vulnerability and needs among these populations are complex and multisectoral, and there are multiple interrelationships between community wellbeing and local drivers of risk*	Patel et al. 2016; Pasteur and McQuistan 2016
Inaccessible locations and lack of basic amenities: *Their location and lack of access to safe housing and reliable basic services like water, sanitation, energy, and education affect the capacity of vulnerable residents to recover. In current digital societies, lack of internet and access to digital services is also considered a basic amenity*	ICLEI 2019; Patel et al. 2016; HABITAT III 2015; Sheppard et al. 2013
Partial treatment: *They suffer from consequences of uneven relief and recovery efforts with obstacles to access entitlements, assistance,* etc	Sheppard et al. 2013
Poor governance and planning: *Their vulnerability to disasters is influenced by social, economic, political, and institutional factors that govern entitlements, including national and global trends that contribute to uncertainty. In addition, the rigid and regulative planning frameworks have no flexibility to consider scenarios other than business, as usual, leading to the marginalization of urban poor*	Pinchoff 2018; Pasteur and McQuistan 2016; Sheppard et al. 2013; Birkmann et al. 2014; Zebardast 2006

(continued)

Table 18.1 (continued)

Key challenges and obstacles in building the resilience of vulnerable communities	*Literature evidence*
Climate change: *The climate crisis is hurting the world's most vulnerable populations first and most severely, adding to violence, food crisis, and displacement. "People who are socially, economically, culturally, politically, institutionally, or otherwise marginalized are especially vulnerable to climate change" and vice-versa*	Matthews & Nell, 2019; IPCC, 2014, p. 4
Financial exclusion: *Vulnerable people are more exposed to financial exclusion. The key barriers include—not being able to meet legal requirements for access, financial requirements for access to services, and absence of required skills or means. The risk of exclusion is very high with elderly and disabled people*	Jérusalmy et al., 2020; UK Parliament, 2017
Social exclusion: *Extreme poverty and marginalization put vulnerable groups among actors who are not being involved in decision-making processes. These processes are either related to disaster risk reduction and adaptation or the conventional urban development trajectories*	Clark-Ginsberg et al. 2017; Keck and Sakdapolrak 2013
Digital exclusion: *Globally, there are still 3.6 billion people who lack an affordable connection to the internet and data. For example, in a developed country like the UK, two million households are without access to the internet and 22 percent of the country's population lack basic digital skills. The children in these households are at risk of falling and being left behind in their education. The adults seeking jobs cannot apply online and are missing opportunities. The families are missing their friends and family while living in isolation during COVID-19*	UN-Habitat 2020
Lack of sustainable livelihoods: *Most vulnerable groups work in informal and temporary jobs without sufficient and decent income. Life for many urban poor migrants, mainly living in the developing world, is an everyday struggle*	Serrat 2017
Lack of knowledge and skills: *Low skills perpetuate poverty and inequality. The vulnerable communities like persons with disabilities, learning disorders, low literacy levels, homeless, migrants/refugees, women and girls, etc., are at a high risk of marginalization due to a lack of requisite skills and knowledge*	World Bank 2016

(continued)

Table 18.1 (continued)

Key challenges and obstacles in building the resilience of vulnerable communities	*Literature evidence*
Inaccessible and high cost of healthcare and insurance*: The high cost of medical services and insurance is the most significant barrier to a healthy life. The elderly, disabled, and people with low income are the worst affected and do not have suitable and timely access to healthcare*	The Commonwealth Fund 2018

requires a multidisciplinary and multipronged approach with a holistic risk assessments framework supported by cross-sector priorities and city-wide context (Marana et al. 2019). The new digital technologies can provide solutions to emerging urban challenges through efficient decision-making, resource optimization, and delivery of urban services (Khansari et al. 2014; Mostashari et al. 2011).

According to UNCTAD (2020) and South Asia Disaster (2015), science, technology, and innovation can contribute immensely to the resilience building of the most vulnerable populations. The technology solutions can be used to empower these communities and give them a voice. They can be used effectively to develop early warning systems and extend access to basic services like education, sanitation, and health, better monitoring of environmental risks, and connecting people and community. The digital innovations that drive economic diversification allow economies to adapt to shocks and thrive, further decoupling economic development from environmental degradation and promoting environmental sustainability. In addition, it is asserted that the smart city concept has the potential to democratize the resilience planning process and create conditions for inclusive, just cities in which voices and aspirations across social groups are heard and considered (McPhearson et al. 2016).

Smart planning is a key for the implementation of resilient projects in cities (Moraci et al. 2018). The advantages of a smart city, such as better monitoring and planning, speed and spread, savings and quality of living, have a close relationship and linkage to urban resilience (Traballesi et al. 2019). Inspired by this view and based on literature review and document analysis, the following few relevant and specific contributions that a smart city can make toward building resilient communities are highlighted in Table 18.2 below.

The above list indicates few representative examples of how the smart city approach and ICT can contribute to resilience building. It is contended that these specific interventions addressing the resilience of vulnerable communities can be integrated into the overall smart city strategy. According to Giffinger et al. (2007), the core platform of the smart city strategy is smart government, smart economy, smart environment, smart living, smart mobility, and smart people.

As a narrative approach, we next analyze the three case studies of Smart London, Smart Bengaluru, and Smart Kampala with relevant examples. Narrative methods can be defined as "real-world measures" that are suitable when "real-life problems"

Table 18.2 Smart city contribution to resilience

Smart city contribution to resilience	*Literature evidence*
Smart & resilience governance: *The use of tech, data, and more innovative ways of working will give the city the ability to govern efficiently, at speed and transparently applying collaborative governance to cross-vertical challenges and citizen needs, particularly before, during, and after the occurrence of disasters. For example, the city governance based on ISO standard for smart city operation focuses on four key areas: the end outcomes, cross-silo governance, citizen-centric service, and data management are beneficial to build resilience*	Marana et al. 2019; ISO 37106 2018
Risk intelligence: *Risk information such as data on vulnerability and exposure to shocks and stress is the key to building a sound resilience strategy. The open data approach in a smart city helps in improving urban resilience. For example, the data is the critical resource to manage and rebuild post-disaster, and the power of the mobile phone is enormous when used as an urban planning and disaster management tool*	HABITAT III 2015; Landry et al. 2016; South Asia Disaster 2015
Smart warning systems: *ICTs contribute a significant part in disaster prevention, mitigation, and management. They play an essential role in the early warning system and track and communication during emergencies and post-disaster periods. For example, ICT is the key backbone of smart city disaster management solutions in Japan. Japanese city resilience is built on its ability to process, store, and share vast amounts of data (big data). It involves multiple observation systems, information gathering capabilities, data gathering and analysis, decision-making systems combined with intelligent warning systems. As a warning in the event of an earthquake, the Japanese mobile network is equipped to send out messages in five languages to mobile users and use the TV broadcast*	Berawi 2018; Japan Meteorological Agency 2013

(continued)

Table 18.2 (continued)

Smart city contribution to resilience	*Literature evidence*
Smart communication channels: *Smartness combined with resilience will contribute to a strong communication network, emergency management, and time for recovery. For example, the social bonds and communication channels between people and communities during the recent COVID-19 situation helped in the timely dissemination of information, thus mitigating the risk levels as much as possible. Most countries set up exclusive hotlines to provide information to the public on COVID-19 and the action taken. The commonly used communication methods included emails and text messages via SMS, WhatsApp, Viber, or others. The resilience portal project in many countries improves communications with relevant stakeholders and includes services like information sharing, establishing a communication structure with stakeholders, engagement and heightening awareness of stakeholders, especially citizens, knowledge sharing, information sovereignty, and usability*	Traballesi et al. 2019; Marana et al. 2019
Participatory planning: *Collective intelligence, collaborative technology, smarter machines, hidden talents, and renewed participatory governance help in building a robust and resilient city. Furthermore, smart cities foster creativity and innovation as citizens generate ideas rather than passive recipients. Urban planners, solution providers, and investors can also exploit digital technologies to practice evidence-based decision-making based on a scientific and rational approaches*	Kubicki 2017
Climate-smart cities: *The United Nations Climate Change sponsored Climate-Smart Cities program aims to "deliver an evidence-based plan for rapid deployment of energy-efficient technologies and investment in climate-resilient infrastructure at the local level." These smart cities can contribute to lowering greenhouse gas emissions from the transport sector and minimizing electricity and heat production*	United Nations 2021

(continued)

Table 18.2 (continued)

Smart city contribution to resilience	*Literature evidence*
Knowledge and skills empower vulnerable communities: *Smart and digital tools provide knowledge and skills at a cheaper cost and more access. For example, online education of poor children during COVID-19*	OECD 2020
Safe and secure city: *Technology tools such as surveillance cameras, IoT sensors, and smartphones support safety applications with fast and quick alarms of any harassment and violence of women*	ARUP and The Rockefeller Foundation 2015
Smart digital entrepreneurs: *Digital innovations lead to diversification of the economy. If utilized appropriately with the private sector, data that is considered new gold or new oil may lead to several business innovations opening plenty of new opportunities for city dwellers. This may also contribute to increased start-ups and new entrepreneurship, especially attracting the youth. For example, today, Africa with more than 50 percent of its population owning mobile phone benefits from a huge mobile phone penetration rate. Many entrepreneurs take this opportunity to create new businesses and tackle social issues with innovative solutions*	OECD 2020
Smart banking & financial inclusion: *Digital tools are increasingly used for financial inclusion mainly due to reduction in cost and its reach and accessibility. For example, mobile money is being quickly picked up in developing countries in Asia, Africa, and Latin America*	Bloomberg 2021; OECD 2020

are examined. Accordingly, a selection of three narratives is provided as examples to showcase how technology innovations can address the challenge of the resilience of vulnerable communities.

The three city examples from Europe, Asia, and Africa highlight different smart solutions used to tackle the unique and specific local challenges, along with a focus on improving and building the resilience of vulnerable communities. These case examples represent different societies with unique and distinct challenges at different growth and development levels, including their motivation and adaptation of ICT and the smart city approach. The three cities also represent the varied urban landscapes from high-income, middle-income, and low-income countries, respectively. Small innovations and simple technology tools have immense potential to build resilience across diverse economies in varied geographical contexts.

18.5 Smart London: A Data-Driven Strategy Toward Building the Resilience of Vulnerable Communities

London, the capital city of the United Kingdom, is a thriving global hub for business, education, culture, and international affairs. "London is the tech capital of Europe by size, level of investment, and the presence of more than a third of Europe's billion-dollar 'unicorn' companies" (Morningstar 2020). It is one of the first cities to adopt a technological approach to development. The Smart London Plan (2013; 2016) proposes the adoption of technology and a new form of collaboration between Londoners, government, businesses, and academics to approach London's challenges in an integrated way in effort to make it a more competitive and liveable city of the world. The thrust areas of Smart London include city-wide collaboration, data, digital skills and capacity, connectivity, and inclusive technology.

The London City Resilience Strategy (2020–2050) focuses on long-term resilience challenges to improve the ability of the city to manage crises up to 2050. The resilience vision of London identifies opportunities across three cross-cutting themes: resilience for people, resilience for place, and resilience for process. The vision further states that a successful resilience is achieved through systems approach and flexibility to deal with unpredictable events and recover to a new, better, normal when things go wrong.

The London City Resilience Strategy further proposes to mobilize its collective intelligence to improve societal wellbeing for current and future generations. The "London Datastore" project is one such project toward building collective intelligence of the city. This unique smart city initiative aims to build resilience, sustainable growth, development of the city, and make available vast amounts of free data for multiple stakeholders, including citizens. This data can be used to tackle crucial challenges of a city like improving air quality, easing road congestion, etc. Availability of this data led to several other smart projects addressing the urban challenges of London, including meeting the needs and requirements of many vulnerable groups and communities. The key features of London Datastore are detailed in Table 18.3 below.

The initial budget for the setup of the Datastore was less than £100,000. At present, the Datastore has only two permanent employees, and the development costs are split across multiple clients. In 2018, the Datastore moved beyond open data and initiated facilitating access to data across the data spectrum.

The London Datastore today contributes extensively toward the resilience building of its citizens, particularly the vulnerable groups and communities. Table 18.4 below highlights some of the relevant projects and contributions:

Table 18.3 Key features of London Datastore (*Source* London Datastore)

Amount of free data	6000 + data sets
Domain wise listing of datasets	Education (75), Crime and Community Safety (74), Demographics (173), Environment (179), Transparency (152), Business and Economy (131), Employment and Skills (148), Planning (127), Health (103), Housing (111), Transport (95), Young People (60), Income, Poverty and Welfare (32), COVID-19 Data and Analysis (33), Art and Culture (30), Sport (26), London 2012 (5), and Championing London (19)
Data format	The data is available in 23 formats, including pdf, excel, and ZIP file
Number of data publishers	116 agencies and includes Greater London Authority (GLA), Transport for London, Office for National Statistics, London Fire Brigade Census Information Scheme, and NHS
Software	The Datastore uses CKAN, an open-source data portal software, which allows much better management of publishing organizations, each of which can have one superuser with administrator rights
Number of users	60,000/month

18.6 Smart Bengaluru: A Safe and Secure City for Women

Bengaluru is the capital of the state of Karnataka, one of the southern states in India. It is one of India's fastest-growing cities, a major economic center with a billion-dollar economy, and the leading earner of foreign exchange from services and exports (Reserve Bank of India). Bengaluru city, often known as the Silicon Valley of India, is home to some of the largest and prominent IT companies in the world. Over 400 of the fortune 500 companies have an office in Bengaluru, and it contributes to almost 60 percent of the state's GDP. Additionally, it has emerged as a global destination for "new" service sector economies like IT and biotechnology. The most significant attraction toward the city comes from its high economic growth rate, the premier academic institutions, and availability of skilled workforce. It has the highest per capita income among metro cities in India and the largest number of Research & Development labs. Bengaluru has highly educated youths and constitutes 40 percent of the urban educated voters. Furthermore, Bengaluru is claimed as the start-up Capital of India, with nearly 1,200 to 1,500 start-up companies registered so far.

In recent years, Bengaluru has made significant advances in mobility, energy, environment, safety and security, civic participation, and grievance redressal. For example, the Smart City Plan of Bengaluru (2015) was proposed with a goal of "Smart Communities build Smart City." The city's vision is for "Livable Bengaluru" through *"Connected, Vibrant and Healthy Communities"* which is sustainable on three fronts: Environment, Economy, and Equity. In essence, the strategic focus and blueprint highlight a conscious directive for sustainable choice, informed decisions through continued civic participation in city management and services, and building on public–private partnerships for infrastructure projects in the city.

Table 18.4 List of London Datastore projects contributing toward building the resilience of vulnerable communities

Project name	*Details*
Monitoring COVID-19 cases and vaccinations	Use resident data to locate and support the most at-risk population like the elderly or disabled. This data is shared with the grassroots organization to coordinate and manage the infected cases and vaccination
Economic Impact of COVID-19 on London's Small and Medium-sized Enterprises (SMEs):	Data analysis of the impacts of COVID-19 on London's SMEs by sectoral, financial, employment, and risk indicators
COVID-19 Labor Market Analysis	Analysis of the labor market covering all the impacts of the pandemic
Equality, Diversity, and Inclusion Evidence Base Report and Measures	This dataset assesses the level of equality in London across various policy areas and supports the Mayor's equality, diversity, and inclusion strategy. It brings together data relevant to equality that is publicly available
London Job Posting Analytics	This workbook summarizes Job Posting Analytics related to delivery drivers in London
Life in Generation Z	This project focuses on the key issues and concerns affecting young people aged 16–24, examines the challenges they face, and the required action to take in order to to help them thrive across the capital
Universal Credit and Disabled Londoners	The briefing provides an insight into the ecosystem of organizations in London, offering support to disability benefit claimants migrating to Universal Credit (UC)
"Climate Just" data	The "Climate Just" Map Tool shows the geographical map of England highlighting vulnerability to climate change at a neighborhood scale. It shows the places that are at disadvantage due to climate impacts. It aims to heighten awareness on social vulnerability and exposes knowledge on flooding and heat that leads to uneven impacts in different neighborhoods. The Climate Just Map Tool comprises of maps on flooding, heat, and fuel poverty
SafeStats	It shows nearly 15 million records of datasets covering issues of crime and public safety in London. The data is provided by the Metropolitan Police Service, Transport for London, London Ambulance Service, British Transport Police, Fire Brigade, and Hospital Emergency Departments

(continued)

Table 18.4 (continued)

Project name	*Details*
Better Environment, Better Health	It offers borough wise tailored information on seven important environmental factors that impact residents' health. These factors are healthy food, green spaces, surface water and flood risk, active travel and transport, air quality, overheating and fuel poverty
The Planning London Datahub	It shows live data fed from boroughs and applicants on how the city and environment is changing and how planning policies are impacting that change and so on

One of the main goals of the Bengaluru smart city is to work toward the safety and security of women and enhance neighborhood safety. This takes prominence because the sexual assaults on women and girls in India are rapidly increasing. The average rate of reported rape cases is about 6.3 per 100,000 of the population. According to a recent news report, about 99 percent of cases of sexual violence go unreported. As per the last National Crime Records Bureau (NCRB, India) data, Bengaluru is one of the most unsafe cities for women. Among the recent crime statistics of 19 metropolitan cities in the country, Bengaluru is in the top three positions for four crimes such as: murder, crimes against women, crimes against children, and thefts. The statistics of "Crime in India – 2017" released by NCRB raises concern on Bengaluru's crime graph which is on steep upswing compared to the two previous years.

To manage this challenge, Bengaluru has adapted the use of ICT for city-wide surveillance, with the project being called the Safe City project. Essentially, it is aimed at increasing 24/7 safety for women in Bengaluru by installing CCTV surveillance cameras, panic buttons, up-gradation of dial 100, and GIS mapping of crimes in the city. The Government of India sponsors the Safe City project under the Nirbhaya Fund scheme (named after the victim of the 2012 Delhi gang-rape case). In addition to Bengaluru, this scheme is implemented in Delhi, Mumbai, Kolkata, Chennai, Hyderabad, Lucknow, and Ahmedabad.

Based on GIS mapping of crimes, the Bengaluru police will install nearly 16,120 high-tech cameras in sensitive areas with a panic button attached to the streetlight pole, which will make a loud noise when prompted. The women can go to the nearest "Suraksha deep" (safety light) and press the panic button. The panic button will then alert the city's command and control center (CCC), and the CCC will notify the local police patrol cars to proceed to the spot. The Safe City Project also has a provision to upgrade the helpline 100 and child helpline 1098.

In addition, the Bengaluru City Police recently launched an innovative and easy-to-use mobile App, SURAKSHA, meaning "safety" in English. This Bengaluru City Police SOS is a fully integrated personal safety app with policing. During emergency, this App turns the user's smartphone into a discreet personal safety device with a call of service to police that can be triggered by simply activating the SOS button's icon on the user's cell phone.

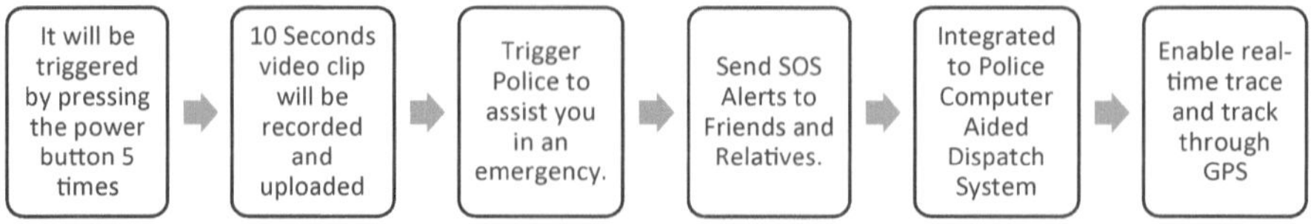

Fig. 18.1 The key features of the SURAKSHA App in Bengaluru (Government of Karnataka, 2020, para. 2)

Furthermore, the SURAKSHA App, which has a panic button, can trigger the police's control room. The police can be reached in times of distress, and the nearest patrol vehicle can address the issue. While the app was first launched in 2017 by the city police and since undergone modifications, more than 200,000 women in Bengaluru city alone have downloaded this app. The key features of SURAKSHA are shown in Fig. 18.1 below.

In line with this, India's ride-sharing company "Ola" has integrated its in-app emergency button with the Bengaluru police's safety App, SURAKSHA. The integration will allow users to share their ride information, including driver and car details, with the city police to help them track the particular trip. Whenever a customer presses the "emergency button," or dial 100, the police will get real-time access to the customer's ride information along with the vehicle's GPS coordinates. The nearest police team is then able to track and trace the car. To attend to the distress calls, nearly 270 police patrol vehicles are stationed at sensitive zones and over 1,200 Cheetah motorcycle patrol squads are available on 24/7 duty.

18.7 Smart Kampala: Building Urban Resilience Through Inclusive Digital Financial Services

Kampala, the capital city of Uganda, is the largest city in the country with a size of 189 km^2. It is the 13th fastest-growing city and ranked the best city to live in East Africa. The resident population in the city is estimated at 1.5 million, while the day population is estimated at 4 million. By 2040, the population is projected to be 10 million, and the city's contribution to Uganda's GDP is almost 60 percent. Due to its widespread scale, the city is divided into five divisions that oversee local planning.

The Kampala Capital City Authority (KCCA) Smart City Strategic Plan strives to holistically weave technology throughout the urban fabric to improve and enhance community quality of life. KCCA Smart City focuses on data-driven decision-making, stakeholder engagement with relevant and timely information, with user feedback into service and program design for a better experience and outcomes. Overall, the vision of Smart Kampala is to be a vibrant, attractive, and sustainable city, while its mission is to provide quality services to the city with core excellence, client care, integrity, teamwork, and innovativeness. The core objectives are to improve institutional effectiveness, enhance the client experience, promote collaboration experience, and develop service delivery through ICT innovation.

Today, Kampala is one of the lead cities in the Africa Smart Towns Network. The Smart Kampala plan is developed on core SMART values of Sustainability, Mobility, Accessibility, Resiliency, and Transparency. Additionally, the significant services planned under the smart city are smart payments, smart education, smart planning, smart communities and social services, and smart transport.

Despite these planned services, Kampala faces significant challenges in terms of building urban resilience particularly in slums and informal settlements where there is huge challenge of integrating the informal economy. In general, access to formal financial services is still low in Uganda. Lack of access to financial institutions forces several informal workers and businesses to depend on precarious informal and short-term funding offered at high-interest rates. Particularly, the informal vendors are more vulnerable and have fewer opportunities to formalize their businesses. It further leads to unregulated, uncontrolled, and unsafe work environment and markets and services. However, the trend changed dramatically in 2009 with the introduction of mobile money that has helped to expand formal financial services to previously excluded populations such as these.

It is often argued that the development of innovative financial products on mobile platforms has tremendous impact in improving financial inclusion outcomes across many African countries. Uganda is one country where the access to and use of mobile money has expanded rapidly over the last few years. At present the digital financial services (DFS) market in Uganda is dominated by mobile money service providers (MMSPs). The mobile money ecosystem comprises of mobile network operations, commercial banks, non-banking financial institutions, the Bank of Uganda (BoU), technology providers, and third-party operations.

At present, seven MMSPs provide a wide range of financial services across the country and, as a result, mobile money services enhance the financial inclusion of vulnerable populations. For example, digitalized cash-based transfers are currently made to millions of refugees, as Uganda hosts nearly 1.2 million forcibly displaced persons and refugees; the most significant number in Africa and the third largest in the world. Mobile money services are increasingly used to transfer social benefits. For example, the social assistance grants for empowerment (SAGE) scheme aimed at senior citizens and vulnerable families uses the digital cash transfer initiative of the Government of Uganda.

In line with being a crucial financial tool for the most vulnerable, a mobile phone is interestingly found to be an economic tool that liberates women entrepreneurs from poverty and empowers them with knowledge. In Kampala mobile phone usage is high among women entrepreneurs the majority of who are micro-entrepreneurs. Women entrepreneurs rose especially in mainstream operations like customer service delivery, marketing and sales, information sharing leading to increased business processes, and profits, further facilitating increase in productivity and enterprise transformation.

Furthermore, mobile money is providing an increased access to finance SMEs (small and medium-sized enterprises), which have traditionally been poorly served by existing lending institutions. At present, banking through mobile phones allows for the real-time transfer and the receipt of small funds at a low cost. The mobile money has opened up new opportunities in small businesses to access financial services by reducing the costs of processing and administering small loans, thereby alleviating a significant disincentive for lenders to extend credit to SMEs. Overall, mobile technologies for development (M4D) represent a force for change in the market and women's economic activities, challenging but also reinforcing the informality of the Kampala markets. Based on this success and building on the ongoing collaboration with UNCDF in implementing the "Leaving No One Behind in the Digital Era" strategy, Airtel Uganda Limited is building a new partnership model with two financial technology firms that aim to empower rural customers to access digital services.

18.8 Discussion

Many cities link their resilience initiatives with established planning systems and development trajectories, including actions to enhance the inclusion of all relevant actors in decision-making, transparency of the processes of planning, empowering the local environment through digital innovations, and enhancing fair access to basic infrastructures and services for vulnerable groups, among others. These strategies predispose cities to resist change and bounce back to a former condition(coping), adjust to change without modifying established structures (incremental adaptation), and adapt to change and transform through training and self-organization (transformational adaptation) in the context of rapid urbanization, climate change, and the associated consequences (Torabi et al. 2018; Béné et al. 2017; Ziervogel et al. 2016). A resilient approach to development is expected to improve governance challenges and highlight the weak links, possible vulnerabilities, and other loopholes in the system (HABITAT III 2015). The rapidly increasing urbanization is shifting the focus toward building more resilient cities, thus saving more lives, resources, and the environment (UN-Habitat 2017). However, the question of what contributes to urban resilience and what kind of interactions are needed to enhance it remains on the agenda (Asadzadeh et al. 2017).

More recently, many scientific and academic attempts have made the case that establishing long-term resilience to increasing complexities and uncertainties necessitates innovative and novelties for thinking, doing, and organizing (Hölscher et al. 2019; Van Winden and Carvalho 2017). In other words, science, technology, and innovation contribute immensely to resilience building (UNCTAD 2020). Harnessing the power of technology can mitigate the challenges of urbanization and can help catalyze urban resilience. The foundational technologies for cities, including mobile communications, cloud computing, data analytics, and social applications, have a critical role in supporting urban resilience. Without the deliberate use and integration of technology in approaching resilience, the cities will be ill-equipped to address the shocks and stresses that are rising with the phenomena of rapid urbanization, climate change, and globalization.

The smart city approach, particularly the use of technology, helps in building the resilience of cities. The smart city is a term most often associated with innovative technology solutions in urban environments. However, the smart cities movement is much more than just technology. It combines urban planning, public administration, and community engagement, all powered by innovation and expressed through actual use cases that improve government operations or service delivery (100 Resilient Cities). In order to realize this potential, many city governments started using ICT to address various challenges, which later became more focused and identified through smart city initiatives.

Across the globe, many technologies are used to improve government services, including disaster risk management and urban resilience. Smart urban governance, therefore, is increasingly considered as a core factor facilitating progress on all activities related to urban resilience. Similarly, open and updated risk information and communication contribute to transparent and evidence-based decision-making. The three specific real-world cases discussed in this chapter highlighted the use of ICT in different ways for building the resilience of vulnerable communities addressing the local challenges. The London Datastore effectively uses the data to address the city's challenges, particularly vulnerable communities, through maps and apps open to the public and other stakeholders. The Bengaluru Safe City Project aims to address the challenges of women in cities who often experience violence and untoward incidents in their daily life. And finally, the case of Kampala discussed building urban resilience of vulnerable communities, including women, through inclusive digital financial services.

Digging a little deeper, the London Datastore highlights how innovation and digital technologies can be implemented to build resilience and improve overall business function and operational performance, including social and environmental impact. This data-based evidence guides and influences GLA, including local boroughs, to make informed and scientific decisions and accordingly design policy innovations and allocate budgets. The open and transparent data encourages more collaborative and constructive work among various stakeholders and reinforces London's resilience action plan. This project addresses two significant resilience development challenges: (1) lack of access to information, and public participation in decision-making (2) need for efficient communication and exchange among the relevant stakeholders when shaping the resilience initiatives.

Regarding the second case study, the Bengaluru Safe city project highlights the use of technology in building a safe, secure, and empowering environment for women in public places. Thus, enabling them to pursue all opportunities equally without the threat of gender-based violence or harassment. Inspired by the success of this project, recently, the city police have decided to launch a pilot project that will identify the areas where women generally do not go because it is considered unsafe. Consequently, in association with a local civil society group, the Bengaluru police have identified eight spots, including roads, parks, and other public spaces where women do not feel safe, turning them into more friendly spaces. Also, in the first of

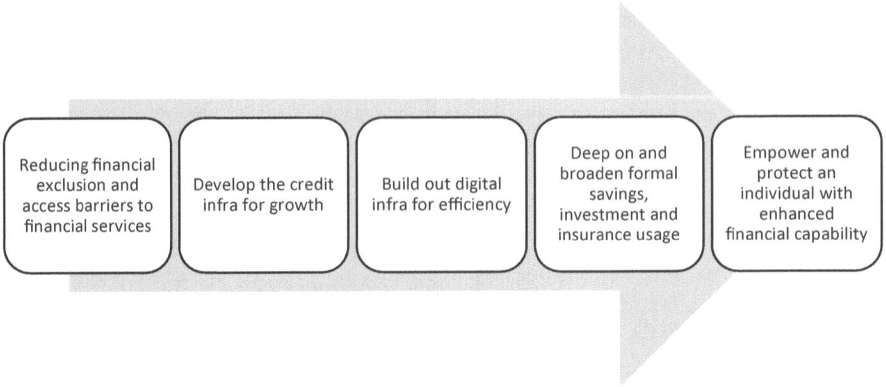

Fig. 18.2 National financial inclusion strategy of Uganda (2017)

its case, policewomen will patrol the streets of select Bengaluru areas such as where usually, only police officers patrol the streets on motorbikes.

Lastly, the Kampala project is an example to highlight how everyone in society can access essential financial services regardless of their income or savings. It focuses on providing financial solutions to the economically underprivileged who are most vulnerable to economic shocks. The mobile phones and mobile money in Kampala have enabled women to have the ability and tools to manage and save their money and also empower them with the right skills and knowledge to make the right financial decisions. This will further lead to equality both within the community and within the family. Inspired by the success stories, Uganda launched a national financial inclusion strategy in 2017. This strategy focuses on five key areas, as shown in Fig. 18.2 below.

Urbanization, demographic shift, climate changes, and new technologies are reshaping how city leaders are looking at sustainability and resilience, particularly the design and delivery of public services to address these new dynamics. Based on the recent COVID-19 crisis alone, city governments have shifted their focus on the need to build long-term resilience into their recovery strategies. Smart cities, which are established upon complex and intelligent approaches of universal digital networks, can better connect citizens, governments, and objects seamlessly. This will enable the right environment and can optimize the performance of urban basic infrastructures and connect to citizens using integrated command and control centers where the centralized units help manage and administer vital services efficiently during and after a crisis.

The current digital age with increasing use of ICT applications can promote the simultaneous improvement of smart cities and urban resilience, while also helping to establish a new "smart urban resilience" model. As a result, the insights supported by open data can deliver more resilient infrastructure and deliver a more sustainable future. This, in turn, brings the most high-level technology into a centralized data environment, enabling actors to maximize performance, minimize cost and delay, and assure effective services.

On the flip side, increasing digital dependency invites potential attacks on ICT infrastructure. The cyber-attacks on cities' physical and digital infrastructure often could be more severe and disruptive than ever before. Therefore, cyber resilience must become another dimension to tackle when building a "smart urban resilience" model.

18.9 Conclusion

Cities are the key hubs of economy and growth. Strengthening urban resilience is a critical element for global sustainable development. Resilient cities can better withstand, adjust, and adapt to social, economic, political, and environmental shocks and stresses. The vulnerable communities such as elderly people with disabilities, women, children, poor, migrants, refugees, and other minority groups are more prone and exposed to natural and human-made disasters, including other life threats, shocks, and stresses. Therefore, resilience initiatives recognize the importance of promoting procedural justice by the inclusion of the needs and interests of all relevant actors, including vulnerable and marginalized populations.

Such attitudes toward resilience call for innovations and novelties in planning processes and practices since urbanization and climate change dynamics are globally rising, and cities are at the forefront of these phenomena. More recently, international scientific and policy discourses on sustainable development stress the need to incorporate smart city concepts into urban resilience trajectories. The city managers and planners can leverage the benefits of ICT and use the same for building a resilient and fool-proof city. In addition to this, the use of technology, such as early warning systems for risk information and communication, enables efficient delivery of services during and after disasters. Smart technologies predispose local governance and agencies to set more preventive measures that enable them to respond in a time of emergencies and plan for long-term resilience and sustainability.

Further, the smart city approach enhances opportunities for developing innovative solutions for building the resilience of vulnerable communities. The use of data and communication systems will increase efficiency, collaboration, transparency, and evidence-based decision among multiple actors working at city resilience strategy.

Consequently, the three case studies discussed in this chapter present different use cases where the technology can contribute to resilience building. The London Datastore is a classic example to highlight the use of data for evidence-based decision-making. As described, London is using the data to measure and tackle the risk exposure of vulnerable and marginalized populations with appropriate strategies and interventions. Similarly, the case of Bengaluru highlights the use of emerging technology solutions to build a safe and secure city environment for women. Finally, the case example of Kampala highlights how technology is contributing to enhancing financial inclusion with increased business opportunities for the informal sector.

While the integrated concept of resilience thinking and smart city approach predisposes decision-makers and planners to shape sustainable and just cities, more empirical insights are needed to understand how and to what extent different smart city tools and paradigms contribute to the realization of different pathways to resilience. However, it is debatable whether cities can effectively merge resilience and smartness without the right policy framework, institutional and people capacity, and consistent disaster risk reduction plans at both governance and operational level. Further, an additional layer posed is the challenge of how to run and maintain the ICT services during and after disasters without causing any disruption. Overall, while these considerations are relevant and highly significant to the discourse, the value that smart cities provide in security, community inclusion, and resilience cannot go ignored, as it offers modest insight in how to navigate a world that is seeing an increase in both complexity and uncertainty.

References

10 years of the London Datastore & thinking on city data for the next decade (2020) Smart London Medium. Retrieved 26 April 2021, from https://smartlondon.medium.com/10-years-of-the-london-datastore-thinking-on-city-data-for-the-next-decade-b634ae62dc3c

15 – Urban Resilience (2015) Habitat Iii Issue Papers. Retrieved 26 April 2021, from http://uploads.habitat3.org/hb3/Habitat-III-Issue-Paper-15_Urban-Resilience-2.0.pdf

Ageing cities (2021) HelpAge International. Retrieved 26 April 2021, from https://www.helpage.org/what-we-do/ageing-cities/

AR5 synthesis report: climate change 2014—IPCC. IPCC (2014) Retrieved 26 April 2021, from https://www.ipcc.ch/report/ar5/syr/

ARUP, The Rockefeller Foundation (2015) City resilience index. http://publications.arup.com/Publications/C/City_Resilience_Framework.aspx

Asadzadeh A, Kötter T, Salehi P, Birkmann J (2017) Operationalizing a concept: the systematic review of composite indicator building for measuring community disaster resilience. Int J Disaster Risk Reduct 25:147–162. https://doi.org/10.1016/j.ijdrr.2017.09.015

Balogun A, Marks D, Sharma R, Shekhar H, Balmes C, Maheng D, Arshad A, Salehi P (2020) Assessing the potentials of digitalization as a tool for climate change adaptation and sustainable development in urban centres. Sustain Cities Soc 53:101888. https://doi.org/10.1016/j.scs.2019.101888

Baron M (2012) Do we need smart cities for resilience. J Econ Manag 10:32–46.

Béné C, Mehta L, McGranahan G, Cannon T, Gupte J, Tanner T (2017) Resilience as a policy narrative: potentials and limits in the context of urban planning. Clim Dev 10(2):116–133. https://doi.org/10.1080/17565529.2017.1301868

Berawi MA (2018) The role of technology in building a resilient city: managing natural disasters. Int J Technol 9(5):862–865.

Birkmann, Garschagen M, Setiadi N (2014) New challenges for adaptive urban governance in highly dynamic environments: revisiting planning systems and tools for adaptive and strategic planning. Urban Clim 7(July):115–133.https://doi.org/10.1016/j.uclim.2014.01.006

Bizzotto M, Huseynova A, Vital Estrada V (2019) Resilient cities, thriving cities: the evolution of urban resilience. ICLEI-local governments for sustainability. Retrieved 26 April 2021, from https://e-lib.iclei.org/publications/Resilient-Cities-Thriving-Cities_The-Evolution-of-Urban-Resilience.pdf

Bloomberg. Bloomberg.com (2021) Retrieved 26 April 2021, from https://www.bloomberg.com

Borie M, Pelling M, Ziervogel G, Hyams K (2019) Mapping narratives of urban resilience in the global south. Glob Environ Chang, 54(August 2018):203–213. https://doi.org/10.1016/j.gloenvcha.2019.01.001

Case study: London datastore. Greater Manchester (2019) Retrieved 26 April 2021, from https://greatermanchester-ca.gov.uk/media/3540/london-datastore.pdf

Challenges of urban resilience in India. South Asia disaster (2015) Retrieved 26 April 2021, from https://www.preventionweb.net/files/43556_urbanriskandresilience.pdf

Chan J, Zhang Y (2019) Urban resilience in the smart city. ResearchGate. Retrieved 26 April 2021, from https://www.researchgate.net/publication/334398312_Urban_Resilience_in_the_Smart_City

Chauhan N (2019) How gender-based violence in India continues to rise. YourStory.com

Clark-Ginsberg A, Aguirre B, Birkland T, Browne KE, Cutter SL, Oliver-Smith A, Perrow C, Ben Wisner (2017) Urban challenges and opportunities for FEMA during the Trump administration. Retrieved from http://www.aaroncg.me/wp-content/uploads/2017/01/Urban-challenges-and-opportunities-for-FEMA-during-the-Trump-administration.pdf

Climate Smart Cities in Emerging Economies. Unfccc.int (2021) Retrieved 26 April 2021, from https://unfccc.int/climate-action/momentum-for-change/activity-database/momentum-for-change-climate-smart-cities-in-emerging-economies

Coaffee J, Therrien MC, Chelleri L, Henstra D, Aldrich DP, Mitchell CL, Tsenkova S, Rigaud É (2018) Urban resilience implementation: a policy challenge and research agenda for the 21st century. J Contingencies Cris Manag 26(3):403–410. https://doi.org/10.1111/1468-5973.12233

Cole J (2013, October) Measuring the resilience of cities: the role of big data. In Proc Conf Meas Resil Cities: Role Big Data (vol 25)

Correa E, Ramírez F, Sanahuja H (2011) Populations at risk of disaster: a resettlement guide. World Bank

Covid-19 accelerates the adoption of smart city tech to build resilience. Smart Cities World 2020 Retrieved 26 April 2021, from https://www.smartcitiesworld.net/news/news/covid-19-accelerates-the-adoption-of-smart-city-tech-to-build-resilience--5259

Crime In India | National Crime Records Bureau. Ncrb.gov.in (2019) Retrieved 26 April 2021, from https://ncrb.gov.in/en/crime-india

Dominici D (2019) A set of good practices and recommendations for smart city resilience engineering and evaluation

Demirgüç-Kunt A, Klapper L (2013) Measuring financial inclusion: explaining variation in use of financial services across and within countries. Brook Pap Econ Act 2013(1):279–340

Edwards A (2016) Global forced displacement hits record high. UNHCR. Retrieved 26 April 2021, from https://www.unhcr.org/uk/news/latest/2016/6/5763b65a4/global-forced-displacement-hits-record-high.html

Fekete A, Asadzadeh A, Ghafory-Ashtiany M, Amini-Hosseini K, Hetkämper C, Moghadas M, Ostadtaghizadeh A, Rohr A, Kötter T (2020) Pathways for advancing integrative disaster risk and resilience management in Iran: needs, challenges and opportunities. Int J Disaster Risk Reduct 49. https://doi.org/10.1016/j.ijdrr.2020.101635

Financing the resilient city—a demand driven approach to development, disaster risk reduction, and climate adaptation. ICLEI Global Reports (2019) Retrieved 26 April 2021, from https://resilientcities2019.iclei.org/wp-content/uploads/Report-Financing_Resilient_City-Final.pdf

Giffinger R, Fertner C, Kramar H, Meijers E (2007) City-ranking of European medium-sized cities. Cent Reg Sci Vienna UT, 1–12

Government of Karnataka (2020) SURAKSHA-Bengaluru city police. https://apps.karnataka.gov.in/app/94/en

Gu D (2019) Exposure and vulnerability to natural disasters for world's cities. United Nations, Department of Economics and Social Affairs, Population Division, Technical Paper No. 4

HM Chaitanya S (2020) In a first, women cops to patrol Bengaluru streets on motorbikes. Deccan Herald. Retrieved 26 April 2021, from https://www.deccanherald.com/city/in-a-first-women-cops-to-patrol-bengaluru-streets-on-motorbikes-808487.html

Hölscher K, Frantzeskaki N, McPhearson T, Loorbach D (2019) Capacities for urban transformations governance and the case of New York City. Cities 94(May):186–199. https://doi.org/10.1016/j.cities.2019.05.037

Home: Sustainable Development Knowledge Platform. Sustainabledevelopment.un.org (2021) Retrieved 26 April 2021, from https://sustainabledevelopment.un.org/index.html

House of Lords—Tackling financial exclusion: A country that works for everyone?—Select Committee on Financial Exclusion. Publications.parliament.uk (2017) Retrieved 26 April 2021, from https://publications.parliament.uk/pa/ld201617/ldselect/ldfinexcl/132/13209.htm

Hyogo Framework for Action 2005 2015: ISDR International Strategy for Disaster Reduction International Strategy for Disaster Reduction. Unisdr.org (2005) Retrieved 26 April 2021, from https://www.unisdr.org/2005/wcdr/intergover/official-doc/L-docs/Hyogo-framework-for-action-english.pdf

Inclusion A (2019) Uganda's journey to inclusive finance through digital financial services. Issuu. Retrieved 26 April 2021, from https://issuu.com/afiglobal/docs/afi_ms_uganda_aw_digital

Inclusive Growth and Social Cohesion in Cities: An OECD Approach. Urban20.org (2020) Retrieved 26 April 2021, from https://www.urban20.org/wp-content/uploads/2020/11/U20_WP_Inclusive_growth_in_cities.pdf

Information Systems Strategic Plan 2020–2026. Kampala Capital City Authority (KCCA) (2020) Retrieved 26 April 2021, from https://www.kcca.go.ug/media/docs/Information_Systems_Strategic_Plan%202020-2026.pdf

International Monetary Fund (2020, October 15) World Economic Outlook. International Monetary Fund. Retrieved 15 May 2021, from https://libraryguides.vu.edu.au/apa-referencing/7Webpages

ISO 37106:2018, Sustainable cities and communities. ISO (2018) Retrieved 26 April 2021, from https://www.iso.org/standard/62065.html

Jérusalmy O, Fox P, Hercelin N, Mao L (2020) Financial exclusion: making the invisible visible. A study on societal groups encountering barriers to accessing financial services in the EU. Finance-watch.org. Retrieved 26 April 2021, from https://www.finance-watch.org/wp-content/uploads/2020/03/FW-Report_Vulnerable_Groups_March2020.pdf

Jovanovic A (2018) Innovative approaches to assessing infrastructure resilience: developing indicators and data for smart infrastructures. http://www.oecd.org/gov/risk/Aleksandar-Jovanovic-Session-2B.pdf

Kampala strives to improve resilience. Undrr.org (2016) Retrieved 26 April 2021, from https://www.undrr.org/news/kampala-strives-improve-resilience

Karnataka Government App Store. Apps.karnataka.gov.in (2021) Retrieved 26 April 2021, from https://apps.karnataka.gov.in/app/94/en

Keck M, Sakdapolrak P (2013) What is social resilience? lessons learned and ways forward. Erdkunde 67(1):5–19. https://doi.org/10.3112/erdkunde.2013.01.02

Khan N (2020) Making women reclaim public spaces: Bengaluru police identify 8 'black spots' for pilot project—WATCH. Times Now News. Retrieved 26 April 2021, from https://www.timesnownews.com/mirror-now/in-focus/article/making-women-reclaim-public-spaces-bengaluru-police-identify-8-black-spots-for-pilot-project-watch/552622

Khansari N, Mostashari A, Mansouri M (2014) Impacting sustainable behavior and planning in smart city. Int J Sustain Land Use Urban Plan 1(2)

Komunte M (2015) Usage of mobile technology among women entrepreneurs: a case Study of Uganda

Kubicki S (2017) Beyond smart cities, welcome to resilient cities

Lall SV, Deichmann U (2010) Density and disasters: economics of urban hazard risk. The World Bank

Landry JN, Webster K, Wylie B, Robinson P (2016) How can we improve urban resilience with open data. Open Data Institute

Larsson CW, Svensson J (2018) Mobile phones in the transformation of the informal economy: stories from market women in Kampala, Uganda. J East Afr Stud 12(3):533–551

London City Resilience Strategy 2020. Greater London Authority (2020) Retrieved 26 April 2021, from https://www.london.gov.uk/sites/default/files/london_city_resilience_strategy_2020_digital.pdf
London Datastore—Greater London Authority. Data.london.gov.uk (2021) Retrieved 26 April 2021, from https://data.london.gov.uk/
London Resilience Strategy Preliminary Resilience Assessment. Greater London Authority (2019) Retrieved 26 April 2021, from https://www.london.gov.uk/sites/default/files/london_urban_resilience_strategy_pra_final_0.pdf
Make Kampala better! How can we build urban resilience and integrate the informal economy? | UNDP in Uganda. UNDP (2020) Retrieved 26 April 2021, from https://www.ug.undp.org/content/uganda/en/home/presscenter/pressreleases/2020/make-kampala-better--how-can-we-build-urban-resilience-and-integ.html
Making Health Care Work Better for Vulnerable Patients. The Commonwealth Fund (2018) Retrieved 26 April 2021, from https://www.commonwealthfund.org/blog/2018/making-health-care-work-better-vulnerable-patients
Marana P, Eden C, Eriksson H, Grimes C, Hernantes J, Howick S, Labaka L, Latinos V, Lindner R, Majchrzak TA, Pyrko I, Serrano N (2019) Towards a resilience management guideline—cities as a starting point for societal resilience. Sustain Cities Soc 48:101531
Matthews N, Nel D (2019) Climate change hits vulnerable communities first and hardest. IISD. Retrieved 26 April 2021, from https://www.iisd.org/articles/climate-change-hits-vulnerable-communities-first-and-hardest
Mawejje J, Lakuma EC (2017) Macroeconomic effects of mobile money in Uganda (No. 677–2017–2261)
McPhearson T, Iwaniec DM, Bai X (2016) Positive visions for guiding urban transformations toward sustainable futures. Curr Opin Environ Sustain 22:33–40. https://doi.org/10.1016/j.cosust.2017.04.004
Melkonyan A, Koch J, Lohmar F, Kamath V, Munteanu V, Alexander Schmidt J, Bleischwitz R (2020) Integrated urban mobility policies in metropolitan areas: a system dynamics approach for the Rhine-Ruhr metropolitan region in Germany. Sustain Cities Soc 61(February). https://doi.org/10.1016/j.scs.2020.102358
MG C (2019). Bengaluru city crime graph up. The New Indian Express. Retrieved 26 April 2021, from https://www.newindianexpress.com/cities/bengaluru/2019/oct/22/bengaluru-city-crime-graph-up-2051110.html
Moglia M, Cork SJ, Boschetti F, Cook S, Bohensky E, Muster T, Page D (2018) Urban transformation stories for the 21st century: insights from strategic conversations. Glob Environ Chang 50(August 2017):222–237. https://doi.org/10.1016/j.gloenvcha.2018.04.009
Moraci F, Errigo MF, Fazia C, Burgio G, Foresta S (2018) Making less vulnerable cities: resilience as a new paradigm of smart planning. Sustainability 10(3):755
Morningstar M (2020, February 25) London. About smart cities. https://www.aboutsmartcities.com/smart-city-london/
Mostashari A, Arnold F, Maurer M, Wade J (2011, November) Citizens as sensors: the cognitive city paradigm. In 2011 8th International Conference & Expo on Emerging Technologies for a Smarter World (pp 1–5). IEEE
Moudgal S (2019) Safe city project: to make Bengaluru safe for women, 16,000 cameras with panic buttons will be installed. The Times of India. Retrieved 26 April 2021, from https://timesofindia.indiatimes.com/city/bengaluru/safe-city-project-to-make-bengaluru-safe-for-women-16000-cameras-with-panic-buttons-will-be-installed-across-city/articleshow/71705019.cms
OECD (2013) What does 'resilience' mean for donors? OECD
OECD (2018) System thinking for critical infrastructure resilience and security. http://www.oecd.org/gov/risk/systems-thinking-for-critical-risks-resilience-and-security-2018-agenda.pdf
OECD (2020, October 9) Skills Summit 2020: skills strategies for a world in recovery. [Virtual Meeting]. OECD. https://www.oecd.org/officialdocuments/publicdisplaydocumentpdf/?cote=COM/DELSA/ELSA/EDU/EDPC/RD(2020)1&docLanguage=En

Pasteur K, McQuistan C (2016) From risk to resilience: a systems approach to building long term, adaptive wellbeing for the most vulnerable Rugby. Practical Action Publishing

Patel R, Phelps L, Sanderson D, King J (2016) What are the practices to identify and prioritize vulnerable populations affected by urban humanitarian emergencies? A systematic review protocol

Pelling M, O'Brien K, Matyas D (2015) Adaptation and transformation. Climatic Change 133(1):113–127. https://doi.org/10.1007/s10584-014-1303-0

Pinchoff J (2018) Building resilience in communities most vulnerable to environmental stressors

Population Reference Bureau – Inform, Empower, Advance. Prb.org (2021) Retrieved 26 April 2021, from https://www.prb.org/

Prospects: The 2018 Revision (ST/ESA/SER.A/420). United Nations. Education. Unicef.org (2021) Retrieved 26 April 2021, from https://www.unicef.org/education

Retrieved 26 April 2021, from https://yourstory.com/socialstory/2019/09/gender-violence-india

Romero-Lankao P, Bulkeley H, Pelling M, Burch S, Gordon DJ, Gupta J, Munshi D (2018) Urban transformative potential in a changing climate. Nat Clim Chang 8(9):754–756. https://doi.org/10.1038/s41558-018-0264-0

Rouganne M (2018) How mobile phones are shaping the future of Africa. Digital Corner. Retrieved 26 April 2021, from https://www.digitalcorner-wavestone.com/2018/02/how-mobile-phones-are-shaping-the-future-of-africa/

Serrat O (2017) The sustainable livelihoods approach. In Knowledge solutions (pp 21–26). Springer

Sheppard A, Tatham P, Fisher R, Gapp R (2013) Humanitarian logistics: enhancing the engagement of local populations. J HumItarian Logist Supply Chain Manag

Shulla K, Voigt BF, Cibian S, Scandone G, Martinez E, Nelkovski F, Salehi P (2021) Effects of COVID-19 on the sustainable development goals (SDGs). Discov Sustain 2(1):1–19

Skills Development. World Bank (2021) Retrieved 26 April 2021, from https://www.worldbank.org/en/topic/skillsdevelopment

Smart cities and Inclusive growth. OECD (2020) Retrieved 26 April 2021, from http://www.oecd.org/cfe/cities/OECD_Policy_Paper_Smart_Cities_and_Inclusive_Growth.pdf

Smart cities and infrastructure. Unctad.org (2016) Retrieved 26 April 2021, from https://unctad.org/system/files/official-document/ecn162016d2_en.pdf

Smart City Challenge Round 3: Smart City Proposal Bengaluru | Smartnet. Smartnet.niua.org (2017) Retrieved 26 April 2021, from https://smartnet.niua.org/content/c8fdaabb-767c-4854-befc-9e9966c28282

Smart City Resilience Learning from Emergency Response and Coordination in Japan. Japan Meteorological Agency (2013) Retrieved 26 April 2021, from https://www.gsma.com/iot/wp-content/uploads/2013/02/cl_SmartCities_emer_01_131.pdf

Tesh J, Cole J (2013) Measuring the resilience of cities—identifying the risks to resilience. Rusi.org. Retrieved 26 April 2021, from https://rusi.org/sites/default/files/201405_op_resilient_cities.pdf

The role of science, technology and innovation in building resilient communities, including through the contribution of citizen science. Unctad.org (2019) Retrieved 26 April 2021, from https://unctad.org/system/files/official-document/ecn162019d3_en.pdf

Torabi E, Dedekorkut-Howes A, Howes M (2018) Adapting or maladapting: building resilience to climate-related disasters in coastal cities. Cities 72(December 2016):295–309. https://doi.org/10.1016/j.cities.2017.09.008

Traballesi A, Bertocchi G, Bologna S, Carrozzi L, Inzerilli P, Franchina L, Di Ludovico D, Dominici D (2019) A set of good practices and recommendations for smart city resilience engineering and evaluation. https://doi.org/10.13140/RG.2.2.11440.81926

Tzioutziou A, Xenidis Y (2021) A study on the integration of resilience and smart city concepts in urban systems. Infrastructures 6(24). https://doi.org/10.3390/infrastructures6020024

Ulrichs M, Slater R (2016) How can social protection build resilience? Insights from Ethiopia, Kenya and Uganda. BRACED Working Paper. ODI, London

UN-Habitat (2004) The challenge of slums: global report on human settlements 2003. Manag Environ Qual: Int J 15(3):337–338.

UN-Habitat (2017) Trends in urban resilience. https://unhabitat.org/trends-in-urban-resilience-2017

UN-Habitat (2020) People-centered smart cities

UN-SDG (2015) Transforming our world: the 2030 agenda for sustainable development. General assembley 70 session. https://doi.org/10.1007/s13398-014-0173-7.2

UNCTAD (2020) World investment report 2020: international production beyond the pandemic. https://doi.org/10.1057/s42214-020-00078-2

UNICEF (2012) The state of the world's children 2012: children in an urban world. Esocialsciences

UNISDR (2018) United Nations Office for Disaster Risk Reduction 2018 Annual Report. United Nations. 64454_unisdrannualreport2018eversionlight.pdf (reliefweb.int)

United Nations (2019, January 24) Extreme weather hit 60 million people in 2018, no part of the world spared. https://news.un.org/en/story/2019/01/1031182

United Nations, Department of Economic and Social Affairs, Population Division (2018) World urbanization prospects: the 2018 revision (ST/ESA/SER.A/420). United Nations

United Nations (2021) Theme report on energy transition. United Nations. 2021-twg_2–062321.pdf (un.org)

Urban Resilience Hub. UN-Habitat (2021) Retrieved 26 April 2021, from https://urbanresilience hub.org/what-is-urban-resilience/

Van Winden W, Carvalho L (2017) How digitalization changes cities: innovation for the urban economy of tomorrow Business models of smart city coalitions View project. https://doi.org/10.13140/RG.2.2.27447.83362

Vermiglio C, Kudo H, Zarone V (2020) Making a step forward towards urban resilience. The contribution of digital innovation. Smart Innovation, Systems and Technologies. https://doi.org/10.1007/978-3-030-52869-0_10

What is the Sendai Framework? UNDRR (2015) Retrieved 26 April 2021, from https://www.undrr.org/implementing-sendai-framework/what-sendai-framework

World Bank Group (2016) Investing in urban resilience: protecting and promoting development in a changing World. World Bank

World Health Organization & World Bank (2011) World report on disability 2011. World Health Organization. https://apps.who.int/iris/handle/10665/44575

World Health Organization (2011) World report on disability 2011. World Health Organization

Zebardast E (2006) Marginalization of the urban poor and the expansion of the spontaneous settlements on the Tehran metropolitan fringe. Cities 23(6):439–454. https://doi.org/10.1016/j.cities.2006.07.001

Ziervogel G, Cowen A, Ziniades J (2016) Moving from adaptive to transformative capacity: building foundations for inclusive, thriving, and regenerative urban settlements. Sustainability 8(9). https://doi.org/10.3390/su8090955

Chapter 19
Advancing Community Resilience Through Smart Approaches: A Resource for Canadian Communities

Ewa Jackson, Amy Jones, and Pourya Salehi

Abstract In response to the severe climate change impacts faced by Canada, Open Smart Cities are increasing in number across the country. Open Smart Cities facilitate collaboration between different sectors and actors regarding the use of data and technology to address social, economic and environmental challenges in an equitable and just manner. This chapter rigorously examines how open smart approaches can increase urban resilience and draws upon initiatives implemented by the City of Surrey in British Colombia to contextualize the theoretical underpinnings of resilience and smart solutions, within a case study example.

Keywords Open Smart Cities · Resilience · Canada

19.1 Introduction

"Now, perhaps more than ever, Canadian communities are being both encouraged (by, for example, the Smart Cities Challenge) and required (due to climate change, urbanization, and other emerging trends threatening to disrupt daily life) to find innovative solutions to the challenges facing their communities. The growing prevalence of "smart" cities reflects this shift. While commonly understood as one that uses data and connected technology to accelerate innovation within local communities (Sodhi et al. 2018), we approach smart cities through a newer concept, called the Open Smart Community.

E. Jackson
ICLEI – Local Governments for Sustainability (Canada Office), Toronto, Canada
e-mail: pourya.salehi@iclei.org

E. Jackson
e-mail: ewa.jackson@iclei.org

A. Jones · P. Salehi
ICLEI—Local Governments for Sustainability (World Secretariat), Bonn, Nordrhein-Westfalen, Germany
e-mail: amy.jones@iclei.org

A. Sharifi and P. Salehi (eds.), *Resilient Smart Cities*, The Urban Book Series,
https://doi.org/10.1007/978-3-030-95037-8_19

First developed by Open North in the Open Smart Cities Guide V1.0 and most recently refined in the State of Open Smart Communities in Canada,[1] the Open Smart Community is defined as: "one where all sectors, including residents, collaborate to mobilize data and technologies to develop their community through fair, ethical and transparent government practices that balance economic development, social progress and environmental responsibility" (Pembleton et al. 2019, p. 5). Critical to this concept are principles like equity, engagement and transparency. We believe that these elements can help to mitigate the potential risks associated with data and connected technology initiatives, from data privacy, to accessibility, to the risks of outsourcing public services (Sodhi et al. 2018).

The concept of the Open Smart Community is now being operationalized in the context of the Community Solutions Network through four domains: Data, Governance, Hardware & Software, and People & Engagement.

This resource is an outcome of the British Columbia Idea Camp held in the City of Victoria, British Columbia, on October 30, 2019. Idea Camps are a series of regional events that took place in 5 regions across the country in 2019. Each event brought together mid- and senior-level municipal staff and Indigenous leaders for peer-to-peer exchange and intensive instruction from experts to build capacity in areas related to open data, governance and public engagement".

"The British Columbia Idea Camp explored the connection between open smart approaches and their ability to support community resilience goals (particularly around climate change) and improve the lives of residents. The event hosted 23 participants representing 13 municipalities and 2 provinces, including Councillors, municipal managers and community planners.

This resource has been developed to assist Canadian municipal and community officials and administrators. It brings together key learnings from communities that are already using open smart approaches to resilience initiatives, as well as resources and guiding principles to help communities identify and implement resilience initiatives and strategies".

19.2 The State of Climate Resilience in Canada

Climate change impacts are being felt across Canada in significant ways. With observed increases in average temperature and precipitation over the last six decades, including especially rapid rates of warming in the North, climate change is already affecting Canada's environment and economy, as well as the safety, physical, mental, cultural, and spiritual health and well-being of Canadians. With these impacts projected to intensify in the coming decades, it is essential that Canadians act now to adapt and build resilience to climate change.

[1] https://drive.google.com/file/d/1sWySv6IDQE-Fyqm-FuU73NcIYqVJ8khi/view.

"According to the recently released *Canada's Changing Climate Report*,[2] Canada has been warming over the last six decades, where temperature has increased in all regions of Canada and in the surrounding oceans" (Bush and Flato 2019). "From 1948 to 2020, there is a trend in annual average temperature departures, showing 1.8°C of warming over that period" (Government of Canada 2021, para. 3). "Canada has also generally become wetter over the past several decades, with a shift toward less snowfall and more rainfall. Annual and winter precipitation is projected to increase everywhere in Canada over the twenty-first century. In the future, a warmer climate will intensify some weather extremes. Extreme hot temperatures will become more frequent and more intense. This will increase the severity of heatwaves, and contribute to increased drought and wildfire risks. While inland flooding results from multiple factors, more intense rainfalls will increase urban flood risks (Bush and Flato 2019).

Extreme weather events such as storms, floods, and windstorms are projected to increase in severity and frequency under a future climate scenario. These types of events have already impacted Canadian communities; the costs of recent major Canadian weather events are evidence of the consequences that these types of events can have:

- Windstorms in Toronto, Eastern Ontario, and Quebec in May 2018 brought claims worth CAD $600 million (Gambrill 2019).
- Heavy rains in May 2017 caused flooding and damage resulting in more than CAD $233 million in insured damage in Eastern Ontario and Western Quebec" (Insurance Bureau of Canada 2017).
- In April 2019, the Canadian Government announced that "annual precipitation is projected to increase in all regions of Canada and a warmer climate is expected to intensify some weather extremes. Projected increases in severe precipitation are expected to increase the potential for future urban flooding" (Insurance Bureau of Canada 2019, para. 2).
- "Insured damage for severe weather events across Canada reached $2.4 billion" in 2020 (Insurance Bureau of Canada 2021, para. 1).
- "Wildfires in British Columbia in 2017 (...) cost the province CAD $568 million, equal to more than double the historical average annual cost of CAD $214 million" (Lindsay 2018).

While communities across Canada prepare strategies and implement actions to prepare for climate risks, they are, at the same time, embracing data and connected technology approaches in an effort to lower operating costs, enhance public services and improve the quality of life of residents. Examples can be identified in transportation, where data from public transportation fare cards pinpoint commuter hotspots and determine bus fleet needs, where bike sharing programs with self-service bike stations are promoting active transportation for residents, and where street light management is enabling the accommodation of real-time traffic conditions. In the realm of health and wellness, the availability of telemedicine to account for increasing

[2] https://changingclimate.ca/CCCR2019/

demands for healthcare services in underserved communities and populations is being increasingly explored, piloted and implemented.

Similarly, communities are exploring how smart approaches [could] link to climate resilience. This resource emerged from the discussions held at the British Columbia Idea Camp that explored how communities can and are advancing resilience strategies, and how open smart approaches to data and technologies can be leveraged to help address resilience challenges.

19.3 Making the Connection: Climate Resilience and Open Smart Approaches

Impacts of a changing climate are already being felt in towns and cities across Canada. Municipalities are becoming increasingly vulnerable to a variety of impacts, including increased incidences of extreme weather events, sea level rise, increasing temperatures and changes in precipitation. Indeed, there is a growing acknowledgement that it is now adaptation and resiliency strategies at the local level that are essential to support infrastructure building and maintenance, land use planning and the broad range of services provided by local governments. While climate change adaptation has received increasing attention over the past several years, the focus has largely been on the planning process (e.g. conducting vulnerability and risk assessments, developing different types of asset management plans, storm water management plans, community adaptation plans, etc.) and less on tangible on-the-ground actions. Adaptation planning is a process that is often capable of delivering positive outcomes; however, it must be supplemented with action; otherwise, vulnerability is not reduced and a community is still exposed to climate risk. For many communities, bridging this gap between planning and implementation is an ongoing barrier towards becoming more resilient".

Big data is proving to be a highly effective decision-making tool. This is because analysing data from different sources such as climate, social media, traditional media, NGOs and citizen-led groups offers a level of precision for tackling global challenges. In addition to analysis, big data creates a space of dialogue as it can be used to educate or warn communities about various shocks and stresses (Sarker et al. 2020).

"The connection between open smart approaches and local government action to reduce climate impacts are increasingly being considered and explored, but it is a newly emerging practice across Canada. At this intersection, for example, geographic information systems (GIS) can be used in local hazard mapping and analysis to help identify evacuation routes and locate vulnerable housing. Through enhanced connectivity and the proliferation of mobile technology, people living in vulnerable areas can use mobile phones to participate in collective mapping exercises and help identify local priorities for action. They can also provide ongoing feedback on the effectiveness of risk-reduction initiatives that have been deployed in their communities. Many of the methods for reducing climate impacts rely on sensors either to record and relay real-time consumption information or to detect the activities of residents. In many parts of Canada, fixed and mobile telecommunications networks are ideally suited for a lot of these tasks. The detection and collection of key data, when used appropriately, can help local governments make informed decisions that contribute to climate reduction.

To date, much of the work on resilience and smart cities has been siloed; the two initiatives are not frequently thought of as mutually beneficial and complimentary, or even as existing within the prevue of local government. Instead, smart cities approaches have typically come from outside of local government. In order for the two concepts to be mainstreamed, local governments will also need to co-create solutions from within, working with residents, entrepreneurs, academics, not-for-profits and others to develop innovative prevention and risk reduction tools that help to address identified community challenges and needs".

Cities have traditionally functioned as a collection of individual silos which provide essential services to citizens but are not always built around their needs. Smart cities, however, value holistic, citizen-centric models which facilitate the cross pollination of ideas between different departments and avoid operating in silos (BSI Group, n.d).

Resilient cities tend to be discussed as an abstract and multi-facetted concept (Cummings et al. 2005) which differs to smart cities, which are driven by technocratic solutions (Kitchin et al. 2017). This is reflected in the way that resilience frames cities as a socio-ecological system of metabolic flows (Kaye et al. 2006), whereas smart cities are seen as technical systems where challenges can be overcome with technological solutions (Kitchin et al. 2017).

However, there is a nexus between resilient and smart cities; both resilient and smart cities require a degree of urban transformation and the uptake of solutions into mainstream practice. In fact, smart cities have been described as a mainstream form of experimental urbanism (Cugurullo 2018) which can also apply to resilient cities as an integral component of achieving urban resilience is allowing cities to develop strategies for transitioning to more sustainable systems which offer an alternative to mainstream practices (Evans 2011).

"Some examples of the nexus between resilience and smart approaches are:

The role of big data in climate resilience, where anonymized cell-phone use records can be made available to track movements during extremes weather events. In work done by Professor James Ford (McGill University and University of Leeds), big data can also be used to georeference data around vulnerability of high risk groups—such as household income, demographics, housing characteristics—data which is often non-existent, outdated or unreliable.

The creation of warning systems that incorporate social media data to trigger emergency response measures (e.g., heat or flood alert systems). In this way, smartphones could be equipped with sensors that allow for the monitoring of human movement before, during, and after an extreme event to aid with disaster response (Ford et al. 2016). Integrating detailed considerations on data, technology and connectivity into the planning process for climate change strategies (alongside the people who will be responsible for their implementation) can be one of the ingredients for success in moving forward with open smart approaches.

Some recent examples of how Canadian cities are using data and connected technology as part of their climate adaptation strategies include:

- The City of Windsor used thermal imaging to understand how increased temperatures were affecting playgrounds in the city. Through imaging, the City found that temperatures on surfaces below certain play structures ranged from 50–70 C.
- Libraries of Things: The concept of borrowing a neighbour's drill, lawnmower, or tent is not new. However, increasing rates of social isolation have virtually eliminated this practice. In response, Libraries of Things have emerged to not only help neighbors interact, but also to address the consumption of resources (a significant contributor to climate change). Libraries of Things, enabled by software platforms, have become community hubs which increase entrepreneurship, social innovation and local economic development.
- Many cities are adopting technology-based water monitoring systems that not only broadcast risk alerts, but also help to control and record behaviours in maintaining a satisfactory water flow (see the River Watch mobile app,[3] developed by Civic Tech Fredericton and now serving all New Brunswickers). Such early flood detection and warning systems are reliable and have stable structures in terms of communications and accuracy.

As a result of the British Columbia Idea camp, three key opportunities were identified for connecting smart approaches to resilience action:

- Raising awareness among Canadian communities around smart approaches as a tangible and locally relevant response to (...) climate change challenges.
- Knowledge exchange and sharing around smart approaches—digitization, data integration, connected technology – that can be applied in remote or rural, small-to medium- sized communities.
- Building relationships across a network of communities and innovation partners centred around collaboratively advancing smart approaches as solutions to local resilience challenges".

19.4 Theoretical Foundation

19.4.1 Resilience

The concept of resilience evolved from the natural sciences in the 1970s. Since then, different disciplines have adopted the concept (Alexander 2013) and it has become key for sustainability and disaster risk reduction (Davoudi 2016). Resilient urbanism is now a prominent topic for discussion within governments. Definitions of resilience

---

[3] https://www2.gnb.ca/content/gnb/en/news/public_alerts/river_watch.html.

vary between disciplines; however, there is general agreement that a resilient city can be defined as being "prepared to absorb and recover from any shock or stress while maintaining its essential functions, structures, and identity as well as adapting and thriving in the face of continual change" (ICLEI 2019, p. 5). A resilient approach to urban development is synonymous with a holistic approach as it considers a whole system's ability to bounce back after a shock and its adaptive capacity (Sarker et al. 2020). It is important for cities to be resilient because in this way, they can be in a state of constantly adjusting to the consequences of the climate crisis and revising the mainstream practices that take place.

19.4.2 Smart Cities

Digital cities were heavily discussed in the early 2000s until 2009 where greater attention was given to smart cities and it has continued to gain traction to the point where academic work referred to the concept more than sustainable cities in 2012 (de Jong et al. 2015). Smart cities are prominent in the New Urban Agenda, which calls for a commitment "to adopting a smart-city approach that makes use of opportunities from digitalization, clean energy and technologies, as well as innovative transport technologies, thus providing options for inhabitants to make more environmentally friendly choices and boost sustainable economic growth and enabling cities to improve their service delivery".

The concept of smart cities began with a focus on "technology assisted—through sensors, surveillance cameras, control centres, autonomous driving, and connected infrastructure and communities" (Yigitcanlar et al. 2019). Since then, the concept has evolved as technology has become more advanced and sophisticated (Kitchin et al. 2017). There is not yet a single standardized agreed upon definition of smart cities. McKinlay (2017) argues that this debate is helpful for local governments to refine their approach to smart cities. This paper recognizes, however, that a smart city can be generally defined as a city that uses technology to increase the efficiency of urban systems and strive for a high quality of life for its citizens (Batty et al. 2012; Harrison et al. 2010; Xie et al. 2019; Yigitcanlar and Cugurullo 2020). Smart cities arose from the belief that smart solutions are crucial for addressing the main challenges in sustainability and that a technocratic approach can alter the trajectory of the climate crisis (Kitchin et al. 2017).

Kitchin et al. (2020) argue that smart city solutions advocate for collaborative working between the public sector and the private sector. However, as smart strategies become more sophisticated and evolve over time, a gap is created within the knowledge of decision makers and this can lead to the potential of implementing these solutions being overlooked. The limitation created by asymmetrical knowledge between different groups can lead to inaction and prevent the benefits of smart solutions being harnessed.

19.4.3 Open Smart Cities

Open Smart Cities aim to overcome the challenge of asymmetric knowledge in different groups by building trust within administration processes and being transparent about potential opportunities and risks of increased data use with all stakeholders (Pembleton et al. 2019). Specifically, an open smart city is one in which "residents, civil society, academics, and the private sector collaborate with public officials to mobilize data and technologies when warranted in an ethical, accountable and transparent way in order to govern the city as a fair, viable and liveable commons and balance economic development, social progress and environmental responsibility" (Sodhi et al. 2018, p. 10). Open Smart Cities are able to combat some of the challenges faced by smart cities such as the lack of regulations in place to prevent data privacy leaks, ransomware attacks and biases within algorithms.

19.4.4 Vulnerability

The vulnerability of a city depends on its exposure, sensitivity and adaptive capacity. Exposure refers to the extent to which people, livelihoods and assets could potentially be harmed. Sensitivity refers to the extent to which communities or individuals are affected by a climate change impact. Adaptive capacity refers to the extent to which people possess the relevant knowledge, capacity and opportunities to adapt to and limit climate change impacts (Hardoy and Pandiella 2009). Bridging the gap between planning and implementing capacity building initiatives in the community increases resilience and decreases vulnerability.

19.4.5 Disseminating Smart Solutions into Mainstream Practice

The private sector has heavily influenced the development of what constitutes a smart city. IBM was one of the first companies to advocate for smart cities. In 2008, IBM announced its smart city vision within its Smarter Planet Initiative which argued the importance of investing in urban digital systems. However, the incentive was to increase efficiency and lessen the risk of economic decline (Wiig 2015). Canadian government have invested in smart cities, i.e. Industry Canada invested $60 million into its nationwide "Smart Communities" initiative (Hollands 2008), and since then, there has been more research into smart cities and interest at the government level has increased further.

Put simply, smart cities require a change in dynamic between technology and society (Soderstrom et al. 2014). Within academic literature, increasing importance is being placed upon increasing the level of citizen engagement. In Canada, there is a growing demand at the municipal level for this citizen engagement to be smart. The open element of Open Smart Cities means that government data is accessible to communities and this results in more transparent and accessible forms of governance (Johnson et al. 2020).

Johnson et al. (2020) identify a gap in the research to explore how smart approaches to citizen engagement can be achieved through the planning stages of projects. This paper contributes to this gap in the literature as it investigates how the City of Surrey's initiatives have disseminated smart solutions throughout communities.

19.5 Methodology

The Victoria Idea Camp took place on 30 October 2019 and lasted 5.5 h. The idea camp was open to LCF delegates and other communities; it was designed to provide a space of dialogue to municipalities from British Columbia so they could "connect, share, and ask questions around smart cities, which achieve meaningful outcomes for residents through the use of data and connected technology". The conversation explored "the connection between smart cities approaches and how digitalization and information technology can provide opportunities to achieve broader resilience goals" (University of British Columbia 2019, para. 1).

Pre-consultation calls were offered by ICLEI Canada and Evergreen to those attending in order to advance the understanding of the issues and questions facing communities in British Columbia and the divergence between their current capacity to navigate the smart cities landscape and the capacity needed (University of British Columbia 2019). This contributed to the Idea Camp being able to precisely target the main challenges faced by communities and build capacity.

A case-study approach was taken to carry out an in-depth assessment on the City of Surrey and its strategy for becoming a smart city. The City of Surrey was chosen because of its prominent position within the sustainability arena that it has held for over a decade. In addition, the city is aiming to become a leader in innovation via its "Smart Surrey Strategy". This "seeks to guide Surrey's growth from an innovation and technological perspective". The case-study approach "is used to generate an in-depth, multi-faceted understanding of a complex issue in its real-life context" (Crowe et al. 2011, p. 1). This method has been used to rigorously analyse how the City of Surrey has implemented smart city solutions.

19.6 Case Study: The City of Surrey, BC

"The City of Surrey is located in the Metro Vancouver region of British Columbia. Surrey has been a leader on a variety of sustainability initiatives for decades and has been working on climate change since 2010, with their *Community Climate*

Action Strategy[4] being endorsed by Council in 2013. Likewise, it has been working on advancing smart approaches through a variety of initiatives under the banner of the Surrey's strategy *Smart Surrey: Leading Through Innovation.*[5] Surrey began by creating a local definition: "A 'Smart' City creates sustainable economic development and high quality of life by considering innovation and technological advancements as a key ingredient in its decision making and strategy development" (City of Surrey 2020, p. 7).

There are several examples of initiatives that highlight how Surrey has operationalized *Smart Surrey: Leading Through Innovation.* These include:

Surrey mobile apps

An example of a citizen information system, providing access to a variety of mobile apps developed by Surrey to help residents access the following City of Surrey services:

- Building Inspection Request App[6]
- MySurrey App[7]
- Surrey Libraries App[8]
- Surrey RCMP App

Data Driven Decision Making (D3M)

Surrey's D3M process launched in 2018 in partnership with the consulting company, CGI. Through this process, 80 different technologies were identified as being used by City departments. Staff were asked to work through problem statements to get to a common "wish list" that included:

- Reducing data lags and "Self-service" access to data
- Effective forecasting
- Data catalogue
- More public-facing data

The City's focus on creating and improving city information systems, while at the same time moving towards more integrated data-driven decision making is helping to build resilience in Surrey, both within their corporate operations as well as in the broader community. The variety of mobile apps that are offered and downloaded by residents are vital communications platforms that can be used before, during, or after an emergency event to communicate vital information regarding the level of risk and where residents can turn for help. Advancing D3M is an important process for city staff to use to design better, more targeted, and equitable resilient actions and solutions".

---

4 https://www.surrey.ca/about-surrey/sustainability-energy-services/climate-adaptation-strategy.

5 https://www.surrey.ca/sites/default/files/media/documents/SmartSurreyStrategy.pdf.

6 https://www.surrey.ca/renovating-building-development/building/building-inspections/building-inspection-request-app.

7 https://www.surrey.ca/services-payments/online-services/mobile-apps/mysurrey-app.

8 https://www.surrey.ca/services-payments/online-services/mobile-apps/surrey-libraries-app.

19.7 Discussion

19.7.1 Key Considerations for Local Governments

A number of lessons learned emerged from the City of Surrey case study and the broader discussions held at the British Columbia Idea Camp. These lessons provide insight around some guiding elements that local governments can keep in mind when considering and/or moving forward with the implementation of open smart approaches to help address climate resilience challenges:

- **Early capacity building and knowledge sharing:** Staff knowledge and capacity must be developed early on in order to reduce reliance on external organizations and providers.
- **Aligning around a shared vision:** Changing the way we make decisions around resilience and technology will take time. Establishing the trust needed for collaborative decision-making can help to leverage the potential in communities and advance a collective vision for an equitable and resilient community. Aligning around a shared vision of what a resilient community looks like, and the role of technology in it, will lead to long-term progress and shared ownership for implementation.
- **Meaningful public and stakeholder engagement is crucial:** Ensuring comprehensive and meaningful public and stakeholder engagement is key to building and maintaining trust with residents, and opens the door for co-creation. The shared vision of your resilient community should drive your engagement strategy and process. By making decisions and developing policies collaboratively, and ensuring ongoing transparency and accountability, the foundation will be laid for meaningful and long-term change.
- **Quality of life > technology for innovation's sake:** It is easy to be swept up in the promises of new technologies and the benefits and solutions they claim to enable. Understanding the needs and wants of your residents and maintaining focus on improving their quality of life can help you to avoid the potential pitfalls of prioritizing innovation and newness. A few questions to ask are: What problem are we trying to solve with this technology? How certain are we that the problem will be solved by this technology?
- **Time and resources are real constraints:** Many communities lack dedicated resources to take on new data and technology initiatives (few Canadian communities have a dedicated smart city or open data branch or department). In most cases, data and connected technology initiatives are undertaken by existing departments that are already limited in resources.
- **Understand the risks:** While smart approaches promise many benefits, they also have the potential to pose significant threats to individual rights. From invasion of privacy due to increasing and new forms of surveillance, to the ability of data or connected technology initiatives to reflect or even exacerbate existing social and economic inequalities, there is reason for exercising caution. This is particularly

significant when approaching resilience initiatives; resilience, at its core, is meant to promote equity and social connectedness. Understanding the risks associated with a certain intervention and the purpose behind it, is a crucial early step.

Discussions at the British Columbia Idea Camp also triggered important questions for local governments to consider when applying data and technology approaches to climate resilience:"

1. "Do you have reliable and consistent access to broadband internet?

   Generally, smart approaches to data and technology rely on continuous and reliable broadband internet and connectivity. Many parts of Canada, especially in Northern, rural, and remote areas, struggle with connectivity issues and do not have adequate or consistent access to broadband internet. To meet this need, governments at all levels have acknowledged this as a fundamental challenge and are taking action to close the digital divide by creating funding mechanisms and infrastructure to provide reliable internet access for all Canadians.

   For more information, go to:

- *Government of Canada Universal Broadband Fund*[9]
- *Internet speed test with Canadian Internet Registration Authority (CIRA)*[10]
- *Conference Board of Canada Progress Report on Connectivity*[11]
- *Community Solutions Network"*[12]

2. "Do you have a comprehensive community and stakeholder engagement strategy?

When considering the application of data and connected technology to advance resilience, it is important to engage with communities and stakeholders more broadly. Better engagement helps to identify resident needs and how open smart approaches can help to meet those needs. When building these relationships, it is important to identify *all* stakeholders: governments, utilities, IT, finance, private businesses, etc. Additionally, it is important to identify opportunities for a variety of partnerships, both public/public and public/private as prospects for funding. Models to help connect innovators with practitioners and funders (e.g. the *Innovate BC* model, which connects innovators with funding, tools, resources, etc.) should be pursued when considering open smart approaches.

For more information, go to:

- *Innovate BC*[13]
- *Future Cities Canada: Getting to the Open Smart City*[14]

---

[9] https://www.ic.gc.ca/eic/site/139.nsf/eng/h_00006.html.

[10] https://performance.cira.ca/

[11] https://www.conferenceboard.ca/hcp/Details/Innovation/connectivity.aspx.

[12] https://www.evergreen.ca/our-projects/community-solutions-network/

[13] https://www.innovatebc.ca/

[14] https://futurecitiescanada.ca/downloads/2018/Getting_to_Open_Smart_City.pdf.

- *Municipal World: Smart(er) cities focus on open data and community engagement*[15]"

3. "Do you understand the unique needs and challenges of your community's equity-seeking communities? Have you addressed them in your approach to resilience?

Climate change impacts are disproportionately impacting Indigenous people, low-income families, and people of colour. Climate change worsens gaps that historically (and presently) prevent inclusion of whole communities in building resilience. We need to make sure that open smart approaches are people-centred (and equitable) by ensuring that all are included or considered in conversations on data and technology. Likewise, technological responses must be catered to a perceived and/or identified local need or gap and implemented actions must reflect people and place. Fundamentally, communities need to make sure that solutions include (or at a minimum do not discriminate) against those for whom technology and connectivity are not as readily available.

For more information, go to:

- *City of Calgary Resilience Strategy*[16]"

4. "Are you aware of and considering the differences for large versus small communities?

In large communities it can be hard to "act locally" due to large bureaucracy, and many pre-identified and competing priorities. In small communities, however, municipal staff may not necessarily see open smart technologies as a priority area and also may not have the resources to take on something new. To overcome these considerations, smart city approaches need to be framed as a solution to resilience challenges (in any type of community) leading to better uptake and long-term implementation.

For more information, go to:

- *Do not forget about Smart Towns*[17]"

19.8 Conclusion

"Communities across Canada are actively working on resilience – from vulnerability and risk assessments, to creating resilience strategies/plans, to implementing resilience interventions. In most cases these initiatives are not being approached through a data and connected technology lens; digitization, data-driven decision making, automation, and technological innovation are often not central components

[15] https://www.municipalworld.com/feature-story/smarter-mid-sized-cities-focus-open-data-community-engagement/

[16] https://www.calgary.ca/CS/Documents/ResilientCalgary/ResilienceStrategyBooklet.PDF.

[17] https://link.springer.com/article/10.1007/s12599-018-0536-2#citeas.

of resilience strategies and plans. In many cases, we are relying on the ways we have always planned cities to plan for a new and uncertain future.

Planning for this uncertain future requires a paradigm shift in the way communities are planned and designed. Innovative and sustainable approaches to many aspects of community life, such as public transportation, urban agriculture, construction, housing and smart building technologies can advance resilience, create jobs, and help communities stand up to a changing climate. Open smart approaches can have a role to play in advancing these future-ready solutions. By better integrating principled data and connected technology approaches into resilience strategies in a manner that is equitable, transparent and participatory, we can ensure our communities are taking long-term action that is inclusive and responds to local risks and stressors".

It is clear that Open Smart Cities have great potential to help build resilience. To implement successful smart solutions, it is important to take a holistic approach to urban development and look beyond cities as just technical systems, and instead, view them as socio-ecological systems as this allows planners to consider all aspects of urban areas when implementing smart initiatives.

Acknowledgements This paper has been reproduced from "*Advancing Community Resilience Through Smart Approaches: A Resource for Canadian Communities*",[18] a toolkit published on the Community Solutions Portal in March 2020. Please note that all sentences and paragraphs that have been directly taken from the original report are in quotation marks.

"This resource was created for the Community Solutions Network, a program of Future Cities Canada. As the program lead, Evergreen is working with Open North and partners to help communities of all sizes across Canada navigate the smart cities landscape.

This toolkit was written by Ewa Jackson, Managing Director of ICLEI Canada. It is an outcome of the lessons learned by Canadian communities that attended the British Columbia Idea Camp in Victoria, British Columbia on October 30, 2019. It provides guidance and resources to Canadian communities seeking to build resilience through data and connected technology approaches.

The Community Solutions Network toolkit series was curated and edited by Carly Livingstone, Senior Program Manager, Evergreen and Sanchita Rajvanshi, Program (…) [Officer], Evergreen. Special thanks to Zoya Sodhi, (…) Program Manager, Evergreen, who provided additional feedback on this paper.

The Community Solutions Network is supported with funding provided by Infrastructure Canada. The views expressed in this publication are the views of the author and Evergreen and do not necessarily reflect those of the Government of Canada.

We would like to thank all of the Canadian community members who have participated in the Community Solutions Network and the Idea Camps throughout this program".

References

Alexander DE (2013) Resilience and disaster risk reduction: an etymological journey. Nat Hazards Earth Syst Sci 13(11):2707–2716. https://doi.org/10.5194/nhess-13-2707-2013

[18] https://portal.futurecitiescanada.ca/resources/advancing-community-resilience-through-smart-approaches/

Batty M, Axhausen KW, Giannotti F, Pozdnoukhov A, Bazzani A, Wachowicz M, Ouzounis G, Portugali Y (2012) Smart cities of the future. Eur Phys J: Spec Top 214(1):481–518. https://doi.org/10.1140/epjst/e2012-01703-3
BSI Group (n.d.) Sustainable cities and communities – ISO 37106 Summary Guide. https://www.bsigroup.com/en-GB/smart-cities/iso-37106-guide/
Bush E, Flato G (2019) Executive summary, in Canada's changing climate report. In Bush E, Lemmen DS (eds) Government of Canada, Ottawa, Ontario, pp 7–23
City of Surrey (2020) Smart surrey strategy. https://www.surrey.ca/files/SmartSurreyStrategy.pdf
Crowe S, Cresswell K, Robertson A, Huby G, Avery A, Sheikh A (2011) The case study approach. BMC Med Res Methodol 11(1):1–9
Cugurullo F (2018) Exposing smart cities and eco-cities: Frankenstein urbanism and the sustainability challenges of the experimental city. Environ Plan A: Econ Space 50(1):73–92 https://doi.org/10.1177/0308518X17738535
Cumming GS, Barnes G, Perz S, Schmink M, Sieving KE, Southworth J, Binford M, Holt RD, Stickler C, Holt T Van (2005) An exploratory framework for the empirical measurement of resilience. https://doi.org/10.1007/s10021-005-0129-z
Davoudi S (2016) Resilience and governmentality of unknowns. Governmentality after Neoliberalism (pp 152–170). Routledge. https://doi.org/10.4324/9781315685083-9
De Jong M, Joss S, Schraven D, Zhan C, Weijnen M (2015) Sustainable-smart-resilient-low carbon-eco-knowledge cities; making sense of a multitude of concepts promoting sustainable urbanization. J Clean Prod 109, 25–38. https://doi.org/10.1016/j.jclepro.2015.02.004
Evans JP (2011) Resilience, ecology and adaptation in the experimental city. Trans Inst Br Geogr 36(2):223–237. https://doi.org/10.1111/j.1475-5661.2010.00420.x
Ford JD, Tilleard SE, Berrang-Ford L, Araos M, Biesbroek R, Lesnikowski AC, Bizikova L (2016) Opinion: big data has big potential for applications to climate change adaptation. Proc Natl Acad Sci 113(39):10729–10732
Gambrill D (2019, January 22) Why the global protection gap is the lowest it's been in 14 years. Canadian Underwriter. https://www.canadianunderwriter.ca/insurance/why-the-global-protection-gap-is-the-lowest-its-been-in-14-years-1004151048/
Government of Canada (2021, May 26) Temperature change in Canada. Government of Canada. https://www.canada.ca/en/environment-climate-change/services/environmental-indicators/temperature-change.html
Hardoy J, Pandiella G (2009) Urban poverty and vulnerability to climate change in Latin America. Environ Urban 21(1):203–224.
Harrison C, Eckman B, Hamilton R, Hartswick P, Kalagnanam J, Paraszczak J, Williams P (2010) Foundations for smarter cities. IBM J Res Dev 54(4). https://doi.org/10.1147/JRD.2010.2048257
Hollands RG (2008) Will the real smart city please stand up? Intelligent, progressive or entrepreneurial? City 12(3):303–320
ICLEI (2019) Resilient cities, thriving cities: the evolution of urban resilience
Insurance Bureau of Canada (2017, September 1) Spring flooding in Ontario and Quebec caused more than 223 million in insured damage. Insurance Bureau of Canada. http://www.ibc.ca/on/resources/media-centre/media-releases/spring-flooding-in-ontario-and-quebec-caused-more-than-223-million-in-insured-damage
Insurance Bureau of Canada (2019, April 2) Canada's changing climate report confirms increase in extreme rainfall. Insurance Bureau of Canada. http://www.ibc.ca/on/resources/media-centre/media-releases/canada%E2%80%99s-changing-climate-report-confirms-increase-in-extreme-rainfall
Insurance Bureau of Canada (2021, January 18) Severe Weather caused $2.4 billion in insured damage in 2020. Insurance Bureau of Canada. http://www.ibc.ca/on/resources/media-centre/media-releases/severe-weather-caused-$2-4-billion-in-insured-damage-in-2020
Johnson PA, Acedo A, Robinson PJ (2020) Canadian smart cities: are we wiring new citizen-local government interactions? Can Geogr/Le Géographe Can 64(3):402–415

Kaye JP, Groffman PM, Grimm NB, Baker LA, Pouyat RV (2006) A distinct urban biogeochemistry? Trends Ecol Evol 21(4):192–199. Elsevier Current Trends. https://doi.org/10.1016/j.tree.2005.12.006

Kitchin R, Coletta C, Evans L, Heaphy L, Donncha D Mac (2017) Smart cities, urban technocrats, epistemic communities and advocacy coalitions. http://progcity.maynoothuniversity.ie/

Kitchin R, Moore-Cherry N (2020) Fragmented governance, the urban data ecosystem and smart city-regions: the case of Metropolitan Boston. Regional Studies, 1–11

Lindsay B (2018, August 21) How British Columbia budgets for wildfire: choose an arbitrary number, put it in the books. CBC News. https://www.cbc.ca/news/canada/british-columbia/how-b-c-budgets-for-wild%EF%AC%81res-choose-an-arbitrary-number-put-it-in-the-books-1.4793981

McKinlay P (2017) Make way for smart cities: opportunities, challenges and capacities of New Zealand local governments. Asia Pac J Public Adm 39(4):297–303

Pembleton C, Ahmed N, Lauriault T, Landry J, Planchenault M (2019) The state of open smart communities in Canada. OpenNorth. https://www.opennorth.ca/publications/#the-state-of-open-smart-communities-2019

Sarker NI, Yang B, Lv Y, Huq E, Kamruzzaman MM (2020) Climate change adaptation and resilience through big data. IJACSA Int J Adv Comput Sci Appl 11(3). www.ijacsa.thesai.org

Söderström O, Paasche T, Klauser F (2014) Smart cities as corporate storytelling. City 18(3):307–320. https://doi.org/10.1080/13604813.2014.906716

Sodhi Z, Flatt J, Landry JN (2018) Getting to the open smart city. Future cities Canada. https://futurecitiescanada.ca/downloads/2018/Getting_to_Open_Smart_City.pdf

University of British Columbia (2019) Community solutions network: Victoria idea camp. The University of British Columbia. https://scarp.ubc.ca/school/weekly-digest/2019/08/community-solutions-network-victoria-idea-camp

Wiig A (2015) IBM's smart city as techno-utopian policy mobility. City 19(2–3):258–273

Xie J, Tang H, Huang T, Yu FR, Xie R, Liu J, Liu Y (2019) A Survey of blockchain technology applied to smart cities: research issues and challenges. IEEE Commun Surv Tutor 21(3):2794–2830. https://doi.org/10.1109/COMST.2019.2899617

Yigitcanlar T, Cugurullo F (2020) The sustainability of artificial intelligence: an urbanistic viewpoint from the lens of smart and sustainable cities. Sustainability 12(20):8548. https://doi.org/10.3390/su12208548

Yigitcanlar T, Han H, Kamruzzaman M, Ioppolo G, Sabatini-Marques J (2019) The making of smart cities: are Songdo, Masdar, Amsterdam, San Francisco and Brisbane the best we could build? Land Use Policy 88. https://doi.org/10.1016/j.landusepol.2019.104187

Chapter 20
Toward Integrating Resilience Thinking in Smart City Planning and Development

Pourya Salehi and Ayyoob Sharifi

Abstract Along with sustainability, resilience and smartness have arguably been the most dominant terms in science and policy debates about cities in the past 2–3 decades. Much has been written on the integration of sustainability and smartness as well as sustainability and resilience. By comparison, integration of smartness and resilience has received limited attention. This is despite the fact that, on the one hand, smart city solutions and technologies can provide numerous benefits for enhancing urban resilience, and, on the other hand, resilience thinking can contribute to planning and implementation of more efficient smart city projects that are able to survive and thrive during different types of crises. Different theoretical and empirical issues related to the integration of resilience thinking into smart city planning were discussed in this book that has contributed to gaining a better understanding of the concept of resilient smart cities. In addition to a summary of the main issues discussed in the previous chapters, this concluding chapter provides recommendations for further integration of the concepts of smart city and urban resilience.

Keywords Smart city · Urban resilience · Integrated approaches · Urban planning · Resilient cities

20.1 The Importance of Smart and Resilient Cities

Resilience is a beneficial concept for urban areas aiming to address the impacts of human-made and natural disasters and respond to stressors such as climate change and pandemics with a proactive approach. The United Nation's Sendai Framework for Disaster Risk Reduction 2015–2030 states that disaster prevention should span

P. Salehi
ICLEI—Local Governments for Sustainability (World Secretariat), Bonn, Nordrhein-Westfalen, Germany
e-mail: pourya.salehi@iclei.org

A. Sharifi (✉)
Graduate School of Humanities and Social Sciences, Hiroshima University, Higashihiroshima, Japan
e-mail: sharifi@hiroshima-u.ac.jp

A. Sharifi and P. Salehi (eds.), *Resilient Smart Cities*, The Urban Book Series,
https://doi.org/10.1007/978-3-030-95037-8_20

across multiple sectors in order to maximize its efficacy. The framework comprises of four key aims: firstly, to understand the risk of disasters occurring, secondly, to govern risks at different scales, thirdly, to invest in reducing the risk of disasters occurring, and fourthly, to increase preparedness in order to improve the response to disasters, as well as building back better if a disaster does occur. The first three aims are top-down strategies whereas the fourth aim is to empower communities through building resilient critical infrastructure (Meriläinen 2019). Specifically, the framework sets out the goal of increasing the resilience of critical infrastructure such as schools and hospitals by 2030. The Sendai Framework is critical for achieving the Sustainable Development Goals (SDGs) relating to building resilience. With regard to Goal 9 (Industry, Innovation, and Infrastructure), the framework aims to improve the process of building and re-building by implementing design, construction, and retrofitting principles to withstand hazards. In addition, the framework encourages investment in innovation and technology to carry out research in disaster risk management. The framework contributes to Goal 11 (Sustainable Cities and Communities), by aiming to mainstream the presence of disaster risk assessments in land-use policies, urban planning, land degradation assessments, and informal housing (UN 2015).

Similarly, the smart cities concept is increasingly included in recent sustainability agendas; as an example, smart solutions feature in the New Urban Agenda (NUA), specifically, digitalization. More specifically, in the NUA, paragraph 66 highlights an alliance to the smart city approach including digitalization, paragraph 151 illuminates the importance of administrative and technical capacity via the digitalization of accounting processes and records, paragraph 156 encourages the use of citizen-centric digital governance tools in addition to geospatial information systems (GIS), and paragraph 157 discusses digital innovation for urban planning and policy making (UN-Habitat III 2016). As another example, in Addis Ababa Action Agenda, paragraph 114 advocates for the commercial banking sector to remove barriers in order to increase the accessibility of financial services across the world, by encouraging the use of microfinance and digital banking tools which are considered smart solutions (United Nations 2015).

20.2 Defining Resilient Cities and Smart Cities

In different parts of this volume, efforts have been made to further clarify the underlying principles of resilient cities and smart cities and define the concept of smart city resilience. Resilience is a multidimensional and multifaceted concept used for measuring the capacity of a city to respond to shocks and stresses. While there is no agreement among scholars where the concept of resilience originates from, most believe that the concept was derived from physics where the terminology was adopted in the second half of the seventeenth century, to convey an object's reaction to the impact of an intrinsic force (Peng et al. 2017). Subsequently, the term gained traction as a popular term in the 1960s and 1970s and was used in ecology, firstly to measure the time taken for a system to recover after a shock, and secondly, to ascertain the size of shock needed to cause a disruption in the system (Perrings 2006). It should also be noted that this definition of resilience has been expanded upon by scholars such as Perrings (1998), who argues that the resilience of a system equates to its ability to withstand uncertainties. The engineering profession also uses the concept of resilience and defines it as the ability of physical infrastructure to withstand disruptions (Martin-Moreau and Ménascé 2018). The extensive use of the term resilience by different disciplines demonstrates its malleability and usefulness for a variety of stakeholders. Urban areas are no exception to this, as various academics have connected the concept with the behavior of socio-ecological systems. Urban resilience is widely recognized as a positive feature of cities, but some schools of thought have questioned whether a system returning to its original state before a shock is wholly positive. For example, returning to conditions such as high levels of poverty would not be desirable. For this reason, Gleeson (2008) advocates that resilient cities should be able to withstand disturbances, adapt from one state to another, and transform radically.

It is important to consider that the different ways in which disciplines refer to resilience, as well as the different weightings of importance that researchers and institutes place on the various aspects of resilience, are nuanced. By and large, there is no globally agreed definition for resilience which is partly due to the different technical language used by different disciplines and the differing perspectives. Some academics value characteristics such as redundancy, diversity, efficiency, autonomy, strength, interdependence, adaptability, and collaboration, and other academics consider these characteristics in addition to social capital, innovation, tight feedback, ecosystem services, and overlap in governance (Tasan-Kok et al. 2013). However, despite the differences between eminent definitions, there are some unifying recognized characteristics of urban resilience from various scholars which are as follows: robustness, diversity, redundancy, connectivity, flexibility, resourcefulness, agility, efficiency, adaptive/learning capacity, modularity, creativity (innovation), equity, inclusiveness, and foresight capacity, which suggests commonalities for local leaders to prioritize in policy making.

In a way that is comparable to resilience, there is no agreed-upon definition of smart cities (Ahvenniemi et al. 2017; Angelidou 2015; Ismagilova et al. 2019). Scholars believe that this is due to the various lenses through which the smart city is viewed by different disciplines. For this reason, the body of research on smart cities is both technological and holistic in approach.

There are various definitions of the smart city. Early definitions view smart cities as being synonymous with the use of ICT and business-oriented objectives (Harrison et al. 2010; Lombardi et al. 2012; Paul et al. 2011). However, later definitions incorporate a deeper understanding of smart cities. For instance, Marsal-Llacuna et al. (2015) argue that the purpose of smart cities is not only limited to improving urban performance through the use of technology but smartness is also instrumental in providing efficient services to citizens, increasing economic collaboration, and encouraging business models that involve both public and private stakeholders. In addition, it is argued by Giffinger et al. (2007) and Hollands (2008) that technological solutions should be coupled with social, human, cultural, and governance factors.

The smart city concept originates from information, digital, and intelligent cities (Yigitcanlar et al. 2018). The early literature on smart cities is predominantly focused on one dimensional, technocentric, and top-down solutions but the literature recognizes that since this early research, there has been a shift toward slowly incorporating the elements of citizen engagement and human capital (citizens, businesses, and government) to cultivate innovation for social, economic, environmental, and cultural growth. This shift has resulted in the development of the widely agreed-upon four pillars for smart cities, namely, physical infrastructure, social infrastructure, institutional infrastructure, and economic infrastructure (Silva 2018). Also, as was discussed in Chapter 3, seven major smart city dimensions, namely, economy, people, governance, environment, living, mobility, and data have been frequently used in the literature to define smart cities (Sharifi 2019, 2020a).

The concept has been gaining momentum in recent years, with the number of smart city projects increasing at a rapid pace, and research inputs growing from the electronics and machine learning community. Concurrently, there has been an increase in the uptake of Artificial Intelligence (AI) for data mining and analysis, which is supporting decision-makers in making increasingly informed choices (Batty 2013; Allam and Dhunny 2019). This increase in usage and research has occurred simultaneously with the evolving development of new digital technologies and the opinion that smart city initiatives are becoming increasingly essential in achieving sustainable urban development and helping cities transform to more sustainable systems.

In line with this, there is a strong school of thought that smart cities require smart citizens, which places citizens at the center of the concept and places great importance on education to amass a skilled workforce to further support smart city initiatives (Winters 2011). It is worth noting that there is a geographical divide regarding the approach to smart cities. European research tends to be aligned with a holistic view of smart cities, whereas North American research tends to take a technocentric and innovation-focused approach (Mora et al. 2017). As well as this divide, there is a prominent debate between top-down and bottom-up approaches and the mono-dimensional versus multidimensional approaches to smart cities.

20.3 Synergies Created by the Resilient Smart City

There are positive impacts of both smart cities and resilient cities, and these are amplified due to the way in which the two concepts complement each other. As explained in Chapter 5, the two concepts of smart city and resilient city should not be viewed in isolation and their synergistic benefits should be utilized. As noted in Chapter 5: "In the resilient-smart city framework, smart city solutions and technologies provide technical support to ensure a resilient city can deal with disasters and emergencies in an efficient manner, while the resilient city provides positive feedback for smart city in resisting external interferences and disturbances." Connections can be drawn between smart solutions such as sensors used to monitor smart cities in real-time and respond immediately to various situations, with the need for resilient cities to be able to access data to assist with informed decision-making. In addition, the ability of resilient cities to protect people, ecosystems, assets, resources, and services is essential for smart cities as well. This extends to the use of technologies to increase the ability of cities to prevent disasters, as opposed to only placing the emphasis on recovery (Sharifi 2016). This overlap demonstrates that both concepts rely on a similar set of principles and as such, the success of one can benefit the other.

The resilient smart city centers around the idea that resilience cannot be detached from technology, and smart cities can and should enhance the resilience of urban areas. In other words, smart cities also increase resilience by providing the necessary technical support for resilient initiatives and infrastructure. There are a variety of resilient practices that can be enhanced and improved by technology such as natural disaster monitoring and assessment, urban disaster relief, and disaster prevention and mitigation systems. For instance, the COVID-19 pandemic illuminated the ability of the combination of ICT, big data, and geospatial and temporal information to enable the monitoring of the situation and consequently better manage the impacts (Sharifi et al. 2021). This same concept of building resilience through monitoring can also be applied to infrastructure, transportation, buildings, electricity, and water to ensure high quality of resources and resource-use efficiency. However, due attention must be given to the potential risks associated with smart city initiatives as these can lower resilience by, for example, increasing a city's vulnerability to cyber-attacks. For the above-mentioned reasons, scholars have generated frameworks enhancing urban resilience with the utilization of smart city solutions. Most of these frameworks emphasize that resilience is a process rather than a goal (Ilmola 2016).

20.4 Key Considerations for Decision-Makers

It is known that the act of resilience thinking is conducive for decision-makers to incorporate a range of possible future scenarios into their planning that ultimately allows the city to implement initiatives that will be the most effective for the local context. Local authorities should begin the process of resilient planning by assessing prominent risks and vulnerability of their jurisdiction to various hazards and incorporate tailor-made solutions to combat potential disruptions. In addition, decision-makers must consider the dynamics of socio-technical systems, and multilevel governance structures (Burch et al. 2014), as it illustrates in greater depth the root causes of unsustainable practices and uncovers the current development path. However, as the world continues to change at an accelerating rate and becomes increasingly complex, conventional practices to resilience building are falling short. This is well manifested by the COVID-19 pandemic when cities and local governments are finding it challenging to match past performance results while responding to the impacts of the COVID-19 pandemic and at the same time adapting to accelerating changes.

Against this backdrop, and knowing that the smart cities concept is bringing unconventional and innovative approaches to the table, it is necessary to ensure that decision-makers further leverage the offerings of smart solutions and frameworks when developing an overarching resilience strategy for their jurisdictions.

In essence, resilient and smart approaches require a shift away from unsustainable trajectories; it is, therefore, essential for decision-making bodies to go beyond focusing on climate policy, by changing the underlying development pathway of their local area. This change should be carried out in a way that capitalizes on the synergistic benefits generated by embedding resilience thinking, as well as deploys smart solutions in their transformation pathways.

Overall, the theoretical and empirical insights presented in the volume can be utilized for better integration of the concepts of smart city and resilience in science and policy debates. This, in turn, could lead to developing innovative and transformative plans and solutions that would enable cities to better prepare for, absorb, recover from, and adapt to various types of adverse events that are expected to increase in frequency and intensity due to climate change.

Despite the contributions of this volume, the concept of smart city resilience is still in its infancy and is expected to further develop in the coming years. Among various issues that can be prioritized in research and practice, we recommend the following items:

- It is essential for decision-makers to engage citizens in innovative and transformative solutions toward smartness and resilience. Participatory practices should be facilitated by local leaders to ensure local expertise and grass-root knowledge are harnessed and leveraged. Participation also ensures public buy-in which is integral to the success of resilient smart cities agendas. Participatory approaches are only successful when diversity is considered. A wide range of perspectives and local knowledge should be harnessed to aid the implementation of tailor-made resilient smart cities agendas and ensure buy-in from the whole community. For this reason, local leaders should prioritize making the process accessible for all members of the community and ensure all groups are engaged in the conversation of transformative change based on smart solutions toward enhancing community resilience.
- Contributions of smart city solutions and technologies to resilience against adverse events are increasingly recognized. However, such contributions have mainly been associated with technical aspects of smart cities. It is essential to recognize that smartness is a multidimensional concept and is beyond just technical solutions. Hence, potential resilience benefits that can be achieved through enhancing non-technical aspects of smart cities should also be taken into account.
- Similarly, it can be argued that a more holistic approach toward resilience is needed. Traditionally, research and practice on urban resilience have paid more attention to recovery from adverse events (Sharifi 2020b). As a result, short- and medium-term approaches, and measures related to recovery to pre-disaster

equilibrium states are currently dominant. As discussed across different chapters of this volume, resilience is a multidimensional concept and entails continuous efforts that go beyond absorption and recovery and include planning and adaptation. Through data analytic tools and approaches, smart city solutions can provide opportunities for a paradigm shift toward more holistic approaches. Such approaches will also be conducive to a better understanding of the actual and/or potential contributions of smart and resilient cities to achieving different Sustainable Development Goals.

- As expected, there is limited knowledge of the empirical contributions of smart cities to resilience. This could be explained by the fact that the concept is still in its infancy and there are still ongoing debates on the theoretical aspects. The case studies presented in this volume could pave the way for more studies on the operationalization of the concept of smart city resilience.
- Finally, it should be noted that what we define as a smart city approach and framework is constantly evolving due to the increasing complexities in the world and urban systems. What may have been considered a smart approach a decade ago is no longer considered a smart approach now. The concept of resilience is no different, as it too is constantly evolving which draws attention to the need for new definitions and types of resilience. It is very apparent that resilience thinking needs to be embedded into sustainability planning and action at the local level. At the same time, having a tailor-made smart cities agenda that supports local governments in providing basic services for their communities while achieving sustainable development in their jurisdiction seems necessary. These issues should be duly considered in future efforts aimed at revisiting and updating the underlying principles and definitions of smart city and resilient city.

References

Ahvenniemi H, Huovila A, Pinto-Seppä I, Airaksinen M (2017) What are the differences between sustainable and smart cities? Cities 60:234–245

Allam Z, Dhunny ZA (2019) On big data, artificial intelligence and smart cities. Cities 89:80–91

Angelidou M (2015) Smart cities: a conjuncture of four forces. Cities 47:95–106

Batty M (2013) Big data, smart cities and city planning. Dialogues Hum Geogr 3(3):274–279

Burch S, Shaw A, Dale A, Robinson J (2014) Triggering transformative change: a development path approach to climate change response in communities. Climate Policy 14(4):467–487

Giffinger R, Pichler-Milanović N (2007) Smart cities: ranking of European medium-sized cities. Centre of Regional Science, Vienna University of Technology

Gleeson B (2008) Critical commentary. Waking from the dream: an Australian perspective on urban resilience. Sage J 45(13):2653–2668. https://doi.org/10.1177/0042098008098198

Harrison C, Eckman B, Hamilton R, Hartswick P, Kalagnanam J, Paraszczak J, Williams P (2010) Foundations for smarter cities. IBM J Res Dev 54(4):1–16

Hollands RG (2008) Will the real smart city please stand up? intelligent, progressive or entrepreneurial? City 12(3):303–320

Ilmola L (2016) Approaches to measurement of urban resilience. Urban resilience (pp 207–237). Springer

Ismagilova E, Hughes L, Dwivedi YK, Raman KR (2019) Smart cities: advances in research—an information systems perspective. Int J Inf Manag 47:88–100
Lombardi P, Giordano S, Farouh H, Yousef W (2012) Modelling the smart city performance. Innov: Eur J Soc Sci Res 25(2):137–149
Marsal-Llacuna M, Colomer-Llinàs J, Meléndez-Frigola J (2015) Lessons in urban monitoring taken from sustainable and livable cities to better address the smart cities initiative. Technol Forecast Soc Chang 90:611–622
Martin-Moreau M, Ménascé D (2018) Urban resilience: introducing this issue and summarizing the discussions. Field Actions Science Reports, 18. http://journals.openedition.org/factsreports/4629
Meriläinen E (2019) The dual discourse of urban resilience: Robust city and self-organised neighbourhoods. Disasters 44:125–151
Mora L, Deakin M, Reid A (2017) Smart-city development paths: insights from the first two decades of research. Paper presented at the International Conference on Smart and Sustainable Planning for Cities and Regions 403–427
Paul A, Cleverley M, Kerr W, Marzolini F, Reade M, Russo S (2011) Smarter cities series: understanding the IBM approach to public safety. IBM Corporation
Peng C, Yuan M, Gu C, Peng Z, Ming T (2017) A review of the theory and practice of regional resilience. Sustain Cities Soc 29:86–96
Perrings, C. (1998). Resilience in the dynamics of economy-environment systems. Environ Resour Econ 11:503–520. https://doi.org/10.1023/A:1008255614276
Perrings C (2006) Resilience and sustainable development. Environ Dev Econ 11(4):417–427
Sharifi A (2016) A critical review of selected tools for assessing community resilience. Ecol Indic 69:629–647. https://doi.org/10.1016/j.ecolind.2016.05.023
Sharifi A (2019) A critical review of selected smart city assessment tools and indicator sets. J Clean Prod 233:1269–1283. https://doi.org/10.1016/j.jclepro.2019.06.172
Sharifi A (2020a) A typology of smart city assessment tools and indicator sets. Sustain Cities Soc 53. https://doi.org/10.1016/j.scs.2019.101936
Sharifi A (2020b) Urban resilience assessment: mapping knowledge structure and trends. Sustainability (Switzerland) 12(15). https://doi.org/10.3390/SU12155918
Sharifi A, Khavarian-Garmsir AR, Kummitha RKR (2021) Contributions of smart city solutions and technologies to resilience against the COVID-19 pandemic: a literature review. Sustainability 13:8018
Silva BN, Khan M, Han K (2018) Towards sustainable smart cities: a review of trends, architectures, components, and open challenges in smart cities. Sustain Cities Soc 38:697–713
Tasan-Kok T, Stead D, Lu P (2013) Conceptual overview of resilience: history and context. In Ayda Eraydin and Tuna Tasan-Kok (eds) Resilience thinking in urban. Planning. Springer
United Nations (2015) Addis Ababa action agenda of the third international conference on financing for development. Conference Report 53(9). https://doi.org/10.1017/CBO9781107415324.004
UN (2015) Sendai framework for disaster risk reduction 2015–2030. https://www.preventionweb.net/files/43291_sendaiframeworkfordrren.pdf
UN-Habitat III (2016) New Urban Agenda. General Assembly (Issue January)
Winters JV (2011) Why are smart cities growing? Who moves and who stays. J RegNal Sci 51(2):253–270
Yigitcanlar T, Kamruzzaman M, Buys L, Ioppolo G, Sabatini-Marques J, da Costa EM, Yun JJ (2018) Understanding 'smart cities': intertwining development drivers with desired outcomes in a multidimensional framework. Cities 81:145–160

Correction to: Digital Solutions for Resilient Cities: A Critical Assessment of Resilience in Smart City Initiatives in Melbourne, Victoria

Leila Irajifar and Khanh N. Vu

Correction to:
Chapter 11 in: A. Sharifi and P. Salehi (eds.), *Resilient Smart Cities*, The Urban Book Series, https://doi.org/10.1007/978-3-030-95037-8_11

The original version of this chapter has been revised by adding Khanh N. Vu as a co-author. The chapter has been updated with the change.

The updated version of this chapter can be found at
https://doi.org/10.1007/978-3-030-95037-8_11

A. Sharifi and P. Salehi (eds.), *Resilient Smart Cities*, The Urban Book Series,
https://doi.org/10.1007/978-3-030-95037-8_21

Index

A. Sharifi and P. Salehi (eds.), *Resilient Smart Cities*, The Urban Book Series,
https://doi.org/10.1007/978-3-030-95037-8